The Statistical Analysis of
Failure Time Data

The Statistical Analysis of Failure Time Data

Second Edition

JOHN D. KALBFLEISCH

ROSS L. PRENTICE

WILEY-INTERSCIENCE

A JOHN WILEY & SONS, INC., PUBLICATION

Published by John Wiley & Sons, Inc., Hoboken, New Jersey.
Published simultaneously in Canada.

For general information on our other products and services please contact our Customer Care Department within the U.S. at 877-762-2974, outside the U.S. at 317-572-3993 or fax 317-572-4002.

Wiley also publishes its books in a variety of electronic formats. Some content that appears in print, however, may not be available in electronic format.

Library of Congress Cataloging-in-Publication Data Is Available

ISBN 0-471-36357-X

Printed in the United States of America.

10 9 8 7 6 5 4 3

To Sharon and Didi

Contents

Preface

As in the first edition of this book, the purpose of this revision is the collection and unified presentation of statistical models and methods for the analysis of failure time data. The motivation for this effort continues to derive primarily from biomedical contexts and, to a lesser extent, industrial life-testing purposes.

A voluminous literature on failure time analysis and the closely related event history analysis has developed in the more than 20 years since the publication in 1980 of the first edition of this book. The theoretical underpinnings of the methods described previously have been strengthened in the interim, and many important generalizations and related developments have taken place. Counting process methods and related martingale convergence results have led to precise and general asymptotic results for tests and estimators under key classes of failure time models and important censoring and truncation mechanisms. These developments have also contributed to the formulation of broader classes of models and methods.

An important challenge in developing this revision was to preserve the feature of a fairly elementary and classical likelihood-based presentation of failure time models and methods while integrating the counting process notation and related theory. This we have done by using classical notation and descriptions throughout the first four chapters of the revision while introducing the reader to key estimating functions and estimators in notation involving counting processes and stochastic integration. These chapters deal with survivor function estimation and comparison of survival curves (Chapter 1); statistical models for failure time distributions, including parametric and semiparametric regression models (Chapter 2); testing and estimation in parametric regression models under right censoring and other selected censoring schemes (Chapter 3); and testing and estimation under the semiparametric Cox regression model (Chapter 4). These chapters, along with parts of Chapters 6 to 8, can form the basis for an introductory graduate-level biostatistics or statistics course. We have tried to keep a solid contact with the first edition in many places and, for example, have retained illustrations from that edition where they still seemed to make the relevant points well.

A new Chapter 5 provides a more systematic introduction to counting processes and martingale convergence results and describes how they can be applied to yield asymptotic results for many of the statistical methods discussed in the first four chapters. The treatment is somewhat less formal than in some more specialized books, but presents the reader with a development and summary of the main ideas and a good basis for further investigation and study.

The remainder of the book uses the notation from counting processes and stochastic integrals where it is helpful, but continues to emphasize the likelihood basis for testing and estimation procedures. Like Chapter 5 in the first edition, Chapter 6 is devoted to general concepts of likelihood and partial likelihood construction, especially in relation to time-dependent and evolving covariate histories. We also provide an example in which martingale methods do not allow the development of asymptotic results because the conditioning events are not nested in time. Like our previous Chapter 6, Chapter 7 is devoted to the semiparametric log-linear or accelerated failure time model. Over the past two decades much effort has been devoted to regression estimation under this model, to the point where it can provide a practical alternative to the Cox model. Like our previous Chapter 7, Chapters 8 through 10 are devoted to aspects of multivariate failure time data analysis, including competing risk and multistate failure time modeling and estimation (Chapter 8), recurrent event modeling and estimation (Chapter 9), and correlated failure time methods (Chapter 10). Aside from a part of Chapter 8, most of the material in these chapters reflects developments since the first edition was published. Martingale convergence results are applicable to some of the estimating functions considered in these chapters, but others rely on empirical process methods. The latter methods can largely subsume the martingale methods, but we have not attempted comprehensive coverage here. Chapter 11 is devoted to more specialized topics. We have retained some of the material from our original Chapter 8 while providing a description of methods for such topics as risk set sampling, missing covariate data, mismeasured covariate data, sequential testing and estimation, and Bayesian methods, mostly in the context of the Cox model. The revision as a whole can serve as the textbook for a more advanced graduate course in biostatistics or statistics.

With the vast literature that has developed on failure time analysis, we have had to be selective in both the scope and depth of our coverage. We have chosen not to provide in-depth coverage of probability theory that is relevant to the asymptotic methods and results discussed, nor, except for some general comments in Appendix B, have we attempted to include a description of how available statistical software packages can or cannot be used to implement the various methods. We have chosen to emphasize some statistical models and approaches that seem to us to be of particular importance, to stress the ideas behind their development and application, and to provide some worked examples that illustrate their use.

To augment the usefulness of this revision as a graduate text, we have included a set of exercises at the end of each chapter. A number of these problems introduce the reader to additional pertinent failure time literature. As before, we have used references sparingly, especially in the early chapters, and bibliographic notes are provided at the close of each chapter. For historical reasons we have retained most of bibliographic notes from the original version, but we have augmented them with important recent references for each failure time topic.

There are a number of books on failure time methods that nicely complement this work and provide more comprehensive coverage of specific topics. For example, Lawless (1982) provides extensive coverage of parametric failure time models and estimation procedures; Cox and Oakes (1984) provide a concise and readable account of a range of failure time data topics; Fleming and Harrington (1991) provide a rigorous presentation of Cox regression methods and selected other failure time topics with considerable attention to model checking procedures; Andersen et al. (1993) give a comprehensive compendium of failure time and event history analysis methods with emphasis on counting processes. Andersen et al. (1993) provide additional material on a number of the topics discussed here. Books by Collett (1994) and Klein and Moeschberger (1997) provide relatively less technical accounts of the methods for key failure time topics. Collett includes a presentation of computer software options. Therneau and Grambsch (2000) discuss the implementation of failure time methods using SAS and S-Plus and provide a number of detailed illustrations with particular attention to model building and testing. Hougaard (2000) presents the first book dedicated to multivariate failure time methods. His book nicely complements our Chapters 8 through 10, with a greater emphasis on random effects or frailty models.

We would like to express our thanks to colleagues and to former and current students who have helped to shape our understanding of failure time analysis issues and methods. Their ideas and efforts have helped to inform this presentation.

JOHN D. KALBFLEISCH
ROSS L. PRENTICE

February 2002

CHAPTER 1

Introduction

1.1 FAILURE TIME DATA

We consider methods for the analysis of data when the response of interest is the time until some event occurs. Such events are generically referred to as *failures*, although the event may, for instance, be the performance of a certain task in a learning experiment in psychology or a change of residence in a demographic study. Major areas of application, however, are biomedical studies and industrial life testing.

We assume that observations are available on the failure time of n individuals usually taken to be independent. A principal problem examined is that of developing methods for assessing the dependence of failure time on explanatory variables. Typically, such explanatory variables will describe prestudy heterogeneity in the experimental material or differential allocations of treatments resulting from the study design. A secondary problem involves the estimation and specification of models for the underlying failure time distribution.

Additional problems arise in the analysis of multivariate failure times and failure types. These problems entail assessing the frequency of recurrent failures and estimating the correlation among failure times and types. There are a number of reasons why special methods and special treatment is required for failure time data, and it is convenient to illustrate some of the distinguishing features through the following examples.

1.1.1 Carcinogenesis

Table 1.1 gives the times from insult with the carcinogen DMBA to mortality from vaginal cancer in rats. Two groups were distinguished by a pretreatment regimen. We might consider comparing the two regimes using the t-test (presumably to transformed data) or one of several nonparametric tests. Such procedures cannot be applied immediately, however, because of a feature very prevalent in failure time studies. Specifically, four failure times in Table 1.1 are *censored*. For these four rats, we can see that the failure times exceed 216, 244, 204, and 344 days,

Table 1.1 Days to Vaginal Cancer Mortality in Rats

Group 1	143,	164,	188,	188,	190,	192,	206,	209,	213,	216,	220
	227,	230,	234,	246,	265,	304,	216*,	244*			
Group 2	142,	156,	163,	198,	205,	232,	232,	233,	233,	233,	233
	239,	240,	261,	280,	280,	296,	296,	323,	204*,	344*	

Source: Pike (1966).

* These four items are right censored.

respectively, but we do not know the failure times exactly. In this example, the (right) censoring may have arisen because these four rats died of causes unrelated to application of the carcinogen and were free of tumor at death, or they may simply not have died by the time of data analysis. The necessity of obtaining methods of analysis that accommodate censoring has been a principal motivating factor for the development of specialized models and procedures for failure time data.

A larger set of animal carcinogenesis data is given in Appendix A (data set V). Two groups of male mice were given 300 rads of radiation and followed for cancer incidence. One group was maintained in a germ-free environment. The new feature of these data is that more than one failure mode occurs. It is of interest, for example, to evaluate the effect of a germ-free environment on the incidence rate of reticulum cell sarcoma while accommodating the competing risks of developing thymic lymphoma or other causes of failure.

1.1.2 Randomized Clinical Trial

Table 1.2 gives some data from a randomized clinical trial on 64 patients with severe aplastic anemia. Prior to the trial, all the patients were treated with high-dose cyclophosphamide followed by an infusion of bone marrow from an HLA-identical family member. Patients were then assigned to each of two treatment groups: cyclosporine and methotrexate (CSP + MTX) or methotrexate alone (MTX). One endpoint of interest was the time from assignment until the diagnosis of a life-threatening stage (≥ 2) of acute graft versus host disease (AGVHD). The times are given in days. Also included are two covariates measured at the outset: the patient's age in years at the time of transplant and an indicator of whether or not the patient was assigned to a laminar airflow (LAF) isolation room. Storb et al. (1986) report on the subset of 46 patients who were randomly assigned to treatment, with stratification by age group and LAF. For purposes of illustration, we shall treat the data as though all 64 patients had been randomly assigned. In this trial, only 20 of the 64 patients actually reached the endpoint; the remaining 44 patients were right censored.

Appendix A (data set II) gives a part of the data from a much larger clinical trial carried out by the Radiation Therapy Oncology Group. The full study included patients with squamous cell carcinoma of 15 sites in the mouth and throat, with 16 participating institutions, although only the data on three sites in the oropharynx

Table 1.2   Time in Days to Severe (Stage $\geq$ 2) Acute Graft Versus Host Disease (AGVHD), Death, or Last Contact for Bone Marrow Transplant Patients Treated with Cyclosporine and Methotrexate (CSP + MTX) or with MTX Only [a]

CSP + MTX						MTX					
Time	LAF	Age	Time	LAF	Age	Time	LAF	Age	Time	LAF	Age
3*	0	40	324*	0	23	9	1	35	104*	1	27
8	1	21	356*	1	13	11	1	27	106*	1	19
10	1	18	378*	1	34	12	0	22	156*	1	15
12*	0	42	408*	1	27	20	1	21	218*	1	26
16	0	23	411*	1	5	20	1	30	230*	0	11
17	0	21	420*	1	23	22	0	7	231*	1	14
22	1	13	449*	1	37	25	1	36	316*	1	15
64*	0	20	490*	1	37	25	1	38	393*	1	27
65*	1	15	528*	1	32	25*	0	20	395*	0	2
77*	1	34	547*	1	32	28	0	25	428*	0	3
82*	1	14	691*	1	38	28	0	28	469*	1	14
98*	1	10	769*	0	18	31	1	17	602*	1	18
155*	0	27	1111*	0	20	35	1	21	681*	0	23
189*	1	9	1173*	0	12	35	1	25	690*	1	9
199*	1	19	1213*	0	12	46	1	35	1112*	1	11
247*	1	14	1357*	0	29	49	0	19	1180*	0	11

[a] Asterisks indicate that time to severe AGVHD is right censored; that is, the patient died without severe AGVHD or was without severe AGVHD at last contact.

reported by the six largest institutions are given. Patients entering the study were randomly assigned to one of two treatment groups: radiation therapy alone or radiation therapy together with a chemotherapeutic agent. One objective of the study was to compare the two treatment policies with respect to patient survival.

Approximately 30% of the survival times are censored, owing primarily to patients surviving to the time of analysis. Some patients were lost to follow up because the patient moved and was unable to continue, but these cases were relatively rare. From a statistical point of view, a key feature of these data is the considerable lack of homogeneity between individuals being studied. Of course, as a part of the study design, certain criteria for patient eligibility had to be met which eliminated extremes in the extent of disease, but still many factors are not controlled. This study included measurements of many covariates that would be expected to relate to survival experience. Six such variables are given in the data of Appendix A (sex, T staging, N staging, age, general condition, and grade). The site of the primary tumor and possible differences between participating institutions require consideration as well.

The TN staging classification gives a measure of the extent of the tumor at the primary site and at regional lymph nodes. T_1 refers to a small primary tumor, 2 cm or less in largest diameter, whereas T_4 is a massive tumor with extension to adjoining tissue. T_2 and T_3 refer to intermediate cases. N_0 refers to the absence of clinical evidence of a lymph node metastasis and N_1, N_2, and N_3 indicate, in increasing magnitude, the extent of existing lymph/node involvement. Patients with classifications T_1N_0, T_1N_1, T_2N_0, or T_2N_1 or with distant metastasis were excluded from study.

The variable "general condition" gives a measure of the functional capacity of the patient at the time of diagnosis (1 refers to no disability, whereas 4 denotes bed confinement; 2 and 3 refer to intermediate levels). The variable grade is a measure of the degree of differentiation of the tumor (the degree to which the tumor cell resembles the host cell) from 1 (well differentiated) to 3 (poorly differentiated).

In addition to the primary question of whether the combined treatment mode is preferable to the conventional radiation therapy, it is of considerable interest to determine the extent to which the several covariates are related to subsequent survival. In answering the primary question, it may also be important to adjust the survival times for possible imbalance that may be present in the study with regard to the other covariates. Such problems are similar to those encountered in the classical theory of regression and the analysis of covariance. Again, the need to accommodate censoring is an important distinguishing point. In many situations, nonparametric and robust procedures are desirable since there is frequently little empirical or theoretical work to support a particular family of failure time distributions.

1.1.3 Heart Transplant Data

Crowley and Hu (1977) give survival times of potential heart transplant recipients from their date of acceptance into the Stanford heart transplant program. These data are reproduced in Appendix A, data set IV. One problem of considerable interest is to evaluate the effect of heart transplantation on subsequent survival.

For each study subject the explanatory variables "age" and "prior surgery" were recorded. There were also donor–recipient variables that may be predictive of post-transplant survival time. The main new feature here is that patients change treatment status during the course of the study. Specifically, a patient is part of the control group until a suitable donor is located and transplantation takes place, at which time he or she joins the treatment group. Correspondingly, some explanatory variables, such as waiting time for transplant, are observed during the course of the study and depend on the time elapsed to transplant. This study is examined in some detail in Chapter 6 using the ideas of time-dependent covariates and time-dependent stratification.

The existence of covariates that change over time is yet another unusual feature of failure time data that requires special methods and attention to model characteristics and implications. Transplant studies, such as the heart transplant study, provide a class of examples where such covariates arise because of the very nature of the treatment. Alternatively, we can imagine a system operating under stress where the stress factor is varied as time elapses. In such a situation, it would be common to examine the relationship between the stress applied now and the current risk of failure. Other examples arise in clinical studies, such as, for example, measures of immune function taken at regular intervals for leukemia patients in remission. One may wish, in this instance, to study the relationship between changes in immune function and corresponding propensity to relapse. Such examples are also discussed in Chapter 6. In comparative trials, time-dependent covariates such as measures of immune function can be *responsive*; that is, they can be affected by the treatments under investigation. Responsive covariates have the potential to be useful in examining the mechanism of a treatment effect (does the treatment work by improving immune function?) or even in serving as a surrogate for the primary failure time outcome. If, however, they are treated as ordinary covariates in a regression model to investigate the effect of treatments, they can mask a treatment effect.

1.1.4   Accelerated Life Test

Nelson and Hahn (1972) present data on the number of hours to failure of motorettes operating under various temperatures. The name *accelerated life test* for this type of study derives from the use of a stress factor, in this case temperature, to increase the rate of failure over that which would be observed under normal operating conditions. The data are presented in Table 1.3 and exhibit severe censoring, with only 17 of 40 motorettes failing. Note that the stress (temperature) is constant for any particular motorette over time. The principal interest in such a study involves determination of the relationship between failure time and temperature for the purpose of extrapolating to usual running temperatures. Of course, the validity of such an extrapolation depends on the constancy of certain relationships over a very wide range of temperatures. For this study, the failure time distribution at the regular operating temperature of 130°C was of interest.

As in earlier examples, the censoring here is *type I* or *time censoring*. That is, censored survival times were observed only if failure had not occurred prior to a predetermined time at which the study was to be terminated. Experiments of this type, where considerable control is available to the experimenter, offer the possibility of other censoring schemes. For instance, in the study above it might have been decided in advance to continue the study until specified numbers of motorettes had failed at each of the temperatures (e.g., until one, three, five, and seven motorettes had failed at 150°C, 170°C, 190°C, and 220°C, respectively). Such censoring is usually referred to as *type II* or *order statistic censoring*, in that the study terminates as soon as certain order statistics are observed. With certain models, some

Table 1.3   Hours to Failure of Motorettes

150°C	All 10 motorettes without failure at 8064 hours
170°C	1764, 2772, 3444, 3542, 3780, 4860, 5196
	3 motorettes without failure at 5448 hours
190°C	408, 408, 1344, 1344, 1440
	5 motorettes without failure at 1680 hours
220°C	408, 408, 504, 504, 504
	5 motorettes without failure at 528 hours

inferential procedures (e.g., exact significance tests) are simpler for type II than for type I censoring. It should be noted, however, that type II censoring usually does not allow an upper bound to be placed on the total duration of the study and is generally not a feasible study design if there is staggered entry to the study.

Some of the examples above are considered further throughout the book. We turn now, however, to mathematical representations of failure times and consider the very simplest case of an independent sample from a homogeneous population (no explanatory variables) with a single failure mode.

1.2 FAILURE TIME DISTRIBUTIONS

Let T be a nonnegative random variable representing the failure time of an individual from a homogeneous population. The probability distribution of T can be specified in many ways, three of which are particularly useful in survival applications: the survivor function, the probability density function, and the hazard function. Interrelations among these three representations are given below for discrete and continuous distributions.

The *survivor function* is defined for discrete and continuous distributions by the probability that T exceeds a value t in its range; that is,

$$F(t) = P(T > t), \qquad 0 < t < \infty.$$

Note that F in some settings refers to the cumulative distribution function, $P(T \leq t)$, and therefore gives the probabilities in the left tail rather than in the right tail of the distribution. The right tail, however, is the important component for the incorporation of right censoring, so it is more convenient to concentrate on the survivor function in dealing with failure time distributions. Clearly, $F(t)$ is a nonincreasing right-continuous function of t with $F(0) = 1$ and $\lim_{t \to \infty} F(t) = 0$.

1.2.1 *T* (Absolutely) Continuous

The *probability density function* (PDF) of T is

$$f(t) = -dF(t)/dt.$$

The range of T is $[0, \infty)$, and this should be understood as the domain of definition for functions of t. It is convenient to remember that $f(t)$ gives the density of probability at t and for h small has the interpretation

$$f(t)h \simeq P(t \leq T < t + h) = F(t) - F(t + h),$$

provided that $f(t)$ is continuous at t. We note also that $f(t) \geq 0$, $\int_0^\infty f(t)\,dt = 1$, and

$$F(t) = \int_t^\infty f(s)\,ds.$$

The *hazard function* is defined as

$$\lambda(t) = \lim_{h \to 0^+} P(t \le T < t + h \mid T \ge t)/h \tag{1.1}$$

and specifies the instantaneous rate at which failures occur for items that are surviving at time t. The hazard function fully specifies the distribution of t and so determines both the density and the survivor functions. From (1.1) and using the definition of the density function, it follows that

$$\lambda(t) = f(t)/F(t)$$
$$= -d \log F(t)/dt.$$

Now integrating with respect to t and using $F(0) = 1$, we obtain

$$F(t) = \exp\left[-\int_0^t \lambda(s)\, ds\right]$$
$$= \exp[-\Lambda(t)], \tag{1.2}$$

where $\Lambda(t) = \int_0^t \lambda(s)\, ds$ is called the *cumulative hazard function*. The PDF of T can be obtained by differentiating (1.2) to find that

$$f(t) = \lambda(t)\exp[-\Lambda(t)]. \tag{1.3}$$

Examination of (1.2) indicates that any nonnegative function $\lambda(t)$ that satisfies

$$\int_0^t \lambda(s)\, ds < \infty$$

for some $t > 0$ and

$$\int_0^\infty \lambda(s)\, ds = \infty$$

can be the hazard function of a continuous random variable.

Other representations of the failure time distribution are occasionally useful. An example is the *expected residual life* at time t,

$$r(t) = E(T - t \mid T \ge t),$$

which uniquely determines a continuous survival distribution with finite mean. To see this, note that

$$r(t) = \frac{\int_t^\infty (s - t)f(s)\, ds}{F(t)}$$

and integrate by parts to obtain

$$r(t) = \frac{\int_t^{\infty} F(s)\, ds}{F(t)},$$ (1.4)

where we have used the fact that $E(T) < \infty$ implies that $\lim_{t \to \infty} tF(t) = 0$. Substituting $t = 0$ in (1.4) gives the useful result

$$E(T) = r(0) = \int_0^{\infty} F(s)\, ds.$$ (1.5)

Taking the reciprocal of both sides of (1.4), we obtain

$$\frac{1}{r(t)} = -\frac{d}{dt} \log \int_t^{\infty} F(s)\, ds,$$

so that

$$\int_0^t \frac{ds}{r(s)} = -\log \int_t^{\infty} F(s)\, ds + \log r(0).$$

This leads finally to the expression

$$F(t) = \frac{r(0)}{r(t)} \exp\left[-\int_0^t \frac{du}{r(u)} \right]$$

for the survivor function.

1.2.2 T Discrete

If T is a discrete random variable taking values $a_1 < a_2 < \cdots$ with associated probability function

$$f(a_i) = P(T = a_i), \qquad i = 1, 2, \ldots,$$

the survivor function is

$$F(t) = \sum_{j|a_j > t} f(a_j).$$

The hazard at a_i is defined as the conditional probability of failure at a_i given that the individual has survived to a_i,

$$\lambda_i = P(T = a_i \mid T \geq a_i) = \frac{f(a_i)}{F(a_i^-)}, \qquad i = 1, 2, \ldots,$$

where $F(a^-) = \lim_{t \to a^-} F(t)$. Corresponding to (1.2) and (1.3), the survivor function and the probability function are given by

$$F(t) = \prod_{j \mid a_j \le t} (1 - \lambda_j) \qquad (1.6)$$

and

$$f(a_i) = \lambda_i \prod_{j=1}^{i-1} (1 - \lambda_j). \qquad (1.7)$$

As in the continuous case, the discrete hazard function $(\lambda_i, i = 1, 2, \ldots)$ uniquely determines the distribution of the failure time variable T.

The results in (1.6) and (1.7) are quite easily deduced by considering the failure time process unfolding over time and a sequence of trials, each of which may or may not result in a failure. For example, the result in (1.7) follows from noting that an individual fails at time a_i if and only if:

- The individual survives in sequence each of the preceding discrete failure times $a_1, \ldots, a_{i-1}$ with corresponding (conditional) probabilities $(1 - \lambda_1), \ldots, (1 - \lambda_{i-1})$.
- Having survived to a_i, the individual fails at a_i with (conditional) probability λ_i.

1.2.3 *T* has Discrete and Continuous Components

More generally, the distribution of T may have both discrete and continuous components. In this case, the hazard function can be defined to have the continuous component $\lambda_c(t)$ and discrete components $\lambda_1, \lambda_2, \ldots$ at the discrete times $a_1 < a_2 < \cdots$. The overall survivor function can then be written

$$F(t) = \exp\left[-\int_0^t \lambda_c(u)\,du\right] \prod_{j \mid a_j \le t} (1 - \lambda_j).$$

The discrete, mixed, and continuous cases can be combined. The cumulative hazard function,

$$\Lambda(t) = \int_o^t \lambda_c(u)\,du + \sum_{j \mid a_j \le t} \lambda_j,$$

is a right-continuous nondecreasing function. From $\Lambda(t)$ we define the differential increment

$$\begin{aligned}
d\Lambda(t) &= \Lambda(t^- + dt) - \Lambda(t^-) \\
&= P[T \in [t, t + dt) \mid T \ge t] \\
&= \begin{cases} \lambda_i, & t = a_i, \quad i = 1, 2, \ldots \\ \lambda_c(t)\,dt, & \text{otherwise.} \end{cases}
\end{aligned}$$

which specifies the hazard of failure over the infinitesimal interval $[t, t + dt)$.

The survivor function in the discrete, continuous, or mixed cases can then be written as

$$F(t) = \mathscr{P}_0^t[1 - d\Lambda(u)], \qquad (1.8)$$

where the *product integral* $\mathscr{P}$ is defined by

$$\mathscr{P}_0^t[1 - d\Lambda(u)] = \lim \prod_{k=1}^{r} \{1 - [\Lambda(u_k) - \Lambda(u_{k-1})]\}.$$

Here $0 = u_0 < u_1 < \cdots < u_r = t$ and the limit is taken as $r \to \infty$ and $\max(u_i - u_{i-1}) \to 0$. In the continuous case ($\lambda_i = 0$ for all i), it can be shown that this reduces to

$$F(t) = \mathscr{P}_0^t[1 - d\Lambda(u)] = \mathscr{P}_0^t[1 - \lambda_c(u)\,du] = \exp\left[-\int_0^t \lambda_c(u)\,du\right].$$

In the discrete case [$\lambda_c(t) = 0$ for all t], it is easily seen that

$$\mathscr{P}_0^t[1 - d\Lambda(u)] = \prod_{j|a_j \leq t} (1 - \lambda_i).$$

This unification shows that failure time data can be considered to arise in essentially the same way in both the discrete and continuous cases. The product representation in (1.8) can be thought of as describing a coin-tossing experiment in which the probability of heads varies over time. The coin is tossed repeatedly and failure corresponds to the first occurrence of a tail. Thus, in general, the survival probability at time t is obtained by taking the product of the conditional survival probabilities $1 - d\Lambda(u)$ over infinitesimal intervals up to time t. This way of viewing a failure mechanism has led to many developments in the area and is crucial in understanding many of the ideas and techniques. In effect, it is possible to examine survival experience by looking at the survival experience over each interval conditional upon the experience to that point. Simple arguments for estimating the survivor function (Section 1.4) or for constructing censored data tests (Section 1.5) depend on this idea. It also underlies failure time analysis by counting processes and martingales (Chapter 5), the construction of the likelihood under independent censoring (Section 6.2), the construction of partial likelihood in the Cox model (Section 4.3), and the analysis of multivariate failure times and life-history processes (Chapter 9).

Note that $f(t)$ and $F(t)$ [or more usually, the cumulative distribution function $\bar{F}(t) = 1 - F(t)$] are common representations of the distribution of a random variable. The hazard function $\lambda(t)$ is a more specialized characterization but is particularly useful in modeling survival time data. In many instances, information is

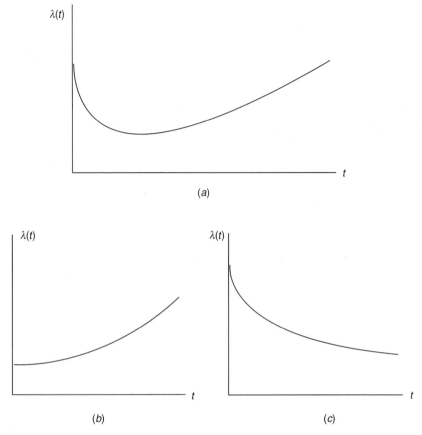

Figure 1.1 Examples of hazard functions: (*a*) hazard for human mortality; (*b*) positive aging; (*c*) negative aging.

available as to how failure rates change with the amount of time on test. This infor-mation can be used to model $\lambda(t)$ and easily translated into implications for $F(t)$ and $f(t)$ using the formulas above. For example, in modeling age at death of human populations, it is clear that initially, $\lambda(t)$ is elevated, owing to infant mortality and childhood diseases. This is followed by a period of relatively low mortality, after which the mortality rate increases very rapidly (see Figure 1.1*a*). In other applica-tions, monotone increasing hazards (positive aging) or decreasing hazards (negative aging) may be suggested (Figure 1.1*b* and *c*). Such qualitative information on $\lambda(t)$ can be useful in selecting a family of probability models for T. In Chapter 2 we discuss and examine some commonly used models for failure time and their associated hazard functions.

In the discussion above, we have specified models for a homogeneous popula-tion in which all individuals independently experience the same probability laws governing their failure. As noted earlier, there are many applications where we wish to incorporate measured covariates into the model. With covariates x measured at the time origin of the study, we can then think of models for the corresponding hazard function

$$\lambda(t;x) = \lim_{h \to 0} P\{T \in [t, t+h]|T \geq t, x\}/h,$$

which applies to those individuals with covariate value x. Corresponding to this, there are density and survivor functions, written $f(t;x)$ and $F(t;x)$, respectively.

1.3   TIME ORIGINS, CENSORING, AND TRUNCATION

In considering failure time data, it is important to have a clear and unambiguous definition of the time origin from which survival is measured. In some instances, time may represent age, with the time origin the birth of the individual. In other instances, the natural time origin may be the occurrence of some event, such as randomization or entry into a study or diagnosis of a particular disease. In like manner, one must have a clear definition of what constitutes failure. For example, in a trial to compare treatments of heart disease, one might take previous documented occurrence of a heart attack as providing eligibility for study. The time origin might be admission and randomization to the study, and failure may correspond to the recurrence of a heart attack. One would need to define carefully the clinical medical conditions that correspond to failure (and eligibility for the study). We will not talk about this further, but the clear identification of an origin and an endpoint are crucial applied aspects of failure time studies.

As noted earlier, failure time data often include some individuals who do not fail during their observation period; the data on these individuals are said to be *right censored*. In some situations, right censoring arises simply because some individuals are still surviving at the time that the study is terminated and the analysis is done. In other instances, individuals may move away from the study area for reasons unconnected with the failure time endpoint, so contact is lost. In yet other instances, individuals may be withdrawn or decide to withdraw from the study because of a worsening or improving prognosis. As is intuitively apparent, some censoring mechanisms have the potential to introduce bias into the estimation of survival probabilities or into treatment comparisons.

A right-censoring mechanism is said to be *independent* if the failure rates that apply to individuals on trial at each time $t > 0$ are the same as those that would have applied had there been no censoring. We discuss this idea more thoroughly in Chapter 6, but a brief discussion here is useful to set the stage. Suppose that the failure rate at time t that applies in the absence of censoring for an individual selected at random from a group with covariate value x is $\lambda(t;x)$. Here, as before, x consists of measurements taken on the individual at the time that he or she enters the study, such as age, sex, measures of physical condition, and so on. Suppose that within this group, individuals are to be censored according to a specific mechanism.

Consider the subset of individuals who are at risk of failure (neither failed nor censored) at some time $t > 0$. The censoring mechanism or scheme is independent if for an individual selected at random from this subset, the failure rate is $\lambda(t; x)$. Thus we require that at each time t,

$$\lim_{h \to 0} \frac{P\{T \in [t, t+h) | x, T \geq t\}}{h} = \lim_{h \to 0} \frac{P\{T \in [t, t+h) | x, T \geq t, Y(t) = 1\}}{h}, \quad (1.9)$$

where $Y(t) = 1$ indicates that the individual has neither failed nor been censored prior to time t (is at risk of failure at time t). If the censoring scheme is independent, it can be shown that an individual who is censored at time t contributes the term $P(T > t; x) = F(t; x)$ to the likelihood. Thus the information that the individual is censored at time t tells us only that the time to failure exceeds t.

As mentioned, independent censoring is examined more fully in Chapter 6. It is interesting to note, however, that some standard censoring schemes are independent. Consider, for example, a random censorship model where the ith individual has a time T_i to failure and a time C_i to censoring. Given the covariate value x_i, we suppose that C_i and T_i are independent random variables. Further, conditional on the x_i's, (T_i, C_i) are independent, $i = 1, \ldots, n$, where n is the number of subjects in the study. The time T_i to failure is observed if $T_i \leq C_i$. Otherwise, the individual is censored at C_i. For this case, it is easy to see that

$$\lim_{h \to 0} \frac{P\{T_i \in [t, t+h) | x_i, T_i \geq t\}}{h} = \lim_{h \to 0} \frac{P\{T \in [t, t+h) | x_i, T_i \geq t, C_i \geq t\}}{h},$$

which is equivalent to the condition (1.9). Type II censoring, in which individuals are put on trial until the kth item fails, for some fixed k, was discussed briefly Section 1.1.4. This censoring scheme is also independent.

In general, a censoring scheme is independent if the probability of censoring at each time t depends only on the covariate x, the observed pattern of failures and censoring up to time t in the trial, or on random processes that are independent of the failure times in the trial. Mechanisms in which the failure times of individuals are censored because the individuals appear to be at unusually high (or low) risk of failure are not independent. For these mechanisms, the condition (1.9) is violated, and the basic methods of survival analysis are not valid. Because of this, it is very important to follow the individuals entered into a study as completely as possible, so that the possibility of dependent censoring is minimized.

In some studies, individuals are not identified for observation at their respective time origin, but rather, at the occurrence of a subsequent event. Thus, there is a larger group of individuals who could have been observed, but the study is comprised of a subset of those in the cohort who experience some intermediate event. For these individuals, we observe the time origin and the follow-up time until they fail or are censored. For example, suppose that is the chosen time variable, so that time of birth is the time origin. Interest centers on the group of individuals who were exposed to some environmental risk, and individuals are identified for study at the time they respond to an advertisement. Any individuals who died prior to the advertisement are not observed, and in fact may not even be known to exist. Those who are observed are subject to *delayed entry* or *left truncation*. There is a condition similar to (1.9) for independent left truncation which requires that the failure rates of individuals under observation at time t are representative of those in the study population. Many of the methods and analyses that we discuss extend easily to allow for independent left truncation as well as independent right censoring.

Individuals can also be subject to *left censoring*, which occurs if the individual is observed to fail prior to some time t, but the actual time of failure is otherwise unknown. In this case, we observe that $T \in [0, t]$, which is analogous to right censoring, where we observe that $T \in (t, \infty)$. Left censoring should not be confused with left truncation, as discussed in the preceding paragraph. With left censoring, we know the individual exists and failed prior to the time t. With left truncation, the existence of an individual who fails before the beginning of observation is hidden from us.

Other types of censoring also arise. For example, in some situations individuals are interval censored, so we observe only that the failure time falls within some interval $T \in (a, b)$. One might also have situations in which individuals are subject to right truncation. That is, an individual is observed if and only if its failure time is less than some given time t. Exercise 1.13 gives an example. We discuss these more general censoring schemes in Chapter 3 in the context of parametric analyses. Most of our attention, however, is focused on independent right censoring and extensions to allow independent delayed entry or left truncation.

1.4 ESTIMATION OF THE SURVIVOR FUNCTION

1.4.1 Kaplan–Meier or Product Limit Estimator

The *empirical distribution function,*

$$\bar{F}_n(x) = \frac{\text{no. sample values} \leq x}{n}$$

is a simple estimate of the distribution function $\bar{F}(x) = P(X \leq x)$ and is a familiar and convenient way to summarize and display data. A plot of $\bar{F}_n(x)$ versus x visually represents the sample and provides full information on the percentile points, the dispersion, and the general features of the sample distribution. Besides these obvious descriptive uses, it is an indispensable aid in studying the distributional shape of the population from which the sample arose; in fact, the empirical distribution function can serve as a basic tool in constructing formal tests of goodness of fit of the data to hypothesized probability models (see, e.g., Cox and Hinkley, 1974, pp. 69ff.).

In the analysis of survival data, it is very often useful to summarize the survival experience of particular groups of patients in terms of the empirical survivor function. If an uncensored sample of n distinct failure times is observed from a continuous homogeneous population, the sample survivor function $F_n(t) = 1 - \bar{F}_n(t)$ is a step function that decreases by n^{-1} at each failure time observed. As noted earlier, survival data very often involve right censoring, and in this case a convenient method for estimating $F(t)$ is required.

Let $t_1 < t_2 < \cdots < t_k$ represent the observed failure times in a sample of size $n = n_0$ from a homogeneous population with (unknown) survivor function F. Suppose that d_j items fail at t_j and m_j items are censored in the interval $[t_j, t_{j+1})$ at times $t_{j1}, \ldots, t_{jm_j}, j = 0, \ldots, k,$ where $t_0 = 0$ and $t_{k+1} = \infty$. Let $n_j = (m_j + d_j) + \cdots + (m_k + d_k)$ denote the number of items at risk at a time just prior to t_j. The probability of failure at t_j is

$$P(T = t_j) = F(t_j^-) - F(t_j).$$

We assume that the contribution to the likelihood of a censored survival time at t_{jl} is

$$P(T > t_{jl}) = F(t_{jl}).$$

Here we are assuming that the observed censoring time t_{jl} tells us only that the unobserved failure time is greater than t_{jl}. This is appropriate provided that the censoring is independent, as discussed in Section 1.3.

The probability of the data is then of the form

$$L = \prod_{j=0}^{k} \left\{ \left[F(t_j^-) - F(t_j) \right]^{d_j} \prod_{l=1}^{m_j} F(t_{jl}) \right\},$$

which, given the data, can be viewed as a likelihood function on the space of all survivor functions F. The (nonparametric) maximum likelihood estimate (MLE) is the survivor function $\hat{F}$ that maximizes L.

Clearly, $\hat{F}(t)$ is discontinuous at the failure times observed (i.e., places some positive probability mass at each t_j) since otherwise, $L = 0$. Further, since $t_{jl} \geq t_j$, $F(t_{jl})$ is maximized by taking $F(t_{jl}) = F(t_j)$ $(j = 1, \ldots, k; l = 1, \ldots, m_j)$. The required MLE, $\hat{F}(t)$, is therefore a discrete survivor function with hazard components $\hat{\lambda}_1, \ldots, \hat{\lambda}_k$ at $t_1, \ldots, t_k$, respectively. Thus

$$\hat{F}(t_j) = \prod_{l=1}^{j} (1 - \hat{\lambda}_l) \tag{1.10}$$

and

$$\hat{F}(t_j^-) = \prod_{l=1}^{j-1} (1 - \hat{\lambda}_l), \tag{1.11}$$

Table 1.4   Kaplan–Meier Survivor Function Estimates for Carcinogenesis Data

\multicolumn{5}{c}{Group 1}					\multicolumn{5}{c}{Group 2}				
t_i	d_i	n_i	$\hat{F}(t_i)$	$\widehat{\mathrm{var}}\,(\hat{F})$	t_i	d_i	n_i	$\hat{F}(t_i)$	$\widehat{\mathrm{var}}\,(\hat{F})$
143	1	19	0.947	0.00262	142	1	21	0.952	0.00216
164	1	18	0.895	0.00496	156	1	20	0.905	0.00410
188	2	17	0.789	0.00875	163	1	19	0.857	0.00583
190	1	15	0.737	0.01021	198	1	18	0.810	0.00734
192	1	14	0.684	0.01137	205	1	16	0.759	0.00885
206	1	13	0.632	0.01225	232	2	15	0.658	0.01109
209	1	12	0.579	0.01283	233	4	13	0.455	0.01240
213	1	11	0.526	0.01312	239	1	9	0.405	0.01208
216	1	10	0.474	0.01312	240	1	8	0.345	0.01148
220	1	8	0.414	0.01311	261	1	7	0.304	0.01067
227	1	7	0.355	0.01264	280	2	6	0.202	0.00814
230	1	6	0.296	0.01170	296	2	4	0.101	0.00459
234	1	5	0.237	0.01029	323	1	2	0.051	0.00243
246	1	3	0.158	0.00873					
265	1	2	0.079	0.00530					
304	1	1	0.000						

where the $\hat{\lambda}_l$'s are chosen to maximize the function

$$\prod_{j=1}^{k}\left[\lambda_j^{d_j}\prod_{l=1}^{j-1}(1-\lambda_l)^{d_j}\prod_{l=1}^{j}(1-\lambda_l)^{m_j}\right]=\prod_{j=1}^{k}\lambda_j^{d_j}(1-\lambda_j)^{n_j-d_j}, \qquad (1.12)$$

obtained by substituting (1.10) and (1.11) in L. Clearly, $\hat{\lambda}_j=d_j/n_j(j=1,\ldots,k)$ and the Kaplan–Meier or *product limit estimate* of the survivor function is

$$\hat{F}(t)=\prod_{j|t_j\le t}\frac{n_j-d_j}{n_j}. \qquad (1.13)$$

In the product limit estimate, we are in effect making the estimated hazard or conditional probability of failure at each t_j agree exactly with the observed proportion (d_j/n_j) of the n_j individuals at risk who fail at t_j. Again we are viewing the survival experience sequentially and at each failure time estimating the hazard of failure to be the observed proportion of failures. It should be noted that $\hat{F}(t)$ never reduces to zero if $m_k>0$. In this instance, the largest time recorded is censored and it is usual to take $\hat{F}(t)$ as undefined for $t>t_{km_k}$.

The estimate $\hat{F}(t)$ is the direct generalization of the sample survivor function for censored data. It was first derived by Kaplan and Meier (1958), and as a consequence, is often referred to as the *Kaplan–Meier estimate*. Table 1.4 and Figure 1.2 exemplify the Kaplan–Meier estimate (1.13) for the carcinogenesis data of Section 1.1.1.

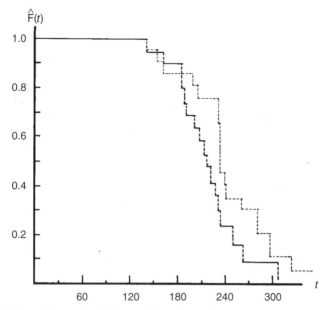

Figure 1.2 Kaplan–Meier survivor funtions estimates for carcinogenesis data: solid line, group; dashed line, group 2.

We consider now the asymptotic distribution of $\hat{F}(t)$ at a prespecified value of t. A heuristic derivation of an asymptotic variance can be obtained by regarding (1.12) as a parametric likelihood in the parameters $\lambda_1, \ldots, \lambda_k$. Standard likelihood methods, reviewed in Section 3.4, would yield an estimate $d_j(n_j - d_j)/n_j^3$ for the asymptotic variance of $\hat{\lambda}_j$ and hence for

$$\log \hat{F}(t) = \sum_{j|t_j \leq t} \log (1 - \hat{\lambda}_j),$$

an asymptotic variance estimate of

$$\widehat{\text{var}}\left[\log \hat{F}(t)\right] = \sum_{j|t_j \leq t} (1 - \hat{\lambda}_j)^{-2} \widehat{\text{var}} (1 - \hat{\lambda}_j)$$

$$= \sum_{j|t_j \leq t} \frac{d_j}{n_j(n_j - d_j)}.$$

The induced expression for the asymptotic variance of $\hat{F}(t)$ is then

$$\hat{V}_F(t) = \widehat{\text{var}}\left[\hat{F}(t)\right] = \hat{F}^2(t) \sum_{j|t_j \leq t} \frac{d_j}{n_j(n_j - d_j)}. \tag{1.14}$$

Expression (1.14), known as *Greenwood's formula* (Greenwood, 1926), was first derived as the asymptotic variance of the classical life-table estimator, which is discussed below. The derivation above would be valid if the distribution of t were discrete with finitely many mass points. Proper treatment of the asymptotic properties of the Kaplan–Meier estimator in the continuous case can be based on counting process formulations and related martingale theory. We discuss these topics in Chapter 5, and asymptotics for the Kaplan–Meier and related estimates are discussed further in Section 1.7. Essentially, under reasonably mild conditions on the censoring and large n, the results justify the use of a normal approximation of the distribution of $\hat{F}(t)$ with mean $F(t)$ and variance estimate (1.14). These results hold whether T is discrete or continuous or mixed with discrete and continuous components.

An approximate 95% confidence interval for $F(t)$ is $\hat{F}(t) \pm 1.96[\widehat{\text{var}}\,\hat{F}(t)]^{1/2}$. At extreme values of t (e.g., $t \le 188$ or $t > 246$ for the group 1 data of Table 1.3, such an approximate confidence interval may include impossible values outside the range $[0, 1]$. This problem can be avoided by applying the asymptotic normal distribution to a transformation of $F(t)$ for which the range is unrestricted. For example, the asymptotic variance of

$$\hat{v}(t) = \log\left[-\log \hat{F}(t)\right]$$

is, from Greenwood's formula and asymptotic theory (Section 3.4), estimated by

$$\hat{s}^2(t) = \widehat{\text{var}}\left[\log \hat{F}(t)\right] / \left[\log \hat{F}(t)\right]^2.$$

An asymptotic 95% confidence interval of $\hat{v}(t) \pm 1.96\hat{s}(t)$ for $v(t) = \log[-\log F(t)]$ gives a corresponding asymptotic 95% confidence interval for $F(t)$ of

$$\left[\hat{F}(t)\right]^{\exp[\pm 1.96\hat{s}(t)]},$$

which takes values in $[0, 1]$. Application of this method to the group 1 data of Table 1.1 gives an approximate 95% confidence interval for $F(t)$ at $t = 150$ of $(0.679, 0.992)$. A normal approximation to the distribution of $\hat{F}(150)$, in contrast, gives $(0.846, 1.047)$, a clearly unsatisfactory result.

It should be noted that many authors consider first the cumulative hazard function $\Lambda(t)$, which is most naturally estimated using the *Nelson–Aalen estimator*,

$$\hat{\Lambda}(t) = \sum_{t_i \le t} d_i/n_i = \sum_{t_i \le t} \hat{\lambda}_i, \tag{1.15}$$

which is a right-continuous step function whose increments are the empirical hazard estimates. Note that since the estimated distribution is discrete, the

Nelson–Aalen and Kaplan–Meier estimators are related in the way that one should expect [see (1.6) and (1.8)]

$$\hat{F}(t) = \mathscr{P}_0^t[1 - d\hat{\Lambda}(u)] = \prod_{t_i \leq t}(1 - \hat{\lambda}_i).$$

1.4.2   Life-Table and Related Estimates

Many other estimators of the survivor function have been considered. The oldest is that formed from the life table (see, e.g., Chiang, 1968). A *life table* is a summary of the survival data grouped into convenient intervals. In some applications (e.g., actuarial), the data are often collected in such a grouped form. In other cases, the data might be grouped to get a simpler and more easily understood presentation. Suppose, for example, that the data are grouped into intervals $I_1, \ldots, I_k$ such that

$$I_j = (b_0 + \cdots + b_{j-1}, b_0 + \cdots + b_j)$$

is of width b_j with $b_0 = 0$. The life table then presents the number of failures and censored survival times falling in each interval.

Suppose that m_j censored times and d_j failure times fall in the interval I_j, and let $n_j = \sum_{l \geq j}(d_l + m_l)$ be the number of individuals at risk at the start of the jth interval. The standard life-table estimator of the conditional probability of failure in I_j given survival to enter I_j, is $\hat{q}_j = 1$ if $n_j = 0$ and

$$\hat{q}_j = \frac{d_j}{n_j - m_j/2}$$

otherwise. The $m_j/2$ term in the denominator is used in an attempt to adjust for the fact that not all the n_j individuals are at risk for the whole of I_j. The corresponding life-table estimator of the survivor function at the end I_j is

$$\tilde{F}(b_1 = \cdots + b_j) = \prod_{l=1}^{j}(1 - \hat{q}_l). \qquad (1.16)$$

Greenwood's formula (1.14), with n_j replaced by $n_j - m_j/2$, provides an estimator of the variance of $\tilde{F}$.

The life-table method is designed primarily for situations in which actual failure and censoring times are unavailable and only the d_j's and m_j's are given for the jth interval. A simple modification of the life-table method utilizes the additional information when the (continuous) failure times are known. Suppose, for example, that $t_{j1}, \ldots, t_{jr_j}$ are the observed times in I_j of which m_j are censored and d_j are failures, $r_j = d_j + m_j(j = 1, \ldots, k)$. Suppose that the hazard function $\lambda(t)$ is taken to be a step function having constant value λ_j in the interval I_j. In this case, it can be shown that the maximum likelihood estimate of λ_j is

$$\hat{\lambda}_j = d_j/S_j,$$

where

$$S_j = \sum_{l=1}^{r_j} \left(t_{jl} - \sum_{0}^{j-1} b_i \right) + n_{j+1} b_j$$

is the total observed survival time in the interval I_j. The corresponding estimator of the survivor function is for $t \in I_j$,

$$\hat{F}(t) = \exp\left[-\hat{\lambda}_j \left(t - \sum_{l=0}^{j-1} b_l \right) - \sum_{i=1}^{j-1} \hat{\lambda}_i b_i \right]. \tag{1.17}$$

Unlike the preceding estimators, this is a continuous function of t and so relatively easier to view and to interpret shape. There is, however, an arbitrariness in the choice of intervals and in the piecewise constant model. Nonetheless, for exploratory purposes, the estimator (1.17) can be very useful.

1.5   COMPARISON OF SURVIVAL CURVES

Often, it is of interest to determine whether two or more samples could have arisen from identical survivor functions. One approach would involve the use of the asymptotic results for $\hat{F}(t)$ mentioned above to devise a test for equality of the survivor functions at some prespecified time t. Such a procedure, however, would not usually make efficient use of the data available, and attention has turned instead to test statistics that attempt to evaluate differences between survivor function estimators over the entire study period. The most commonly used statistics of this type can be viewed as censored data generalizations of such familiar nonparametric rank tests as the Wilcoxon test and the Savage (1956) or exponential scores test.

In this section, a heuristic derivation of the *log-rank test* is given. This test is a censored data generalization of the Savage test and is particularly good when the ratio of hazard functions in the populations being compared is approximately constant. It can also be advocated on the basis of ease of presentation to nonstatistical personnel since the test statistic is particularly simple in form. It amounts to the difference between the number of failures observed in each group and a quantity that, for most purposes, can be thought of as the corresponding expected number of failures under the null hypothesis.

Suppose that one wishes to test the hypothesis that the survivor functions $F_0(t), \ldots, F_p(t)$ are equal on the basis of samples from each of $p + 1$ populations.

Table 1.5   Frequency of Failures and Survivals at the Observed Failure Time t_j

	Sample 0	$\cdots$	Sample i	$\cdots$	Sample p	Total
Failures	d_{0j}	$\cdots$	d_{ij}	$\cdots$	d_{pj}	d_j
Survivors	$n_{0j} - d_{0j}$	$\cdots$	$n_{ij} - d_{ij}$	$\cdots$	$n_{pj} - d_{pj}$	$n_j - d_j$
At risk	n_{0j}	$\cdots$	n_{ij}	$\cdots$	n_{pj}	n_j

Let $t_1 < \cdots < t_k$ denote the failure times for the sample formed by pooling the $p + 1$ samples. Suppose that d_j failures occur at t_j and that n_j study subjects are at risk just prior to t_j ($j = 1, \ldots, k$). Let d_{ij} and n_{ij} be the corresponding numbers in sample i ($i = 0, \ldots, p$). The data at t_j can be summarized in the form of a $2 \times (p + 1)$ contingency table, as illustrated in Table 1.5. Conditional on the failure and censoring experience up to time t_j, the joint probability function of $d_{0j}, \ldots, d_{pj}$ is simply the product of independent binomial terms,

$$\prod_{i=0}^{p} \binom{n_{ij}}{d_{ij}} \lambda_j^{d_{ij}} (1 - \lambda_j)^{n_{ij} - d_{ij}},$$

where λ_j is the conditional failure probability (or hazard) at t_j, which under the null hypothesis is common for each of the $p + 1$ samples. The conditional distribution for $d_{0j}, \ldots, d_{pj}$ given d_j is then the multivariate hypergeometric distribution with probability function

$$\prod_{i=0}^{p} \binom{n_{ij}}{d_{ij}} \binom{n_j}{d_j}^{-1}. \tag{1.18}$$

The conditional mean and variance of d_{ij} from (1.18) are, respectively,

$$e_{ij} = n_{ij} d_j n_j^{-1}$$

and

$$(W_j)_{ii} = n_{ij}(n_j - n_{ij}) d_j (n_j - d_j) n_j^{-2} (n_j - 1)^{-1}. \tag{1.19}$$

The conditional covariance of d_{ij} and d_{lj} is

$$(W_j)_{il} = -n_{ij} n_{lj} d_j (n_j - d_j) n_j^{-2} (n_j - 1)^{-1}. \tag{1.20}$$

Thus, the statistic $w_j' = (d_{1j} - e_{1j}, \ldots, d_{pj} - e_{pj})$ has conditional mean 0 and $p \times p$ variance matrix W_j. Summing over the k failure times yields the log-rank statistic

$$w = \sum_{j=1}^{k} w_j = O - E, \tag{1.21}$$

where $O = (O_1, \ldots, O_p)'$, $E = (E_1, \ldots, E_p)'$, $O_i = \sum_{j=1}^{k} d_{ij}$, and $E_i = \sum_{j=1}^{k} e_{ij}$, $i = 1, \ldots, p$. Note that O is the vector of observed numbers of failures and E can informally be thought of as a vector of "expected" failures. This is informal only in that E is the sum of conditional expectations and its elements are random variables.

If the k contingency tables were independent, the variance of the log-rank statistic w would be $W = W_1 + \cdots + W_k$, and an approximate test of equality of the $p + 1$ survival distributions could be based on an asymptotic χ_p^2 distribution for

$$w'W^{-1}w. \tag{1.22}$$

Note that any of the $p + 1$ samples might be chosen as sample 0 and the log-rank statistic computed on the remaining p samples relabeled $1, \ldots, p$. It can be shown that the value of the statistic (1.22) is unchanged under any such relabeling.

Application of the log-rank method to a comparison of the two groups ($p = 1$) of survival data in Section 1.1.1 gives a log-rank statistic (1.21), $w = 19 - 23.763 = -4.763$, with corresponding variance estimate $W = 7.263$. The approximate χ_1^2 statistic has value $(4.763)^2(7.263)^{-1} = 3.12$, which is just significant at the 10% level. The slight evidence of a difference that this test shows suggests improved survival for the group 2 rats. This is exhibited in the log-rank statistic, in which we see that the observed number (19) of failures in this group is less than the expected number (23.763).

The derivation of the log-rank test above is similar to that given by Mantel (1966). It is difficult, however, to formalize the distribution theory from this development since the contingency tables over failure times are clearly not independent. It can, however, be shown that the w_j's are uncorrelated and that W provides an estimate of the covariance matrix of w. The chi-squared limiting distribution of (1.22) can be shown to hold under fairly general conditions. The asymptotic results are most easily established using counting processes and martingale limit theorems, as outlined in Chapter 5.

There are two important extensions of the log-rank procedure which can be mentioned at this stage. The first is stratification, and the second concerns the inclusion of weights.

1.5.1 Stratified Log-Rank Test

A simple means of testing equality of several survival curves while allowing for heterogeneity in the populations to be compared involves stratification on auxiliary variables. An overall test statistic is obtained by summing the log-rank statistics (1.21) and corresponding variances obtained within each of the independent strata. Specifically, if the strata are indexed by h, and $w^{(h)}$ and $W^{(h)}$ are the corresponding log-rank and variance statistics based on the data in stratum h, the stratified log-rank test is based on the statistic

$$\left(\sum_{h=1}^{s} w^{(h)} \right)^T \left(\sum_{h=1}^{s} W^{(h)} \right)^{-1} \left(\sum_{h=1}^{s} w^{(h)} \right). \tag{1.23}$$

Under the null hypothesis, (1.23) typically has an asymptotic χ_p^2 distribution. It should be noted that this test will be most sensitive to differences among the $p+1$ treatment groups that are similar across the strata. Examination of the individual log-rank tests in each of the strata can also provide some insights into possible treatment by strata interactions. This method can provide a valuable means of initial analysis and presentation for many data sets. As well, it is often a useful tool for communicating the results of a more complex analysis to nonstatistical personnel.

1.5.2 Weighted Log-Rank Test

The log-rank statistic as formulated above is most sensitive to departures from the null hypothesis in which the hazard ratios among the samples are roughly constant over time. In some instances, there may be reason to expect that any differences in the failure rates would occur early and that after the treatment has been in place for some time, treated and untreated individuals would show little difference. Conversely, there may be situations where any differences in failure rates between treatment groups might be expected to be small to begin and then larger later. Consider the weighted log-rank statistic

$$w(g) = \sum_{j=1}^{k} g_j w_j, \tag{1.24}$$

where $g_1, \ldots, g_k$ are weights chosen in specific applications to emphasize or deemphasize in an appropriate way the differences measured by the w_j's. The g_j's may be functions of time or of j, or they may depend on the past failure and censoring experience in the study. For example, one might consider the weights $g_j^{(G)} = n_j$, which yields the Gehan–Breslow generalization of the Wilcoxon or Kruskal–Wallis statistic. Alternatively, the weights $g_j^{(P)} = \prod_{i \leq j}[1 - d_i/(n_i + 1)]$ yield the Peto and Prentice generalization of the Wilcoxon. Note that $g_j^{(P)}$ is a survivor function estimate, close to the Kaplan–Meier estimator at t_j. Both of these weighting schemes emphasize early differences in the failure rates.

Under the null hypothesis, arguments similar to those outlined above show that the weighted log-rank statistic (1.24) has mean 0 and variance estimated by $W(g) = \sum g_j^2 W_j$. This again yields a simple asymptotic χ_p^2 statistic,

$$w(g)'W(g)^{-1}w(g).$$

These statistics are considered much more comprehensively in Chapter 7.

1.6 GENERALIZATIONS TO ACCOMMODATE DELAYED ENTRY

The methods of survivor function estimation and log-rank and related tests are easily generalized to accommodate independent left truncation or delayed entry into the study sample. In fact, there are essentially no changes involved in the formula and results given. As individuals enter the study, they become at risk of failure and so are included in the n_j or n_{ij}'s. With right censoring only, the number at risk in each sample will decrease over time as individuals fail or are censored. With left truncation, however, each new entry increases the number at risk in the appropriate group.

As a brief example, the Atomic Bomb Casualty Commision/Radiation Effects Research Foundation in Japan has, since 1950, followed a lifespan study cohort of over 100,000 persons who resided in Hiroshima or Nagasaki as of October 1, 1950. Data on this cohort are used to assess the effects of ionizing radiation exposure on mortality. The cohort includes a subsample who were residents of, but not in, either city at the time of the 1945 bombings. Key analyses from this cohort use date of bombing in the respective cities as the time origin, since mortality risk as a function of radiation exposure and time since exposure is of interest from both the public health and radiation biology perspectives. Data on time from exposure to death in this cohort are subject to left truncation since the cohort was not assembled until 1950. One can, however, estimate failure rates just as before as d_j/n_j, where d_j is the number of deaths at the jth chronological death time t_j and n_j is the number of cohort members alive and without censoring just prior to t_j. Similar changes generalize the log-rank procedures to this case. In this example it is not possible to estimate the failure rates or survival distribution for early times because no individuals who die early are included in the data. Typically, however, one can estimate the survival experience after some threshold time. Thus we can estimate that

$$F(t \mid T > a) = P(T > t \mid t > a) = F(t)/F(a)$$

for some suitably chosen a, where a might be October 1, 1950 or later in the illustration above.

In other instances, the data are subject to right truncation. In this case, the condition for study membership is that the event of interest occurs before some time of recruitment. Appendix A (data set III) gives data on transfusion-related AIDS cases in the United States. This study contains those individuals who were diagnosed with AIDS prior to 1988 and for whom the mode of infection was determined to be by blood transfusion. The distribution of the time from infection to diagnosis of AIDS (the incubation period) is of interest. In this study, individuals whose diagnosis occurs after the end of the study period are not included in the study, and the times included in the study are subject to very strong selection favoring the shorter incubation times. Right truncation is more difficult than left truncation to incorporate. Right truncation and this example are discussed further in Exercise 1.13.

1.7 COUNTING PROCESS NOTATION

Counting processes provide an alternative very compact notation for describing many of the results discussed above, and the related martingale theory provides a framework for deriving asymptotic properties. The theoretical framework and some of the asymptotic results are discussed in Chapter 5. In this section, some of the counting process notation is introduced and the estimators, tests, and variance formulas are reexpressed in these terms. The counting process notation is widely used in the literature on failure time analysis, and a general acquaintance with it is important.

1.7.1 Kaplan–Meier and Related Estimators

As in Section 1.4, suppose that n individuals from a homogeneous population are put on study at time 0. Let F be the survivor function and Λ be the cumulative hazard function; these may be discrete, continuous, or mixed. For the ith individual, let $N_i(t)$ count the number of failures observed in the interval $(0, t]$ and let $N_i(0) = 0$. Note that N_i is right continuous and takes value 0 until a failure is observed to occur, at which time it jumps to 1. Let Y_i be the at-risk process defined such that $Y_i(t) = 1$ if the individual is without failure and uncensored just prior to time t, and $Y_i(t) = 0$ otherwise. By convention, Y_i is taken to be left continuous. Let $N.(t) = \sum_{i=1}^{n} N_i(t)$ and $Y.(t) = \sum_{i=1}^{n} Y_i(t)$, $0 < t < \infty$. Clearly, $Y.(t)$ is the number of individuals in the entire study group that are at risk at time t, and $N.(t)$ is the total number of observed failures in the interval $(0, t]$. In the notation of Section 1.4, $N.(t) = \sum_{t_i \leq t} d_i$ is a right-continuous step function with a jump of d_i at t_i, $i = 1, \ldots, k$ and $Y.(t), 0 < t < \infty$ is a left-continuous step function that specifies the number of individuals who are uncensored and surviving at time t. Note that $Y.(t_i) = n_i, i = 1, \ldots, k$.

The Nelson–Aalen estimator of the cumulative hazard (1.15) can be written as the stochastic integral

$$\hat{\Lambda}(t) = \int_0^t \frac{J(u)}{Y.(u)} \, dN.(u), \tag{1.25}$$

where $J(u) = I[Y.(u) > 0]$ with the convention that $0/0$ is interpreted as 0. Note that $J(u)$ is used as a device to account for the possibility that at time u^-, there may be no items at risk. The Kaplan–Meier estimator of the survivor function is

$$\hat{F}(t) = \prod_{u \leq t} [1 - d\hat{\Lambda}(u)] = \mathscr{P}_0^t [1 - \frac{J(u)}{Y.(u)} \, dN.(u)]. \tag{1.26}$$

We had previously considered the Kaplan–Meier and Nelson–Aalen estimators to be undefined for t values greater than the maximum observed time if that time corresponded to a censoring. The convention being used in (1.25) and (1.26), however, takes the estimates as defined at all t, but constant following the maximum observed time. The former convention is more appropriate in most contexts, but the latter is convenient for some theoretical arguments.

A variance estimator for the Nelson–Aalen estimator (1.15) or (1.25) is

$$\hat{V}(t) = \int_0^t \frac{J(u)}{[Y.(u)]^2} \left[1 - \frac{\Delta N.(u)}{Y.(u)} \right] dN.(u)$$

$$= \sum_{t_j \leq t} \frac{d_j(n_j - d_j)}{n_j^3}, \tag{1.27}$$

where $\Delta N.(u) = N(u) - N(u^-)$. Large-sample properties of the Nelson–Aalen estimator can be shown to hold under relatively mild conditions, as outlined in Section 5.5. If for given t, $Y.(u) \to \infty$ for all $u \in (0, t]$ as $n \to \infty$, it is shown that $\hat{\Lambda}(t) \xrightarrow{\mathscr{P}} \Lambda(t)$ and

$$[\hat{\Lambda}(t) - \Lambda(t)]/\hat{V}(t)^{0.5} \xrightarrow{\mathscr{D}} N(0, 1),$$

where $\xrightarrow{\mathscr{P}}$ and $\xrightarrow{\mathscr{D}}$ indicate convergence in probability and convergence in distribution, respectively.

Greenwood's variance formula (1.14) can be written

$$\widehat{\text{var}}\,[\hat{F}(t)] = [\hat{F}(t)]^2 \int_0^t \frac{1}{Y.(s)[Y.(s) - \Delta N.(s)]} \, dN.(s). \qquad (1.28)$$

1.7.2 Log-Rank and Related Tests

Consider the experimental situation described in Section 1.5, where n_{i0} items are placed on test in the ith group at time 0, and let $N_{il}(t), t > 0$ be the counting process for the number of failures observed in $(0, t]$ for the lth individual in the ith group, $l = 1, \ldots, n_{i0}; i = 0, \ldots, p$. The corresponding at risk processes are $Y_{i\ell}(t)$, and again we assume independent censoring. Let $N_{i.}(t) = \sum_{l=1}^{n_{i0}} N_{il}(t)$ record the number of observed failures in the ith group and $Y_{i.}(t) = \sum Y_{i\ell}(t)$ specify the number at risk at time t. The ith component of the log-rank statistic (1.21) can now be written as

$$w_i = \int_0^\infty dN_{i.}(u) - \frac{Y_{i.}(u)}{Y..(u)} \, dN..(u), \qquad (1.29)$$

where $N..(t) = \sum_{i=0}^p N_{i.}(t)$. With some algebra, it can be verified that

$$w_i = \sum_{\ell=0}^p \int_0^\infty \left[\delta_{i\ell} - \frac{Y_{i.}(u)}{Y..(u)} \right] dN_{\ell.}(u), \qquad i = 1, \ldots, p, \qquad (1.30)$$

where $\delta_{i\ell} = \mathbf{1}(i = \ell)$. The variance and covariance formulas can also be expressed in counting process notation, as discussed further in Section 5.6.

BIBLIOGRAPHIC NOTES

Some useful references to life-table estimation are those by Berkson and Gage (1952), Cutler and Ederer (1958), Chiang (1960,1968), and Gehan (1969). The Kaplan–Meier or product limit estimator appears first to have been proposed as a limit of the life-table estimator by Böhmer (1912). It was not followed up, however, and was reintroduced in the important paper by Kaplan and Meier (1958), who showed that the estimate was a nonparametric MLE through an argument similar to that given in Section 1.4.1. Efron (1967) showed that the estimate satisfied a certain self-consistency property and discussed asymptotic properties. Breslow and Crowley (1974) first derived the asymptotic results for the Kaplan–Meier estimator under a random censorship model. More recent references for asymptotic results utilizing counting processes and martingales are reviewed in the notes to Chapter 5. The estimates based on the life table will tend to be slightly biased due to the grouping, and this will also typically be true for the piecewise continuous estimate (1.17). The nonparametric maximum likelihood approach and the self-consistency ideas of Efron (1967) were extended by Turnbull (1974, 1976) to include left and right truncations and interval censoring. Some of this work is reviewed in Section 3.9.1, and additional references and discussion on interval censoring are given in the bibliographic notes for Chapter 3.

The Nelson–Aalen estimate was first proposed by Nelson (1969,1972) as the basis for simple graphical checks for hazard shape in industrial life testing. Its large-sample properties were studied by Breslow and Crowley (1974) and by Aalen (1976). Altshuler (1970) also derived the Nelson–Aalen estimator and gave a related estimate of the survivor function. The product integral was introduced in the statistical literature by Cox (1972) as a compact description of the relationship between the hazard and the survivor function. A useful summary can be found in Dollard and Friedman (1979). See Gill and Johansen (1990) for a comprehensive account of product integration in relation to failure time data.

The adequacy of the asymptotic approximations to the Kaplan–Meier and Nelson–Aalen estimators has received some attention in the literature. It is evident that transformations to improve the asymptotic approximation in the tail is a useful technique and this has been explored by Klein (1991), who suggests a logistic rather than a $\log(-\log)$ transformation. Thomas and Grunkemeier (1975) developed a generalized likelihood ratio test of an hypothesized value for $F(c)$ at a given c (see Exercise 1.8) and argued that a χ_1^2 asymptotic distribution should apply and gave some simulations. This approach has received some attention in the literature, and asymptotic results have been derived by Li (1995a,b), Li et al. (1996), and Murphy (1995) for nonparametric likelihood ratio tests in various contexts. This approach is essentially that of empirical likelihood, and the recent book by Owen (2001) gives references and a good summary of asymptotic results.

We have given a derivation of the log-rank test in Section 1.5 that is essentially the same as that given originally by Mantel (1966). The test has been widely used in the literature, and both it and the weighted log-rank test arise in various contexts. The name log-rank was coined by Peto and Peto (1972) and the motivation of the term is not entirely clear to all — some say to apply it one first logs the data and then ranks them. The weighted log-rank test has been considered by many authors. Tarone and Ware (1977) first considered the general class. Harrington and Fleming (1982) considered a family of weight functions indexed by a parameter ρ that included the Wilcoxon and log-rank tests as special cases. They derive the asymptotic null distribution of the maximum weighted log-rank statistic in the class.

Fleming and Harrington (1991, Chap. 7) give an extensive discussion of log-rank and weighted log-rank procedures and have collected numerous references. References for counting processes and associated asymptotics are collected in Chapter 5.

EXERCISES AND COMPLEMENTS

1.1 Consider the mouse carcinogenesis data of Appendix A (data set V). Compute the product limit (Kaplan–Meier) estimates (1.10) of the survivor function for the endpoint, reticulum cell sarcoma, for the control and germ-free groups by:

(a) Ignoring failures from thymic lymphoma and other causes (i.e., eliminate mice dying by these causes before carrying out calculations).

(b) Regarding failure times from lymphoma or other causes as right censored.

Comment on the relative merits of parts (a) and (b). (*Hint:* Try to understand what is being estimated in both cases.) On the basis of the survivor function plots, does the germ-free environment appear to reduce the risk of reticulum cell sarcoma?

1.2 Plot on a single graph the logarithms of the estimates obtained from the life table (1.16), product limit (1.10), and the continuous (1.17) estimates of the survivor function for the thymic lymphoma data in the germ-free group. Regard failures from reticulum cell sarcoma and other causes as censored. Use grouping intervals of width 50 days for (1.16) and (1.17).

1.3 Show that the Kaplan–Meier estimate reduces to $\hat{F}(t) = $ (no. observations $> t)/n$ when there is no censoring. Show that Greenwood's formula (1.14) reduces in this case to the usual estimate of the variance of a binomial proportion. That is,

$$\widehat{\mathrm{var}}\,[\hat{F}(t)] = n^{-1}\hat{F}(t)[1 - \hat{F}(t)].$$

1.4 Let T be a discrete failure time variable taking values on the points $x_1, x_2, \ldots$ with survivor function $F(t)$. Show that the area under the survivor function, $\int_0^\infty F(t)\,dt = E(T)$. (*Note:* A simple geometric proof of this is obtained by partitioning a plot of the survivor function into rectangles with bases along the vertical axis.)

1.5 Let T be a discrete, continuous, or mixed random variable with survivor function $F(t)$. Show that $E(T) = \int_0^\infty F(t)\,dt$.

1.6 An electronic system is at continuous risk of failure with a constant hazard of λ events per hour. In addition, power surges occur each hour (i.e., at times 1,2,...), and at each power surge there is a 10% chance that the system will fail immediately. Obtain expressions for the survivor and cumulative hazard functions. Find the mean of T.

1.7 Let the survival time $T > 0$ be an integer-valued random variable with finite mean r_0 and let

$$r_i = E(T - i \mid T > i)$$

be the expected residual life at time i, $i = 1, 2, \ldots$. Show that the survivor function for integer t is

$$F(t) = P(T > t) = \prod_{i=1}^{t} \frac{r_{i-1} - 1}{r_i}.$$

Thus, in the discrete case also, the residual mean lifetime specifies the distribution of T. (*Note*: The geometric argument in Exercise 1.4 can be used to show that $r_j = [r_0 - F(0) - \cdots - F(j-1)]/F(j)$, for $j = 1, 2, \ldots$.)

1.8 As in Section 1.3, let $t_1 < t_2 < \cdots < t_k$ represent the observed failure times in a sample of size n_0 from a homogeneous population with survivor function $F(t)$. Suppose that d_j items fail at t_j and that n_j items are at risk at t_j^-.

(a) Let b be a prespecified time $(b > t_1)$ and c be a constant $(0 \le c \le 1)$. Show that subject to the constraint $F(b) = c$, the nonparametric maximum likelihood estimate of $F(t)$ is

$$\tilde{F}(t) = \prod_{j \mid t_j \le t} (1 - \tilde{\lambda}_j),$$

where $t_0 = \tilde{\lambda}_0 = 0$ and $\tilde{\lambda}_j = d_j/(n_j + a)$ if $t_j \le b$ and d_j/n_j if $t_j > b$, $j = 1, \ldots, k$. The value a is chosen to satisfy $\tilde{F}(b) = c$. Note that if $b \le t_1$, the constrained estimate is not unique for $t < b$. An arbitrary convention would assign a hazard $1 - c$ at $t = \epsilon$ for some small positive $\epsilon < b$.

(b) Show that the log-likelihood ratio statistic for the hypothesis $F(b) = c$ can be written

$$R = \sum_{i \mid t_i \le b} \left[(n_i - d_i) \log\left(1 + \frac{a}{n_i - d_i} \right) - n_i \log\left(1 + \frac{a}{n_i} \right) \right].$$

(c) Thomas and Grunkemeier (1975) show that the usual asymptotic properties apply and that $-2R$ is asymptotically χ_1^2 under the hypothesis. Use this result to establish a 95% confidence interval for $F(b)$. Compare these results with those obtained in Section 1.3 for the carcinogenesis data (Table 1.1) with $b = 150$.

1.9 Suppose that censored samples are available on two populations with survivor functions $F_1(t)$ and $F_2(t)$. Consider the hypothesis $F_2(b) = F_1(b)$ at some prespecified time b. Extend the results in Exercise 1.8 to obtain the nonparametric likelihood ratio statistic for this hypothesis. Apply this approach to test for equality of the survivor functions at $b = 250$ for the carcinogenicity data (Table 1.1).

1.10 Show that the mean vector and variance matrix for $(d_{1j}, \ldots, d_{pj})$ in the distribution (1.18) are as asserted.

1.11 Consider again the mouse carcinogenesis data (data set V, Appendix A). Use the log-rank test (1.16) to test the hypothesis that germ-free isolation does not affect overall mortality.

1.12 Suppose that $T_1, \ldots, T_n$ are independent exponential variates with respective failure rates $\lambda_1, \ldots, \lambda_n$. Let $\gamma_1, \ldots, \gamma_m$ be the distinct elements of $\lambda_1, \ldots, \lambda_n$. Let $S = \sum_{i=1}^{n} T_i$.

(a) Show that the survivor function of S may be written as

$$F_S(t) = P(S > t) = \sum_{j=1}^{m} p_j(t) e^{-\gamma_j t},$$

where the p_j's are polynomials in t.

(b) Let $\lambda_S(t)$ be the hazard function of S and show that $\lambda_S(t) \leq \lambda_{\min}$ for all t and that $\lim_{t \to \infty} \lambda_S(t) = \lambda_{\min}$, where $\lambda_{\min} = \min(\gamma_1, \ldots, \gamma_m)$.

1.13 Consider the transfusion-related AIDS data in data set III, Appendix A. As discussed in Section 1.6, these data are subject to right truncation in that a condition for study membership is that diagnosis of AIDS takes place prior to the end of the study period. Let T represent the number of months from transfusion to AIDS diagnosis, and $F(t) = P(T > t)$ be the corresponding (discrete) survivor function. Let t_i be the month of diagnosis, and let a_i be the total months elapsed to the end of the study period for the ith subject, $i = 1, \ldots, n$.

(a) Under what conditions would the likelihood function be of the form $\prod_{i=1}^{n} \{[F(t_i) - F(t_i^-)]/[1 - F(a_i)]\}$.

(b) Explain why F can only be estimated up to a constant of proportionality.

(c) Let $a = \max(a_1, \ldots, a_n)$, and find the maximum likelihood estimate of the conditional survivor function $G(t) = F(t)/[1 - F(a)] = P(T > t \mid T \leq a)$.

(d) What additional information would you need to estimate the median time from transfusion infection to diagnosis with AIDS? (Lagakos et al., 1988; Kalbfleisch and Lawless, 1989)

1.14 Consider the data of Table 1.2. Apply the log-rank test to compare the two treatment groups in the trial. Consider dividing the data into three strata consisting of patients in the age groups ≤ 15, 16–25, and ≥ 26, respectively. Apply a log-rank test separately in each stratum and the stratified log-rank test. Discuss the results.

CHAPTER 2

Failure Time Models

2.1 INTRODUCTION

Although a primary focus of this book concerns the relationship between failure time and explanatory variables, it is a useful first step to consider failure time distributions for homogeneous populations. Throughout the literature on failure time data, certain parametric models have been used repeatedly; exponential and Weibull models, for example, are often used. These distributions admit closed-form expressions for tail area probabilities and thereby simple formulas for survivor and hazard functions. Log-normal and gamma distributions are generally less convenient computationally but are still applied frequently. We also consider more general parametric models (e.g., the log F and generalized gamma) that are able to adapt to a diverse range of distributional shapes.

In Section 2.2 we discuss some of the standard failure time models for homogeneous populations. In the presentation we concentrate on model interpretation both through hazards and through models for the logarithm of failure time. The general properties and theoretical bases of these distributions are considered here only briefly. More detailed discussion can be found in various sources, and we mention, in particular, Johnson and Kotz (1970a,b), Mann et al. (1974), and Lawless (1982) as good reference sources. In Section 2.3 we consider extensions of the parametric models to include regressor variables and identify two general classes of models: the relative risk or Cox models and the accelerated failure time models. In Section 2.4 we consider discrete regression models and their relationships to continuous models.

2.2 SOME CONTINUOUS PARAMETRIC FAILURE TIME MODELS

As before, $T > 0$ is a random variable representing failure time, and t represents a point in its range. We use $Y \equiv \log T$ to represent the log failure time and summarize the failure time distributions in terms of both T and Y. Shape comparisons among the parametric models are often simpler in terms of Y than T.

2.2.1   Exponential Distribution

The *one-parameter exponential distribution* is obtained by taking the hazard function to be constant, $\lambda(t) = \lambda > 0$, over the range of T. The instantaneous failure rate is independent of t, so that the conditional chance of failure in a time interval of specified length is the same regardless of how long the individual has been on study; this is referred to as the *memoryless property* of the exponential distribution. The survivor function and density functions of T are, respectively,

$$F(t) = e^{-\lambda t} \qquad \text{and} \qquad f(t) = \lambda e^{-\lambda t}$$

from (1.2) and (1.3). Figure 2.1 graphs the functions $\lambda(t), f(t)$, and $F(t)$ for the exponential distribution. An empirical check of the appropriateness of the exponential model for a set of survival data is provided by plotting the log of a survivor function estimate versus t. Such a plot should approximate a straight line through the origin. As can be deduced from Table 1.1, the carcinogenesis data of Section 1.1.1 are not well described by a single-parameter exponential model, owing primarily to the absence of vaginal cancer deaths within the first 140 days.
   The PDF of $Y \equiv \log T$ is

$$\exp(y - \alpha - e^{y - \alpha}), \qquad -\infty < y < \infty,$$

where $\alpha = -\log \lambda$. Letting $Y = \alpha + W$, the PDF is

$$\exp(w - e^w), \qquad -\infty < w < \infty, \tag{2.1}$$

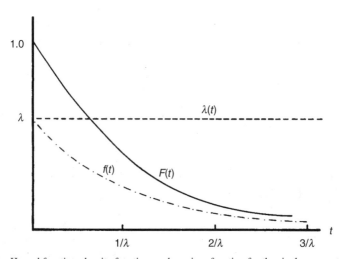

Figure 2.1   Hazard function, density function, and survivor function for the single-parameter exponential model. Note that λ may exceed 1.

which is an extreme value (minimum) distribution. This distribution derives its name from its appearance as the limiting distribution of a standardized form of the minimum of a sample selected from a continuous distribution with support on $(-\infty, a)$ for some $a \leq \infty$. Details of the derivation are given, for example, by Johnson and Kotz (1970a, Chap. 2.1). In a similar manner, the exponential distribution arises as the limiting distribution of a standardized form for the minimum of a sample from certain densities with support $(0, \infty)$ (see Exercise 2.1). This can sometimes be taken as theoretical justification for its use in survival studies in which a complex mechanism fails when any one of its many components fails. The extreme value distribution (2.1) is a unimodal distribution with skewness -1.14 and kurtosis 2.4. The mean of (2.1) is $-0.5722\ldots$, the negative of Euler's constant, and the variance is $\pi^2/6 = 1.6449\ldots$. The moment generating function is

$$M_W(\theta) = E(e^{\theta W}) = \Gamma(\theta + 1), \qquad \theta > -1,$$

where

$$\Gamma(k) = \int_0^\infty x^{k-1} e^{-x} dx$$

is the gamma function.

2.2.2 Weibull Distribution

An important generalization of the exponential distribution allows for a power dependence of the hazard on time. This yields the two-parameter Weibull distribution with hazard function

$$\lambda(t) = \lambda\gamma(\lambda t)^{\gamma-1}$$

for λ, $\gamma > 0$. This hazard (see Figure 2.2) is monotone decreasing for $\gamma < 1$, increasing for $\gamma > 1$, and reduces to the constant exponential hazard if $\gamma = 1$. The PDF is

$$f(t) = \lambda\gamma(\lambda t)^{\gamma-1} \exp[-(\lambda t)^\gamma]$$

and the survivor function is

$$F(t) = \exp[-(\lambda t)^\gamma].$$

Clearly,

$$\log[-\log F(t)] = \gamma(\log t + \log \lambda),$$

so that an empirical check for the Weibull distribution is provided by a plot of $\log[-\log \hat{F})]$ versus $\log t$, where $\hat{F}$ is a sample (Kaplan–Meier) estimate of the survivor function. The plot should give approximately a straight line, the slope and x intercept of which provide a rough estimate of γ and $-\log \lambda$, respectively.

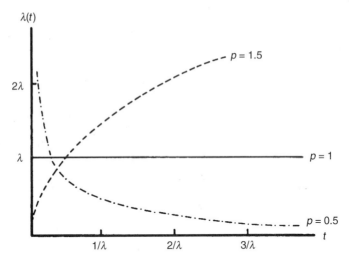

Figure 2.2 Hazard functions for the two-parameter Weibull model with shape parameter $\gamma = p$.

The PDF of the log failure time Y is

$$\sigma^{-1}e^{(y-\mu)/\sigma}\exp(-e^{(y-\mu)/\sigma}), \qquad -\infty < y < \infty,$$

where $\sigma = \gamma^{-1}$ and $\alpha = -\log \lambda$. More simply, we can write $Y = \alpha + \sigma W$, where W has the extreme value PDF (2.1). The shape of the density for Y is fixed because λ and γ affect only the location and the scaling of the distribution. The Weibull distribution can also be developed as the limiting distribution of the minimum of a random sample from certain distributions (Exercise 2.1).

2.2.3 Log-Normal Distribution

Again, the model for $Y = \log T$ is of the form $Y = \alpha + \sigma W$, but W is a standard normal variate with density

$$\phi(w) = \frac{1}{\sqrt{2\pi}}e^{-w^2/2}, \qquad -\infty < w < \infty. \tag{2.2}$$

The density function for T can be written

$$f(t) = (2\pi)^{-1/2}\gamma t^{-1}\exp\left[\frac{-\gamma^2(\log \lambda t)^2}{2}\right],$$

where, as before, $\alpha = -\log \lambda$ and $\sigma = \gamma^{-1}$. The survivor and hazard functions involve the normal distribution function $\Phi(w) = \int_{-\infty}^{w} \phi(u)\,du$. The survivor function is

$$F(t) = 1 - \Phi(\gamma \log \lambda t)$$

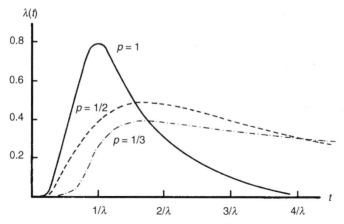

Figure 2.3 Hazard functions of the log-normal distribution with shape parameter $\gamma = p$.

and the hazard function is $f(t)/F(t)$. The hazard function has value 0 at $t = 0$, increases to a maximum and then decreases, approaching zero as t becomes large (see Figure 2.3). The log-normal model is particularly simple to apply if there is no censoring, but with censoring the computations become more difficult. The log-logistic distribution of Section 2.2.6 provides a good approximation to the log-normal distribution, and since it has a closed form for the survivor and hazard function, may frequently be a preferable survival time model.

2.2.4 Gamma Distribution

As noted above, the Weibull distribution is a two-parameter generalization of the exponential model; another such generalization is the gamma distribution with density function

$$f(t) = \frac{\lambda(\lambda t)^{k-1} e^{\lambda t}}{\Gamma(k)},$$

where $k, \lambda > 0$. The model for $Y = \log T$ can be written $Y = \alpha + W$, where $\alpha = -\log \lambda$ and W has the density

$$\frac{\exp(kw - e^w)}{\Gamma(k)}. \tag{2.3}$$

The error quantity W has a negatively skewed distribution with skewness decreasing with increasing k. When $k = 1$, at the exponential model, W has the extreme value distribution (2.1). More generally, the moment generating function of W is

$$M(\theta) = \frac{\Gamma(\theta + k)}{\Gamma(k)},$$

from which it is easily shown that the mean and variance of (2.3) are the digamma function

$$\psi(k) = \frac{d \log \Gamma(k)}{dk}$$

and the trigamma function

$$\psi^{(1)}(k) = \frac{d^2 \log \Gamma(k)}{dk^2},$$

respectively. These are discussed and tabulated, for example, in Abramowitz and Stegun (1965).

It is of interest to note that W, suitably standardized, has a limiting normal distribution as $k \to \infty$: From Stirling's formula,

$$\log \Gamma(k) = -k + \left(k - \frac{1}{2}\right) \log k + \log \sqrt{2\pi} + O(k^{-1}),$$

it is easily seen that $\psi(k) = \log k + O(k^{-1})$ and $\psi^{(1)}(k) = k^{-1} + O(k^{-2})$. We consider, then, the standardized variate

$$W^* = \sqrt{k}(W - \log k)$$

and obtain

$$\lim_{k \to \infty} M_{W^*}(\theta) = \exp(\theta^2/2),$$

where $M_{W^*}(\theta)$ is the moment generating function of W^*. Thus as $k \to \infty$, the distribution of W^* converges to that of a standard normal variate.

The survivor and hazard functions of the gamma distribution involve the incomplete gamma integral,

$$I_k(s) = \frac{\int_0^s x^{k-1} e^{-x} \, dx}{\Gamma(k)},$$

and are, respectively,

$$F(t) = 1 - I_k(\lambda t)$$

and

$$\lambda(t) = \frac{\lambda(\lambda t)^{k-1} \exp(-\lambda t)}{\Gamma(k)[1 - I_k(\lambda t)]}.$$

The hazard function is monotone increasing from 0 if $k > 1$, monotone decreasing from ∞ if $k < 1$, and in either case approaches λ as t becomes large. If $k = 1$, the gamma distribution reduces to the exponential distribution. With integer k, the gamma distribution is sometimes called a *special Erlangian distribution*.

The gamma distribution with integer k (and the exponential distribution, $k = 1$) can be derived as the distribution of the waiting time to the kth emission from a Poisson source with intensity parameter λ. As a side result, it is apparent from this and the properties of the Poisson process that the sum of k independent exponential variates with failure rate λ has a gamma distribution with parameters λ and k.

2.2.5 Generalized Gamma Distribution

The gamma family of Section 2.2.4 can be generalized by incorporating a scale parameter σ in the model for $Y = \log T$ to give $Y = \alpha + \sigma W$, where W has the distribution with density (2.3). This three-parameter model was introduced by Stacy (1962) and includes as special cases the exponential ($\sigma = k = 1$), the gamma ($\sigma = 1$), and the Weibull ($k = 1$). The log-normal is also a limiting special case as $k \to \infty$.

The PDF for T can be written

$$f(t) = \frac{\lambda\gamma(\lambda t)^{\gamma k-1}\exp[-(\lambda t)^{\gamma}]}{\Gamma(k)}, \qquad t > 0,$$

where $\lambda = \exp(-\alpha)$ and $\gamma = \sigma^{-1}$. The hazard function incorporates a variety of shapes, as indicated by the special cases. The distribution is most easily visualized in terms of Y, the log survival time.

2.2.6 Log-Logistic Distribution

Other failure time models can be constructed by selecting different distributions for the error variable W in $Y = \alpha + \sigma W$. One such is the log-logistic distribution for T obtained if W has the logistic density

$$\frac{e^w}{(1 + e^w)^2}. \tag{2.4}$$

This is a symmetric density with mean 0 and variance $\pi^2/3$ with slightly heavier tails than the normal density function, the excess in kurtosis being 1.2. The probability density function of t is then

$$f(t) = \lambda\gamma(\lambda t)^{\gamma-1}[1 + (\lambda t)^{\gamma}]^{-2},$$

where again $\lambda = \exp(-\alpha)$ and $\gamma = \sigma^{-1}$. Although this model is used less frequently in life-testing applications, it has the advantage (like the Weibull and exponential models) of having simple algebraic expressions for the survivor and hazard functions. It is therefore more convenient than the log-normal distribution in handling censored data, while providing a good approximation to it except in the extreme tails. The survivor and hazard functions are, respectively,

$$F(t) = \frac{1}{1 + (\lambda t)^\gamma}$$

and

$$\lambda(t) = \frac{\lambda \gamma (\lambda t)^{\gamma - 1}}{1 + (\lambda t)^\gamma}.$$

This hazard function is identical to the Weibull hazard, aside from the denominator factor $1 + (\lambda t)^\gamma$; it is monotone decreasing from ∞ if $\gamma < 1$ and is monotone increasing from λ if $\gamma = 1$. If $\gamma > 1$, the hazard resembles the log-normal hazard in that it increases from zero to a maximum at $t = (\gamma - 1)^{1/\gamma}/\lambda$ and decreases toward zero thereafter.

2.2.7   Generalized F Distribution

The final parametric model to be discussed incorporates all the foregoing distributions as special cases; the primary value of this model may be its use for discriminating between competing models such as the Weibull and log-logistic distributions for a given set of data (see Section 3.8). It also has the advantage that it can adapt to a wide variety of distributional shapes.

Once again we consider a location and scale model for the log failure time Y in which the error distribution is now assumed to be that of the logarithm of an F variate on $2m_1$ and $2m_2$ degrees of freedom. That is, $Y = \mu + \sigma W$, where the PDF of W is

$$\frac{(m_1/m_2)^{m_1} \exp(wm_1)(1 + m_1 e^w/m_2)^{-(m_1 + m_2)}}{B(m_1, m_2)}, \tag{2.5}$$

where $B(m_1, m_2) = \Gamma(m_1)\Gamma(m_2)/\Gamma(m_1 + m_2)$ is the beta function. The resulting model for T is the generalized F distribution. It can be seen that the distributions discussed in Sections 2.2.1 through 2.2.6 are special cases of the generalized F. If $(m_1, m_2) = (1, 1)$, then (2.5) reduces to the logistic model and T has the log-logistic distribution. The Weibull model is obtained as $(m_1, m_2) = (1, \infty)$, for which (2.5) is the extreme value error density (2.1); if, in addition, $\sigma = 1$, the exponential model is obtained. The generalized gamma distribution corresponds to $m_2 = \infty$ and, as before, reduces to the gamma distribution for $\sigma = 1$. The log-normal distribution arises by allowing $(m_1, m_2) \to (\infty, \infty)$, as discussed below.

The densities (2.5) are positively skewed for $m_1 > m_2$, negatively skewed for $m_2 > m_1$, and symmetric along $m_1 = m_2$. The $\log F$ distribution is tabulated in the Fisher–Yates tables (1938).

The moment generating function can be used to establish the limiting special cases noted above:

$$M_W(\theta) = \frac{1}{B(m_1, m_2)} \left(\frac{m_2}{m_1}\right)^\theta \int_0^\infty \frac{x^{\theta + m_1 - 1}}{(1+x)^{m_1 + m_2}} \, dx$$

$$= \frac{\Gamma(m_1 + \theta)\Gamma(m_2 - \theta)}{\Gamma(m_1)\Gamma(m_2)} \left(\frac{m_2}{m_1}\right)^\theta,$$

where a change of variables to $x = (m_1/m_2)e^w$ and

$$B(a, b) = \int_0^\infty \frac{x^{a-1}}{(1+x)^{a+b}} \, dx$$

have been used. The mean and variance of W are easily obtained as

$$E(W) = \frac{d}{d\theta} \log M_W(\theta)|_{\theta=0} = \psi(m_1) - \psi(m_2) + \log \frac{m_2}{m_1}$$

and

$$\text{var}(W) = \frac{d^2}{d\theta^2} \log M_W(\theta)|_{\theta=0} = \psi^{(1)}(m_1) + \psi^{(1)}(m_2),$$

where, as before, ψ and $\psi^{(1)}$ are the digamma and trigamma functions.

All the special cases of the generalized F distribution listed above are apparent upon inserting the Stirling approximation into $M_W(\theta)$ except the convergence to the normal as $(m_1, m_2) \to (\infty, \infty)$. It is easily seen that $E(W) \to 0$ as $(m_1, m_2) \to (\infty, \infty)$ and further that

$$\text{var}(W) = m_1^{-1} + m_2^{-1} + O(m_1^{-2}) + O(m_2^{-2})$$

by Stirling's approximation. If $W^* = \sqrt{m_1 m_2/(m_1 + m_2)}\,W$, then $\text{var}(W^*) \to 1$ as $(m_1, m_2) \to (\infty, \infty)$ and it can be shown further (with rather tedious algebra) that as $(m_1, m_2) \to (\infty, \infty)$, $M_{W^*}(\theta) \to e^{\theta^2/2}$, the moment generating function of a standard normal variate.

The generalized F distribution is discussed in more detail in Section 3.8, where parameter transformations are considered to obtain the properties of regular estimation on the boundaries $m_1 = \infty$ or $m_2 = \infty$. Its use in discrimination among submodels is also considered there.

2.2.8 Other Distributions and Generalizations

There are, of course, many other distributions that have been or could be used as models for survival data. We have attempted only to outline some of the more commonly used models along with some of their extensions and generalizations. In modeling adult human mortality, a more rapidly increasing hazard function than that represented by, say, the Weibull distribution is necessary. In fact, a relationship in which the hazard function is an exponential function of follow-up time (age at death) has been found to be descriptive in many investigations, at least for ages greater than 35. Such a relation leads to the Gompertz (1825) hazard $\lambda(t) = \lambda \exp(\gamma t)$. Sometimes the exponential term is generalized to a polynomial function of t. The Makeham (1860) generalization adds a constant to the hazard to give $\lambda(t) = \alpha + \lambda \exp(\gamma t)$.

Failure time is sometimes modeled to include an initial threshold parameter (or guarantee parameter) Δ before which it is assumed that failure cannot occur. The models given above could all be modified in this way simply by replacing T with $T' = T - \Delta$. When such a Δ is known to exist, it should be incorporated in the modeling. But it would be rare that Δ would be known to exist without its value being known. For this reason, and also because of analytical difficulties in estimating Δ, such a threshold is usually not included as a free parameter to be estimated in the methods of Chapter 3. The exercises at the end of Chapter 3 outline some of the standard results for models with thresholds.

All the models discussed above are appropriate for continuous failure time variables. As noted earlier, however, failure time can be discrete and, correspondingly, discrete models are required. Some examples of discrete failure time models are given in the exercises at the end of the chapter. Discrete models in a regression framework are discussed in Section 2.4.

2.3 REGRESSION MODELS

In Section 2.2, several survival distributions were introduced for modeling the survival experience of a homogeneous population. Usually, however, there are explanatory variables upon which failure time may depend. It therefore becomes of interest to consider generalizations of these models to take account of concomitant information on the individuals sampled.

Consider failure time $T > 0$ and suppose that a vector of basic covariates $x' = (x_1, x_2, \ldots)$ is available on each individual, their measurements having been taken at or before time 0. Thus, x may contain information on treatment group, various physical measurements, time on study, and so on, and aspects of x are expected to be predictive of subsequent failure time. We consider models for the failure time that depend on a vector of derived covariates $Z' = (Z_1, \ldots, Z_p)$ which are obtained as functions of the basic covariates x. Note that Z may include both quantitative variables and qualitative variables such as treatment group; the latter can be incorporated through the use of indicator variables. It may also, for example, include interactions between elements of x or quadratic terms in elements of x. The principal problem dealt with in this book is that of modeling and determining the relationship between T and derived covariates Z. Certain of the covariates are usually of primary interest, such as those specifying treatment groups or specific risk factors of interest. One primary aim is then to evaluate treatment effects, or examples, while accounting for heterogeneity among the individuals sampled.

2.3.1 Exponential and Weibull Regression Models

The exponential distribution can be generalized to obtain a regression model by allowing the failure rate to be a function of the derived covariates Z. The hazard at time t for an individual with basic covariate vector x can be written

$$\lambda(t; x) = \lambda(Z).$$

Thus the hazard for a given x is a constant characterizing an exponential failure time distribution, but the failure rate depends on the derived covariates Z. The $\lambda(\cdot)$ function may be parameterized in many ways. If the effect of the components of Z is only through a linear function, $Z'\beta$, one has

$$\lambda(t; x) = \lambda \, c(Z'\beta),$$

where $\beta' = (\beta_1, \ldots, \beta_p)$, of regression parameters, λ is a constant, and c is a specified functional form. The choice of c may depend on the particular data being considered. Three specific forms have been used (e.g., Feigl and Zelen, 1965): (1) $c(s) = 1 + s$, (2) $c(s) = (1 + s)^{-1}$, and (3) $c(s) = \exp(s)$. The first two of these, corresponding to (1) the failure rate and (2) the mean survival time, being linear functions of Z. They both suffer from the disadvantage that the set of β values considered must be restricted to guarantee that $c(Z'\beta) > 0$ for all possible Z. In many ways, (3) is the most natural of the forms since it takes only positive values. We use the form $c(s) = \exp(s)$ here and elsewhere, although it should be kept in mind that other forms may be more appropriate in specific settings and could be used without adding unduly to numerical or analytical computations.

Consider then the model with hazard function

$$\lambda(t; x) = \lambda \exp(Z'\beta). \tag{2.6}$$

The conditional density function of T given x is then

$$f(t; x) = \lambda \exp(Z'\beta) \exp[-\lambda t \, \exp(Z'\beta)].$$

The model (2.6) specifies that the log failure rate is a linear function of the covariate Z. In terms of the log survival time, $Y = \log T$, the model (2.6) can be written

$$Y = \alpha - Z'\beta + W, \tag{2.7}$$

where $\alpha = -\log \lambda$ and W has the extreme value distribution (2.1). The model (2.6) is a log-linear model; it is a linear model for Y with the error variable W having a specified distribution.

The Weibull distribution can be generalized to the regression situation in essentially the same way. If the conditional hazard is

$$\lambda(t;x) = \gamma(\lambda t)^{\gamma - 1} \exp(Z'\beta),$$

the conditional density of T is

$$f(t;x) = \lambda \gamma(\lambda t)^{\gamma - 1} \exp(Z'\beta) \exp[-(\lambda t)^{\gamma} \exp(Z'\beta)]. \tag{2.8}$$

The effect of the covariates is again to act multiplicatively on the Weibull hazard. Alternatively, in terms of $Y = \log T$, the model (2.8) is the linear model

$$Y = \alpha + Z'\beta^* + \sigma W, \tag{2.9}$$

where $\alpha = -\log \lambda$, $\sigma = \gamma^{-1}$, and $\beta^* = -\sigma\beta$.

The forms of the exponential and Weibull regression models suggest two distinct generalizations. First, the effect of the covariates in either (2.6) or (2.8) is to act multiplicatively on the hazard function. This relationship suggests a general model called the *relative risk or Cox model*. Second, both of these models are log-linear models; that is, the covariates act additively on Y (or multiplication on T). From this we obtain a general class of log-linear models called the *accelerated failure time model*.

2.3.2 Relative Risk or Cox Model

Again, let $\lambda(t;x)$ represent the hazard function at time t for an individual with basic covariates x. The relative risk model (Cox, 1972) specifies that

$$\lambda(t;x) = \lambda_0(t) \exp(Z'\beta), \tag{2.10}$$

where $\lambda_0(\cdot)$ is an arbitrary unspecified baseline hazard function for continuous T. In this model, the covariates act multiplicatively on the hazard function. If $\lambda_0(t) = \lambda$, (2.10) reduces to the exponential regression model (2.6); the Weibull model (2.8) is the special case $\lambda_0(t) = \lambda \gamma(\lambda t)^{\gamma - 1}$.

The conditional density function of T given x corresponding to (2.10) is

$$f(t;x) = \lambda_0(t) \exp(Z'\beta)\exp\left[-\exp(Z'\beta) \int_0^t \lambda_0(u)\, du\right]. \tag{2.11}$$

The conditional survivor function for T given Z is

$$F(t;x) = [F_0(t)]^{\exp(Z'\beta)} \tag{2.12}$$

where

$$F_0(t) = \exp\left[-\int_0^t \lambda_0(u)\,du\right].$$

Thus the survivor function of t for a covariate value, x, is obtained by raising the baseline survivor function $F_0(t)$ to a power. The class of models produced by this process is sometimes referred to as the *Lehmann class* (Lehmann, 1953).

If $\lambda_0(\cdot)$ is arbitrary, this model is sufficiently flexible for many applications. There are, however, two important generalizations that do not substantially complicate the estimation of β. First, the nuisance function $\lambda_0(t)$ can be allowed to vary in specific subsets of the data. Suppose that the population is divided into r strata and that the hazard $\lambda_j(t;x)$ in the jth stratum depends on an arbitrary shape function $\lambda_{0j}(t)$ and can be written

$$\lambda_j(t,x) = \lambda_{0j}(t)\,\exp(Z'\beta), \tag{2.13}$$

for $j = 1,\ldots,r$, where $Z = (Z_1,\ldots,Z_p)$ is again a vector of derived covariates. Such a generalization is useful, for instance, if some explanatory variable or variables do not appear to have a multiplicative effect on the hazard function. The range of such variables can then be divided into strata with only the remaining regression variables contributing to the exponential factor in (2.13).

The second important generalization allows the covariates Z to depend on time. Such regression variables arise, for example, in the heart transplant example of Section 1.1.3, where the treatment group itself is time-dependent, as are certain donor–recipient matching variables. In other instances, the covariate $Z(t)$ may be thought of as a stress factor affecting the individuals under study at time t, as for example pollution levels where the failure corresponds to a severe asthmatic attack. In other instances, components of $Z(t)$ may simply reflect interactions between covariates x measured at baseline with time. With such time-dependent covariates, the relative risk model is of the form

$$\lambda[t;X(t)] = \lambda_0(t)\,\exp[Z(t)'\beta],$$

where $X(t) = \{x(u) : 0 \leq u < t\}$ is the history prior to time t of basic, possibly time-dependent, covariates $x(u)$. The covariates modeled are defined as suitable functions of $X(t)$. The use and analysis of time-dependent covariables are examined in Chapters 4 and 6 in various contexts.

We have avoided the fairly common nomenclature *proportional hazards model* for the model (2.10). With fixed covariates, the hazards within a stratum are proportional under the model. The important generalization to time-dependent covariates, however, suggests that *relative risk* (or *Cox*) *model* is a more appropriate name.

2.3.3 Accelerated Failure Time Model

The multiplicative effect of the regression variables on the hazard as specified in (2.10) has a clear and intuitive meaning. Without restriction on $\lambda_0(\cdot)$, however, this model postulates no direct relationship between Z and the time to failure T itself. In Section 2.2.1 it was noted that the exponential and Weibull regression models are linear in $Y = \log T$ [see (2.7) and (2.9)]. In this section we obtain a second semiparametric class of survival models, the class of log-linear models for T.

Suppose that $Y = \log T$ is related to the derived covariate Z via a linear model $Y = Z'\beta + W$, where W is an error variable with density f. Exponentiation gives $T = \exp(Z'\beta)S$, where $S = \exp(W) > 0$ has hazard function $\lambda_0(s)$, say, that is independent of β. It follows that the hazard function for T can be written in terms of this baseline hazard $\lambda_0(\cdot)$ according to

$$\lambda(t;x) = \exp(-Z'\beta)\lambda_0[te^{-Z'\beta}]. \qquad (2.14)$$

The survivor function is

$$F(t;x) = \exp\left\{ -\int_0^t \exp(-Z'\beta)\lambda_0[ue^{-Z'\beta}]\,du \right\}$$
$$= \exp\{-\Lambda_0[te^{-Z'\beta}]\}, \qquad (2.15)$$

where $\Lambda_0(t) = \int_0^t \lambda_0(u)\,du$. The density function is the product of (2.14) and (2.15).

Although the interpretation of the model (2.14) in terms of $\log T$ is straightforward, it is also easily seen that this model specifies that the effect of the covariate is multiplicative on t rather than on the hazard function. That is, there is a baseline hazard function $\lambda_0(t)$ which applies when $Z = 0$, and the effect of the regression variables is to alter the rate at which an individual proceeds along the time axis. Equivalently, it is supposed that the role of Z is to accelerate (or decelerate) the time to failure. The accelerated failure time model (2.14) is discussed further in Chapters 3 and 7. An extension of the model to incorporate time-dependent covariates is given in Chapter 7. Figure 2.4 compares the relative risk and the accelerated failure time models for a simple $Z = 0$ or $Z = 1$ covariate which acts multiplicatively on the hazard, or multiplicatively on the failure time.

All the parametric models discussed in Section 2.2 lead to linear models for Y, so that, in many ways, the log-linear regression model is their most natural generalization. The exponential and Weibull regression models can be considered as special cases of either the accelerated failure time model (2.14) or the relative risk model (2.10). Note, however, that log-linear models derived from the other parametric models are not special cases of (2.10). For example, log-normal hazard functions (Section 2.2.3) with different location parameters α_1 and α_2, are not proportional to one another. In the next section it is shown that the only log-linear models that are also relative risk models are the exponential and Weibull regression models of Section 2.3.1.

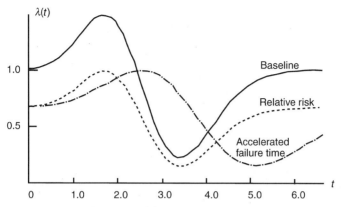

Figure 2.4 The baseline hazard function $\lambda_0(u)$ corresponding to $Z = 0$ is compared to the hazard for $Z = 1$ ($\beta = -\log 1.5$) under a relative risk model and to $z = 1$ ($\beta = \log 1.5$) under an accelerated failure time model.

2.3.4 Comparison of Regression Models

Consider now the intersection of the relative risk and accelerated failure time models; or equivalently, consider the subset of log-linear models in which the regression variable acts multiplicatively on the hazard function. Using subscripts 1 and 2 for the respective models and changing the sign of the accelerated failure time regression parameter, we require that

$$\lambda_{01}(t)\,\exp(Z'\beta_1) = \lambda_{02}[t\,\exp(Z'\beta_2)]\,\exp(Z'\beta_2)$$

for all (t, Z). The value $Z = 0$ gives $\lambda_{01}(\cdot) = \lambda_{02}(\cdot) = \lambda_0(\cdot)$, say, while $z = (-\log t / \beta_{21}, 0, \dots, 0)$ gives, at that t,

$$\lambda_0(t)t^{\beta_{11}\beta_{21}^{-1}} = \lambda_0(1)t^{-1},$$

where β_{11} and β_{21} are the first components of β_1 and β_2, respectively. It follows that for all t,

$$\lambda_0(t) = \lambda\gamma(\lambda t)^{\gamma - 1},$$

where $\gamma = \beta_{11}\beta_{21}^{-1}$ and $\lambda = [\lambda_0(1)/\gamma]^{1/\gamma}$. Note also that $\beta_1 = \gamma\beta_2$. The Weibull (and exponential) log-linear regression models are then the only log-linear models in (2.10). Also, the discussion above leads to a characterization of the two-parameter Weibull model as the unique family that is closed under both multiplication of failure time and multiplication of the hazard function by an arbitrary nonzero constant.

Both general classes of models described would provide sufficient flexibility for many purposes if methods for estimating β were available that did not require undue restrictions on the nuisance function $\lambda_0(\cdot)$. The remarkable feature of the relative risk or Cox model is that suitable methods of inference are available without any restriction on $\lambda_0(\cdot)$. This is discussed in Chapter 4. Parametric procedures for estimation in the accelerated failure time family are given in Chapter 3. In Chapter 7 we discuss rank-based methods of inference under (2.14) that can provide consistent regression parameter estimates regardless of the error density f.

2.4 DISCRETE FAILURE TIME MODELS

2.4.1 General

All the models discussed to this point are appropriate for failure time data arising from continuous distributions. As remarked earlier, however, failure time data are sometimes discrete either through the grouping of continuous data due to imprecise measurement or because time itself is discrete. The latter case arises, for example, when the response time represents the number of episodes that occur prior to a terminal event. A concrete example would arise if the response were the number of standardized blows required to fracture a piece of pavement.

Any of the continuous failure time models discussed in Section 2.2 can be used to generate a discrete model by introducing a grouping on the time axis. For example, suppose that the underlying continuous failure time S has a Weibull distribution with survivor function

$$\exp[-(\lambda s)^\gamma]$$

and times are grouped into unit intervals so that the discrete observed variable is $T = [S]$ (where $[c]$ represents "integer part of c"). The probability function of T can be written

$$f(t) = P(T = t) = P(t \leq S < t + 1)$$
$$= \theta^{t^\gamma} - \theta^{(t+1)^\gamma}, \qquad t = 0, 1, 2, \ldots, \tag{2.16}$$

where $0 < \theta = \exp(-\lambda^\gamma) < 1$. The special case $\gamma = 1$ is the geometric distribution with probability function $\theta^t(1 - \theta)$. The hazard function corresponding to (2.16) is

$$\lambda(t) = P(T = t | T \geq t)$$
$$= 1 - \theta^{(t+1)^\gamma - t^\gamma},$$

which is monotone increasing, monotone decreasing, or constant for $\gamma > 1$, $\gamma < 1$, or $\gamma = 1$, respectively. This can be generalized to a regression model by applying the same grouping to the Weibull regression model with density (2.8).

In the next two sections we discuss general discrete regression models which, like the continuous relative risk or accelerated failure time models, allow an arbitrary baseline hazard function which can be estimated from the data.

2.4.2 Discrete Regression Models

A discrete analog of the relative risk model with fixed covariates can be obtained by applying the survivor function relationship (2.12) directly to a discrete model. Let the failure time T given basic covariates x have a discrete distribution with mass points at $0 < a_1 < a_2 < \cdots$. Let $Z = (Z_1, \ldots, Z_p)$ be a vector of derived covariates as before, and let $F_0(t)$ represent the baseline survivor function for $Z = 0$. The corresponding survivor function for covariates x is

$$F(t; x) = F_0(t)^{\exp(Z'\beta)}, \tag{2.17}$$

as in (2.12). If the hazard function corresponding to F_0 has contribution λ_i at a_i, then

$$F_0(t) = \prod_{i|a_i \leq t} (1 - \lambda_i)$$

and from (2.17),

$$F(t; x) = \prod_{i|a_i \leq t} (1 - \lambda_i)^{\exp(Z'\beta)}. \tag{2.18}$$

The hazard at a_i for covariate Z is then

$$1 - (1 - \lambda_i)^{\exp(Z'\beta)}. \tag{2.19}$$

It is of some interest to note that the discrete model (2.19) can also be obtained by grouping the continuous model (2.10). Thus if continuous failure times arising from the relative risk model (2.10) are grouped into disjoint intervals $[0 = c_0, c_1), [c_1, c_2), \ldots, [c_{k-1}, c_k = \infty)$, the hazard of failure in the ith interval for an individual with covariate Z is

$$P\{T \in [c_{i-1}, c_i) | T \geq c_{i-1}\} = 1 - (1 - \lambda_i)^{\exp(Z'\beta)},$$

where $\lambda_i = 1 - \exp[-\int_{c_{i-1}}^{c_i} \lambda_0(u)\, du]$. This discrete model is then the uniquely appropriate one for grouped data from the continuous relative risk model.

If the discrete baseline cumulative hazard function is written as

$$\Lambda_0(t) = \sum_{a_i \leq t} \lambda_i, \tag{2.20}$$

as in Section 1.2.3, we see that the hazard function corresponding to covariate vector Z is

$$d\Lambda(t; x) = 1 - [1 - d\Lambda_0(t)]^{\exp(Z'\beta)}. \tag{2.21}$$

Note that if $d\Lambda_0(t)$ is replaced with a continuous hazard $\lambda_0(t)\,dt$ so that $\Lambda_0(t)$ is the cumulative hazard of a continuous failure time variate, then (2.21) gives precisely $\lambda(t;x) = \lambda_0(t)\exp(Z'\beta)$, as in (2.10). If $\Lambda_0(t)$ is the baseline cumulative hazard function $(Z = 0)$ for a discrete, continuous, or mixed random variable, (2.21) defines a regression model for which the relationship between the survivor and hazard functions is

$$
\begin{aligned}
F(t;Z) &= \mathscr{P}_0^1[1 - d\Lambda(u;x)] \\
&= \mathscr{P}_0^1[1 - d\Lambda_0(u)]^{\exp(Z'\beta)},
\end{aligned}
\tag{2.22}
$$

where $\mathscr{P}$ is the product integral of Section 1.2.3. The expression (2.22) reduces to (2.18) in the discrete case and to (2.12) in the continuous case. We shall refer to the discrete model in (2.21) as the *grouped relative risk model*. The form of (2.21) suggests a procedure whereby many different discrete models might be generated.

In terms of hazard relationships, perhaps the simplest discrete, mixed, or continuous model is described by the hazard relationship

$$
d\Lambda(t;x) = \exp(Z'\beta)\,d\Lambda_0(t),
\tag{2.23}
$$

which retains the multiplicative hazard relationship. Here again, $\Lambda_0(t)$ is a discrete, continuous, or mixed baseline cumulative hazard function. In the discrete or mixed case, there will be some range restrictions on the parameter β induced by the fact that the discrete hazard components are less than or equal to unity. We shall refer to the model (2.23) as the *discrete and continuous relative risk model*. The model has the advantage of retaining the relative risk interpretation of the multiplicative factor $\exp(Z'\beta)$.

Another discrete failure time regression model was proposed by Cox (1972) and specifies a linear log odds model for the hazard probability at each potential failure time. Thus if $\Lambda_0(t)$ is an arbitrary discrete or continuous cumulative hazard function, the hazard for an arbitrary Z is $d\Lambda(t;Z)$, where

$$
\frac{d\Lambda(t;x)}{1 - d\Lambda(t;x)} = \frac{d\Lambda_0(t)}{1 - d\Lambda_0(t)}\,\exp(Z'\beta).
\tag{2.24}
$$

This is a linear binary logistic model with an arbitrary location parameter corresponding to each discrete failure time point. We refer to this as the *discrete logistic model*. The effect of the covariates is to act multiplicatively, not on the discrete hazards but on the discrete odds. Thus the interpretation of $\exp(Z'\beta)$ is as an odds ratio rather than a relative risk. If the cumulative hazard $\Lambda_0(t)$ is continuous, the denominator terms reduce to 1, and this model again yields the continuous proportional hazards model (2.10). The three discrete models (2.20), (2.23), and (2.24) are therefore very similar if all of the discrete hazard contributions λ_i are small.

Any of these three discrete models could be generalized to allow time-dependent covariates. Discrete models are discussed further in Section 4.8.

BIBLIOGRAPHIC NOTES

Properties of the exponential, Weibull, log-normal, gamma, generalized gamma, and log-logistic distributions are discussed by Johnson and Kotz (1970a,b), who also give extensive bibliographies on these distributions. Some of these distributions are also discussed by Cox (1972), Mann et al. (1974), Gross and Clark (1975), Lawless (1982), and Klein and Moeschberger (1997). The generalized gamma distribution was introduced by Stacy (1962) and has been discussed by Stacy and Mihram (1965), Parr and Webster (1965), Harter (1967), Hagar and Bain (1970), and Prentice (1974). The generalized F distribution is discussed by Prentice (1975).

Exponential, Weibull, and log-normal regression models have received considerable use in the literature. Since most of this work has been concerned with estimation, a list of references for these and other parametric models is deferred to the bibliographic notes at the end of Chapter 3.

The relative risk model was introduced by Cox (1972); the two-sample special case with censored data was considered by Peto and Peto (1972). This model has been discussed extensively with regard to inferential problems, and references are given in Chapter 4. The grouped relative risk model was given by Kalbfleisch and Prentice (1973), and Cox (1972) suggested the linear logistic model for the discrete case. The discrete relative risk model (2.23) was proposed and discussed by Prentice and Kalbfleisch (2002). Additional references are given in Chapter 5, where inference from discrete models is considered more fully. The accelerated failure time model was introduced by Cox (1972) and considered by Prentice (1978) and by many authors since. A bibliographic summary is given in Chapter 7, where general inference in this class is considered. Methods of inference for discrete failure time distributions are considered in Section 4.8, and additional references are given there.

EXERCISES AND COMPLEMENTS

2.1   **(a)** Let $T_1, \ldots, T_n$ be a random sample from a distribution with survivor function $F(t)$ and suppose that as $t \to 0$,

$$F(t) = 1 - \lambda t + o(t)$$

for some $\lambda > 0$. Show that the limiting distribution of $X_n = n \min (T_1, \ldots, T_n)$ is exponential with failure rate λ. As noted in Section 2.2.1, this property is sometimes taken as justification for the choice of an exponential model.

[Note that a function $g(t)$ is said to be $o(t^r)$ as $t \to 0$ if $\lim_{t \to 0} g(t)/t^r = 0$.]

  (b) Suppose now that for t near 0,

$$F(t) = 1 - (\lambda t)^\gamma + o(t^\gamma),$$

where $\lambda > 0$ and $\gamma > 0$. Show that the limiting distribution of

$$Y_n = n^{1/r} \min(T_1, \ldots, T_n)$$

is Weibull with shape γ and scale λ.

2.2 (a) Show that the exponential distribution is the only continuous distribution for which the mean residual lifetime $r(t) = $ constant for all $t > 0$.

  (b) Show that if $\lambda(t) > 0$ for all t and $\lambda(t) \to c \in [0, \infty)$ as $t \to \infty$, then $r(t) \to c^{-1}$ as $t \to \infty$.

  (c) Examine the form of $r(t)$ for the log-normal and gamma distributions.

2.3 Show that the moment generating function of the logistic distribution (2.4) is

$$\Gamma(1 + \theta)\Gamma(1 - \theta)$$

and that the $2r$th cumulant is

$$k_{2r} = \frac{(2r)!\,\zeta(2r)}{r}, \qquad r = 1, 2, \ldots,$$

where $\zeta(n)$ is the Riemann zeta function

$$\zeta(n) = \sum_{i=1}^{\infty} i^{-n}.$$

Hence show that the excess in kurtosis for the extreme value density is 1.2. [Note that $\zeta(2) = \Pi^2/6$ and $\zeta(4) = \Pi^4/90$.]

2.4 Let W_1 and W_2 be independent with the extreme value density $\exp(w - e^w)$. Show that $V = W_1 - W_2$ has the logistic density

$$f(v) = e^v(1 + e^v)^{-2}.$$

Derive the variance and kurtosis of the logistic distribution from those of the extreme value distribution.

2.5 Consider a two-sample situation $(Z = 0, 1)$, in which the hazard is exponential:

$$\lambda(t; Z) = \lambda \exp(Z'\beta)$$

for the continuous failure variable T. The time axis is grouped into disjoint intervals $I_j = [c_{j-1}, c_j), j = 1, 2, \ldots,$ where $c_0 = 0 < c_1 < \cdots$ and $c_k \to \infty$ as $k \to \infty$. Define the discrete variable $Y = j$ for $T \in I_j, j = 1, 2, \ldots.$

(a) Verify that the resulting discrete model is of the form (2.21) and that the same parameter β measures the sample differences.

(b) Show that if the grouping intervals are constant so that $c_j - c_{j-1} = c/\lambda$, $j = 1, 2, \ldots$, where c is a positive constant, the discrete model also has the logistic relationship (2.24). For the latter case, note that the log odds ratio is β^*, where

$$\exp(\beta^*) = \frac{1 - \exp(-ce^\beta)}{1 - \exp(-c)}.$$

This is close to β if c is small.

(c) Again with constant grouping intervals, show that the model can also be written in the form (2.23) as a mixed discrete and continuous Cox model. Express the relative risk in the discrete model in terms of the relative risk in the continuous model.

CHAPTER 3

Inference in Parametric Models and Related Topics

3.1 INTRODUCTION

As mentioned previously, one important reason for specialized statistical models and methods for failure time data is the need to accommodate right censoring in the data. It is usually the case that censoring greatly complicates the exact distribution theory for the estimators even when the censoring mechanism is simple and well understood. In other cases, complex censoring mechanisms may make such computations impossible even in principle. This fact leads, in most instances, to a reliance on asymptotic methods for estimation and testing.

The purpose of this chapter is to consider estimation and testing with parametric regression models. We begin with some discussion of exact methods within the exponential distribution, but note that at least in the medical setting, the necessary control on the experiment to allow exact confidence intervals and tests is rarely available. We therefore consider the asymptotic methods based on the likelihood as the primary means for inference.

We delay general discussions of censoring to Chapter 6, and in the present context, consider random censorship models. The derivations are first given in terms of random right censorship, and extensions are considered to allow for more general random censoring and truncation. Our discussion begins with an analysis of independent right-censored data arising from a parametric model, and asymptotic results are first discussed in that context.

3.2 CENSORING MECHANISMS

We consider survival studies in which n items or individuals are put on test and data of the form (T_i, δ_i, x_i), $i = 1, \ldots, n$, are collected. Here δ_i is an indicator variable ($\delta_i = 0$ if the ith item is censored; $\delta_i = 1$ if the ith item failed) and T_i is the corresponding failure or censoring time. As before, $x_i' = (x_{i1}, x_{i2}, \ldots)$ is a vector of basic covariates associated with the ith individual and $Z_i = (Z_{i1}, \ldots, Z_{ip})$ is a vector of modeled or derived covariates that will be incorporated into the failure time model, which is presumed specified up to a parameter vector θ. The survivor function for the ith individual is $P(\tilde{T}_i > t; \theta, x_i) = F(t; \theta, x_i)$ with the corresponding density $f(t; \theta, x_i)$, where $\tilde{T}_i$ is the underlying uncensored failure time variable. For the time being, we restrict attention to absolutely continuous failure times $\tilde{T}_i$.

To obtain the likelihood function for θ, it is necessary to consider the nature of the censoring mechanism. For most of this chapter, we make the rather restrictive assumption of *random censoring*. Specifically, we assume that the censoring time C_i for the ith individual is a random variable with survivor and density functions G_i and g_i, respectively ($i = 1, \ldots, n$), and that given $x_1, \ldots, x_n$, the C_i's are stochastically independent of each other and of the independent failure times $\tilde{T}_1, \ldots, \tilde{T}_n$. Note that the random censorship model includes the special case of *type I censoring*, where the censoring time of each individual is fixed in advance, as well as the case where items enter the study at random over time and the analysis is carried out at a prespecified time (Exercise 3.3). The latter situation provides a good model for some medical studies. It should be noted, however, that the random censorship formulation is not sufficiently general to include some censoring schemes that are commonly used in certain areas of application, as discussed below.

Let $T_i = \tilde{T}_i \wedge C_i$ and $\delta_i = \mathbf{1}(T_i = \tilde{T}_i)$. Thus, T_i is the observed, possibly censored failure time. For random censorship

$$P[T_i \in (t, t + dt), \delta_i = 1; x_i, \theta] = P[\tilde{T}_i \in (t, t + dt), C_i > t; x_i, \theta]$$
$$= G_i(t)f(t; \theta, x_i)\, dt$$

and

$$P[T_i \in (t, t + dt), \delta_i = 0; x_i, \theta] = P[C_i \in (t, t + dt), \tilde{T}_i > t; x_i, \theta]$$
$$= g_i(t)F(t; \theta, x_i)\, dt.$$

Given $x_1, \ldots, x_n$, the pairs (T_i, δ_i), $i = 1, \ldots, n$, are independent. Thus, if the censoring is *noninformative* [i.e., $G_i(t)$ does not involve θ], the likelihood on the data $(T_i = t_i, \delta_i, x_i)$, $i = 1, \ldots, n$, conditional on $x_1, \ldots, x_n$ is

$$L(\theta) \propto \prod_{i=1}^{n} f(t_i; \theta, x_i)^{\delta_i} F(t_i; \theta, x_i)^{1-\delta_i}. \tag{3.1}$$

This likelihood is of the form $L(\theta) = \Pi\, L_i(\theta)$, where $L_i(\theta)$ is $f(t_i; \theta, x_i)$ for a failure and $F(t_i; \theta, x_i)$ for a censored time. The contribution to the likelihood of an individual censored at t_i is just $P(\tilde{T}_i > t_i; x_i, \theta)$.

In fact, the likelihood (3.1) is much more generally correct than the discussion above would suggest. As described explicitly in Section 6.2, there is a class of censoring mechanisms called *independent censoring* for which this likelihood is appropriate. Briefly, the censoring procedure or rules may depend arbitrarily during the course of the study on previous failure times, on previous censoring times, on random mechanisms external to the study, or on values of covariates included in the model. The likelihood is then built up by considering the experience of the entire study group as it evolves over time. This leads to an expression of the form

$$L(\theta) \propto \prod_{i=1}^{n} \lambda(t_i : \theta, x_i)^{\delta_i} \exp\left[-\int_0^{\infty} \sum_{\ell \in R(u)} \lambda(u; \theta, x_\ell)\, du \right], \tag{3.2}$$

where $R(u)$ is the *risk set* at time u and comprises the set of individuals who are alive and still under observation (not censored) at time u^-, just prior to time u. Thus $R(u) = \{i : t_i \geq u\}$. It is left as an exercise to verify that (3.2) and (3.1) are equivalent. Many of the censoring schemes commonly discussed in the literature are independent. For example, the study may continue until the dth smallest failure time occurs, at which time all surviving items are censored (*type II censoring*), or a specific fraction of individuals at risk may be censored at each of several ordered failure times (*progressive type II censoring*).

The censoring scheme is not independent if individuals are censored selectively or withdrawn from study because they appear to be at an unusually high (or low) risk of failure compared to others on study with the same covariates. Some restriction of this type is clearly necessary since it would be impossible to obtain meaningful survival data if, for example, individuals were withdrawn from study whenever they appeared to be in imminent danger of failure.

3.3   CENSORED SAMPLES FROM AN EXPONENTIAL DISTRIBUTION

A simple example illustrates the relationship between the censoring mechanism and the complexity of exact (frequentist) inferences. Suppose that failure times arise from a homogeneous exponential distribution with failure rate λ. In the notation above, $\theta = \lambda$, $F(t; Z, \theta) = F(t; \lambda) = e^{-\lambda t}$ and $f(t; Z, \theta) = \lambda e^{-\lambda t}$. Based on an uncensored sample $t_1, \ldots, t_n$, the likelihood function is

$$L(\lambda) = \lambda^n \exp(-\lambda v),$$

where $v = \sum t_i$ and the maximum likelihood estimate (MLE) is $\hat{\lambda} = n/v$. The likelihood function is determined by the observed value v of $V = \sum \tilde{T}_i$ so that V is sufficient for λ. That is, inference on λ can be based on the value of V and its distribution. The distribution of V is simply obtained from its moment generating function (MGF). The MGF of $\lambda \tilde{T}_i$ with parameter ξ is $(1 - \xi)^{-1}$, so that of λV is $(1 - \xi)^{-n}$. This is the MGF of a gamma variate (2.2.4) with shape $k = n$ and scale $\lambda = 1$. Equivalently, $2\lambda V$ has a χ^2 distribution with $2n$ degrees of freedom. This result leads to simple significance tests and confidence intervals for λ.

Inference on λ is equally simple with type II censoring. Suppose that n individuals are placed simultaneously on test and the study terminates when the dth failure occurs. Denote the ordered failure times by $t_{(1)} < t_{(2)} < \cdots < t_{(d)}$. As in the general expression (3.1), the likelihood contribution from each of the $(n - d)$ items censored at $t_{(d)}$ is $\exp(-\lambda t_{(d)})$, since the corresponding failure times are known only to exceed $t_{(d)}$, and the contribution of the failure $t_{(i)}$ is $\lambda \exp(-\lambda t_{(i)})$, $i = 1, \ldots, d$. The likelihood is then

$$L(\lambda) = \lambda^d \exp(-\lambda v), \tag{3.3}$$

for which $\hat{\lambda} = d/v$. In this case, v is the observed value of the total survival time $V = \sum_1^d \tilde{T}_{(i)} + (n - d)\tilde{T}_{(d)}$, which is again sufficient for λ. The joint density of $\tilde{T}_{(1)}, \ldots, \tilde{T}_{(d)}$ is required to derive the density for V. For this purpose, we divide the time axis into intervals $[0, t_{(1)}), [t_{(1)}, t_{(1)} + dt_{(1)}), [t_{(1)} + dt_{(1)}, t_{(2)}), \ldots, [t_{(d)}, t_{(d)} + dt_{(d)}), [t_{(d)} + dt_{(d)}, \infty)$ and use the multinomial probability of obtaining frequencies $0, 1, 0, \ldots, 1, (n - d)$ in the respective intervals. This gives a probability element

$$\frac{n!}{(n - d)!} \exp[-(n - d)\lambda t_{(d)}] \prod_1^d [\lambda \exp(-\lambda t_{(i)}) \, dt_{(i)}].$$

The corresponding joint PDF of $\tilde{T}_{(1)}, \ldots, \tilde{T}_{(d)}$ is

$$\frac{n! \, \lambda^d \exp(-\lambda v)}{(n - d)!}, \qquad 0 < t_{(1)} < \cdots < t_{(d)},$$

which, of course, gives rise to (and amounts to a derivation of) the likelihood function (3.3) above.

Consider the change of variables to the *normalized spacings* $U_1, \ldots, U_d$, where

$$U_i = (n - i + 1)(\tilde{T}_{(i)} - \tilde{T}_{(i-1)}), \qquad i = 1, \ldots, d,$$

where $\tilde{T}_{(0)} = 0$. The transformation has Jacobian $(n - d)!/n!$ and it is easily seen that $V = \sum U_i$. It follows that the joint density of the U_i's is

$$\prod_1^d (\lambda e^{-\lambda u_i}), \qquad u_1, \ldots, u_d > 0,$$

so that $U_1, \ldots, U_d$ have independent exponential distributions with failure rate λ. As above, it then follows that $2\lambda V = 2\lambda \sum_1^d U_i$ has a χ^2 distribution with $2d$ degrees of freedom and inference proceeds as in the uncensored case. Note that with exponential data, the same estimating efficiency is achieved by following d items until all have failed or a larger number, n, until the dth failure.

With $d = n$, the transformation to the U_i's allows a simple expression for the exponential order statistics since

$$\tilde{T}_{(i)} = \frac{U_1}{n} + \frac{U_2}{n-1} + \cdots + \frac{U_i}{n-i+1}, \qquad i = 1, \ldots, n. \tag{3.4}$$

Since the expectation of each U_i is λ^{-1}, the expectation of $T_{(i)}$ is

$$E(\tilde{T}_{(i)}) = \lambda^{-1} \sum_{j=1}^{i} (n-j+1)^{-1}, \qquad i = 1, \ldots, n.$$

This suggests a simple graphical test for exponentiality with uncensored or type II censored data: A plot of $t_{(i)}$ versus $\sum_{j=1}^{i}(n-j+1)^{-1}$ should be approximately linear if the exponential model is suitable. This procedure is a first-order approximation to that based on a plot of the log survivor function estimator as suggested in Section 2.2.1: The Kaplan-Meier estimator, for $t \in [t_{(i)}, t_{(i+1)})$, can be written

$$\hat{F}(t) = \prod_{j=1}^{i}[1 - (n-j+1)^{-1}],$$

so that

$$\log \hat{F}(t) = \sum_{j=1}^{i} \log[1 - (n-j+1)^{-1}]$$

$$\approx -\sum_{j=1}^{i}(n-j+1)^{-1}.$$

Consider sampling from an exponential model with an arbitrary independent censoring mechanism. The likelihood function can again be written

$$L(\lambda) = \lambda^d \exp(-\lambda v),$$

where $d = \sum \delta_i$ and $v = \sum t_i \delta_i + \sum t_i(1 - \delta_i) = \sum t_i$. Typically, not only V, but also the number of failures D, is random and the sampling distribution of the sufficient statistic (V, D) or that of the MLE $\hat{\lambda} = D/V$ is complicated by the censoring mechanism.

Consider, for example, the simple case of type I censoring where the censoring times $c_1, c_2, \ldots, c_n$ are specified in advance for the n individuals on study. In this case

$$V = \sum_{i=1}^{n} [(1 - \delta_i)c_i + \delta_i \tilde{T}_i],$$

where, as before, $\tilde{T}_i$ is the failure time of the ith individual. Even in this very simple case, the joint distribution of D and V is quite complicated, as is the exact distribution of the MLE $\hat{\lambda} = D/V$. Further, the MLE is no longer sufficient for inference about λ. This complication of the distribution theory and the lack of sufficiency properties leads, in most cases, to the use of asymptotic likelihood arguments for inference about the parameters.

3.4 LARGE-SAMPLE LIKELIHOOD THEORY

In this section we consider the main asymptotic results that typically apply to the likelihood function and the maximum likelihood estimator. Generally speaking, these results are applicable to the likelihoods derived from parametric regression models for failure time data with independent censoring mechanisms. Some discussion of the derivations of the asymptotic results is given in Section 3.8 and in Chapter 5. Our purpose here, however, is to draw together the various methods that are used for inference from a parametric likelihood function. These methods are based on asymptotic approximations to the distribution of the score statistic, the maximum likelihood estimator and the likelihood ratio statistic. These basic procedures also apply to the partial likelihood that arises in analysis of the relative risk or Cox regression model, discussed in Chapter 4.

As discussed in Section 3.2, we suppose that data (t_i, δ_i, x_i), $i = 1, \ldots, n$, are available from a parametric model with parameter $\theta = (\theta_1, \ldots, \theta_p)$ and give rise to the likelihood function

$$L(\theta) = \prod_{i=1}^{n} L_i(\theta) \tag{3.5}$$

$$= \prod_{i=1}^{n} \lambda(t_i; \theta, x_i)_i^{\delta} \exp\left[-\int_0^{\infty} \sum_{\ell \in R(u)} \lambda_\ell(u; \theta, x_\ell) \, du \right], \tag{3.6}$$

where $R(u) = \{\ell : t_\ell \geq u\}$ is the set of individuals at risk (i.e., alive and uncensored) at time $u > 0$. These two expressions describe two ways of viewing survival data. In (3.5), the term $L_i(\theta)$ is the likelihood contribution arising from the ith individual and represents the corresponding density function if the individual is observed to fail, and the survivor function if the individual is censored. This is a natural way to view the problem with random censoring, and asymptotic arguments proceed in fairly standard ways. Further, the approach to asymptotics extends in a natural way to other types of random censoring (interval or left censoring) as well as random truncation. The expression (3.6) arises naturally from viewing the survival data as unfolding sequentially in time and relates to general independent right censoring. In this case, asymptotic results can be obtained through martingale central limit theorems as discussed in Chapter 5. In some respects, however, these methods are more restrictive in that they generally do not extend in natural ways to certain other types of censoring or truncation that arise with some frequency in applications.

With either approach, however, the main result relates to a central limit theorem for the score vector. Other results follow in fairly standard ways from this foundation. In this section we discuss asymptotic results as they relate to the random censorship model and the likelihood expression (3.5).

3.4.1 Score Statistic

Central to asymptotic likelihood arguments are the efficient score vectors

$$U_i(\theta) = \frac{\partial}{\partial \theta} \log L_i(\theta) = \left[\frac{\partial}{\partial \theta_j} \log L_i(\theta) \right]_{p \times 1}, \tag{3.7}$$

$i = 1, \ldots, n$. If the operations of expectation and differentiation with respect to θ can be interchanged, it can be shown that $U_i(\theta)$ has expectation 0 and covariance matrix

$$\mathscr{I}_i(\theta) = E[U_i(\theta)U_i'(\theta)] = -\left(E \frac{\partial^2 \log L_i}{\partial \theta_j \, \partial \theta_k} \right)_{p \times p}. \tag{3.8}$$

It should be noted that although the conditions above leading to (3.8) are quite mild, they are nonetheless sufficient to exclude parameters that define the range of support for the random variable $\tilde{T}$ and thus threshold or guarantee parameters are not covered by these arguments.

With random censorship, $U_1(\theta), \ldots, U_n(\theta)$ are independent. Thus, under certain conditions, a central limit theorem will apply to the total score statistic

$$U(\theta) = \sum_{i=1}^{n} U_i(\theta).$$

As a consequence, $U(\theta)$ is typically asymptotically normal with mean 0 and covariance matrix $\mathscr{I}(\theta) = \sum_{i=1}^{n} \mathscr{I}_i(\theta)$. For convenience, we speak of $U(\theta)$ as being asymptotically normal with mean 0 and covariance $\mathscr{I}(\theta)$, although it is, of course, the standardized version $n^{-1/2}U(\theta)$ that converges. For the central limit theorem to apply, the requirements are basically that the relative information $\mathscr{I}_i(\theta)\mathscr{I}(\theta)^{-1}$ approaches a zero matrix for each i as $n \to \infty$, and the total information $\mathscr{I}(\theta)$ approaches infinity at a suitable rate. Necessary and sufficient conditions for the central limit theorem for sums of independent variables were given by Lindeberg (e.g., Feller, 1971, p. 262; Shorack, 2000, p. 260). The sufficient conditions of Ljapunov (e.g., Feller, 1971, p. 286) are often more easily verified.

Under these conditions, the asymptotic distribution of the score $U(\theta)$ can be used for approximate inference about θ. Specifically, under a hypothesized $\theta = \theta_0$, the score statistic $U(\theta_0)$ is asymptotically normal with mean 0 and variance $\mathscr{I}(\theta_0)$. If $\mathscr{I}(\theta_0)$ is nonsingular, it follows that

$$U'(\theta_0)\mathscr{I}(\theta_0)^{-1}U(\theta_0) \tag{3.9}$$

has an asymptotic χ_p^2 distribution where p is the dimension of θ. The hypothesized value of θ_0 is assessed by comparing the value of (3.9) with the χ^2 tables. Alternatively, an approximate confidence region for θ can be formed as the set of values θ_0 for which (3.9) is less than a specified upper level of the χ^2 distribution. It is also possible to use the score function for tests of certain composite hypotheses as discussed in Section 3.4.4.

The limiting normal distribution of the score function is the fundamental result of asymptotic likelihood theory. It serves as the basis on which the other asymptotic results are built.

3.4.2 Maximum Likelihood Estimator

Simpler methods of interval estimation are available than those based on the score statistic. These involve the asymptotic distribution of the MLE $\hat{\theta}$. If θ is interior to the parameter space, $L(\theta)$ is thrice differentiable, and certain boundedness conditions on the third derivatives are satisfied, it can be shown that for sufficiently large n, $\hat{\theta}$ is the unique solution to $U(\theta) = 0$, that $\hat{\theta}$ is consistent for θ, and that the asymptotic distribution of $\hat{\theta}$ is multivariate normal with mean θ and covariance matrix $\mathscr{I}(\theta)^{-1}$. We can write

$$\hat{\theta} \sim N(\theta, \mathscr{I}(\theta)^{-1}), \tag{3.10}$$

where again, it is a standardized form $n^{1/2}(\hat{\theta} - \theta)$ that converges. Tests of hypotheses about θ and interval estimation can be based on this result.

For example, approximate confidence regions can be specified using the asymptotic χ_p^2 distribution of

$$(\hat{\theta} - \theta)' \mathscr{I}(\theta)(\hat{\theta} - \theta). \tag{3.11}$$

Similarly, if $\theta' = (\theta_1', \theta_2')$, where $\theta_1' = (\theta_1, \ldots, \theta_k)$, and

$$\mathscr{I}(\theta)^{-1} = \begin{pmatrix} \mathscr{I}^{11}(\theta) & \mathscr{I}^{12}(\theta) \\ \mathscr{I}_{21}(\theta) & \mathscr{I}^{22}(\theta) \end{pmatrix},$$

where $\mathscr{I}^{11}(\theta)$ is of dimension $k \times k$, it follows that

$$(\hat{\theta}_1 - \theta_1)' \mathscr{I}^{11}(\theta)^{-1}(\hat{\theta}_1 - \theta_1)$$

has an asymptotic χ_k^2 distribution. In particular, if $k = 1$, this gives a simple asymptotic standard normal variate for estimating θ_1, or equivalently θ_j, as

$$(\hat{\theta}_j - \theta_j)[i^{jj}(\theta)]^{-1/2}, \tag{3.12}$$

where $i^{jj}(\theta)$ is the (j,j) element of $\mathscr{I}(\theta)^{-1}$. Typically, (3.12) involves the unknown θ and to be useful, $\mathscr{I}(\theta)$ must be estimated.

The results above can be modified by replacing the Fisher information $\mathscr{I}(\theta) = E(-\partial^2 L(\theta)/\partial\theta\,\partial\theta')$ with an estimator. For example, under regularity conditions, $\mathscr{I}(\theta)\mathscr{I}(\hat{\theta})^{-1}$ converges in probability to a $p \times p$ identity matrix. As a consequence, $\mathscr{I}(\hat{\theta})$ can replace $\mathscr{I}(\theta)$ in (3.10), (3.11), and (3.12). An even simpler estimator of $\mathscr{I}(\theta)$ is provided by the observed information

$$I(\theta) = \left(\frac{-\partial^2 \log L(\theta)}{\partial\theta_i\,\partial\theta_j}\right)_{p\times p},$$

the expected value of which is the Fisher information. Thus $\mathscr{I}(\theta)$ can be replaced in the results above with $I(\theta)$ or with $I(\hat{\theta})$ without affecting the asymptotic distributions. It should also be noted that $\mathscr{I}(\theta_0)$ in (3.9) may be replaced with $\mathscr{I}(\hat{\theta})$, $I(\theta_0)$, or $I(\hat{\theta})$ while retaining the asymptotic χ^2 result.

3.4.3   Score Tests of Composite Hypotheses

Under the same regularity conditions as for the maximum likelihood methods, score-based procedures can also be used for asymptotic tests of composite hypotheses. Specifically, suppose that the total score is partitioned as

$$U(\theta_1,\theta_2)' = [U^{(1)}(\theta_1,\theta_2)', U^{(2)}(\theta_1,\theta_2)'],$$

where the component vector $U^{(1)}$ is of dimension k corresponding to θ_1, and where θ_2 and $U^{(2)}$ have dimension $p - k$. Let $\tilde{\theta}_2 = \tilde{\theta}_2(\theta_1)$ be the MLE of θ_2 for given θ_1, which is obtained as a solution to

$$U^{(2)}(\theta_1,\theta_2) = 0.$$

Let $\tilde{U}^{(1)}(\theta_1) = U^{(1)}(\theta_1,\tilde{\theta}_2)$ and $\tilde{\mathscr{I}}^{11}(\theta_1) = \mathscr{I}^{11}(\theta_1,\tilde{\theta}_2)$. The score statistic for estimating or testing hypotheses about θ_1 is

$$\tilde{U}^{(1)}(\theta_1)'\tilde{\mathscr{I}}^{11}(\theta_1)\tilde{U}^{(1)}(\theta_1),$$

which has an asymptotic χ_k^2 distribution. For inference about the scalar component parameter θ_j, we could utilize the related asymptotic normal statistic,

$$\tilde{U}^{(j)}(\theta_j)[\tilde{i}^{jj}(\theta_j)]^{1/2}.$$

As in statistics based on the MLE, the Fisher information $\mathscr{I}(\theta)$ in the score statistics above can be replaced by the observed information $I(\theta)$ while retaining the asymptotic results.

3.4.4 Likelihood Ratio Statistic

A third class of likelihood statistics arises from the likelihood ratio

$$R(\theta) = \frac{L(\theta)}{L(\hat{\theta})}$$

and its asymptotic distribution. If the regularity conditions of maximum likelihood theory hold, then under the hypothesis $\theta = \theta_0$, the asymptotic distribution of

$$-2 \log R(\theta_0) \tag{3.13}$$

is that of a χ_p^2 variate where p is the dimension of θ.

If $\theta' = (\theta_1', \theta_2')$ as above, then under the hypothesis $\theta_1 = \theta_1^0$, $-2 \log R[\theta_1^0, \tilde{\theta}_2(\theta_1^0)]$ has a χ_k^2, where k is the dimension of θ_1. This result and the related score test are quite general since by reparametrization it is usually possible to use the likelihood ratio to test any hypothesis of interest about the θ_i's. Basically, we need to be able to reparametrize to a vector of parameters, $\gamma = \gamma(\theta)$, to which maximum likelihood theory applies and for which the hypothesis of interest is equivalent to specifying the values of a subset of γ_i's. Thus, suppose that $\gamma = h(\theta)$ is a k-dimensional parameter of interest, where h is a differentiable mapping from $R^p \rightarrow R^k$, and the matrix $H(\theta) = \partial h / \partial \theta'$ is of full row rank k. Under the hypothesis $\gamma = \gamma_0$, the statistic

$$-2 \log R_M(\gamma_0) = -2 \log \frac{\sup_{h(\theta) = \gamma_0} L(\theta)}{L(\hat{\theta})}$$

has an asymptotic χ_k^2 distribution, where sup denotes supremum.

One desirable feature of any inference procedure is that the conclusions drawn should be independent of the (to some extent) arbitrary parametrization used. Thus if θ is replaced by some 1:1 function, say, $\lambda = \lambda(\theta)$, the conclusions should be unaltered. All tests and inference procedures based on the likelihood ratio statistic have this property. The score test statistic (3.9) also has this property. To see this for (3.9), we suppose that $\lambda = \lambda(\theta)$ is a 1:1 differentiable function of θ and let $\lambda_0 = \lambda(\theta_0)$. If $U^*(\lambda)$ and $\mathscr{I}^*(\lambda)$ are the score statistic and the information for λ, respectively, straightforward calculations show that

$$U^*(\lambda_0) = J_0 U(\theta_0)$$
$$\mathscr{I}^*(\lambda_0) = J_0 \mathscr{I}(\theta_0) J_0',$$

where J_0 is the Jacobian matrix with (i,j) element $(\partial \theta_i / \partial \lambda_j)$ evaluated at $\lambda = \lambda_0$. It then follows that

$$U^{*'}(\lambda_0) \mathscr{I}^*(\lambda_0)^{-1} U^*(\lambda_0) = U'(\theta_0) \mathscr{I}(\theta_0)^{-1} U(\theta_0).$$

Testing and estimation procedures based on the asymptotic distribution of the maximum likelihood estimate do not have the property of functional invariance. Even the score function tests do not possess this property if $\mathscr{I}(\theta_0)$ is replaced with an estimate, as is generally necessary with censored failure time data, owing to the complex sample spaces over which expectations must be taken. As discussed further in Section 3.4.5, use of the expected information can also be criticized since the inference would then depend on potential censoring times even when these have not affected the observations.

The fact that asymptotic theory applied to $\hat{\theta}$ or $\hat{\lambda}$ leads to different results suggests that some care need be exercised in the use of asymptotic results for the maximum likelihood estimator. Consideration should in general be given to selecting a parametrization $\lambda = \lambda(\theta)$ for which a normal approximation to the distribution of $\hat{\lambda}$ is suitable. Some rough guidelines are to select λ so that its components do not have unnecessary range restrictions and so that its asymptotic variance matrix is reasonably stable near $\hat{\lambda}$. It is also possible, in some instances, to choose λ to make the likelihood nearly symmetric in shape by making the third derivatives of the log likelihood, evaluated at $\hat{\lambda}$, small. The asymptotic distribution of $\hat{\theta}$ is obtained from that of $U(\theta)$ by approximating the log likelihood with a quadratic function, and choosing a parametrization that makes the third derivatives small improves this approximation. This suggests that the normal shape of the likelihood is important for the application of large-sample theory for the estimator. It should be noted that if the likelihood can be made very close to normal in shape, the likelihood ratio statistic and the maximum likelihood estimates yield nearly identical inferences.

The discussion above suggests superiority of the asymptotic approximations based on the likelihood ratio, and this same general conclusion has been reached by many authors.

3.4.5 Exponential Sampling Illustration

In this section we consider application of the methods of Section 3.4 to a sample from the exponential distribution with type I censoring as defined at the end of Section 3.3. We suppose, as there, that failure times $\tilde{T}_i$ arise as independent exponential variates with failure rate λ and the ith individual has a censoring time $C_i = c_i$ fixed in advance of follow-up. Note that the same analysis would apply to a random censorship model where independent censoring times C_i, $i = 1, \ldots, n$, are determined according to some distribution free of λ. In the latter case, the analysis is conditioned upon $C_i = c_i$, the censoring times being ancillary statistics for the estimation of λ. Let $T_i = \tilde{T}_i \wedge c_i$ and again let t_i be the observed value of T_i.

As noted before, the contribution of the ith individual to the likelihood is

$$L_i(\lambda) = \left(\lambda e^{-\lambda t_i}\right)^{\delta_i} \left(e^{-\lambda t_i}\right)^{1-\delta_i}$$
$$= \lambda^{\delta_i} e^{-\lambda t_i}, \qquad i = 1, \ldots, n,$$

so that the efficient scores are

$$U_i(\lambda) = \frac{\delta_i}{\lambda} - T_i$$

and $-\partial^2 \log L_i(\lambda)/\partial\lambda^2 = \delta_i/\lambda^2$. It is easily seen that

$$E[U_i(\lambda)] = \frac{E(\delta_i)}{\lambda} - E(T_i)$$
$$= 0,$$

while

$$\mathcal{I}_i(\lambda) = \frac{E(\delta_i)}{\lambda^2} = \frac{1 - p_i}{\lambda^2},$$

where $p_i = \exp(-\lambda c_i)$ is the probability that the ith item is censored.

If, as before, $D = \sum \delta_i$ represents the total number of failures and $V = \sum T_i$, the total accumulated survival, it follows from Section 3.4.1 that the score function

$$U(\lambda) = \frac{D}{\lambda} - V$$

is asymptotically normal with mean 0 and variance

$$\mathcal{I}(\lambda) = \frac{E(D)}{\lambda^2} = \frac{\sum(1 - p_i)}{\lambda^2}. \tag{3.14}$$

The requirement for asymptotic normality of $U(\lambda)$ is simply that the expected number of deaths approaches infinity as $n \to \infty$. This in turn places some very mild restriction on the censoring times to the effect that the c_i's must not converge too rapidly to zero as n becomes large.

Now we have a common situation in the application of asymptotic methods with censored data: (3.14) involves the potential censoring times for individuals that fail. It is certainly not clear, however, that potential but unobserved censoring times should affect the inference even if they are available. For these reasons, the observed information,

$$I(\lambda) = \frac{-\partial^2 \log L(\lambda)}{\partial\lambda^2},$$

or $I(\hat{\lambda})$, is commonly substituted for $\mathcal{I}(\lambda)$, as discussed earlier. Replacing $\mathcal{I}(\lambda)$ with $I(\lambda) = D/\lambda^2$, we find that $D^{-1/2}\lambda U(\lambda)$ has an asymptotic standard normal distribution.

For example, suppose that failure time in excess of 100 days in the group 1 data of Section 1.1.1 is exponentially distributed with failure rate λ. Then $D = d = 17$ and $v = (43 + 64 + \cdots + 144) = 2195$. The results above give an approximate 95% confidence interval for the failure rate, λ, as the set of λ values for which

$$|\lambda d^{-1/2} U(\lambda)| = |d^{1/2} - \lambda d^{-1/2} v| < 1.96$$

since ± 1.96 bracket the central 95% probability from a standard normal distribution. Direct solution gives $(0.00406, 0.01143)$ as the approximate confidence interval. Maximum likelihood results applied to $\hat{\lambda} = d/v$, with the information estimated by $I(\hat{\lambda}) = v^2/d$, give

$$(\hat{\lambda} - \lambda)I(\hat{\lambda})^{1/2} = d^{1/2} - \lambda d^{-1/2} v$$

as an approximate standard normal variate, which, in this case, is precisely the same as the score procedure. Alternatively, we may prefer to apply asymptotic MLE results to $\gamma = \log \lambda$ since γ has unrestricted range and the asymptotic variance of $\hat{\gamma} = \log \hat{\lambda}$ is estimated by $I_*(\gamma)^{-1} = d^{-1}$ a constant. The latter statement follows from direct manipulation of the likelihood for γ, $L_*(\gamma) = L(e^\gamma) = \exp(\gamma d - e^\gamma v)$. It follows that $d^{1/2} \log \hat{\lambda}/\lambda$ has an asymptotic standard normal distribution, giving

$$[a, b] = \left(\log \frac{d}{v} - 1.96 d^{-1/2}, \ \log \frac{d}{v} + 1.96 d^{-1/2} \right)$$

as an approximate 95% confidence interval for γ and (e^a, e^b) as an approximate 95% interval for λ. For the data of Section 1.1.1, the interval is $(0.00481, 0.01246)$. Similarly, the reparametrization $\alpha = \lambda^{1/3}$ may be considered, since the log likelihood for α is very nearly symmetric about $\hat{\alpha} = \hat{\lambda}^{1/3}$. More precisely, the third derivative of this log likelihood evaluated at $\hat{\alpha}$ is zero. This choice of parameter gives an approximate confidence interval for λ of

$$\left(1 \pm \frac{1.96 d^{-1/2}}{3} \right)^3 \frac{d}{v}$$

based on the asymptotic distribution of $\hat{\alpha}$. As expected, this interval is in close agreement with those given earlier for large d. It has value $(0.00462, 0.01204)$ for the carcinogenesis data.

Finally, the likelihood ratio

$$R(\lambda) = \frac{\lambda^d \exp(-\lambda v)}{(d/v)^d e^{-d}} \tag{3.15}$$

can also be used for inference about λ. The statistic $-2 \log R(\lambda_0)$ has an asymptotic χ_1^2 distribution if $\lambda = \lambda_0$, and an approximate 95% confidence interval for λ is obtained as the set of λ values for which (3.15) has a value less than 3.84, the upper 5% point of χ_1^2. This is simply found by plotting $R(\lambda)$ or $-2\log R(\lambda)$ as a function of λ and reading off the appropriate interval or a simple iterative or search technique will give the more accurate results. In the present case the 95% interval is (0.00462, 0.01203), which is nearly indistinguishable from that based on $\hat{\alpha}$ above. The likelihood ratio statistic (3.15) is particularly well suited for tests of hypotheses about λ. It has, as noted before, the advantage of being functionally invariant so that application to α or γ gives the same results. Moreover, it does not involve unobserved censoring times. It is, in fact, the only inference procedure we have considered with both of these properties.

3.5 EXPONENTIAL REGRESSION

3.5.1 Methods

Suppose that failure times arise from an exponential distribution. The basic covariates are again denoted by x and the model for failure rate depends on a regression vector of p derived covariates $Z = (Z_1, \ldots, Z_p)'$ as in (2.6). For ease of notation let the hazard function be written

$$\lambda(t; x) = \exp(Z'\beta),$$

where $Z_1 = 1$, so that $\lambda = \exp(\beta_1)$ is the failure rate when other components of Z have value 0. On assumption of an independent censoring mechanism, the likelihood function of β is written

$$L(\beta) = \prod_1^n e^{\delta_i Z_i' \beta} \exp(-e^{Z_i' \beta} t_i),$$

where, as before, t_i is an observed survival time with corresponding regression vector $Z_i' = (Z_{1i}, Z_{2i}, \ldots, Z_{pi})$ and δ_i indicates failure ($\delta_i = 1$) or censoring ($\delta_i = 0$). The score vector has components

$$U_j(\beta) = \frac{\partial \log L(\beta)}{\partial \beta_j} = \sum Z_{ji}(\delta_i - e^{Z_i' \beta} t_i), \qquad j = 1, \ldots, p, \qquad (3.16)$$

and the (j, k) element of the observed information matrix $I(\beta)$ is

$$\frac{-\partial^2 \log L(\beta)}{\partial \beta_j \partial \beta_k} = \sum Z_{ji} Z_{ki} \exp(Z_i' \beta) t_i. \qquad (3.17)$$

As in the single-sample case discussed above, the information matrix involves taking an expectation of (3.17) with respect to the variables T_i and δ_i and will typically be difficult or impossible to evaluate. As before, we shall use the observed information in constructing asymptotic results.

For the central limit theorem to apply to the score statistic $U(\beta)$, whose components are defined in (3.16), conditions are required on both the regression vectors Z_i and the censoring mechanism. As before, the times of censoring must not converge too rapidly to zero as $n \to \infty$, but also the Z_i's must be such that the relative information from observation i, $\mathscr{I}_i(\beta)\mathscr{I}(\beta)^{-1}$ goes to a zero matrix as $n \to \infty$. This means that the normal approximation to the distribution of $U(\beta)$ is likely to be poorer when the sample contains isolated and extreme Z values, since such values can exert considerable influence upon the β estimate and upon $\mathscr{I}(\beta)$. Of course, observations corresponding to such Z values should be scrutinized routinely for consistency with the assumed model, quite apart from the asymptotic normality considerations.

The asymptotic distribution of $U(\beta)$ is multivariate normal with mean 0 and variance $\mathscr{I}(\beta)$, while that of $\hat{\beta}$ is multivariate normal with mean β and variance $\mathscr{I}(\beta)^{-1}$. As before, to eliminate the role of unobserved censoring times, the expected information $\mathscr{I}(\beta)$ can be replaced with the observed information $I(\beta)$ or its estimate $I(\hat{\beta})$ in either asymptotic result. Note that $\hat{\beta}$ as well as the likelihood ratio statistic will usually require iterative calculation. The Newton–Raphson method generally works very well. Under this technique, an initial value $\beta^{(0)}$ is updated to $\beta^{(1)} = \beta^{(0)} + [I(\beta^{(0)})]^{-1}U(\beta^{(0)})$ iteratively until convergence is achieved. A starting value of $\beta^{(0)} = (v/d, 0, \ldots, 0)'$ usually will suffice.

3.5.2 Comparisons of Two Exponential Samples

A special case in which $\hat{\beta}$ can be calculated explicitly is that of comparing two exponential samples. Suppose that failure times from two groups of individuals are exponential with failure rates λ_1 and λ_2, respectively. Comparison of the two groups then involves comparison of λ_1 and λ_2. Such data can be placed in the regression framework above by defining for the ith individual a regression vector $Z_i = (Z_{1i}, Z_{2i})$, where $Z_{1i} = 1$, as before, and $Z_{2i} = 0$ if individual i is in group 1 and $Z_{2i} = 1$ if individual i is in group 2. Then $e^{\beta_1} = \lambda_1$ and $e^{\beta_1+\beta_2} = \lambda_2$ and equality of λ_1 and λ_2 is equivalent to $\beta_2 = 0$. The score statistic (3.16) can be written

$$U_1(\beta) = d_1 + d_2 - e^{\beta_1} v_1 - e^{\beta_1+\beta_2} v_2$$
$$U_2(\beta) = d_2 - e^{\beta_1+\beta_2} v_2,$$

where (d_j, v_j) is the number of failures and total survival time in sample $j, j = 1, 2$. This gives $\exp(\hat{\beta}_1) = d_1/v_1$ and $\exp(\hat{\beta}_2) = d_2 v_1/d_1 v_2$. The observed information matrix is

$$I(\beta) = \begin{pmatrix} e^{\beta_1} v_1 + e^{\beta_1+\beta_2} v_2 & e^{\beta_1+\beta_2} v_2 \\ e^{\beta_1+\beta_2} v_2 & e^{\beta_1+\beta_2} v_2 \end{pmatrix},$$

so that

$$I(\hat{\beta}) = \begin{pmatrix} d_1 + d_2 & d_2 \\ d_2 & d_2 \end{pmatrix} \quad \text{and} \quad I(\hat{\beta})^{-1} = \frac{1}{d_1 d_2} \begin{pmatrix} d_2 & -d_2 \\ -d_2 & d_1 + d_2 \end{pmatrix}.$$

Consider now a test of the hypothesis $\beta_2 = 0$, which corresponds to equality of the failure time distributions for the two groups. The asymptotic distribution of $\hat{\beta}_2$ is most convenient for this purpose. From the (2,2) element of $I(\hat{\beta})^{-1}$, $\hat{\beta}_2$ has an asymptotic normal distribution with mean β_2 and variance estimated by $(d_1 + d_2)/d_1 d_2$. The corresponding test for $\beta_2 = 0$ then involves a comparison of

$$\hat{\beta}_2 \sqrt{d_1 d_2/(d_1 + d_2)} \tag{3.18}$$

with standard normal tables.

For illustration, suppose that survival times in excess of 100 days are exponential for both groups of rats for the data in Section 1.1.1. Then $d_1 = 17$, $v_1 = 2195$, as before, and $d_2 = 19$, $v_2 = 2923$ so that $\hat{\beta}_2 = \log(d_2/v_2) - \log(d_1/v_1) = -0.1752$ and (3.18) has value -0.5248, which is central to the standard normal distribution and offers no evidence against the hypothesis $\beta_2 = 0$. A normal approximation to the distribution of (3.18) may be expected to be suitable with only moderate values of d_1, d_2 because of the stable variance estimate and the absence of range restrictions on $\hat{\beta}_2$.

The asymptotic distribution of the score statistic can also be used to test $\beta_2 = \beta_2^0$ upon maximizing out β_1. Let $\hat{\beta}_1(\beta_2^0)$ represent the maximum likelihood estimate of β_1, assuming that $\beta_2 = \beta_2^0$. The asymptotic conditional distribution of $U_2(\beta)$ given $U_1(\beta) = 0$ [i.e., given $\beta_1 = \hat{\beta}_1(\beta_2^0)$] is, from the asymptotic distribution of $U(\beta)$ and multivariate normal theory, normal with mean zero and variance estimated by

$$e^{\beta_1 + \beta_2^0} v_2 - \frac{(e^{\beta_1 + \beta_2^0} v_2)^2}{e^{\beta_1} v_1 + e^{\beta_1 + \beta_2^0} v_2}, \tag{3.19}$$

evaluated at $\beta_1 = \hat{\beta}_1(\beta_2^0)$. Under the hypothesis $\beta_2 = 0$, we have from $U_1(\beta)$, $\exp[\hat{\beta}_1(0)] = (d_1 + d_2)/(v_1 + v_2)$, so that $U_2(\beta)$ evaluated at $(\hat{\beta}_1(0), 0)$ becomes

$$d_2 - \frac{(d_1 + d_2)v_2}{v_1 + v_2} = \frac{d_2 v_1 - d_1 v_2}{v_1 + v_2}$$

with variance estimate, from (3.19), of $(d_1 + d_2)v_1 v_2/(v_1 + v_2)^2$. This gives an asymptotic standard normal statistic

$$\frac{d_2 v_1 - d_1 v_2}{[(d_1 + d_2)v_1 v_2]^{1/2}} \tag{3.20}$$

for testing $\beta_2 = 0$. For the carcinogenesis data, (3.20) has value -0.5255, in close agreement with (3.18). Either the MLE or score procedure could be used to form an approximate confidence interval for β_2, though the asymptotic distribution of $\hat{\beta}_2$ is more convenient for this purpose.

The likelihood ratio method could also be used to test $\beta_2 = 0$. The log likelihood is written

$$\log L(\beta) = d_1\beta_1 - e^{\beta_1}v_1 + d_2(\beta_1 + \beta_2) - e^{\beta_1+\beta_2}v_2,$$

so that

$$\log R(\beta) = \log L(\beta) - \log L(\hat{\beta})$$
$$= \log L(\beta) - d_1 \log \frac{d_1}{v_1} + d_1 - d_2 \log \frac{d_2}{v_2} + d_2.$$

For the hypothesis $\beta_2 = 0$, we have $\exp[\hat{\beta}_1(0)] = (d_1 + d_2)/(v_1 + v_2)$ so that $-2\log R(\beta)$ evaluated at $\beta_1 = \hat{\beta}_1(0)$, $\beta_2 = 0$ has the value

$$2\left[d_1 \log \frac{d_1}{v_1} + d_2 \log \frac{d_2}{v_2} - (d_1 + d_2) \log \frac{d_1 + d_2}{v_1 + v_2}\right], \tag{3.21}$$

which, under the hypothesis $\beta_2 = 0$, has an asymptotic χ_1^2 distribution. Again with the carcinogenesis data, (3.21) has value 0.274, which is in good agreement with the approximate χ_1^2 statistics $(-0.5248)^2 = 0.275$ and $(-0.5255)^2 = 0.276$ from the MLE and score procedures. None of the these tests suggests any survival difference between the two groups of rats. Of course, all the tests are based on an assumed exponential model for time in excess of 100 days, which is suspect on the basis of the survival curves of Figure 1.2. The more general parametric regression models of Chapter 2 would be expected to provide an improved fit.

3.6 ESTIMATION IN LOG-LINEAR REGRESSION MODELS

The likelihood methods of Section 3.5 are easily generalized to any specific log-linear regression model, such as those arising from the parametric models of Section 2.2.

Suppose that the PDF for $Y = \log \tilde{T}$ can be written

$$\sigma^{-1}f(w),$$

where $w = (y - Z'\beta)/\sigma$ and $Z = (Z_1, \ldots, Z_p)'$ is a regression vector corresponding to failure time t. Note that σ is a scale constant and that if $Z_1 = 1$ identically, the first component of $\beta' = (\beta_1, \ldots, \beta_p)$ represents the general location of Y. As simple examples, a Weibull regression model is given by $f(w) = \exp(w - e^w)$; the exponential model of Section 3.5 requires, in addition, that $\sigma = 1$.

Again for an independent censoring mechanism where δ_i and y_i, respectively, represent the noncensoring indicator and the logarithm of the minimum of failure and censoring time for the ith individual, the likelihood function may be written

$$L(\beta, \sigma) = \prod_1^n [\sigma^{-1} f(w_i)]^{\delta_i} F(w_i)^{1-\delta_i},$$

where $w_i = (y_i - Z_i\beta)/\sigma$ and $F(w) = \int_w^\infty f(u)\, du$. The score statistic can be written

$$U_j(\beta, \sigma) = \frac{\partial \log L}{\partial \beta_j} = \sigma^{-1} \sum_{i=1}^n Z_{ji} a_i, \qquad j = 1, \ldots, p$$

$$U_{p+1}(\beta, \sigma) = \frac{\partial \log L}{\partial \sigma} = \sigma^{-1} \sum_{i=1}^n (w_i a_i - \delta_i),$$

$$(3.22)$$

where

$$a_i = -\left[\delta_i \frac{d \log f(w_i)}{dw_i} + (1 - \delta_i) \frac{d \log F(w_i)}{dw_i} \right]$$

$$= -\delta_i \frac{d \log f(w_i)}{dw_i} + (1 - \delta_i)\lambda(w_i)$$

and $\lambda(w_i) = f(w_i)/F(w_i)$. The observed information matrix, $I(\beta, \sigma)$, has entries

$$\frac{-\partial^2 \log L}{\partial \beta_j \partial \beta_k} = \sigma^{-2} \sum_{i=1}^n Z_{ji} Z_{ki} A_i,$$

$$\frac{-\partial^2 \log L}{\partial \beta_j \partial \sigma} = \sigma^{-2} \sum_{i=1}^n Z_{ji} w_i A_i + \sigma^{-1} U_j(\beta, \sigma),$$

$$\frac{-\partial^2 \log L}{\partial \sigma^2} = \sigma^{-2} \sum_{i=1}^n (w_i^2 A_i + \delta_i) + 2\sigma^{-1} U_{s+1}(\beta, \sigma),$$

$$(3.23)$$

where $j, k = 1, \ldots, p$ and

$$A_i = \frac{da_i}{dw_i}$$

$$= \delta_i \frac{d^2 \log f(w_i)}{dw_i^2} + (1 - \delta_i)\left[\lambda(w_i) \frac{d \log f(w_i)}{dw_i} + \lambda^2(w_i) \right].$$

$I(\hat{\beta}, \hat{\sigma})$ is somewhat simpler by virtue of the fact that $U(\hat{\beta}, \hat{\sigma}) = 0$. The same criteria as in Section 3.5 will be associated with the suitability of a normal approximation to the distribution of $U(\beta, \sigma)$ and $(\hat{\beta}, \hat{\sigma})$. Of course, convergence to normality would usually be more rapid for error distributions close to normal.

The likelihood derivatives (3.22) and (3.23) depend on (a_i, A_i), $i = 1, \ldots, n$, which are straightforward to obtain provided that $\lambda(w_i)$ can be calculated conveniently. For instance, the Weibull regression model has $f(w_i) = \exp(w_i - e^{w_i})$, so that $\lambda(w_i) = e^{w_i}$, $a_i = e^{w_i} - \delta_i$, and $A_i = da_i/dw_i = e^{w_i}$. Similarly, the log-logistic regression model has $f(w_i) = e^{w_i}(1 + e^{w_i})^{-2}$, so that $F(w_i) = (1 + e^{w_i})^{-1}$, $\lambda(w_i) = e^{w_i}(1 + e^{w_i})^{-1}$, $a_i = -\delta_i + (1 + \delta_i)e^{w_i}(1 + e^{w_i})^{-1}$, and $A_i = (1 + \delta_i)e^{w_i} (1 + e^{w_i})^{-2}$. The log-normal regression model is given by $f(w_i) = (2\pi)^{-1/2} \exp(-w_i^2/2)$ so that $a_i = \delta_i w_i + (1 - \delta_i)\lambda(w_i)$ and $A_i = \delta_i + (1 - \delta_i)\lambda(w_i)[\lambda (w_i) - w_i]$. The likelihood derivatives involve, through $\lambda(w_i)$, the incomplete normal integral. An approximation, such as that given in Abramowitz and Stegun (1965, p. 932, 26.2.19) gives rise to a straightforward computation in the log-normal model.

3.7 ILLUSTRATIONS IN MORE COMPLEX DATA SETS

3.7.1 Accelerated Life Testing

Consider the accelerated life test data of Nelson and Hahn (1972) as given in Table 1.2 (Section 1.1.4). Hours to failure of motorettes are given as a function of operating temperatures of 150°C, 170°C, 190°C, or 220°C. The primary purpose of the experiment was to estimate certain percentiles of the failure time distribution at a design temperature of 130°C. Nelson and Hahn applied a log-normal model to these data with the single regressor variable, $Z = 1000/(273.2 + °C)$. They used a weighted least squares method of estimation which required at least two failures at each test condition, so that the 150°C data had to be excluded. By this method, they obtained an estimate of 10.454 for the log-median lifetime at 130°C (they use base 10 logarithms rather than base e used here) with an associated 90% confidence interval $10.454 \pm 1.645 \ (0.417)$. This gives an estimated median life time of $\exp(10.454) = 34{,}700$ hours, and approximate 90% confidence interval (17,500, 68,900).

For comparative purposes the models of Section 3.6 are fitted to only the 30 observations at test temperatures of 170°C or greater, although elimination of the 150°C data is unnecessary for the maximum likelihood methods. A log-normal regression model and Newton–Raphson iterative technique yield

$$\hat{\alpha} = -10.471, \qquad \hat{\beta} = 8.322, \qquad \hat{\sigma} = 0.6040,$$

at which parameter values the maximum log likelihood is -24.474.

The estimated covariance matrix of $(\hat{\alpha}, \hat{\beta}, \hat{\sigma})$ is

$$\begin{pmatrix} 7.684 & -3.556 & 0.0327 \\ -3.556 & 1.649 & -0.0128 \\ 0.0327 & -0.0128 & 0.0123 \end{pmatrix}.$$

A comparison of $\hat{\beta}$ with its estimated standard error verifies the important effect of temperature on failure time.

The maximum likelihood estimate of the 100pth percentile of the distribution of y (log-failure time) at $Z = z_0$ is simply

$$\hat{y}_p = \hat{\alpha} + z_0\hat{\beta} + \hat{\sigma}w_p, \tag{3.24}$$

where w_p is the 100pth percentile of the error distribution. Also, if $\hat{\Sigma}$ represents the estimated covariance matrix of $(\hat{\alpha}, \hat{\beta}, \hat{\sigma})$, the estimated variance of $\hat{y}_p$ is

$$(1, z_0, w^p)\hat{\Sigma}(1, z_0, w^p)'.$$

At 130°C, $z_0 = 1000/(273.2 + 130) = 2.480$, so that the estimated log-median lifetime is

$$\hat{y}_5 = -10.471 + 8.322(2.480) + 0.6040(0) = 10.170,$$

while var $\hat{y}_5 = (0.433)^2$. This gives approximate 90% confidence interval for $\hat{y}_5$ of $10.170 \pm 1.645\ (0.433)$. The estimate of the median lifetime is exp $(10.170) = 26,100$ with an associate approximate 90% confidence interval $(12,800, 53,200)$.

A Weibull model may equally well be taken for failure time. In fact, if the motorettes are such that failure occurs when the first of any of several essentially independent components fails, there would be some theoretical reason for considering a Weibull model. We find that

$$\hat{\alpha} = -11.891, \qquad \hat{\beta} = 9.038, \qquad \hat{\sigma} = 0.3613$$

and a maximized log likelihood of -22.952. Since the median of an extreme value minimum distribution (2.1) is log(log2), the maximum likelihood estimate of the log-median lifetime at 130°C is $\hat{y}_5 = 10.391$ from (3.24). The standard error of $\hat{y}_5$ is estimated as 0.303. The estimated median lifetime is then $\exp(10.391) = 32,600$ with approximate 90% confidence interval $(19,800, 53,600)$. Note the greater precision of the Weibull analysis over the log-normal and weighted least squares procedures. The Weibull model is to some extent preferable to the log-normal model on account of the larger maximized log likelihood. Further work with these models could, for example, include additional or alternative functions of temperature in the regression vector. Because of the small number of failures involved, further work should also be carried out, perhaps by simulation, to validate the use of asymptotic methods.

3.7.2   Clinical Trial Data

As a further example, consider the Veterans' Administration lung cancer data of Appendix A. In this trial, males with advanced inoperable lung cancer were randomized to either a standard or test chemotherapy. The primary endpoint for therapy comparison was time to death. Only 9 of the 137 survival times were censored. As is common in such studies, there was much heterogeneity between patients in, for example, disease extent and pathology, previous treatment of the disease, demographic background, and initial health status. The data in the appendix include information on a number of covariates measuring some aspects of this heterogeneity:

1. A measure at randomization, of the patient's performance status (Karnofsky rating); 10–30 completely hospitalized, 40–60 partial confinement, 70–90 able to care for self.
2. Time in months from diagnosis to randomization.
3. Age in years.
4. Prior therapy; 0 = no, 10 = yes.
5. Histological type of tumor: squamous, small cell, adeno, large cell.
6. Treatment: 0 = standard, 1 = test.

After preliminary investigations described below, a Weibull regression model was fitted to these data with eight regressor variables; the results are summarized in Table 3.1. Single indicator variables distinguish treatment and prior therapy groups, and three indicator variables for squamous, small cell, and adenocarcinoma permit arbitrary log linear location effects for the four cell-type classes. The other factors enter as indicated in Table 3.1. The asymptotic χ_1^2 statistics given in the table are formed for the ith component as $[\hat{\beta}_i/(\text{estimated standard error of } \hat{\beta}_i)]^2$. The asymptotic χ_3^2 statistic corresponding to cell type differences is calculated as $b' \sum^{-1} b$, where $\mathbf{b}' = (\hat{\beta}_5, \hat{\beta}_6, \hat{\beta}_7)$ is the vector of maximum likelihood estimates and $\sum$ is the estimated covariance matrix for $\mathbf{b}$. This statistic, of course, does not depend on which three cell types are used to define the indicator variables.

From Table 3.1, a strong prognostic effect of initial performance status is indicated as is a difference among survival times in the different cell type groups.

Table 3.1 Asymptotic Likelihood Inference on Lung Cancer Data Using a Weibull Regression Model

Regressor Variable	Regression Coefficient ($\hat{\beta}$)	χ^2 Statistic
Performance status (Karnofsky)	0.0301	38.79
Disease duration (months)	−0.0005	0.00
Age (years)	0.0061	0.51
Prior therapy (0 no, 10 yes)	−0.0044	0.04
Cell type		
Squamous vs. large	0.3977	
Small cell vs. large	−0.4285	22.03
Adeno. vs. large	−0.7350	
Treatment (standard 0, test 1)	0.2285	1.50

This analysis would indicate, however, that patient survival does not differ significantly between treatment groups after taking account of the prognostic effect of other variables. There is, as well, no apparent dependence of survival time on age or disease duration prior to entry to the clinical trial. Even in a randomized study such as this, it is instructive to conduct an analysis that takes account of prognostic factors. Treatment comparisons that do not control for such factors, however, are also valid and typically have a useful population-averaged interpretation.

Weibull and log-normal analyses of these data with only performance status and cell type as factors yielded maximized log likelihoods of -197.10 and -196.75, respectively. It is of some interest to test the adequacy of the exponential regression model relative to the Weibull model. The Weibull model reduces to the exponential at $\sigma = \gamma^{-1} = 1$. The maximum likelihood estimate of σ under the Weibull model is $\hat{\sigma} = 0.928$ with an estimated standard error 0.062. A test of the hypothesis $\sigma = 1$ provides no evidence against the exponential model relative to the encompassing Weibull model. Further results of fitting various models to these data are given in Sections 3.8.2 and 4.5.

Graphical methods can be very useful with such data in preliminary data exploration and in checking the validity of fitted models. For example, if regression variables do not severely dominate, a plot of $\log[-\log \hat{F}(t)]$ versus $\log t$ ($\hat{F}$ is the product limit estimator) may be used to give an overall impression of adequacy of the Weibull model. Such a plot should be approximately linear with slope $\sigma^{-1} = \gamma$ if the data were homogeneous and Weibull and so, in this case, yields also an informal estimate γ of the Weibull shape parameter. Similar plots may be constructed in strata defined by components of the regression vector, for example, by low, medium, and high initial performance status groups in the lung cancer study. The corresponding plots of $\log[-\log \hat{F}(t)]$ versus $\log t$ should each be roughly linear with approximately common slope if the Weibull regression model holds. Further, the distance between these plots should be roughly proportional to the difference between Z values used in forming the strata. Once a Weibull regression model has been fitted, the same type of plot may be used as a check on the model. For such a plot the original survival times t_i are replaced with $t_i' = t_i \exp(-Z_i \hat{\beta})$ and the product limit estimator computed on the basis of these.

An exploratory tool that can be useful if the distribution of the data is nearly exponential is to compute hazard rate estimators (d/v in the notation of Sections 3.3 and 3.5) in various subsets of the data. For example, in Figure 3.1 exponential failure rate estimators are plotted on a logarithmic scale versus performance status for the lung cancer data. A straight-line relationship agrees with a linear modeling of performance status on $\log t$. Other tabulations of exponential failure rates, taking regressor variables one or two at a time, point out the most important prognostic factors and suggest a form for a log-linear modeling of regression variables. A graphical estimator, $\tilde{\gamma}$, of the Weibull shape parameter may be used to bring to bear this procedure when a Weibull, but not an exponential, model is appropriate. Each censored or uncensored failure time t is simply replaced by $\tilde{t} = t^{\tilde{\gamma}}$ before computing hazard rate estimators (d/v). If a Weibull model is appropriate, the $\tilde{t}$ values will have a distribution closely approximated by an exponential regression model. An

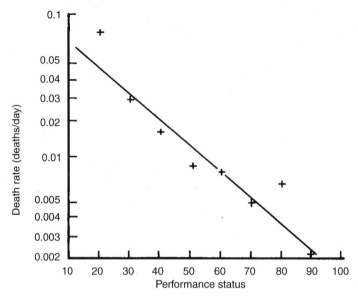

Figure 3.1 Log death rates estimated from an exponential model for the nine performance status groups (VA lung cancer data).

alternative and more generally applicable approach can be based on the logrank test of Sections 1.4 and 4.2.5.

3.8   DISCRIMINATION AMONG PARAMETRIC MODELS

3.8.1   Methods

There are many formal as well as informal methods of assessing the goodness of fit of data to a specified probability model or of selecting a best model among several competitors. One approach, for a log-linear model $y = Z'\beta + \sigma w$, where w has an error distribution of specified form, is to compute the residuals $\hat{w}_i = (y_i - Z_i\hat{\beta})/\hat{\sigma}$ $(i = 1, \ldots, n)$, which should resemble to some extent a (censored) sample from the specified error distribution. As suggested above for the Weibull model, graphical methods based on the Kaplan–Meier estimator computed from these residuals can then provide the tool for an informal assessment of fit. More formally, however, the generalized F model of Section 2.2.7 includes the other parametric models of Chapter 2 as special cases and thus permits their evaluation relative to each other and to a more general model.

Recall that the generalized F is a log-linear model $y = a + Z'\beta + \sigma w$ for $y = \log t$, where the error density $f(w)$ is that of the logarithm of an F variate on $2m_1$ and $2m_2$ degrees of freedom. Its special forms were discussed in Section 2.2.7 and we review these briefly here. The distribution of t is log-logistic for $m_1 = m_2 = 1$, Weibull for $m_1 = 1$, $m_2 \to \infty$, (degenerate) log-normal for

$m_1 = m_2 = \infty$, and generalized gamma as $m_2 \to \infty$; that is,

$$\lim_{m_2 \to \infty} f(w) = \frac{m_1^{m_1} \exp(m_1 w - m_1 e^w)}{\Gamma(m_1)}. \tag{3.25}$$

For specified m_1 and m_2, the results of Section 3.6 can be used to fit this model to the data. If m_1 and m_2 are both finite, then from (2.5),

$$\frac{d \log f(w_i)}{dw_i} = m_1 \frac{(m_1 + m_2)k_i}{1 + k_i}$$

$$\frac{d^2 \log f(w_i)}{dw_i^2} = -\frac{(m_1 + m_2)k_i}{(1 + k_i)^2},$$

where $k_i = m_1 e^{w_i}/m_2$ and $F(w_i) = I(s_i; m_2, m_1)$. Here $s_i = (1 + k_i)^{-1}$ and I represents the incomplete beta ratio that can be calculated using results of Osborn and Madley (1968). Similarly, from (3.25) at finite m_1 and $m_2 = \infty$, $d \log f(w_i)/dw_i = m_1 - m_1 e^{w_i}$; $d^2 \log f(w_i)/dw_i^2 = -m_1 e^{w_i}$ and $F(w_i) = 1 - \mathbf{P}(s_i; m_1)$, where $s_i = e^{m_1 w_i}$ and $\mathbf{P}$ is the incomplete gamma ratio that may be calculated using Abramowitz and Stegun (1965, p.262, 6.5.29). The model at $(m_1 = \infty, m_2)$ can be fit by replacing each w_i by w_i^{-1} and using the method just indicated with m_2 replacing m_1. The log-normal model $(m_1 = \infty, m_2 = \infty)$ can be applied as indicated in Section 3.6. As shown in Prentice (1975), a reparametrization from (m_1, m_2) to $(q, p \geq 0)$ where

$$q = (m_1^{-1} - m_2^{-1})(m_1^{-1} + m_2^{-1})^{-1/2}$$

$$p = 2(m_1 + m_2)^{-1}$$

will lead to a regular maximized log likelihood function (finite, not identically zero likelihood derivatives) everywhere on the boundary $m_1 = \infty$ or $m_2 = \infty (p = 0)$. In the new parametrization (Fig. 3.2) the log-normal, Weibull, log-logistic, reciprocal Weibull, and generalized gamma model occur at respective (q, p) values of $(0, 0)$, $(1, 0)$, $(0, 1)$, $(-1, 0)$, and $(q > 0, 0)$. For inference on (q, p) or, equivalently, on (m_1, m_2) we may calculate the maximized log likelihood over a grid of (q, p) values, and because of the regularity of the log likelihood, we may use the asymptotic distribution of the likelihood ratio statistic to form approximate confidence regions for (q, p) and for evaluating the specific models relative to the generalized F model.

3.8.2　Illustrations

Consider again the two-sample carcinogenesis data of Section 1.1.1. As in Farewell and Prentice (1977), the generalized F model was applied to the variables $t - 100$ at a range of boundary values $(p = 0)$. Note that at $p = 0$,

$$q = \begin{cases} m_1^{-1/2} & \text{at } m_2 = \infty \\ -m_2^{1/2} & \text{at } m_1 = \infty, \end{cases}$$

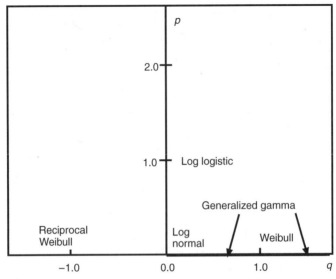

Figure 3.2 Special cases of the log F model. Note that $\{(q,0) : q \geq 0\}$ gives the generalized gamma model.

and that the log-normal and Weibull models previously applied occur at $q = 0$ and $q = 1$, respectively. Figure 3.3 presents a plot of the maximized log relative likelihood, $R(q)$ (maximized log likelihood standardized to have maximal value zero).

Note that the MLE $\hat{q}$ (at $p = 0$) has value 0.87 and that $R(1) = -0.05$ while $R(0) = -2.32$. The asymptotic χ_1^2 distribution for $-2R(q)$ gives an

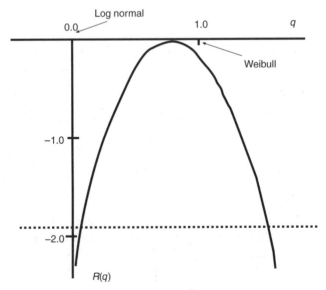

Figure 3.3 Maximized log likelihood assuming a log F model with $p = 0$ and based on the carcinogenicity data of Table 1.1.

approximate 95% confidence interval for q as those values of q for which $R(q) > -1.92$. There is then evidence against the log-normal model relative to this more general class but not against the Weibull model, which, as noted above, indicated an improved survival for the group 2 rats. In fact, the previous calculations of Section 3.7.2 alone, giving maximized log likelihood of -20.618 and -22.890 for Weibull and log-normal models, respectively, are sufficient to provide evidence against the log-normal model. The fact that the difference between log likelihoods $-20.618 + 22.890 = 2.27$ exceeds $1.92 = 3.84/2$ indicates that the log-normal model will be excluded from an approximate 95% confidence interval based on $R(q)$. Note that $P(\chi_1^2 > 3.84) = 0.05$.

The analysis above may be extended to estimate the duration, δ, of the initial failure-free period, rather than specify it as 100 days. Pike (1966) calculates $\hat{\delta} = 98.9$ assuming a Weibull model. Since δ is a threshold parameter, the likelihood function does not possess the required regularity to permit the use of standard asymptotic likelihood results for the estimation of δ. To examine whether inclusion of δ gives rise to a significant improvement in the generalized F, the model with $p = 0$ was applied as above to $y_\delta = \log(t - \delta)$, where t is the time from insult to diagnosis, for several values of δ between 0 and the smallest observed time of 142 days. In each case the log likelihood was maximized over (β, σ, q). Note that the Jacobian factor $[\exp{(y)} - \delta]/\exp{(y)}$ needs to be introduced into the maximized log likelihood $l^*(\delta)$, to describe the change in scale from y to y_δ. Table 3.2 gives values of $R^*(\delta) = l^*(\delta) - l^*(125)$. Apparently, there is little ability to discriminate between values of δ with these data. The values $\hat{q} = \hat{q}(\delta)$ range from 0.36 at $\delta = 0$ to about 1.75 at $\delta = 140$. The generalized F model is sufficiently flexible in this case that the inclusion of a guarantee-type parameter contributes little.

This section ends with a brief discussion of the application of the generalized F regression model to the illustration of Section 3.8. With the accelerated life-test data of Table 1.2, the maximum likelihood estimate of the "skewness" parameter q (subject to a value of zero for the "kurtosis" parameter p) is 1.6. The maximized log relative likelihood yields $R(0) = -1.73$ and $R(1) = -0.21$, so that some doubt is cast on the suitability of the log-normal model, but there is no evidence against the Weibull model relative to this more general model. At $\hat{q} = 1.6$ one obtains an estimated log-median failure time estimate and standard error at 130°C of 10.499 ± 0.252. The corresponding median failure time estimate and approximate 90% confidence interval are then 36,300 and (24,000, 54,900), respectively. The standard error estimate for $\hat{y}_{.5}$ is appropriate assuming that $q = 1.6$ but does not take a account of the correlation between $\hat{y}_{.5}$ and $\hat{q}$.

Table 3.2   Maximized Log Likelihood for a Guarantee Time δ Based on the Data of Section 1.1.1

δ	0	25	50	75	100	125	140
$R^*(\delta)$	-0.55	-0.49	-0.40	-0.32	-0.19	-0.00	-0.16

Table 3.3 Analysis of Veterans Administration Lung Cancer Data Using a Generalized F Regression Model[a]

Regression Variable	Log Normal		Weibull		Log F ($p = 0$)	
	$\hat{\beta}$	S.E.$(\hat{\beta})$	$\hat{\beta}$	S.E.$(\hat{\beta})$	$\hat{\beta}$	S.E.$(\hat{\beta})$
No prior therapy					($\hat{q} = 0.43$)	
(97 patients)						
Performance status	0.030	0.006	0.022	0.006	0.026	0.006
Squamous vs. large	−0.085	0.34	0.175	0.31	0.086	0.32
Small vs. large	−0.762	0.31	−0.521	0.28	−0.669	0.29
Adeno. vs. large	−0.804	0.34	−0.840	0.30	−0.795	0.32
Prior therapy					($\hat{q} = 1.05$)	
(40 patients)						
Performance status	0.059	0.010	0.054	0.009	0.053	0.009
Squamous vs. large	−0.199	0.46	0.428	0.38	0.450	0.38
Small vs. large	−0.388	0.49	−0.044	0.42	−0.033	0.41
Adeno. vs. large	−0.694	0.61	−0.787	0.51	−0.794	0.50

[a] G.E., standard error.

A similar application to the lung cancer data (Appendix A, data set I) with performance status and three cell-type indicators as regressor variables gives $\hat{q} = 0.47$. There is evidence against both Weibull and log-normal models as $R(0) = -2.59$ and $R(1) = -2.94$. Further, the shape of the failure time distribution was found to depend on whether or not the patient had received prior therapy. Data on the 40 patients who had received therapy prior to the start of the study give $\hat{q} = 1.05$, whereas data on the 97 without prior therapy yield $\hat{q} = 0.43$. A likelihood ratio test for equality of the q's gives $\chi_1^2 = 9.0$, which is significant at the 1% level. Separate analyses for the two prior therapy groups are therefore indicated. Table 3.3 gives some results from such analyses. Note the interaction between prior therapy and performance status. Performance status is an important prognostic factor for both groups of patients but is particularly dominating among patients who have received prior therapy.

3.9 INFERENCE WITH INTERVAL CENSORING

In some settings, the failure time for the ith individual may be subject to interval censoring so that T_i is known only to fall in some interval, $(l_i, r_i]$. Such censoring arises, for example, if the individual is subject to occasional inspection times $0 < C_{i1} < C_{i2} < \cdots < C_{im_i} < \infty$, say, and the status as to whether failure has occurred is determined only at those times. Let $C_i = (C_{i1}, \ldots, C_{im_i})$. In such a study, it is observed that $\tilde{T}_i$ falls in one of the intervals $(C_{i,j-1}, C_{ij}]$, $j = 1, \ldots, m_i + 1$, where $C_{i0} = 0$ and $C_{i,m_i+1} = \infty$. Note that right censoring at l_i corresponds to $r_i = \infty$, and left censoring at r_i corresponds to $l_i = 0$.

Censoring schemes of this type occur in clinical studies where subjects are scheduled to return for regular visits that patients may keep, but typically with some random variation, and where it is determined whether or not failure has occurred since the most recent visit. The actual time at which the failure occurs, however, is not observed. They also arise in industrial studies where the status of components of a system might be ascertained only at certain prespecified times. To make valid inferences about the underlying failure distribution, we again need to have certain assumptions about the nature of the censoring. We will assume that pairs $\tilde{T}_i, C_i$ are independent for $i = 1, \ldots, n$ and that the censoring mechanism is such that

$$P\{\tilde{T}_i \in (C_{i,j-1}, C_{ij}] | \tilde{T}_i > C_{i,j-1}, C_{i,j-1}, C_{ij}\} = \frac{F(C_{i,j-1}) - F(C_{ij})}{F(C_{i,j-1})} \tag{3.26}$$

for all $\{i,j\}$. This occurs, for example, if C_i is fixed in advance or if it is distributed independently of $\tilde{T}_i$. It is, however, also satisfied by many other inspection schemes and essentially implies that having observed that the individual is alive at time $C_{i,j-1}$, the timing of the next inspection is distributed independently of the time of failure. We refer to such schemes as independent interval censoring. Note that if there were covariates x_i, we should interpret (3.26) conditional upon them.

A special case of interest, called *current status data*, arises when each individual i is subject to observation only at the single follow-up time C_{1i}, where $0 < C_{1i} < \infty$ so that $m_i = 1$ for all i. It is also of interest to note that the usual case of "exact" observation of failure time in reality corresponds to interval censoring with a small grouping interval. Viewed this way, the continuous-type likelihood arises as a convenient approximation to the likelihood based on interval censoring.

3.9.1   Estimation of the Survivor Function

Consider a sample of n individuals subject to independent interval censoring. Suppose that the underlying failure times are IID (independent and identically distributed) with common survivor function $F(t)$ and that the data consist of observations $T_i \in (l_i, r_i], i = 1, \ldots, n$. We assume that $l_i < r_i$ for all $i = 1, \ldots, n$ and that l_i may be 0 or r_i may be ∞. Under these conditions, the likelihood function is proportional to

$$L = \prod_{i=1}^{n} [F(r_i) - F(l_i)]. \tag{3.27}$$

Consider arranging the distinct values of $l_i^+, r_i, i = 1, \ldots, n$ in increasing order along the time axis where the l's have been shifted slightly to the right to break ties between left and right endpoints. Proceed left to right along the axis and identify all intervals $(L_j, R_j], j = 1, \ldots, m$ (say), where $L_j \in \{l_1, \ldots, l_n\}$, $R_j \in \{r_1, \ldots, r_n\}$ and L_j and R_j are adjacent in the arrangement. The determination of the intervals $(L_j, R_j]$ for a small example is illustrated in Figure 3.4. It can be shown that any nonparametric MLE of F must concentrate all mass on these m

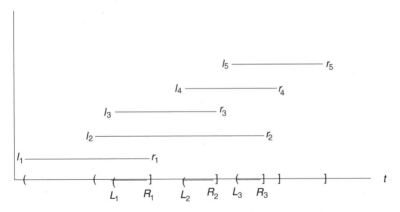

Figure 3.4 Determination of the mass intervals $[L_j, R_j]$ for an interval-censored sample of $n = 5$ individuals. The horizontal lines define the interval censoring and the projection on the x-axis defines the intervals.

intervals. The nonparametric MLE of F is then obtained by maximizing the likelihood

$$\prod_{i=1}^{n}\left\{\sum_{j=1}^{m}\delta_{ij}[F(R_j) - F(L_j)]\right\},$$

where $\delta_{ij} = \mathbf{1}\{(L_j, R_j] \subset (l_i, r_i]\}, i = 1, \ldots, n; j = 1, \ldots, m$. It is convenient to let $p_j = F(R_j) - F(L_j), j = 1, \ldots, m$ and consider maximizing the log likelihood

$$\prod_{i=1}^{n}\left(\sum_{j=1}^{m}\delta_{ij}p_j\right), \tag{3.28}$$

where $0 \le p_j \le 1$ and $\sum p_j = 1$.

Various approaches could be used to maximize (3.28), but the simplest is the expectation-maximization (EM) algorithm (see Appendix B). For this, view the likelihood (3.28) as having arisen from an underlying m-class multinomial distribution for $X_i' = (X_{i1}, \ldots, X_{im})$ with index 1 and probabilities $p_1, \ldots, p_m, i = 1, \ldots, n$ independently. In the corresponding incomplete data problem, the ith outcome is observed only to fall in one of the classes j for which $\delta_{ij} = 1$ and so gives rise to (3.28). Denote the incomplete data by $Y = \{\delta_{ij}, i = 1, \ldots, n, j = 1, \ldots, m\}$ and the complete data by $X = \{X_{ij}, i = 1, \ldots, n, j = 1, \ldots, m\}$. The complete data log likelihood is the multinomial likelihood

$$l_X(p) = \sum_{i=1}^{n}\sum_{j=1}^{m}X_{ij} \log p_j = \sum_{j=1}^{m}X_{\cdot j} \log p_j,$$

where $X._j = \sum_{i=1}^{n} X_{ij}$ and $\sum X._j = n$. Let $p^{(0)} = (p_1^{(0)}, \ldots, p_m^{(0)})'$ be a trial value. The expectation step is

$$E[l_X(p)|Y, p^{(0)}] = \sum_{i=1}^{n} \sum_{j=1}^{m} X_{ij}^{(0)} \log p_j, \tag{3.29}$$

where

$$X_{ij}^{(0)} = E[X_{ij}|Y, p^{(0)}] = \frac{\delta_{ij} p_j^{(0)}}{\sum_{j'=1}^{m} \delta_{ij'} p_{j'}^{(0)}}.$$

The updated estimate of p_j is $p_j^{(1)} = n^{-1} X._j^{(0)}$, $j = 1, \ldots, m$, obtained by maximizing (3.29). The expectation and maximization steps are repeated to convergence to $\hat{p}$. This convergence establishes a self-consistency equation for the $\hat{p}_j$'s:

$$\hat{p}_j = \sum_{i=1}^{n} \frac{\delta_{ij} \hat{p}_j}{\sum_{j'=1}^{m} \delta_{ij'} \hat{p}_{j'}}. \tag{3.30}$$

The EM algorithm simply involves repetitive use of (3.30). It can be shown that the EM algorithm always converges to the MLE from any starting point where all $p_j^{(0)} > 0$ and $\sum p_j(0) = 1$ (Gentleman and Geyer, 1994), although the convergence can be very slow.

This gives rise to a family of MLEs of the survivor function $F(\cdot)$. Specifically, any survivor function $\hat{F}$ that satisfies $\hat{F}(R_j) - \hat{F}(L_j) = \hat{p}_j$, $j = 1, \ldots, m$ and places no mass outside the intervals $(L_j, R_j]$ is a nonparametric MLE. One might adopt various conventions to choose a unique estimate. One such would be to distribute the assigned mass $\hat{p}_j$ uniformly over the interval $(L_j, R_j]$ to obtain a continuous estimate.

Asymptotic properties of such estimators are nonstandard. In the simplest case, there may be a fixed number m of possible intervals $(L_j, R_j]$ as $n \to \infty$. In this case, fairly simple conditions would lead to the usual asymptotic results of the form $n^{1/2}\{\hat{F}(t) - F(t)\}$ approaches a normal limit for any given $t \in \{L_j, R_j, j = 1, \ldots, m\}$. In the more typical situation where the number of intervals increases with sample size, the situation is much more complex. Some investigations have assumed that the endpoints become dense in the neighborhood of a fixed t, and under some conditions, a normal limit holds but with multiplier $n^{1/3}$ instead of $n^{1/2}$. See, for example, Groenboom and Wellner (1992). Some grouping of the data to reduce the number of intervals can achieve $n^{1/2}$ asymptotics. The use of parametric models provides a way to avoid these difficulties.

3.9.2 Analysis of Regression Models with Interval Censoring

Suppose now that a log-linear or accelerated failure time model is postulated for the failure time so that the PDF for $Y = \log \tilde{T}$ is $f_W(w)/\sigma$, where $w = (y - Z'\beta)/\sigma$. Let

$F_W(w) = \int_w^\infty f_W(u)\,du$ be the corresponding survivor function. The log likelihood arising from independent interval censored data $(l_i, r_i], i = 1, \ldots, n$ is

$$l(\beta, \sigma) = \sum_{i=1}^n \log\left[F_W(u_j) - F_W(v_j)\right], \qquad (3.31)$$

where $u_i = (\log r_i - Z_i'\beta)/\sigma$ and $v_i = (\log l_i - Z_i'\beta)/\sigma$. The corresponding score function is of dimension $p + 1$ and given by

$$U(\beta, \sigma) = \frac{1}{\sigma}\sum_{i=1}^n \frac{a_{0i} Z_i}{a_{1i}}. \qquad (3.32)$$

The observed information matrix, again in partitioned form, is

$$I(\beta, \sigma) = \frac{1}{\sigma^2}\sum_{i=1}^n \begin{pmatrix} (b_{0i} + a_{0i}^2)Z_i^{\otimes 2} & (b_{1i} + a_{0i}a_{1i})Z_i \\ (b_{1i} + a_{0i}a_{1i})Z_i' & b_{2i} + a_{1i}^2 \end{pmatrix},$$

where $Z^{\otimes 2} = ZZ'$. In these expressions,

$$a_{li} = \frac{u_i^l f_W(u_i) - v_i^l f_W(v_i)}{F_W(u_i) - F_W(v_i)} \quad \text{and} \quad b_{li} = \frac{u_i^l f_W'(u_i) - v_i^l f_W'(v_i)}{F_W(u_i) - F_W(v_i)} \qquad (3.33)$$

for $l = 0, 1, 2$ and $i = 1, \ldots, n$.

As in the case of right-censored data discussed in Section 3.6, the quantities a_{li} and b_{li} are simply computed in many models. A Newton–Raphson algorithm essentially involves computation of F_W, f_W, and f_W' at each iteration which are the same quantities needed in the right censored case. Thus, these calculations can be extended in a straightforward way to incorporate such flexible error models as the generalized log F.

Under independent interval censoring, standard asymptotic arguments apply. The score components in (3.32) are independent and a central limit theorem applies provided that the censoring mechanism and the covariates are such that the Lindeberg condition holds for the variances of the independent score components. In many instances, it is reasonable to assume that as n becomes large, the average information converges to a positive-definite covariance matrix V. In this case, $n^{-1}I(\beta, \sigma) \xrightarrow{\mathscr{P}} V$ as $n \to \infty$ and

$$n^{-1/2}U(\beta, \sigma) \xrightarrow{\mathscr{D}} N(0, V),$$

and

$$n^{1/2}\begin{pmatrix} \hat{\beta} - \beta \\ \hat{\sigma} - \sigma \end{pmatrix} \xrightarrow{\mathscr{D}} N(0, V^{-1}).$$

As before, V can be estimated by $n^{-1}I(\hat{\beta}, \hat{\sigma})$. This leads to standard inference procedures for all smooth functions of β and σ, including, for example, the survivor function at a given value of Z and t.

Similar results can be obtained for other parametric regression models, such as relative risk models with a specified parametric form for $\lambda_0(t)$. The use of flexible parametric models in the context of interval censoring has much to recommend it. It leads to relatively simple computation and fitting of models and to standard asymptotic results under fairly general and realistic conditions. On the other hand, nonparametric procedures for this estimation problem are generally relatively difficult to implement and have the disadvantage that standard asymptotic results do not apply to estimation of some quantities of interest, such as, for example, the survivor function at a specified time t and covariate Z.

3.9.3   Truncation

Suppose now that individuals are subject to truncation so that individual i enters the study if and only if the corresponding survival time T_i exceeds some threshold value t_{i0}. For individuals who enter, we observe the corresponding t_{i0}; those who do not enter the study are completely unobserved, so that not even their existence is known. The ith individual is subject to interval censoring and we observe $T_i \in (l_i, r_i]$ as above. With a log-linear model, covariate vector Z_i, and independent censoring and truncation, the likelihood function is

$$L(\beta, \sigma) = \prod_{i=1}^{n} \frac{F(u_i) - F(v_i)}{F(r_i)},$$

where u_i and v_i are as defined above and $r_i = (\log t_{i0} - Z_i'\beta)/\sigma$. The same approach can be applied here, and for random censoring and truncation, asymptotic results follow as before. Other truncation schemes (e.g., right truncation or double truncation) can also be accommodated with relatively little additional complication.

3.10   DISCUSSION

Parametric regression models such as Weibull, log-normal, and log-logistic may involve stronger distributional assumptions than it is suitable to make, and the inference procedures mentioned may not be sufficiently robust to departures from these assumptions.

With uncensored data, it can be seen that estimation based on a log-linear model with the right regression form but incorrect error form will give consistent estimates of the regression parameters. Specifically, suppose that the true model is of the form

$$\log T_i = Z_i'\beta + \sigma W_i, \qquad i = 1, \ldots, n,$$

where the W_i's are IID with density $f(w)$ and $Z_i' = (Z_{i1}, \ldots, Z_{ip})$ with $Z_{i1} = 1$ for all i. Suppose, however, that we have adopted a model for inference in which the density of W is assumed to be g instead of f. One cannot, of course, estimate the intercept β_1 or the scale parameter σ consistently. Under the usual regularity conditions, however, it can be shown that estimates of the regression parameters $\beta_2, \ldots, \beta_p$ are consistent as $n \to \infty$; further, adjustments can be made to variance estimation using sandwich-type estimates for estimating equations so as to get robust variance estimates and valid confidence intervals. There is, however, some loss in efficiency due to an incorrect assumption about the error. If the data are subject to independent right censoring or other forms of independent censoring or truncation, however, this consistency result for the regression parameters $\beta_2, \ldots, \beta_p$ no longer holds in general and inconsistent estimates are obtained if the assumed error distribution is incorrect. The extent of such inconsistency has not been examined systematically, but may often be relatively small, provided that the censoring is not too severe. It should also be kept in mind that the regression modeling itself is usually not exactly correct and the practical importance of this inconsistency result requires further investigation. Nonetheless, it is a feature not seen with uncensored data, and it can be argued that it adds some additional motivation to seek techniques that are less dependent on parametric assumptions.

When the primary interest is in the effects of regression variables, a variety of approaches might be considered to achieve greater robustness. The more general parametric models, such as the generalized F, represent one such approach. Another approach would be to extend the M estimation procedures developed for linear regression (e.g., Huber, 1972, 1973; Andrews et al., 1972; Hampel, 1974) to include censoring; that is, to develop specific (pseudo) score functions for which the corresponding regression estimates have both good efficiency and robustness properties. One such generalization was proposed by Hjort (1985). In a counting process notation, the jth component of the score vector arising from a parametric model $\lambda(t; \theta, x)$ can be written

$$U_j(\theta) = \sum_{i=1}^{n} \int_0^\infty \left[\frac{\partial}{\partial \theta_j} \log \lambda(t; \theta, x_i) \right] [dN_i(t) - Y_i(t) \lambda_i(t; \theta, x_i) \, dt],$$

where, as before, $N_i(t)$ is the counting process that records the number of failures observed on the ith individual and $Y_i(t)$ is the at-risk process. Hjort (1985) suggested considering estimates arising from the estimating equation

$$U_j^\dagger(\theta) = \sum_{i=1}^{n} \int_0^\infty K_j(t; \theta, x_i)[dN_i(t) - Y_i(t) \lambda_i(t; \theta, x_i) \, dt] = 0,$$

where $K_j(t)$ is a deterministic function that can be used to emphasize or deemphasize the influence of failures over the time axis. In fact, more general functions are allowable in that K_j can be a predictable process (see Chapter 5 for definitions). Although the suitable choice of K_j could evidently yield quite robust procedures, this suggestion is somewhat different in spirit from the usual M estimation idea in that the estimating function $U^\dagger$ is generally unbiased only if the assumed model is true. Thus, this procedure can again be expected to lead to inconsistent inference if the assumed model is incorrect, although for suitable choices of the weighting function, the degree of inconsistency may be relatively unimportant compared with the uncertainty in regression modeling. There does not seem to have been a systematic study of this suggested generalization of M estimation, although Andersen et al. (1993, pp. 433 ff.) provide some discussion.

A third approach is to consider more general models that are nonparametric, or semiparametric, such as the proportional hazards and accelerated failure time models of Section 2.3. The objective with such models is to develop inference procedures that will be consistent and reasonably efficient regardless of which member of the class obtains. In the next chapter we describe estimation under the important class of relative risk (Cox) regression models. A semiparametric approach based on the accelerated failure time model is considered in Chapter 7.

BIBLIOGRAPHIC NOTES

Inference from type II (order statistic) censored samples has been much discussed in the literature. For example, Epstein and Sobel (1953) gave some basic exponential sampling results. General reference books on order statistic properties include those by Sarhan and Greenberg (1962) and David (1970). Johnson and Kotz (1970a,b) give a comprehensive account of estimation for homogeneous populations assumed to have the parametric models considered in this chapter (exponential, Weibull, log-normal, gamma, log-logistic). Mann et al. (1974) summarize from an industrial life-testing point of view estimation procedures for these as well as other distributions, for both single- and two-sample problems, with censoring. Gross and Clark (1975) give similar results from the biomedical point of view. In general, the methods given are based on asymptotic maximum likelihood procedures with some exact results for order statistic sampling. Bartholomew (1957) provides an early example of a discussion of statistical properties for the MLE with type I (time) censoring. Cox (1953) suggests that under some circumstances, a suitable approximation involves treating type I censored data as if they were type II censored. Some authors (e.g., Gilbert, 1962; Efron, 1967; Breslow, 1970; Breslow and Crowley, 1974) postulate a probability distribution for censoring times in an attempt to derive exact properties or to more easily develop asymptotic distribution theory for estimators. Exact conditional confidence interval estimates have been obtained for type II censored data from the Weibull (or extreme value distribution) by Lawless (1973, 1978, 1982). He compares these exact results with those obtained from asymptotic theory. Similar arguments could be applied to type II censored data from other models in the accelerated failure time class.

There is a very large literature utilizing exponential, Weibull, and log-normal regression models. Some of the earlier papers are the following: Cox (1964), Feigl and Zelen (1965), Zippin and Armitage (1966), Pike (1966), Glasser (1967), Cox and Snell (1968), Sprott and Kalbfleisch (1969), Nelson (1970), Mantel and Myers (1971), Nelson and Hahn (1972), Peto and Lee (1973), Prentice (1973), Myers et al. (1973), Breslow (1974), Byar et al. (1974), Kalbfleisch (1974), and Prentice and

Shillington (1975). Farewell and Prentice (1977) consider regression in the generalized gamma distribution. Summaries of parametric models and examples are given in the books by Lawless (1982), Cox and Oakes (1984), Klein and Moeschberger (1997), and Hougaard (2000). Andersen et al. (1993) give an extensive discussion of parametric inference in the context of multivariate counting processes.

In one of the examples of Section 3.7, the accelerated failure time model is used for extrapolation in an industrial setting. Even more extreme extrapolations beyond dosages (regression variable values) actually considered is often necessary in carcinogenesis testing. Typically, experimental animals are tested at highly accelerated dosages of a suspected carcinogen, and times to tumor incidence are recorded. Downward extrapolation is required to dosages that yield some "safe" level of risk or to dosages comparable to those found in the environment. The regression models considered in this chapter form the basis for proposed methods of extrapolation. The associated literature gives some interesting further insights into biological mechanisms that can give rise, for example, to Weibull or log-normal regression models. For example, "multistate" or "multihit" carcinogenesis theories can lead to the Weibull model (e.g., Armitage and Doll, 1954, 1961; Crump et al., 1976). Rather different assumptions concerning the existence of thresholds leads to lognormal (probit) extrapolations (e.g., Mantel and Bryan, 1961; Mantel et al., 1975). Differences in tail shape for these models generally lead to completely different low-dose risk estimates. Some work in this area (e.g., Hartley and Sielken, 1977) utilizes models more general than the Weibull, but still of a proportional hazards (Section 2.3.2) type.

A summary of asymptotic likelihood theory is given in Cox and Hinkley (1974). These procedures were proposed by Fisher (1922, 1925), and important contributions were made by Neyman and Pearson, Wilks, Wald, Cramér, Bartlett, Le Cam, and many others. In particular, the reader is referred to Le Cam (1970) for a technical discussion of asymptotic normality. Moran (1971) discusses some properties of maximum likelihood estimators of parameters on the boundary of the parameter space. Methods of improving the asymptotic approximation to the distribution of maximum likelihood estimators have been considered by many authors. Variance-stabilizing transformations are wellknown and discussed in several standard texts (e.g., Cox and Hinkley, 1974, p. 275). Transformations to improve symmetry in the log likelihood (by eliminating the cubic term in the Taylor series expansion) have been considered by Anscombe (1964) and by Sprott (1975). The asymptotic arguments referred to in this chapter relate to the use of a central limit theorem for the sum of independent random variables. This Lindeberg–Feller theorem is discussed and proved in many books—Shorack (2000, p. 260) is a recent reference. This approach allows for asymptotic treatment of parametric models under quite arbitrary independent random censoring and truncation, including, for example, interval or left censoring and right truncation. They are therefore quite broad in their application. In certain instances, asymptotic arguments can also be based on counting processes, and martingale central limit theorems as discussed in Section 5.7. This approach was first applied by Aalen and Hoem (1978) and more generally by Borgan (1984). Andersen et al. (1993) give a complete treatment.

This approach applies only to independent right censoring and left truncation and not, for example to random interval censoring or right truncation. It should be noted, however, that the counting process approach does allow general independent (right) censoring mechanisms, and this is more general than independent random censoring considered in this chapter.

Cox (1961, 1962a) and Atkinson (1970) consider general procedures for discriminating among several families of hypotheses. Cox's work involves approximating the distribution of the likelihood ratio. Assuming uncensored data, Dumonceaux and Antle (1973) simulate the distribution of the likelihood ratio statistic in order to discriminate between Weibull and log-normal hypotheses. Hagar and Bain (1970) consider the problem of testing for a Weibull model within the generalized gamma family.

As also mentioned in Chapter 1, Turnbull (1974,1976) first obtained the MLE of the survivor function under interval censoring and truncation, and developed the self-consistency algorithm given in Section 3.9.1. The EM algorithm (see Appendix B) was formally set out by Dempster et al. (1977), who also noted the application to censored data problems. The convergence of the EM algorithm to the NPMLE for interval-censored data was shown by Gentleman and Geyer (1994). Bohning et al. (1996) noted that globally convergent algorithms for fitting nonparametric mixtures can be used to find the NPMLE in interval-censored data. Gentleman and Vandal (2001, 2002) have used graph-theoretic techniques to characterize the NPMLE for general interval-censored models in both univariate and bivariate problems. Asymptotic properties of the NPMLE for interval-censored data is considered in Groenboom and Wellner (1992) and Huang and Wellner (1995), where it is shown that, in general, the survivor function estimates converge at order $n^{-1/3}$ instead of the usual $n^{-1/2}$. The latter paper establishes $n^{-1/2}$ asymptotics for linear functionals of the survivor function. Odell et al. (1992) have considered the use of a Weibull hazard model for interval-censored data. Other work on fitting interval-censored data has used a Cox or relative risk model, and some references are given in the Chapter 4 bibliographic notes.

Current status data have also received some attention in the literature—see, for example, Keiding (1991), Jewell and van der Laan (1995), Rossini and Tsiatis (1996), Sun and Kalbfleisch (1993), and Andersen and Ronn (1995). One place where such data arise is in survival sacrifice experiments with incidental tumor that is clinically unobservable. In this case, tumor status at death constitutes current status data. Some references are Hoel and Walburg (1972), Dinse and Lagakos (1983), McKnight and Crowley (1984), and Dewanji and Kalbfleisch (1986).

EXERCISES AND COMPLEMENTS

3.1 **(a)** Use expression (3.4) to show that if $t_{(n)}$ is the largest of n independent unit exponential variates, the rth cumulant of $t_{(n)}$ is

$$(r-1)! \sum_{i=1}^{n} i^{-r}.$$

(b) Show that the asymptotic distribution of $X_n = \log n - t_{(n)}$ is the extreme value distribution with density (2.1). Note also that the MGF of X_n converges uniformly for $\theta \epsilon(-1, 1)$ to the MGF of (2.1). This implies convergence of the moments of X_n to those of (2.1) (see Rao, 1965, p. 101).

(c) By making use of parts (a) and (b), show that the extreme value density has cumulants

$$\kappa_1 = \psi(1) = \lim_{n \to \infty} \left(\log n - \sum_1^n i^{-1} \right) = -\gamma$$

$$\kappa_r = \psi^{(r-1)}(1) = (-1)^r (r-1)! \, \zeta(r), \qquad r = 2, 3, 4, \ldots,$$

where $\zeta(r)$ is the Riemann zeta function,

$$\psi^{(r-1)}(x) = \frac{d^r}{dx^r} \log \Gamma(x), \qquad r = 1, 2, \ldots,$$

are the polygamma functions with $\psi(x)$ the digamma function, and $\gamma = 0.5772\ldots$ is Euler's constant.

Note that this provides an elementary evaluation of the well-known definite integrals

$$\int_{-\infty}^{\infty} x \exp(x - e^x) \, dx = -\gamma$$

and

$$\int_{-\infty}^{\infty} x^2 \exp(x - e^x) \, dx = \frac{\pi^2}{6} + \gamma^2.$$

3.2 Using the results of Exercise 3.1, determine the expected information matrix for the Weibull distribution with density function

$$\lambda p (\lambda t)^{p-1} \exp[-(\lambda t)^p], \qquad 0 < t < \infty.$$

3.3 Consider the comparison of two type II censored samples where sample i is followed to the observed d_ith failure, $i = 1, 2$.

(a) In the notation of Section 3.3, show that $(\lambda_1 V_1 / d_1)(\lambda_2 V_2 / d_2)^{-1}$ has an F distribution on $2d_1$ and $2d_2$ degrees of freedom.

(b) Suppose that $d_1 = 17$ and $d_2 = 19$ and it is observed that $V_1 = 2195$ and $V_2 = 2923$. Compute a 95% confidence interval for $\beta_2 = \log(\lambda_2 / \lambda_1)$ and compare with the results given in Section 3.5.2 for the same data but with type I censoring.

(c) For large degrees of freedom, the logarithm of an $F_{(n,m)}$ variate is approximately normal with mean $(n^{-1} - m^{-1})$ and variance $2(n^{-1} + m^{-1})$ (Atiquallah, 1962). Use this result to obtain an approximate normal distribution for $\hat{\beta}_2$ and compare it with the asymptotic normal distribution of $\hat{\beta}_2$ in Section 3.5.2.

3.4 Consider again type II censoring in an exponential distribution. As in Section 3.3, suppose that n items are placed on test and followed until the dth failure. Let $U_1, \ldots, U_d$ be the normalized spacings defined by $U_i = (n - i + 1)(\tilde{T}_{(i)} - \tilde{T}_{(i-1)})$, $i = 1, \ldots, d$.

   (a) Obtain the distribution of $Y_1, \ldots, Y_{d-1}$, given $Y_d = y_d$, where $Y_i = \sum_{j=1}^{i} U_j$, $i = 1, \ldots, d$, and show that this distribution is that of the order statistic in a sample size $d - 1$ from the uniform distribution on $(0, y_d)$.

   (b) Develop a test for large d of the exponential assumption using an approximate distribution for $\sum_{1}^{d-1} Y_i$ given $Y_d = y_d$. For what kinds of departures from the exponential would you expect this to be a sensitive test? An insensitive test?

3.5 Consider a type I censored sample from the exponential distribution with censoring time c common to all n individuals on test. Show that the total number of failures, D, has a binomial distribution with parameters n and $p = 1 - \exp(-\lambda c)$. Compare the (asymptotic) efficiency of the MLE of λ from the marginal distribution of D to that from (V, D). Under what conditions on c is it sensible to base inferences about λ upon D alone?

3.6 A laboratory has n test locations for life testing a particular type of electronic equipment. To conserve time, it is decided to place an item in each test location and test all n items simultaneously. As soon as an item fails, it is immediately replaced by a new item and the system is observed until the dth failure occurs. Suppose that the failure times are exponentially distributed. Let S be the random variable representing the time to cessation of testing. Show that S is sufficient for the failure rate λ. Derive the distribution of S and give an exact 95% confidence interval for λ when $n = 25, d = 5$, and the fifth failure occurs 407 hours after the start of the experiment.

3.7 Consider a random censorship model in which failure time $\tilde{T}$ is exponential with failure rate λ and censoring time C is exponential with rate α. Let $T_i = \tilde{T}_i \wedge C_i$, $V = \sum_{i=1}^{n} T_i$, and D be the number of failures. Show that (V, D) is sufficient for (λ, α). Show further that V and D are independent, that D is binomial with parameters n and $\lambda(\lambda + \alpha)^{-1}$, and that $2(\lambda + \alpha)V$ has a χ^2 distribution with $2D$ degrees of freedom. Discuss how inference on λ may be carried out.

3.8 The guaranteed exponential distribution has density function

$$f(t) = \lambda e^{-\lambda(t-G)}, \qquad t > G,$$

where $\lambda, G > 0$ are unspecified parameters. Let $T_{(1)}, \ldots, T_{(n)}$ be the order statistic on a sample of size n.

(a) Show that $U = \sum_2^n T_{(i)}$ and $T_{(1)}$ are jointly sufficient for λ and G and determine the maximum likelihood estimates.

(b) Show that $n(T_{(1)} - G), (n-1)(T_{(2)} - T_{(1)}), (n-2)(T_{(3)} - T_{(2)}), \ldots,$ $(T_{(n)} - T_{(n-1)})$ are independent exponentials with failure rate λ and hence determine the joint distribution of U and $T_{(1)}$.

(c) Establish methods for exact interval estimation of λ and G. (The likelihood ratio statistic gives rise to simple pivotals for these parameters.)

(d) Apply these results to the group 1 data of Table 1.1. For this purpose, omit the censored data points.

3.9 Suppose that the failure times in Exercise 3.8 are type II censored at $T_{(d)}$, where d is fixed in advance. How would this alter the analysis?

3.10 Let $T_1, \ldots, T_n$ and $S_1, \ldots, S_m$ be uncensored samples from two guaranteed exponential distributions with parameters (λ_1, G_1) and (λ_2, G_2), respectively.

(a) Outline a test of the hypothesis $\lambda_1 = \lambda_2$.

(b) Supposing that $\lambda_1 = \lambda_2 = \lambda$ is known, develop a test of the hypothesis $G_1 = G_2$. For this purpose, show that $U = T_{(1)} - S_{(1)}$ has a double exponential distribution with density

$$
f(u) = \begin{cases} \dfrac{nm\lambda}{n+m} \exp(-n\lambda u), & u > 0 \\[2em] \dfrac{nm\lambda}{n+m} \exp(m\lambda u), & u < 0. \end{cases}
$$

(c) Generalize this to a test of $G_1 = G_2$ when $\lambda_1 = \lambda_2 = \lambda$ but λ is unknown. Verify that this is the likelihood ratio test of this hypothesis.

3.11 Freireich et al. (1963) present the following remission times in weeks from a clinical trial in acute leukemia:

Placebo:	1,	1,	2,	2,	3,	4,	4,	5,	5,	8,	8,
	8,	8,	11,	11,	12,	12,	17,	22,	23		
6-MP:	6,	6,	6,	7,	10,	13,	16,	22,	23,	6+,	9+,
	10+,	11+,	17+,	19+,	20+,	25+,	32+,	32+,	34+,	35+	

(a) Test the hypothesis of equality of remission times in the two groups using Weibull, log-normal, and log-logistic models. Which model appears to fit the data best?

(b) Test for adequacy of an exponential model relative to the Weibull model.

(c) As a graphical check on the suitability of exponential and Weibull models, compute the Kaplan–Meier estimators $\hat{F}(t)$ of the survivor functions for the two groups. Plot $\log \hat{F}(t)$ versus t and $\log[-\log\hat{F}(t)]$ versus $\log t$.

3.12 Suppose that uncensored paired failure times $(\tilde{T}_{1i}, \tilde{T}_{2i})$ have regression variables $Z_{1i} = 1$, $Z_{2i} = 0$, $i = 1, \ldots, n$. Suppose also that the hazard function for the jth individual in the ith pair can be written

$$\lambda_i \, \exp(Z_{ji}\beta), \qquad j = 1, 2.$$

(a) Show that the MLE, $\hat{\beta}$, satisfies

$$n - 2 \sum_1^n \frac{w_i e^\beta}{1 + w_i e^\beta} = 0,$$

where w_i is the observed value of $W_i = T_{1i}/T_{2i}$. Does asymptotic likelihood theory apply to $\hat{\beta}$? Show that the usual asymptotic formula, if applicable, would yield an asymptotic variance of $2/n$ for $\hat{\beta}$.

(b) Write down the PDF of W_i, and show that it is independent of $\lambda_i, i = 1, \ldots, n$. Write down the likelihood function based on $w_i, i = 1, \ldots, n$ and show that the corresponding MLE for β, in this case, satisfies precisely the same equation as that given in part (a). Show that asymptotic likelihood results apply to this new (marginal) likelihood function and thereby calculate the asymptotic variance for $\hat{\beta}$. Compare with that given in part (a).

(c) Show that $Y_i = \log W_i$ arises from a linear model with mean $-\beta$. Write down the least squares estimator of β and evaluate its efficiency [ratio of asymptotic variance of $\hat{\beta}$ from part (b) to least squares variance].

(d) Let $\epsilon_i = 1$ if $w_i \leq 1$ and $\epsilon_i = 0$ if $w_i > 1$. Derive the distribution of ϵ_i and compare the efficiency of the MLE based on $\epsilon_i, i = 1, \ldots, n$ to that given in part (b), at $\beta = 0$.

3.13 Suppose that the likelihood function (1.12) arises from a discrete failure time variable with sample space $a_1, \ldots, a_k$. Assume also that censoring can occur only at these discrete times and that, as usual, a censored failure time t means that the underlying survival time exceeds t. Using the asymptotic likelihood methods of Section 3.4, derive the joint asymptotic distribution of the product limit estimators $\hat{F}(a_j), j = 1, \ldots, k$ as defined in (1.14). Compare the asymptotic variance of $\hat{F}(a_j)$ to (1.11).

3.14 Suppose again that failure time is discrete with sample space $a_1, \ldots, a_k$. Suppose that r populations are being compared on the basis of such discrete failure time data. Let λ_{ij} be the conditional probability that an individual in sample i fails at a_j given survival up to a_j. Derive the score test for the hypothesis $\lambda_{ij} = \lambda_j$ all (i, j). Compare with the logrank test of Section 1.5.

3.15 Suppose that

$$\lambda(t, Z) = \lambda h_0(t)e^{Z\beta},$$

where

$$Z = \begin{cases} 0, & \text{group 1} \\ 1, & \text{group 2.} \end{cases}$$

Assuming that the ratio of hazard functions between the two samples is 1.5, and assuming no possibility of censoring, calculate the approximate sample size, common to both groups, required to show a difference between groups at the 0.05 level of significance, with probability (power) 0.80. How does this sample size depend on $h_0(\cdot)$?

3.16 A prior density $p(\theta)$ is said to be conjugate to the density $f(x|\theta)$ if for all $\theta, p(\theta)$ is proportional to a possible likelihood function from $f(x|\theta)$. That is,

$$p(\theta) \propto \prod_{i=1}^{n} f(x_i|\theta)$$

for some n and $x_1, \ldots, x_n$.

(a) Show that the class of gamma distributions is conjugate to the exponential density

$$f(t|\lambda) = \lambda e^{-\lambda t}, \qquad t > 0.$$

(b) If the prior distribution for λ is gamma with parameters γ (scale) and v (shape), show that posterior distribution of λ given data $t_1, \ldots, t_n$ is gamma with parameters $(\gamma + \sum t_i)$ and $(n + v)$. The gamma family of distributions is said to be closed under sampling from the exponential distribution.

(c) Obtain the predictive distribution of the next observation and also the posterior distribution of the reliability parameter $\rho = e^{-\lambda t} = P(T > t|\lambda)$, where t is a specified positive number.

(d) Generalize these results to a censored sample $t_1, \ldots, t_n$ with indicators $\delta_1, \ldots, \delta_n$. Note that in the parametric Bayesian approach, no difficulty is caused by quite complex censoring schemes.

3.17 Beginning at chronological time 0, individuals enter a clinical trial over the interval $(0, a]$ according to a Poisson process with intensity function $\alpha(s) > 0, 0 < s \leq a$. Once entered on study, individuals are followed until chronological time $a + b$ for some $b > 0$, at which time all surviving individuals are censored. The distribution of the time $\tilde{T}$ from entry into the trial until failure occurs is of interest and it is assumed that these times are IID from some distribution with survivor function F.

(a) Let N be the total number of individuals entering the trial. What is $P(N = n)$?

(b) Suppose that $N = n$ is given and consider a random permutation of individuals which are then labeled $1, 2, \ldots, n$, and let $S_1, \ldots, S_n$ be the corresponding times of entry into the study. Show that $S_1, \ldots, S_n$ is a random sample from the density $\alpha(s)[\int_0^a \alpha(u)\, du]^{-1}$.

(c) Explain how the results from this study can be viewed as a random censoring model in which the failure and censoring times $(\tilde{T}_i, C_i)$ are IID and find the common censoring distribution.

3.18 (*continuation*)   An alternative way to look at the situation in Exercise 3.17 is to label the sample $1, 2, \ldots, n$ in the order in which they enter the study. Let T_i and C_i be the corresponding failure and censoring time for the ith entry. Show that in this formulation, $\tilde{T} = (\tilde{T}_1, \ldots, \tilde{T}_n)$ is independent of $C = (C_1, \ldots, C_n)$ but the $C_1, \ldots, C_n$ are not mutually independent. Find the distribution of C.

3.19 (*continuation*)   Suppose that each individual upon entry is independently assigned at random and with equal probabilities to one of two treatment groups. Let F_1 and F_2 be the survivor functions for treatment groups 1 and 2, and let M_1 and M_2 be the respective number of failures observed in the study. Find the joint distribution of (M_1, M_2).

3.20 Show that any nonparametric MLE of F arising from the likelihood (3.27) must concentrate all mass on the m intervals $[L_j, R_j]$, as described in Section 3.9.1.

3.21 (*Current status data*)   Suppose that $T_1, \ldots, T_n$ are IID failure time variables with common survivor function $F(\cdot)$. Suppose that inspection times $a = (a_1, \ldots, a_n)$ are determined independently of $T_1, \ldots, T_n$, the ith item is inspected once at time a_i, and $D_i = \mathbf{1}(T_i \leq a_i)$ is observed. That is, we only observe whether or not the ith individual has failed by time a_i, $i = 1, \ldots, n$. Let $b_1, \ldots, b_k$ be the distinct values of $a_1, \ldots, a_n$.

(a) Place this in the context of interval-censored data and show that any survivor function $\hat{F}$ that satisfies $\hat{F}(b_j) = \hat{q}_j, j = 1, \ldots, k$ is a nonparametric MLE of F where the $\hat{q}_1, \ldots, \hat{q}_k$ maximize the likelihood

$$L = \prod_{j=1}^{k} q_j^{r_j} (1 - q_j)^{s_j}$$

subject to the isotonic constraint $1 \geq q_1 \geq \cdots \geq q_k \geq 0$. In this, r_j and s_j

are, respectively, the number of items inspected at time b_j that are surviving and that have failed at that time.

(b) Suppose that $n = 10$, $a = (1, 2, 3, 4, 5, 6, 7, 8, 9, 10)$, and $D = (0, 1, 0, 0, 1, 0, 1, 0, 1, 1)$. Apply the EM algorithm to find the $\hat{q}_j$'s and describe the class of MLEs in this case.

(c) Use the pool adjacent violators algorithm (see Ayer et al., 1955 or Barlow et al., 1972) to find $\hat{F}$.

Relative Risk (Cox)
Regression Models

4.1 INTRODUCTION

In this chapter, attention is focused on methods of estimation and testing based on data arising from the relative risk or Cox regression model (2.10). In the parametric models discussed in Chapter 3, the failure time distribution is assumed known except for a vector of parameters. The relative risk model, however, has a nonparametric aspect in the sense that it involves an unspecified function in the form of an arbitrary baseline hazard function. The model also incorporates a parametric modeling of the relationship between the failure rate and specified covariates, and is sometimes therefore referred to as *semiparametric*. The relative risk model in its most general form is remarkably flexible, but because of the nonparametric component, nonstandard methods are required for estimation and testing.

The relative risk models that we consider in this chapter incorporate covariates that are fixed (time-independent) or are defined functions of time. This class of models is very rich, and although the potential for flexible modeling of this sort has been known for some time, its flexibility has not been widely used in practice. Indeed, it is often implied that the Cox model is one that specifies proportional hazards for all distinct pairs of covariate values, a relatively strong assumption. One of our aims in the organization and scope of this chapter is to increase awareness of the full breadth of the Cox model for fixed or deterministic covariates. As wide as this class is, further extensions of the model to allow stochastic time-dependent covariates are possible and important. These generalization and some applications of such covariates are considered in Chapter 6.

Let $x = (x_1, x_2, \ldots)'$ be a vector of basic (fixed) covariates that are measured at or before time 0 on individuals under study, and let T be a corresponding absolutely continuous failure time variate. We consider a class of models, termed *relative risk models* or *Cox models*, which are specified by the hazard relationship

$$\lambda(t;x) = \lim_{h \to 0^+} P(t \le T < t + h \mid T \ge t, x)/h$$

$$= \lambda_0(t)r(t,x), \qquad t > 0. \tag{4.1}$$

In (4.1), $\lambda_0(t)$ is an arbitrary unspecified baseline hazard function, and the relative risk function $r(t,x)$ specifies the relationship between the covariates x and the failure rate or hazard function. We comment below on flexibility in the specification of r, but for the majority of this chapter and book, we consider the usual exponential form for the relative risk function, $r(t,x) = \exp[Z(t)'\beta]$, which yields the model

$$\lambda(t;x) = \lambda_0(t)\exp[Z(t)'\beta], \tag{4.2}$$

where $Z(t) = [Z_1(t), \ldots, Z_p(t)]'$ is a vector of derived, possibly time-dependent covariates obtained as functions of t and the basic covariates x. The baseline hazard function $\lambda_0(t)$ corresponds to $Z(t) = (0, \ldots, 0)'$ for all t, and $\beta = (\beta_1, \ldots, \beta_p)'$ is a vector of (unknown) regression parameters. There is wide flexibility in the choice of the regression variables $Z(t)$ in the model and the specification of a suitable model is an important step in data analysis. Once the model is set, inferential problems include estimation of the components of β, the baseline hazard function $\lambda_0(t)$, and functions of β and $\lambda_0(t)$, such as, for example, the survivor function at given values of t and x.

If the failure time T has hazard function (4.2), the corresponding survivor function is

$$F(t;x) = P(T > t|x) = \exp\left\{-\int_0^t \lambda_0(u)\exp[Z(u)'\beta]\,du\right\} \tag{4.3}$$

and the density function is

$$f(t;x) = \lambda(t;x)F(t;x). \tag{4.4}$$

Before proceeding, we consider briefly some examples of relative risk modeling to indicate the breadth of this class of models. We begin by considering the simplest of examples, in which the basic covariate is an indicator $x = 0$ and 1, perhaps representing control and treatment groups, respectively. We consider several different models that might be specified:

- The simplest model is $\lambda(t,x) = \lambda_0(t)\exp(x\beta), x = 0, 1$. The hazards in the two treatment groups are proportional to one another, and the parameter β measures the effect of treatment.
- In many instances, we may wish to extend this model to allow the relative risk to vary with time. To do this, we could specify, for example, $Z_1(t) = x$ and

$Z_2(t) = xt$. The model (4.2) now becomes

$$\lambda(t, x) = \lambda_0(t) \exp[x\beta_1 + (xt)\beta_2],$$

where β_2 measures the effect of the interaction between x and time. This model specifies that the relative risk of $x = 1$ versus $x = 0$ is changing smoothly in time. If $\beta_2 > 0$ ($\beta_2 < 0$), the relative risk function is increasing (decreasing) with time and $\beta_2 = 0$ corresponds to the proportional hazards case above. One important aspect of this extended model is its use for testing the proportional hazards or constant relative risk relationship, but the model also allows a concise description of a useful class of treatment effects. Interactions with other functions of time could also be used, such as $\log t$ instead of t.

- The effect of treatment can also be modulated in other ways. For example, we might suspect that the treatment group has a high initial failure rate, due, for example, to postoperative risk in a surgical study that attenuates over a specified period of, say, 1.5 weeks. To accommodate this, let $Z_1(t) = x$, $Z_2(t) = xt$, and $Z_3(t) = x(1.5 - t)\mathbf{1}(t \leq 1.5)$. The parameter β_3 then allows for the postoperative risk, and having adjusted for that, β_1 and β_2 are interpreted as before.

- In most instances, there will be several measured covariates with $x = (x_1, x_2, \ldots)$. Various models could now be specified that would allow for a smoothly changing effect of any of the basic covariates with time. That is, we can entertain models in which the derived covariates $Z(t)$ incorporate the basic covariates as well as possible interactions among those covariates and interactions between the covariates and functions of time.

Figure 4.1 gives a graphical view of some of the simple relative risk models described above.

Before proceeding further, it is useful to make the following notes:

1. When the covariates in the model are constant so that $Z(t) = Z$ for all t, then $r(t; x) = \exp(Z'\beta)$ is independent of time, and the hazard functions at different covariate values are proportional. The model (4.2) with fixed covariates is often called the *proportional hazards model*. In this case, (4.3) can be written

$$F(t; x) = F_0(t)^{\exp(Z'\beta)}, \tag{4.5}$$

where $F_0(t) = \exp[-\int_0^t \lambda_0(u)\, du]$ is an arbitrary baseline survivor function corresponding to $Z = 0$. The class of models $[F_0(t)^c, \ 0 < c < \infty]$ is sometimes called the *Lehmann family*.

2. The term *proportional hazards model* has been used widely to describe the relative risk or Cox model. It is not a descriptive term, however, except in the special case of fixed covariates, so we avoid its use here. The full flexibility of the relative risk modeling seems often to be ignored in data analysis and in

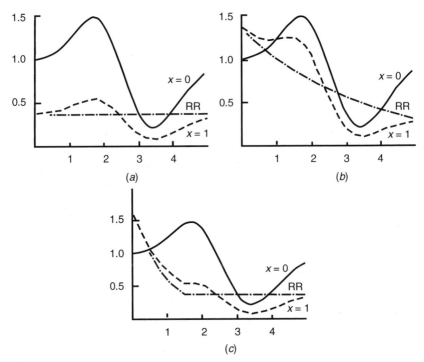

Figure 4.1 Hazards and relative risk functions for two-sample ($x = 0$ or $x = 1$) relative risk models: (a) proportional hazards with RR $= \exp(x\beta)$ and $\beta = 1$; (b) model with interaction with time and RR $= \exp(x\beta_1 + xt\beta_2)$ with $\beta_1 = 0.3$ and $\beta_2 = -0.3$; (c) model with high initial risk and RR $= \exp[x\beta_1 + x(1.5 - t)\mathbf{1}(t \leq 1.5)\beta_2]$ with $\beta_1 = -1.0$ and $\beta_2 = 2$.

the arguments supporting the need for alternative models; in part, this may be due to the proportional hazards misnomer.

3. Time-dependent covariates $Z(t), 0 < t < \infty$ can be much more general than the deterministic type that we consider in this chapter, and they are discussed in more detail in Chapter 6. For example, the covariate may include the output of a stochastic process and may even require the survival of the individual for its existence. The simple deterministic covariates that we consider here, however, allow for useful generalizations of strict proportional hazards and give good flexibility in relative risk modeling. They also lead to simple methods for testing whether individual fixed covariates yield proportional hazards.

4. In the above, the relative risk function $r(t, x)$ in (4.1) has been specified to be of the parametric form $\exp[Z(t)'\beta]$. Other specifications (e.g., $r(t; x) = 1 + Z(t)'\beta$ or $[1 + Z(t)'\beta]^{-1}$) could also be used and may be more appropriate in some instances. We shall, however, use the exponential form in (4.2) as the basis of discussion.

The main problems addressed in this chapter are those related to the estimation of β and the cumulative baseline hazard function defined by $\Lambda_0(t) = \int_0^t \lambda_0(u)\, du$. Estimation of β is considered in Section 4.2, and in Section 4.3 we deal with the estimation of $\Lambda_0(\cdot)$, or equivalently, of $F_0(\cdot)$, where

$$F_0(t) = \exp\left[-\int_0^t \lambda_0(u)\, du\right].$$

Other topics include the extension of (4.2) to include strata, the relationship between the log-rank test and the model (4.2), and the analysis of related discrete models. Although asymptotic results are stated in this chapter, a more detailed discussion can be found in Chapter 5.

Throughout this chapter, unless otherwise specified, failure times are presumed to be subject to arbitrary independent right censoring. Simple extensions apply to incorporate independent left truncation as well, and most formulas apply under both independent left truncation and independent right censoring.

4.2   ESTIMATION OF β

The primary method of analysis is called *partial likelihood*. It formed the basis of the Cox (1972) analysis of the model (4.2), and was discussed further and abstracted in Cox (1975). The presentation in Sections 4.2.1 and 4.2.2 relates closely to that work.

4.2.1   Definition and Some Properties of Partial Likelihood

Suppose that the data consist of an observation on a random vector Y that has density function $f(y; \theta, \beta)$. Here β is the vector of parameters of interest and θ is a nuisance parameter that is typically of very high or infinite dimension. In some applications, θ is in fact a nuisance function, as, for example, the hazard function $\lambda_0(\cdot)$ in the relative risk regression model (4.2). Suppose that Y is transformed into a set of variables $A_1, B_1, \ldots, A_m, B_m$ in a one-to-one manner, and let $A^{(j)} = (A_1, \ldots, A_j)$ and $B^{(j)} = (B_1, \ldots, B_j)$. Suppose that the joint density of $A^{(m)}, B^{(m)}$ can be written

$$\prod_{j=1}^m f(b_j \mid b^{(j-1)}, a^{(j-1)}; \theta, \beta) \prod_{j=1}^m f(a_j \mid b^{(j)}, a^{(j-1)}; \beta). \tag{4.6}$$

The second term in (4.6) is called the *partial likelihood* of β based on $\{A_j\}$ in the sequence $\{A_j, B_j\}$. The number of terms m could be random or fixed. In certain applications one may argue that any information on β in the first term is inextricably tied up with information on the nuisance parameters θ. In these situations, we might choose to base inference on the second term alone, which involves only β.

It should be noted that the partial likelihood

$$L(\beta) = \prod_{j=1}^{m} f(a_j \mid b^{(j)}, a^{(j-1)}; \beta) \tag{4.7}$$

arises as the product of conditional probability statements but is not directly interpretable as a likelihood in the ordinary sense of the word. In fact, (4.7) cannot in general be given any direct probability interpretation as either the conditional or the marginal probability of any event. Nonetheless, in many instances it can be used like an ordinary likelihood for purposes of large-sample estimation in that the usual asymptotic properties formulas and properties associated with the likelihood function and likelihood estimation apply.

We can obtain some intuition about partial likelihood and why it works through consideration of the score components

$$U_j = \frac{\partial \log f(A_j \mid H_j; \beta)}{\partial \beta}, \qquad j = 1, 2, \ldots, m, \tag{4.8}$$

where $H_j = (B^{(j)}, A^{(j-1)})$ is used to specify the conditioning variables for the jth term in (4.8). The total score arising from the partial likelihood (4.7) is

$$U = \frac{\partial \log L}{\partial \beta} = \sum_{j=1}^{m} U_j.$$

Conditionally on H_j, $f(a_j \mid h_j; \beta)$ is a density function. Thus, under the usual regularity conditions, we have $E(U_j \mid H_j = h_j) = 0$. It follows that

$$E(U_j) = EE(U_j \mid H_j) = 0.$$

Further, if $j < k$, the condition $H_k = h_k$ implies that U_j is fixed. Hence, for $j < k$,

$$E(U_j U_k') = EE(U_j U_k' \mid H_k) = E\{U_j E(U_k' \mid H_k)\} = 0.$$

The score contributions $U_1, U_2, \ldots$ thus have mean zero and are uncorrelated.

Again, since $f(a_j \mid h_j; \beta)$ is a (conditional) density,

$$\text{var}(U_j) = E(U_j U_j') = \mathscr{I}_j,$$

where

$$\mathscr{I}_j = -E\left[\frac{\partial^2 \log f(a_j \mid h_j; \beta)}{\partial \beta \, \partial \beta'}\right].$$

From this it follows that the total score U has mean 0 and covariance matrix

$$\text{var}(U) = \mathscr{I} = \sum \mathscr{I}_j.$$

The total score U is thus the sum of m uncorrelated variables, each with mean 0. One might expect a central limit theorem to apply to U as $m \to \infty$ provided that the U_j exhibit a certain degree of independence, the $\mathscr{I}_j$ are not too disparate, and $\sum \mathscr{I}_j$ approaches infinity at a suitable rate. When such a central limit applies, we have the usual basic result for likelihood asymptotics as discussed in Section 3.4: that is, that the score vector has the large-sample approximation

$$U \sim N(0, \mathscr{I}). \tag{4.9}$$

Let $\hat{\beta}$ be the estimator obtained from solving $U = U(\beta) = 0$, and let $I(\beta) = -\partial^2 \log L(\beta)/\partial\beta\,\partial\beta'$ be the observed information matrix from the partial likelihood. If $\hat{\beta}$ is consistent, and $I(\beta).\mathscr{I}^{-1}$ converges to an identity matrix as $m \to \infty$, it would follow that $\mathscr{I}$ in (4.9) can be replaced with $I(\beta)$ or $I(\hat{\beta})$. The other usual asymptotic results based on $\hat{\beta}$ or on the likelihood ratio would then also follow under some additional regularity conditions.

These very informal arguments suggest that at least in some instances, the partial likelihood can be used for large-sample inference, exactly as an ordinary likelihood. The application to the relative risk regression model is given in Section 4.2.2, and in that case, rigorous derivations of the asymptotic results can be based on martingale limit theorems as outlined in Section 5.6.

The argument above deals with the case where the partial likelihood is composed of many terms m, each of which contributes a small amount of information about β. This is the situation that arises in its application to the relative risk regression model discussed in the next section. Sometimes, however, the value m is fixed, and as the sample size increases, each score statistic U_j is based on a large amount of data. This is the situation for some examples with discrete failure times. In that case, a central limit theorem often applies to each score vector U_j individually, and the same asymptotic results as given above apply to the finite sum.

4.2.2 Partial Likelihood for β

The partial likelihood argument applies directly to the relative risk model

$$\lambda(t; x) = \lambda_0(t) \exp[Z(t)'\beta] \tag{4.10}$$

under arbitrary independent right censorship. Suppose that the sample comprises k uncensored failure times $t_1 < \cdots < t_k$ and ignore for the moment the case of ties. The remaining $n - k$ individuals are right censored. Let j denote the individual failing at t_j, and let $Z_\ell(t)$ and x_ℓ denote the covariate vectors for the ℓth individual. In the notation of Section 4.2.1, let B_j specify the censoring information in $[t_{j-1}, t_j)$

plus the information that one individual fails in the interval $[t_j, t_j + dt_j)$. Let A_j specify that item j fails in $[t_j, t_j + dt_j)$. The jth term in the partial likelihood (4.7) is

$$L_j(\beta) = f(a_j \mid b_1^{(j)} a^{(j-1)}; \beta). \tag{4.11}$$

Now, the conditioning event $b^{(j)}, a^{(j-1)}$ specifies all the censoring and failure information in the trial up to time $t_{(j)}^-$ and also provides the information that a failure occurs in $[t_{(j)}, t_{(j)} + dt_{(j)})$. Under independent censoring, it follows that

$$L_j(\beta) = \frac{\lambda(t_j; x_j)\, dt_j}{\sum_{\ell \in R(t_j)} \lambda(t_j; x_\ell)\, dt_j}, \tag{4.12}$$

where $R(t)$ is the set of items at risk of failure at time t^-, just prior to time t. Thus $R(t)$ consists of all individuals who have not failed and are still under observation (uncensored) just prior to time t. It is sometimes convenient to define at risk variables, $Y_i(t) = \mathbf{1}[i \in R(t)]$, where $\mathbf{1}(\cdot)$ is the indicator function. The jth term in the partial likelihood can then be written as

$$L_j(\beta) = \frac{\lambda(t_j; x_j)\, dt_j}{\sum_{\ell=1}^{n} Y_\ell(t_j)\lambda(t_j; x_\ell)\, dt_j}. \tag{4.13}$$

All expressions could be rewritten in this way, and for some purposes, there is an advantage to doing so. At an introductory stage, however, the risk set notation in (4.12) may be more transparent.

Under the relative risk model (4.10), (4.12) simplifies since the baseline hazard $\lambda_0(t_j)\, dt_j$ cancels in the numerator and denominator. The product over j then gives the partial likelihood for β,

$$L(\beta) = \prod_{j=1}^{k} \frac{\exp[Z_j(t_j)'\beta]}{\sum_{\ell \in R(t_j)} \exp[Z_\ell(t_j)'\beta]}. \tag{4.14}$$

The arbitrary baseline hazard function has been eliminated and the resulting likelihood can be used for inferences about β.

In forming the partial likelihood of β in (4.14), we are ignoring any information about β which might be obtained from the observation that no items fail in the intervals (t_{j-1}, t_j), $j = 1, \ldots, k+1$, where $t_0 = 0$ and $t_{k+1} = \infty$. Since $\lambda_0(t)$ is completely unspecified, intuition suggests that the failure-free interval (t_{j-1}, t_j) can yield little information about β since we can account for it simply by taking $\lambda_0(t)$ to be very close to zero over the interval. If one had additional information on $\lambda_0(t)$, for example, a parametric form, there would be contributions to the inference about β from the intervals with no failures.

The maximum likelihood estimate, $\hat{\beta}$, from (4.14), can be obtained as a solution to the vector equation

$$U(\beta) = \partial \log L / \partial \beta = \sum_{j=1}^{k} [Z_j(t_j) - \mathscr{E}(\beta, t_j)] = 0, \tag{4.15}$$

where

$$\mathscr{E}(\beta, t_j) = \sum_{\ell \in R(t_j)} Z_\ell(t_j) p_\ell(\beta, t_j)$$

and

$$p_\ell(\beta, t_j) = \frac{\exp[Z_\ell(t_j)' \beta]}{\sum_{i \in R(t_j)} \exp[Z_i(t_j)' \beta]}, \qquad \ell \in R(t_j). \tag{4.16}$$

Thus, $\mathscr{E}(\beta, t_j)$ is the expectation of $Z_\ell(t_j)$ with respect to the distribution (4.16) on the risk set $R(t_j)$. Similarly, the observed information matrix is

$$I(\beta) = -\frac{\partial^2 \log L}{\partial \beta \, \partial \beta'} = \sum_{j=1}^{k} \mathscr{V}(\beta, t_j), \tag{4.17}$$

where

$$\mathscr{V}(\beta, t_j) = \sum_{\ell \in R(t_j)} [Z_\ell(t_j) - \mathscr{E}(\beta, t_j)]^{\otimes 2} p_\ell(\beta, t_j)$$

is the covariance matrix of $Z_\ell(t_j)$ under the distribution (4.16). (Note that $b^{\otimes 2} = bb'$ for b a vector.] As a consequence, $I(\beta)$ is typically positive definite for all β, the log likelihood is strictly concave, and the estimate $\hat{\beta}$ is typically unique. The value $\hat{\beta}$ that maximizes (4.14) can usually be obtained by a Newton–Raphson iteration utilizing (4.15) and (4.17). A starting value of $\beta_0 = 0$ often suffices. It is important to note, however, that the estimate of a component of β, β_1 say, can be ∞ (or $-\infty$) as outlined in Exercise 4.1. Also, as in ordinary regression, the covariance and information matrices can be singular; see Exercise 4.2.

Examination of the score equation (4.15) shows that $Z_j(t_j)$, the covariate value of the failure at time t_j, is being compared with the expectation, $\mathscr{E}(\beta, t_j)$. It is instructive to think about the scalar case ($p = 1$). In this case, $\mathscr{E}(\beta, t_j)$ is an increasing function of β, and if the observed $Z_j(t_j)$ is one of the larger (smaller) values in the risk set, the jth term in (4.15) favors a positive (negative) value of β. The score equation combines these single-time considerations to find the value of β that best describes the failure experience overall.

Asymptotic results completely analogous to those for parametric likelihoods apply under quite general conditions here, as discussed in Section 5.7. In the absence of ties, the asymptotic distribution of $\hat{\beta}$ is normal with mean β and estimated covariance matrix $I(\hat{\beta})^{-1}$. For example, inference on the lth component β_l of β can be based on the asymptotic result

$$\hat{\beta}_l - \beta_l \approx N(0, \hat{I}^{ll}),$$

where $\hat{I}^{ll}$ is the (l, l) element of $I(\hat{\beta})^{-1}$. Likelihood ratio tests can be based on the partial likelihood, and the score statistic $U(\beta_0)$ from (4.10) can be used to test $\beta = \beta_0$ with the usual χ^2 and normal asymptotic results. Score tests are discussed in Section 4.2.4.

The numerical illustrations in this chapter assume that the conditions for asymptotic normality of the score statistic $U(\beta_0)$ and $\hat{\beta}$ are met and that sample sizes are large enough for asymptotic distributional approximations to be adequate. As in Chapter 3, however, the question of the adequacy with which the asymptotic form of the distribution approximates the actual sampling distribution must be kept in mind in any particular application. The normal approximation to the distribution of the score statistic (4.15) for the special case of comparing several uncensored samples without ties, however, often seems adequate for surprisingly small samples of size 10 or possibly even fewer in each group. Somewhat larger, but probably not appreciably larger, samples sizes are probably necessary for an adequate approximation to the $\hat{\beta}$ distribution provided that the regression variables and censoring mechanism are not too extreme. As with the parametric models, extreme and isolated $Z(t)$ values or very severe censoring increase the total sample size necessary to ensure the adequacy of normal approximations. In small-sample situations where, for example, the results of significance tests are equivocal, some numerical work may be necessary to develop an improved approximation for, or to produce an estimate of, the actual sampling distributions for $U(\beta_0)$ or $\hat{\beta}$.

We have concentrated on the partial likelihood derivation above, but several other approaches to estimating β have been suggested when continuous data arise from the Cox model. Breslow (1974), for example, suggests an approach in which the baseline hazard function is approximated by a step function that is constant between adjacent uncensored failure times, and he shows that the resulting maximized likelihood for β is given by (4.14) (see Exercises 4.4, 4.5, and 4.6). Bayesian approaches have also been suggested and are discussed in Section 11.7. When the covariates are time independent, the estimation of β can also be based on a marginal likelihood arising from the observed rank or generalized rank vector. This approach is discussed in Section 4.7.

4.2.3 Ties in Continuous Failure Time Data

When there are ties among the uncensored failure times, the partial likelihood can be adjusted in various ways. We consider here primarily the case where a continuous model seems appropriate, but there are some ties that occur due to grouping or round-off of the observations.

Perhaps the most natural adjustment for ties is to use the average likelihood that arises through breaking the ties in all possible ways. Suppose, as before, that $t_1 < \cdots < t_k$ are the distinct failure times and suppose that d_j items fail at time t_j, $j = 1, \ldots, k$. Let $D(t_j) = \{j_1, \ldots, j_{d_j}\}$ be the set of labels of individuals that fail at t_j, and let Q_j be the set of $d_j!$ permutations of the labels $\{j_1, \ldots, j_{d_j}\}$. Let $P = (p_1, \ldots, p_{d_j})$ be an element in Q_j and $R(t_j, P, r) = R(t_j) - \{p_1, \ldots, p_{r-1}\}$.

The average partial likelihood contribution at t_j which arises from breaking the ties in all possible ways is then

$$\frac{1}{d_j!}\exp[s_j(t_j)'\beta]\sum_{P\in Q_j}\prod_{r=1}^{d_j}\left\{\sum_{\ell\in R(t_j,P,r)}\exp[Z_\ell(t_j)'\beta]\right\}^{-1},$$

where $s_j(t_j) = \sum_{i=1}^{d_j} Z_{j_i}(t_j)$ is the sum of the covariates of individuals observed to fail at t_j. The corresponding average partial likelihood is proportional to

$$\prod_{j=1}^{k}\left(\exp[s_j(t_j)'\beta]\sum_{P\in Q_j}\prod_{r=1}^{d_j}\left\{\sum_{\ell\in R(t_j,P,r)}\exp[Z_\ell(t_j)'\beta]\right\}^{-1}\right). \qquad (4.18)$$

The result (4.18) is computationally intensive if the number of ties is large at any failure time. If the ties are not too numerous, the expression (4.18) will be well approximated (Peto, 1972b; Breslow, 1974) by

$$L=\prod_{j=1}^{k}\frac{\exp[s_j(t_j)'\beta]}{\{\sum_{\ell\in R(t_j)}\exp[Z_\ell(t_j)\beta]\}^{d_j}}. \qquad (4.19)$$

An alternative approximation suggested by Efron (1977) is

$$L=\prod_{j=1}^{k}\frac{\exp[s_j(t_j)'\beta]}{\prod_{r=0}^{d_j-1}\{\sum_{\ell\in R(t_j)}\exp[Z_\ell(t_j)'\beta]-r\bar{A}(\beta,t_j)\}}, \qquad (4.20)$$

where

$$\bar{A}(\beta,t_j) = d_j^{-1}\sum_{\ell\in D(t_j)}\exp[Z_\ell(t_j)'\beta].$$

Note that (4.14) is a special case of (4.18), (4.19), and (4.20).

It is a straightforward matter to compute the score equations and information matrices that arise from each of these approximate likelihoods. The calculations for (4.19) and (4.20) are particularly simple. For both of these, however, the score has nonzero expectation and the corresponding estimators have some asymptotic bias under a grouped continuous Cox model. In addition, the inverse of the information matrix will not provide an exactly consistent estimator of the variance of $\hat{\beta}$. If the ties arise through grouping of the continuous model and the fraction d_j/n_j of ties at any failure time is large, numerical investigation indicates that (4.18) often still gives good estimates, whereas (4.19) may exhibit a substantial bias for the continuous relative risk parameter. Efron's approximation (4.20) gives an improvement over (4.19) and yields surprisingly good estimates, even when the grouping is quite severe. For applications with relatively few ties, it does not matter much which likelihood is used since asymptotic biases or problems with variance estimation are slight. For estimating the continuous relative risk parameter, there is some preference for using the average likelihood (4.18) or, failing that, the Efron approximation (4.20). If there are many ties in the data, it may be best to consider a discrete model, and these are discussed in some detail in Section 4.8.

The Breslow likelihood (4.19) is often used in practice because its form is so simple, and we have elected to use it in most of the examples in this book. It can be shown that the score equation from the Breslow likelihood is consistent for a relative risk parameter defined in the mixed, discrete, and continuous Cox model

$$d\Lambda(t;x) = \exp[Z(t)'\beta]d\Lambda_0(t) \tag{4.21}$$

of Section 2.3, where $\Lambda_0(t)$ is an unspecified baseline discrete, continuous, or mixed cumulative hazard function. As a consequence, the estimate of β obtained from the Breslow likelihood (4.19) is consistent for β in the model (4.21). This provides additional motivation for basing estimation on the Breslow likelihood. Note that at any mass point where $d\Lambda_0(t) > 0$, the model (4.21) places a constraint on β, since for all relevant x, we must have $d\Lambda(t;x) \leq 1$ (see Section 4.8.2 for additional discussion).

Cox (1972) suggests handling ties through a partial likelihood argument applied to the discrete logistic model,

$$\frac{d\Lambda(t;x)}{1 - d\Lambda(t;x)} = \exp[Z(t)'\beta]\frac{d\Lambda_0(t)}{1 - d\Lambda_0(t)} \tag{4.22}$$

of Section 2.4.3. In this, $d\Lambda_0(t)$ is an unspecified discrete hazard function with mass at the observed failure times $t_1, t_2, \ldots, t_k$. A direct generalization of the partial likelihood argument of Section 4.2.2 can then be used to compute, at each failure time, the probability that the d_j failures should be those observed given the risk set and the multiplicity d_j. A simple computation gives the conditional probability as the jth term in the partial likelihood,

$$\prod_{j=1}^{k} \frac{\exp[s_j(t_j)'\beta]}{\sum_{\ell \in R_{d_j}(t_j)} \exp[s_\ell(t_j)'\beta]}. \tag{4.23}$$

In this expression, $R_{d_j}(t_j)$ is the set of all subsets of d_j items chosen from the risk set $R(t_j)$ without replacement, $\ell = \{\ell_1, \ldots, \ell_{d_j}\}$ is an element of $R_{d_j}(t_j)$, and $s_\ell(t_j) = \sum_{i \in \ell} Z_i(t_j)$. The partial likelihood (4.23) is difficult computationally if the number of ties is large, although there are fairly efficient recursive methods for computation due to Howard (1972). With relatively fewer ties, approximations are afforded by (4.19) or (4.20).

The partial likelihood (4.23) does not give rise to a consistent estimator of the parameter β in (4.6) if the ties arise by the grouping of continuous failure times nor for the relative risk parameter in the extended model (4.21). This inconsistency in the partial likelihood occurs since (4.23) must be thought of as arising from the discrete model (4.22), so the β value that maximizes (4.23) estimates the odds ratio parameter β in that model. Since (4.22) does not arise as a grouping of the continuous model, the two parameters do not have identical interpretations.

4.2.4 Log-Rank Test

Before proceeding to look at an application of these results, it is worth studying the score function tests at $\beta = 0$ that arise from the partial likelihood given above, thereby deriving the important log-rank test of Section 1.4. In this discussion we consider fixed covariates $Z(t) = Z$, although extension to defined covariates is immediate.

First, suppose that there are no ties or censoring in the data so that the observed failure times are $t_1 < t_2 < \cdots < t_n$ with corresponding covariate vectors $Z_1, \ldots Z_n$. The score test statistic for the hypothesis $\beta = 0$ that arises from (4.14) or (4.19) can be written

$$U(0) = \sum_{j=1}^{n} Z_j\{1 - [n^{-1} + (n-1)^{-1} + \cdots + (n-j+1)^{-1}]\}, \qquad (4.24)$$

obtained by substitution in (4.15). This is in the form of a linear rank statistic $\sum_{i=1}^{n} Z_i a_i$, where $a_i = a_{i,n}$ is the score attached to the ith ordered observation or failure time (see Chapter 7). In this case, the ith score is 1 minus the expected ith-order statistic in a sample of size n from a unit exponential distribution. This is the Savage (1956) or exponential scores test.

Generalizations of the Savage test for tied or censored data can be obtained from the score function test corresponding to an exact or approximate partial likelihood. For example, from the approximate Breslow likelihood (4.19), the score test statistic for the global null hypothesis $\beta = 0$ can be written

$$U(0) = \sum_{j=1}^{k} \left(s_j - d_j n_j^{-1} \sum_{\ell \in R(t_j)} Z_\ell \right), \qquad (4.25)$$

where n_j is the number at risk at t_j^-. Note that (4.25) reduces to (4.24) if there are no ties ($d_j = 1$) and no censoring. The special case of the comparison of $p + 1$ survival curves labeled $0, 1, 2, \ldots, p$ arises upon defining $Z_i = (Z_{1i}, \ldots, Z_{pi})$, where Z_{ui} equals 1 or zero according to whether or not the ith study subject is in the uth sample ($i = 1, \ldots, n; u = 1, \ldots, p$). It is then easy to see that (4.25) is precisely the log-rank statistic $U(0) = O - E$ introduced in (1.21), where $O = s_1 + \cdots + s_k$ gives the observed number of failures in each of the samples $1, 2, \ldots, p$ and

$$E = \sum_{j=1}^{k} d_j n_j^{-1} \sum_{\ell \in R(t_j)} Z_\ell$$

is a corresponding vector of summed conditional expected numbers of failures.

This same score statistic (4.25) also arises from the partial likelihood (4.23) as an exact result corresponding to the discrete logistic model (4.22). In fact, the construction of the partial likelihood relates very closely to the hypergeometric arguments of Section 1.4 and the $O - E$ interpretation of the log-rank statistic. The asymptotic results for partial likelihoods show that the log-rank statistic, $O - E$, is asymptotically normal with estimated covariance matrix given by $W_{p \times p}$, whose (h, u) element

$$W_{hu} = \sum_{j=1}^{k} [d_j (n_{uj} n_j^{-1} \delta_{hu} - n_{hj} n_{uj} n_j^{-2})(n_j - d_j)(n_j - 1)^{-1}]$$

is obtained as the negative of the second partial derivative of the logarithm of (4.23). In this expression, n_{uj} is the size of the risk set in sample u just prior to t_j and $\delta_{hu} = 1$ or 0 according to whether or not $h = u$. The appropriate test statistic for testing $\beta = 0$ is then

$$U(0)' W^{-1} U(0) \tag{4.26}$$

which, under the hypothesis, has an asymptotic χ_p^2 distribution. This is exactly the log-rank test given in Section 1.4.

The covariance matrix of $U(0)$ might also be estimated from the approximate likelihood (4.19). This gives $I(0)$ with (h, j) element

$$I_{hu}(0) = \sum_{j=1}^{k} d_j (n_{uj} n_j^{-1} \delta_{hu} - n_{hj} n_{uj} n_j^{-2}).$$

which agrees with W_{hu} if there are no ties ($\delta_i = 1$); this corresponds to the fact that (4.19) is then an exact result. If there are ties, however, the elements of $I(0)$ tend to overestimate the variance of the score statistic.

4.2.5 Some Examples

Example 4.1. Consider the carcinogenesis data and, in the first instance, the simple model $\lambda(t; Z) = \lambda_0(t) \exp(Z\beta)$, where $Z = 0, 1$ is a treatment indicator. The first step in applying the results of this section to data is to order the survival times from smallest to largest, with the additional convention that failure times

Table 4.1   **Relative Risk Model Applied to the Carcinogenesis Data with the Single Covariate Z = Treatment Group**

i	t_i	d_i	Failures, Z_i	Censored, Z_i	Contribution to Likelihood
1	142	1	1		$e^{\beta}/(19 + 21e^{\beta})$
2	143	1	0		$1/(19 + 20e^{\beta})$
3	156	1	1		$e^{\beta}/(18 + 20e^{\beta})$
4	163	1	1		$e^{\beta}/(18 + 19e^{\beta})$
5	164	1	0		$1/(18 + 18e^{\beta})$
6	188	2	0,0		$1/(17 + 18e^{\beta})^2$
7	190	1	0		$1/(15 + 18e^{\beta})$
8	192	1	0		$1/(14 + 18e^{\beta})$
9	198	1	1	1(204)	$e^{\beta}/(13 + 18e^{\beta})$
⋮	⋮	⋮	⋮	⋮	⋮
27	296	2	1,1		$e^{2\beta}/(1 + 4e^{\beta})^2$
28	304	1	0		$1/(1 + 2e^{\beta})$
29	323	1	1	1(344)	$e^{\beta}/2e^{\beta}$

precede censored times in the case of ties. This ordering is presented in Table 4.1 for the carcinogenesis data. Also recorded in the table are the numbers of failures d_i occurring at each distinct failure time t_i, the covariates of the failures, and the covariates of censored times in $[t_i, t_{i+1})$. In general, there is an advantage to beginning the calculation at the last failure time t_k since the risk set at t_i can be formed by adding, to that at $t_{(i+1)}$, the labels of items failing or censored in $[t_i, t_{i+1})$. The contributions to the likelihood (4.19) at a specified β value are then easily computed. With these data, the approximation in (4.19) is not necessary since all items failing at any given time have the same covariates. The ties may thus be broken in any way at all and the likelihood (4.14) used. The approximation (4.19) has been used here for illustration. Only a small difference arises through breaking the ties and using the exact result (4.14).

Three iterations of the Newton–Raphson procedure, with an initial estimate of $\beta = 0$, give the maximum likelihood estimate to three-figure accuracy, as $\hat{\beta} = -0.596$. The information observed is $I(\hat{\beta}) = 8.237$. Thus $\hat{\beta}\sqrt{I(\hat{\beta})} = -1.71$ is, under the assumption that $\beta = 0$, an observation from a $N(0, 1)$ distribution. This gives a significance level of 0.087. There is some indication of a treatment effect, although the evidence is not strong. The log relative likelihood,

$$R(\beta) = \log L(\beta) - \log L(\hat{\beta}),$$

is plotted for these data in Figure 4.2. The close agreement of this plot with the normal log likelihood $-\frac{1}{2}I(\hat{\beta})(\hat{\beta} - \beta)^2$ suggests that the large-sample procedures based on $\hat{\beta}$ are reasonably accurate.

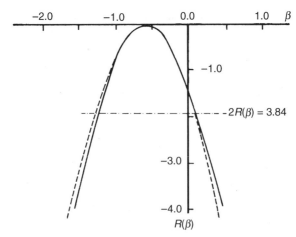

Figure 4.2 Log partial likelihood of β (solid line) and the approximating normal likelihood arising from the carcinogenesis data of Table 1.1.

An alternative test of $\beta = 0$ is based on the likelihood ratio statistic. We find that $-2R(0) = 2.86$, and comparison with the χ_1^2 distribution gives a significance level of 0.091, in close agreement with the test based on the asymptotic distribution of $\hat{\beta}$. The latter gives a χ_1^2 statistic $(1.71)^2 = 2.92$. An approximate 95% confidence interval for β obtained from the likelihood ratio test as $\{\beta | -2R(\beta) \leq 3.84\}$. [Note that $P(\chi_1^2 > 3.84) = 0.05$.] This yields the interval $(-1.27, 0.11)$, compared to $(-1.28, 0.09)$, as the interval based on the approximate normal statistic $(\hat{\beta} - \beta)\sqrt{I(\hat{\beta})}$.

A third procedure for testing $\beta = 0$ is provided by the score test based on $U(0)$. We find that $U(0) = 4.763$ and $I(0) = 7.560$, which gives the test statistic $U(0)^2 I(0)^{-1} = 3.00$. Recall that $U(0) = O - E$ is the log-rank statistic, which was also considered in Section 1.4. The variance estimate $I(0)$ is based on the approximate likelihood (4.19); as discussed above, a better estimate of the asymptotic variance is $V = 7.263$, which arises as the observed information from the likelihood (4.23), and yields a χ_1^2 statistic $U(0)^2 V^{-1} = 3.12$. As is generally the case, the test statistic based on the variance estimate $I(0)$ is smaller and gives a more conservative test, although the differences here are too small to be of practical importance. As noted earlier, since all items failing at any given time have the same covariates, the ties may be broken and the analysis could be based on the log-rank statistic with no ties. This gives $U(0) = 4.584$ and $I = V = 7.653$. The log-rank χ_1^2 statistic is then $(4.584)^2/7.653 = 2.75$. □

One advantage of this analysis of the carcinogenesis data is that the extended initial period with no observed failures is easily handled. Except for the generalized F, all parametric models discussed previously required insertion of a guarantee time to provide an adequate fit to these data. This was usually done by arbitrarily specifying the origin of measurement to be 100 days, although a threshold parameter could be included and estimated from the data. In Section 3.10 we found the generalized F to be sufficiently flexible to be able to account for this failure-free period with no guarantee parameter. The proportional hazards approach, however, is conceptually simpler and involves much less computation.

As noted in Section 4.1, an important use of defined covariates is in checking the proportional hazards assumption within a model involving only fixed covariates. In this two-sample problem, let $Z_1 = 0, 1$ denote the samples and suppose that the hazard function is given by

$$\lambda_0(t) \exp[Z_1 \beta_1 + Z_2(t) \beta_2], \tag{4.27}$$

where $Z_2(t) = Z_1 \log t$ is a defined time-dependent covariate. As discussed earlier, a test of $\beta_2 = 0$ provides a check of the proportional hazards model for the levels of Z_1 versus one in which the hazard ratio of sample 1 to sample 0 is increasing ($\beta_2 > 0$) or decreasing ($\beta_2 < 0$) with time. Various other types of alternatives could be checked by changing the definition of $Z_2(t)$.

The model (4.27) was fitted to the carcinogenesis data in order to check the proportionality assumption implicit in use of the single covariate Z_1 for treatment. On examining the second derivatives of the log partial likelihood, it becomes clear that the estimates of β_1 and β_2 are highly correlated, owing to lack of centering of the $\log t_{(i)}$. There is some numerical advantage to using the equivalent model

$$\lambda_0(t) \exp[Z_1 \beta_1^* + Z_1 (\log t - c) \beta_2],$$

where c is taken to be the average of the $\log t_{(i)}$'s, so that $\beta_1^* = \beta_1 + \beta_2 c$.

Convergence is reached in four iterations from initial values $\beta_1^* = \beta_2 = 0$, and one obtains the estimates $\hat{\beta}_1^* = -0.599$, $\hat{\beta}_2 = -0.230$, with the estimated covariance matrix

$$I^{-1} = \begin{pmatrix} 0.1211 & 0.0487 \\ 0.0487 & 3.3303 \end{pmatrix}.$$

A test of $\beta_2 = 0$ based on the asymptotic distribution of $\hat{\beta}_2$ gives a standard normal statistic $-0.230/\sqrt{3.3303} = -0.1258$, which is not significant. There is no evidence to suggest inadequacy of the proportional hazards assumption (at least in the direction of an increasing or decreasing hazard ratio over time).

The immediate extension of this procedure gives a check on the proportional hazards assumption in the model for fixed covariates, $\lambda_0(t) \exp(x\beta)$. The procedure involves specification of a more general model with a vector of covariates $Z(t)$, where $Z_i(t) = x_i$ and $Z_{s+i}(t) = x_i g_i(t)$ for $i = 1, \ldots, s$. Thus we consider

$$\lambda_0(t) \exp\left[Z(t)' \beta\right],$$

where $g_i(t)$ is some specified function of t, and β_{s+j} is taken to be zero except for components of x for which the proportionality assumption is being questioned. A test of the hypothesis $\beta_{s+j} = 0$ provides a check on proportionality for the jth component of x. This general approach provides a flexible method to evaluate departures from proportionality. It also provides an approach to building a model for the dependence of relative risk on time.

Example 4.2. Consider the data on aplastic anemia presented in Table 1.2. There are three covariates—treatment, age, and laminar airflow isolation—and we consider models involving age and treatment. Laminar airflow exhibits no effect with respect to the endpoint. Treatment is coded in the variable x_1, which takes the value 0 for CSP + MTX and 1 for MTX only. Age is incorporated into the model both as a quantitative variable, with x_2 giving age in years, and also as a grouped variable on three levels (0–$15, 16$–$25, \geq 26$). In the latter case, the groupings were chosen to give approximately equal numbers in each group, and the variable is coded with indicators x_3 and x_4 for the two older groups. There is often some advantage through coding a quantitative variable into groups like this since it limits the effect of extreme values on the fit, and also allows some considerable flexibility in the form of the regression. An alternative approach that yields similar results in this case is to include linear and quadratic terms for age.

Kaplan–Meier estimates of the survivor functions in the two treatment groups are given in Figure 4.3. The estimates suggest a potential treatment benefit to the combined therapy CSP + MTX. The plots also do not suggest a power relationship

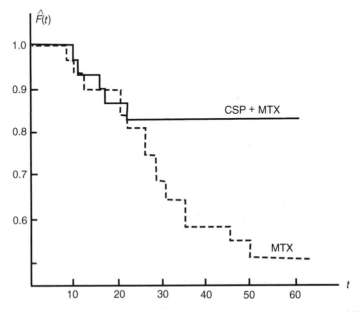

Figure 4.3 Survivor function estimates for the data on acute graft versus host disease of Table 1.2.

Table 4.2 Relative Risk Models for Example 4.2 Fit to the AGVHD Data of Table 1.2 [a]

Model	Treatment	Age	Age Group 16–20	Age Group ≥ 26	Trt $\times$ Time	Log Likelihood
1	—	—	—	—	—	−79.12
2	1.143/0.517 (0.027)	—	—	—	—	−76.14
3	1.388/−0.554 (0.009)	0.057/0.025 (0.025)	—	—	—	−73.71
4	1.165/0.537 (0.030)	—	1.907/0.771 (0.013)	1.678/0.810 (0.038)	—	−71.58
5	−2.053/1.507 —	—	1.933/0.771 (0.012)	1.780/0.810 (0.028)	0.1608/0.0813 (0.048)	−68.19

[a] The numerator is the estimated regression coefficient, the denominator is the estimated standard error, and the number in parentheses is the estimated significance level or p value.

between the two survivor functions and so suggest that the hazards may not be proportional. The use of a time-dependent relative risk function is one way to describe such an effect.

Table 4.2 presents the results of various relative risk models fit to these data. Model 1 gives a baseline log likelihood with no variables in the model, and model 2, with treatment alone, confirms the impression from Figure 4.3 of an overall beneficial effect to the addition of CSP. The estimated relative risk is $\exp(1.143) = 3.14$, with a corresponding confidence interval of $\exp(1.143 \pm 1.96 \times 0.517) = (1.14, 8.64)$. Models 3 and 4 examine the dependence on age. The stratification into three groups (model 4) gives the better fit as measured by the maximized log partial likelihood. This fit suggests that the younger age group 0–15 has a better survival experience than either older group, with little difference between the older groups. Model 5 has a relative risk of the form $\exp[Z(t)'\beta]$, where $Z(t) = (x_1, x_3, x_4, x_1 t)$, allowing an interaction between treatment and time. This interaction term results in an increase in the maximum log likelihood of $71.58 - 68.19 = 3.38$ and a significance level of $P(\chi_1^2 \geq 6.75) < 0.01$. Similar results are found with models that incorporate interactions with other monotone functions of time, such as $g(t) = \log t$.

This likelihood ratio test is preferable to a test based on the maximum likelihood estimate since the likelihood function is positively skewed with respect to this coefficient. In this case, the MLE of the regression coefficient of the treatment by time interaction is $\hat{\beta}_4 = 0.1608$ with a standard error of 0.0813. This gives an approximate significance level of 0.05, which would indicate substantially less evidence against $\beta_4 = 0$ than does the likelihood ratio procedure above. An approximate 95% confidence interval based on the likelihood ratio statistic is $(0.032, 0.362)$, compared with $(0.002, 0.320)$ based on the MLE.

There is a strong evidence here of a net benefit to CSP + MTX and there is evidence suggesting that the size of the effect increases with time since treatment and so tends to delay or avoid late occurrence of AGVHD. From model 5, the estimated relative risk function comparing CSP + MTX versus MTX is

$$\hat{r}(t) = \exp(2.053 - 0.1608t),$$

giving a description of a smoothly changing effect on relative hazard rates as time from treatment increases. The choice of an interaction term $x_1 t$ versus, for example, $x_1 \log t$ cannot be very fully justified from these data, but other such choices lead to qualitatively similar conclusions.   □

4.3 ESTIMATION OF THE BASELINE HAZARD OR SURVIVOR FUNCTION

Consider now the derivation of an estimator of the baseline cumulative hazard function and the baseline survivor function that are analogous to the Nelson–Aalen and Kaplan–Meier estimators obtained in Chapter 1. One way to do this is to embed the continuous-time relative risk model in a more general discrete/mixed/continuous model and use the nonparametric maximum likelihood arguments of the type that led to the Kaplan–Meier estimate in Section 1.4. There are various ways to accomplish this, but perhaps the most natural discrete/continuous model is the grouped relative risk model obtained, with fixed covariates, by grouping the continuous model.

The discussion of the grouped relative risk model in Section 2.4.3 was limited to fixed covariates, and we extend it here, in the obvious way, to include defined time-dependent covariates $Z(t)$. Thus, let $\Lambda_0(t)$ be a baseline cumulative hazard function. The corresponding hazard at basic covariate values x is analogous to (2.21),

$$d\Lambda(t;x) = 1 - [1 - d\Lambda_0(t)]^{\exp[Z(t)'\beta]}, \tag{4.28}$$

which reduces to (4.2) in the continuous case and the discrete model (2.19) in the discrete case $[\Lambda_c(t) = 0]$, except for time-dependent $Z(t)$. The corresponding survivor function for T is

$$F(t;x) = P(T > t;x) = \mathscr{P}_0^t[1 - d\Lambda(u;x)]$$
$$= \mathscr{P}_0^t[1 - d\Lambda_0(u)]^{\exp[Z(u)'\beta]} \tag{4.29}$$

[see also (2.22)].

When $Z(u) = Z$ has only fixed covariates,

$$F(t;x) = F_0(t)^{\exp(Z'\beta)}, \tag{4.30}$$

where the baseline survivor function

$$F_0(t) = \mathscr{P}_0^t[1 - d\Lambda_0(u)] \tag{4.31}$$

corresponds to $Z = 0$. With time-dependent $Z(t)$, however, the relationship between $F(t; x)$ and $F_0(t)$ is summarized in the more complicated (4.29) with $d\Lambda_0(u) = -dF_0(u)/F_0(u)$ at continuity points and $d\Lambda_0(u) = 1 - [F_0(u)/F_0(u^-)]$ at mass points.

Suppose now that data are available from the extended model (4.29) and consider calculation of the nonparametric maximum likelihood estimate of $F_0(t)$. As before, let $t_1, \ldots, t_k$ be the distinct failure times, D_i be the set of labels associated with individuals failing at t_i, and C_i be the set of labels associated with individuals censored in $[t_i, t_{i+1}), i = 0, \ldots, k$, where $t_0 = 0$ and $t_{k+1} = \infty$. The contribution to the likelihood of an individual with covariates x who fails at t_i is, under independent censorship, $F(t_i^-; x) - F(t_i; x)$ and the contribution of a censored observation at time t is $F(t; x)$. The likelihood function can then be written

$$L = \prod_{i=0}^{k} \left\{ \prod_{\ell \in D_i} [F(t_i^-; x_\ell) - F(t_i; x_\ell)] \prod_{\ell \in C_i} F(t_i; x_\ell) \right\}, \qquad (4.32)$$

where D_0 is empty.

As with the Kaplan–Meier estimate, it is clear that L is maximized by taking $F_0(t) = F_0(t_i)$ for $t_i \le t < t_{i+1}$ and allowing probability mass to fall only at the observed failure times $t_1, \ldots, t_k$. These observations lead to the consideration of a discrete model with baseline cumulative hazard function

$$\Lambda_0(t) = \sum_{j=1}^{k} (1 - \alpha_j) \mathbf{1}(t_j \le t),$$

which places a discrete hazard component $1 - \alpha_j$ at each observed failure time $t_j, j = 1, \ldots, k$. Substitution into (4.32) and using (4.29) gives the likelihood function

$$\prod_{i=1}^{k} \left[\prod_{j \in D_i} (1 - \alpha_i^{\exp[Z_i(t_j)'\beta]}) \prod_{\ell \in R(t_i) - D_i} \alpha_i^{\exp[Z_\ell(t_i)'\beta]} \right], \qquad (4.33)$$

which is to be maximized in $\alpha_1, \ldots, \alpha_k$.

The estimation of the survivor function can be carried out by joint estimation of the α's and β in (4.29), as outlined in Exercise 4.11. More simply, however, we can take $\beta = \hat{\beta}$ as estimated from the partial likelihood function and then maximize (4.33) with respect to $\alpha_1, \ldots, \alpha_k$. Differentiating the logarithm of (4.33) with respect to α_i gives the maximum likelihood estimate of α_i as a solution to

$$\sum_{j \in D_i} \exp[Z_j(t_i)\hat{\beta}][1 - \alpha_i^{\exp[Z_j(t_i)\hat{\beta}]}]^{-1} = \sum_{\ell \in R(t_i)} \exp[Z_\ell(t_i)\hat{\beta}]. \qquad (4.34)$$

If only a single failure occurs at t_i, (4.34) can be solved directly for $\hat{\alpha}_i$ to give

$$\hat{\alpha}_i = \left\{ 1 - \frac{\exp[Z_i(t_i)\hat{\beta}]}{\sum_{\ell \in R(t_i)} \exp[Z_\ell(t_i)\hat{\beta}]} \right\}^{\exp[-Z_i(t_i)\hat{\beta}]}.$$

Otherwise, an iterative solution is required; a suitable initial value for the iteration is α_{i_0}, where

$$1 - \alpha_{i_0} = d_i \left\{ \sum_{\ell \in R(t_i)} \exp[Z_\ell(t_i)\hat{\beta}] \right\}^{-1}. \tag{4.35}$$

Note that the $\hat{\alpha}_i$'s can be calculated separately.

The maximum likelihood estimate of the baseline survivor function is

$$\hat{F}_0(t) = \prod_{i|t_i \le t} \hat{\alpha}_i, \tag{4.36}$$

which, like the Kaplan–Meier estimate, is a step function with discontinuities at each observed failure time t_i. The corresponding estimate of the cumulative hazard function is

$$\hat{\Lambda}_0(t) = \sum (1 - \hat{\alpha}_i) I(t_i \le t).$$

The estimated survivor function for covariate function $Z_0(t)$ corresponding to basic covariate x_0 is

$$\hat{F}(t; x_0) = \mathscr{P}_0^t [1 - d\hat{\Lambda}_0(u)]^{\exp[Z_0(u)'\hat{\beta}]}, \tag{4.37}$$

which reduces to $\hat{F}(t; x_0) = [\hat{F}_0(t)]^{\exp(Z'\hat{\beta})}$ when Z is a constant covariate vector.

A closely related estimator is the Nelson–Aalen estimator (also sometimes called the *Breslow estimator*) of the cumulative hazard function, which is given by $\tilde{\Lambda}(t) = \int_0^t d\tilde{\Lambda}_0(u)$, where $d\tilde{\Lambda}_0(t)$ is 0 except at the observed failure times t_i, where it takes the value

$$d\tilde{\Lambda}_0(t_i) = d_i \left\{ \sum_{\ell \in R(t_i)} \exp[Z_\ell(t_i)'\hat{\beta}] \right\}^{-1}. \tag{4.38}$$

This estimator corresponds to (4.35), the suggested initial value for the iteration above. This estimate is particularly simple and can be derived from the discrete and continuous relative risk model as a straightforward moment estimator. It does, however, have some unsatisfactory features; for example, the estimated hazard contribution (4.38) can exceed 1. Nonetheless, it is a reasonable estimate when the hazards are relatively small and one is looking at events relatively early in the failure time distribution. A product integral may seem the natural way to form the corresponding estimate of the survivor function, but this does not yield an estimator that satisfies the relationship (4.37) for different values of x_0. Most authors suggest using

$$\tilde{F}(t; x_0) = \exp\left\{-\int_0^t \exp[Z_0(u)'\hat{\beta}]d\tilde{\Lambda}_0(u)\right\} \qquad (4.39)$$

as the corresponding estimate of the survivor function. This does not reduce to the Kaplan–Meier estimate and does not have the usual discrete relationship with $\tilde{\Lambda}(t; x_0)$ that should be expected. In most applications, however, it provides a good approximation to the maximum likelihood estimator (4.37).

Many other estimates of the survivor function have been proposed. One such, a modified life-table estimate, involves partitioning the time axis into intervals $I_1, \dots, I_k$ and supposing the baseline hazard $\lambda_0(t)$ to be constant within each

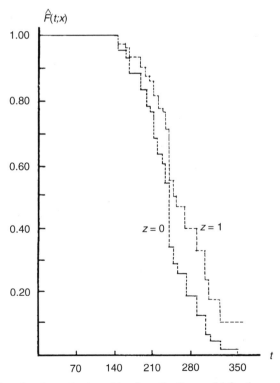

Figure 4.4 Survivor function estimates arising from the Cox model for the carcinogenesis data of Table 1.1.

interval. A simple estimation of the hazard function and consequently of $F_0(t)$ is then available. This approach is analogous to that giving (1.11) in the single-sample problem of Section 1.3 and is outlined in Exercise 4.7. The maximum likelihood approach of Exercise 4.11 gives yet another estimator. All these estimates are typically in reasonably close agreement for particular data sets, and since the use of such estimators is largely descriptive, it probably does not matter much which is used. Large-sample properties of the estimates (4.38) and (4.39) are discussed in Section 4.8.2.

Figure 4.4 gives the estimated survivor functions from (4.37) for the carcinogenesis data for each of the two samples $(Z = 0, 1)$ under the simple model $\lambda(t; Z) = \lambda_0(t)e^{Z\beta}$. To carry out these calculations, the data are again ordered as in Table 4.1 and the hazard contribution at each observed failure time t_i calculated from (4.34). Note that the assumed model constrains the estimates so that one survivor function dominates the other. Such graphs can give a misleading impression that one of the treatments is consistently preferable and suggest significant differences even when they are not present. In this example, a better description is given by the separate Kaplan–Meier estimates in Figure 1.2.

4.4 INCLUSION OF STRATA

The relative risk regression model (4.1) specifies a model which involves all elements of the covariate vector $Z(t)$. In some instances, however, there are components of $Z(t)$, or equivalently of x, for which we do not wish to specify a particular model and in these cases it is possible to let the different levels of that covariate specify strata. For example, if we are using a model with fixed covariates only, the model (4.1) effectively assumes that for any two covariate sets Z_1 and Z_2, the hazards satisfy

$$\lambda(t; Z_1) \propto \lambda(t; Z_2), \qquad 0 < t < \infty.$$

Sometimes there are important factors, the different levels of which produce hazard functions that differ markedly from proportionality. For such a factor, we could model interactions with functions of time through defined time-dependent covariates and model the dependence of hazard ratios on time. But in many instances, especially where the nature of the time dependence in the relative risk is not of particular interest, stratification on these factors provides a simpler and better approach.

Suppose that we wish to stratify on a factor that occurs on q levels. We define the hazard function for an individual in the jth stratum (or level) of this factor as

$$\lambda_j(t; x) = \lambda_{0j}(t) \exp[Z(t)'\beta] \tag{4.40}$$

for $j = 1, 2, \ldots, q$, where $Z(t)$ is the vector of covariates for which a relative risk model is descriptive. The baseline hazard functions, $\lambda_{01}(\cdot), \ldots, \lambda_{0q}(\cdot)$, for the q

strata are allowed to be arbitrary and assumed completely unrelated. In this more general situation, the (approximate) partial likelihood of β is the product of terms like (4.19), one arising from each stratum. In general,

$$L(\beta) = \prod_{j=1}^{q} L_j(\beta), \tag{4.41}$$

where $L_j(\beta)$ is the partial likelihood of β arising from the jth stratum alone.

The maximization of the likelihood (4.41) is easily accomplished; the first and second derivatives are merely sums over strata of those computed earlier [see (4.22) and (4.23)], and a Newton–Raphson technique generally leads to quick convergence to the estimate of $\hat{\beta}$. Although some loss of efficiency is encountered in the estimate of β when stratification is used unnecessarily, it is shown in Section 4.7 that this loss is generally not severe.

Once an estimate of β is obtained, the methods of Section 4.3 can be used to give estimates of the survivor functions in each of q strata separately. With fixed covariates, this provides a graphical check of the appropriateness of a proportional hazards model for those factors used in defining strata. If $\hat{F}_{0j}(t)$ is the estimate of the survivor function for the jth level of a factor, a check to determine whether the corresponding hazards are approximately proportional is afforded by plotting $\log[-\log \hat{F}_{0j}(t)], j = 1, \ldots, q$, versus $\log t$. Such plots for any two values of j should exhibit approximately constant differences over time. Should the differences change systematically over time, the lack of proportionality could perhaps be accounted for by incorporating an interaction with some function of time. If describing the nature of the time dependence explicitly is not so important, an alternative simpler and fully satisfactory approach is to incorporate that factor as defining strata.

If the hypothesis $\beta = 0$ is of interest, the score function test from (4.41) can be used for inference. It is left as an exercise to show that this leads to the stratified log-rank test (1.23).

4.5 ILLUSTRATIONS

We first look at some results of applying the proportional hazards model to the lung cancer data discussed in Section 3.8, and listed in Appendix A, for comparison with the parametric methods of Sections 3.8 and 3.9.2. Table 4.3 summarizes the maximum likelihood estimates and asymptotic χ^2 statistics based on the marginal likelihood (4.8) for the proportional hazards model (4.1) and for the Weibull analysis of Section 3.8. The Weibull model is a special case of the proportional hazards model, and the extremely good agreement between the χ^2 statistics even in the presence of strong prognostic factors suggests that little efficiency is lost in using the semiparametric model (4.1) relative to the fully parametric regression model. Theoretical efficiency comparisons are considered in Section 4.7. Table 4.3 also shows good

Table 4.3   Asymptotic Likelihood Inference on Lung Cancer Data

Regressor Variable	Weibull Model		Proportional Hazards Model	
	Coefficient $(-\hat{b})$	χ^2 Statistic	Coefficient $(\hat{\beta})$	χ^2 Statistic
Performance status	−0.030	38.79	−0.033	35.11
Disease duration (months)	0.000	0.00	−0.000	0.00
Age (years)	−0.006	0.51	−0.009	0.84
Prior therapy	0.004	0.04	0.007	0.10
Cell type				
Squamous	−0.398		−0.400	
Small	0.428	22.03	0.457	18.15
Adeno.	0.735		0.789	
Treatment	0.228	1.50	0.290	1.96

agreement between the absolute values of the regression coefficient estimates. If a Weibull model with regression coefficient b and shape parameter $\gamma = \sigma^{-1}$ holds, the proportional hazards regression parameter is $\beta = -b/\sigma$. The fact that $\hat{\sigma} = 0.928$ is close to unity accounts for the close correspondence between $\hat{\beta}$ and $-\hat{b}$ for these data.

As a second example, consider the clinical trial discussed in Section 1.1.2. In that study patients with primary tumors at any of four sites in the head and neck were randomly assigned to a test or standard treatment policy. The data for one of the sites are given in Appendix A, data set I. Each treatment policy dictated the treatment to be administered during a 90-day period. After this, each patient received medical care as deemed prudent by the participating institution. No restrictions, except a prohibition of the study treatment, were placed on post-90-day care. The primary purpose of the study was to compare patient survival for the two treatment policies in the four primary disease sites. The data considered here are those collected by the eight institutions with the largest patient accession. There were 438 patients entered by these institutions, 217 assigned to the standard, and 221 to the test treatment groups.

As noted in Section 1.1.2 and exemplified in the data of Appendix A, there are many covariates measured on individuals under study and available for consideration. Institution is one such covariate and is of particular importance here since there was considerable variability in patient treatment following the 90-day study period. In addition, the *TN* staging classification, the grade or degree of differentiation of the primary tumor, the site of the primary tumor, age, sex, and general condition are covariates considered here. These covariates are discussed in Section 1.2.3.

In the analysis of such an extensive set of failure time data, the first step is exploratory, its purpose being the identification of which covariates correlate with subsequent survival. One approach makes extensive use of the log-rank test to check for a dependence of failure time on each covariate taken one at a time.

The log-rank test with stratification allows an additional check for possible interactions with other covariates. In the present case, the failure time distribution is reasonably close to an exponential distribution, and as is discussed in Section 3.8, estimated failure rates based on assumed exponential distributions were computed in two-way tables, with covariates being examined in pairs. This examination identified four covariates (i.e., sex, general condition, and TN staging) as being highly related to subsequent survival.

As a preliminary model, these four factors were included as covariates in (4.40), and the eight institutions were allowed to define the strata with baseline hazards $\lambda_{0j}(t)$. To check whether an institution might reasonably be incorporated in the regression portion of (4.40), the corresponding survivor function estimates $\hat{F}_{0j}(t)$ were obtained and $\log[-\log \hat{F}_{0j}(t)]$ was plotted against time for each j. Although any of the estimates of the survivor function discussed in Section 4.3 would be adequate, we have used the modified life-table estimate outlined in Exercise 4.7. Figure 4.5 gives the resulting plots, and the curves are seen to have approximately constant differences over time. This suggests that an institution might be incorporated as a covariate in (4.40). Modeling of the other covariates can be checked in similar ways. Figure 4.6 gives the estimated survivor functions when general condition × sex forms the strata and the survivals are adjusted for T, N, and institution differences. Again there is close correspondence to constant separation. On examining the differences between males and females within the levels of general conditions it is further apparent, that the factors of sex and general condition operate approximately additively on the $\log(-\log)$ survivor function. This suggests that the linear modeling in the exponential factor of (4.40) is appropriate for these variables.

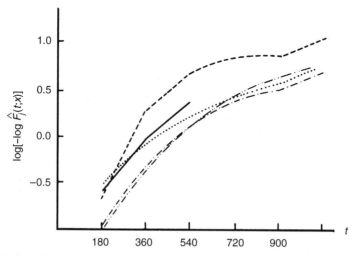

Figure 4.5 Log minus log plot for five largest institutions. Survival curve estimates standardized to male, general condition 1, T classification 3, N classification 2.

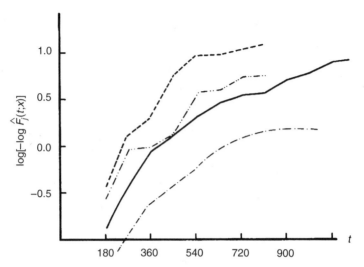

Figure 4.6 Log minus log plot to check for covariate inclusion of sex and general condition. Survival curve estimates standardized to T classification 3, N classification 2. —, male, good condition; – – –, male, poor condition; – · – · – ·, female, good condition; – ·· – ··, female, poor condition.

In investigating the factors of treatment and region, the covariates in (4.40) were sex, general condition, *TN* staging, and institution, while region × treatment formed the strata. Figures 4.7 and 4.8 give the survivor curves for each of the four regions for the two treatment groups, and Figures 4.9 through 4.12 compare the treatments for each of the four regions. Figures 4.7 and 4.8 show considerable departures from constant separation and suggest that the site of primary tumor is probably best not

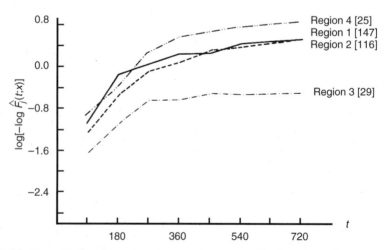

Figure 4.7 Log minus log plot to check for covariate inclusion of region (standard treatment group). Bracketed numbers are sample sizes.

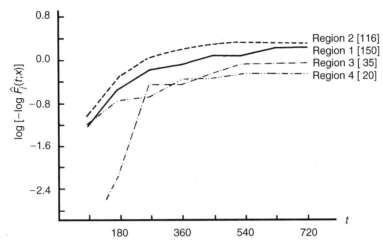

Figure 4.8 Log minus log plot to check for covariate inclusion of region (test treatment group). Bracketed numbers are sample sizes.

included in the regression portion of (4.40). Figures 4.9 through 4.12 suggest that with the possible exception of region 3, the proportional hazards specification for treatment is reasonable within region. In what follows, treatment is included in the regression portion for all regions. However, some further investigation of region 3 may be useful. This could be done by incorporating an interaction of treatment with time within region 3. The comparisons in Figures 4.9 through 4.12 are not constrained by the proportional hazards relationship since for their construction,

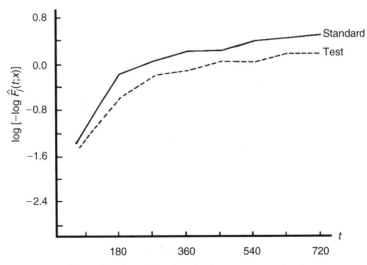

Figure 4.9 Log minus log plot for treatment, region 1.

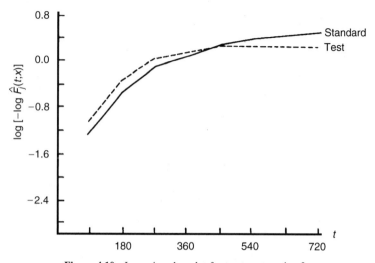

Figure 4.10 Log minus log plot for treatment, region 2.

treatments are determining the strata. It would be possible, once treatment is incorporated as a covariate, to produce estimates of the survivor function that are constrained in this way. Such plots would, however, tend to accentuate treatment differences and are subject to the same criticism as that of Figure 4.2 in the carcinogenesis example. The model incorporating treatment as a regression variable is very useful, however, from an inferential point of view, in that simple tests for treatment differences are then available.

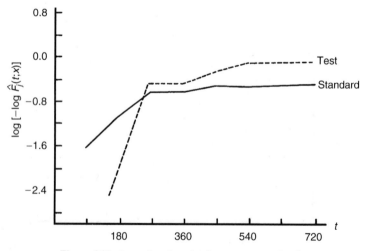

Figure 4.11 Log minus log plot for treatment, region 3.

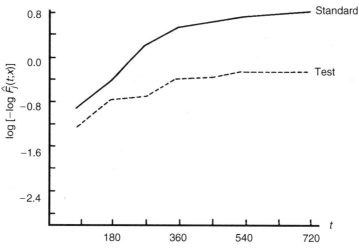

Figure 4.12 Log minus log plot for treatment, region 4.

These considerations lead to the tentative model for these data with hazard function

$$\lambda_j(t; Z, i, x) = \lambda_{0j}(t) \exp(Z\beta + \gamma_i + \alpha_j x), \qquad j = 1, \ldots, 4, \tag{4.42}$$

where Z is the vector giving sex, general condition, the T and the N classifications, j denotes the region of the primary tumor, i is the institution number (with $\gamma_1 = 0$), and x takes values 0 or 1 for the standard and test treatments, respectively. The coefficient α_j gives a measure of treatment differences within the jth region. Table 4.4 gives the estimates for the regression parameters (except the $\hat{\gamma}_i$'s corresponding to institution) and the estimated standard errors of the estimates. The calculations were done using a Newton–Raphson routine with initial values 0 for all parameters.

Table 4.4 Regression Coefficients and Estimated Variances for the Model (4.26)

Variable	Estimated Regression Coefficients	Estimated Standard Error of the Estimate
Sex	$\hat{\beta}_1 = -0.446$	0.154
General condition	$\hat{\beta}_2 = 0.483$	0.103
T classification	$\hat{\beta}_3 = 0.358$	0.098
N classification	$\hat{\beta}_4 = 0.267$	0.055
Treatment region		
1	$\hat{\alpha}_1 = -0.313$	0.267
2	$\hat{\alpha}_2 = 0.102$	0.162
3	$\hat{\alpha}_3 = -0.101$	0.349
4	$\hat{\alpha}_4 = -0.656$	0.369

From Table 4.4 it is easily seen that sex, general condition, and the *TN* classification are all important in evaluating survival prognosis. The treatment effects, however, are not significant. Only in region 4 is there any evidence of dependence on treatment, and there the significance level is at best marginal. In this case, four independent tests for treatment differences have been made. Thus the nominal significance level of about 7% for the treatment effect in region 4 must, to some extent, be discounted, owing to these multiple comparisons.

Once the model (4.42) has been fit, it is possible to check its appropriateness by forming residuals and carrying out residual plots such as in ordinary linear regression. In the model (4.42), let t_{ji} be the survival time of the ith individual in the jth stratum and define

$$e_{ji} = \Lambda_{0j}(t_{ji})e^{Z_{ji}\beta},$$

where $\Lambda_{0j}(t) = \int_0^t \lambda_{0j}(u)\,du$. The e_{ji}'s are a censored sample from the exponential distribution with failure rate 1. If Λ_{0j} and β are replaced with estimates, we obtain estimates of the e_{ji}'s or residuals:

$$\hat{e}_{ji} = \hat{\Lambda}_{0j}(t_{ji})e^{Z_{ji}\hat{\beta}}.$$

In forming these residuals in the present case, the continuous estimate $\hat{\Lambda}_{0j}(t)$ obtained above and the marginal likelihood estimate $\hat{\beta}$ have been used, but other estimates of $\Lambda_{0j}(t)$, for example, the Nelson–Aalen estimator, could also be used. If the model is appropriate, the $\hat{e}_{ji}$'s should be similar to a censored exponential sample (an $\hat{e}_{ji}$ is taken as censored if the corresponding t_{ji} is censored). Survival curve estimates based on the residuals should, when plotted on a log scale, yield

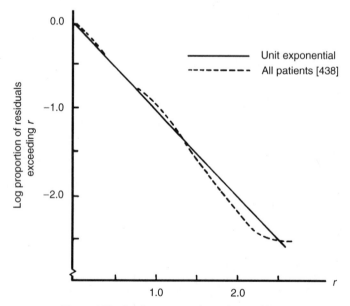

Figure 4.13 Survivor curve estimate from residuals.

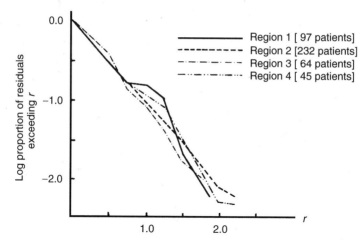

Figure 4.14 Survivor curve estimates from residuals subdivided by region.

approximately a straight line with slope -1. Alternatively, a plot of the cumulative hazard should yield an approximate straight line through the origin of slope $+1$.

The overall adequacy of the model can be partially checked by plotting the survivor curve estimate arising from the residuals, as illustrated in Figure 4.13. The correspondence with the anticipated line is extremely good. A check that the adjustment for covariates has been adequate in the four regions under study is provided by plotting the residual survivor curve for each region (Figure 4.14). Again, close correspondence to the expected line is observed. The modeling of the covariates can be checked in a similar way. Figure 4.15 gives the residual survivor curves for gender

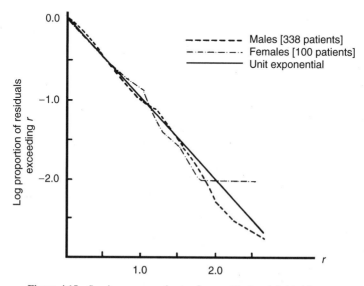

Figure 4.15 Survivor curve estimates from residuals subdivided by sex.

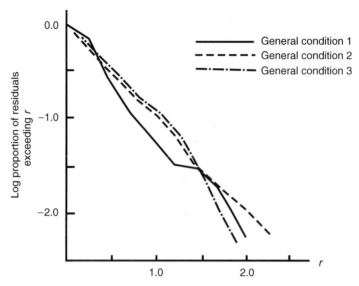

Figure 4.16   Survivor curve estimates subdivided by general condition.

and Figure 4.16 for general condition. Again, the fit seems adequate. It should be noted, however, that it is not apparent what kinds of departures one would expect to see in the residuals if the model is incorrect or even to what extent agreement with the anticipated line should be expected. As pointed out by Crowley and Hu (1977), the many free parameters being fitted to the data may lead to misleadingly good approximations.

The Cox–Snell residuals used above are one of several types that have been proposed for model checking and model formulation. Some aspects of this quite extensive literature are discussed in Section 6.5. A more formal approach to the problems of checking goodness of fit in the proportional hazards model is provided by testing expanded models obtained by incorporating interactions with functions of time, as illustrated in Examples 4.1 and 4.2.

4.6   COUNTING PROCESS FORMULAS

As in earlier chapters, counting process notation is commonly used to describe key aspects of the partial likelihood and related analyses of the relative risk regression model. As before, let $N_i(t)$ be the right-continuous counting process for the number of observed failures on $(0, t]$ for the ith individual, and let $Y_i(t)$ be the left-continuous at-risk process, so that $Y_i(t) = 1$ indicates that the individual is under observation at time t and has not yet failed. Let x_i and $Z_i(t)$ be the basic and derived covariates as above. In addition, let $N.(t) = \sum N_i(t)$ and $Y.(t) = \sum Y_i(t)$. It is convenient to use the following notation:

$$S^0(\beta, t) = \sum_{i=1}^{n} Y_i(t) \exp[Z_i(t)'\beta],$$

$$p_\ell(\beta, t) = \frac{Y_\ell(t) \exp[Z_\ell(t)'\beta]}{\sum_{i=1}^{n} Y_i(t) \exp[Z_i(t)'\beta]},$$

and

$$\mathscr{E}(\beta, t) = \sum_{\ell=1}^{n} Z_\ell(t) p_\ell(\beta, t),$$

where $0 \leq l \leq n$, $0 < t < \infty$, and $0/0$ is interpreted as 0. Note that $\mathscr{E}(\beta, t)$ represents a mean covariate vector in the risk set at time t, the expectation being taken over the distribution $p_\ell(\beta, t), l = 1, ..., n$. Note that $p_\ell(\beta, t) = 0$ if $l \notin R(t)$, and this notation is consistent with that used before.

Many expressions of interest can now be written as stochastic integrals with respect to the counting processes. For example, the log partial likelihood from (4.14) can be written

$$l(\beta) = \sum_{i=1}^{n} \int_0^\infty Z_i(t)'\beta \, dN_i(t) - \int_0^\infty \log \left\{ \sum_{i=1}^{n} Y_i(t) \exp[Z_i(t)'\beta] \right\} dN_.(t)$$

$$= \sum_{i=1}^{n} \int_0^\infty \{Z_i(t)'\beta - \log[S^0(\beta, t)]\} dN_i(t), \qquad (4.43)$$

and the score function (4.15) can be written

$$U(\beta) = \sum_{i=1}^{n} \int_0^\infty [Z_i(t) - \mathscr{E}(\beta, t)] \, dN_i(t). \qquad (4.44)$$

In a similar manner, the observed Fisher information obtained from the partial likelihood (4.43) is

$$I(\beta) = -\frac{\partial^2 l(\beta)}{\partial \beta \, \partial \beta'} = \int_0^\infty \mathscr{V}(\beta, t) \, dN_.(t), \qquad (4.45)$$

where

$$\mathscr{V}(\beta, t) = \sum_{i=1}^{n} [Z_i(t) - \mathscr{E}(\beta, t)]^{\otimes 2} p_i(\beta, t)$$

is the covariance matrix of $Z_i(t)$ taken over the risk set at time t.

The Nelson–Aalen estimator (4.38) of the baseline cumulative hazard function $\Lambda_0(t)$ can be written

$$\hat{\Lambda}_0(t) = \int_0^t [S^{(0)}(\hat{\beta}, u)]^{-1} dN.(u),$$

where it has been assumed that t is less than or equal to the largest observed failure or censoring time. Under regularity conditions, the asymptotic variance of $n^{1/2}[\hat{\Lambda}_0(t) - \Lambda_0(t)]$ is consistently estimated by

$$V_{\hat{\Lambda}}(\hat{\beta}, t) = n \left\{ \int_0^t [S^{(0)}(\hat{\beta}, u)]^{-2} dN.(u) + \left[\int_0^t \mathscr{E}(\hat{\beta}, t)[S^{(0)}(\hat{\beta}, u)]^{-1} dN.(t) \right]^2 [I(\hat{\beta}, t)]^{-1} \right\},$$

where $I(\beta, t) = \int_0^t V(\beta, u) \, dN.(u)$ is the observed information in the partial likelihood up to time t.

We return to this notation and to derivations of various asymptotic results in Chapter 5, where the associated martingales are also introduced.

4.7 RELATED TOPICS ON THE COX MODEL

4.7.1 Marginal Likelihood for β

In this section an alternative approach to inference about β is based on the marginal distribution of the rank statistic or on simple generalizations of that distribution. This serves to illustrate the close connection between inferences about β and rank tests and provides a link between the proportional hazards model and the rank based analyses of the accelerated failure time model outlined in Chapter 7.

Suppose that the model (4.1) holds with the covariates being fixed $[Z(t) = Z]$, so that

$$\lambda(t; x) = \lambda_0(t) e^{Z'\beta}. \tag{4.46}$$

Suppose that n individuals are observed to fail at $t_1, \ldots, t_n$ with corresponding covariates $Z_1, \ldots, Z_n$. For the moment we assume that all failures are distinct and that no censoring is present in the data. Central to our discussion will be the order statistic $O(t) = [t_{(1)}, \ldots, t_{(n)}]$ and the rank statistic $r(t) = [(1), \ldots, (n)]$. The order statistic refers to the t_i's ordered from smallest to largest (i.e., $t_{(1)} < t_{(2)} < \cdots < t_{(n)}$) and the notation (i) in the rank statistic refers to the label attached to the ith element of the order statistic. For example, if $n = 4$ and $t_1 = 5, t_2 = 17, t_3 = 12, t_4 = 15$ are observed, then $O = [5, 12, 15, 17]$ and $r = [1, 3, 4, 2]$.

At the risk of some confusion and for this section only, some specialized notation is introduced. Specifically, the bracketed subscripts are used to indicate the ordered data. This enables discussion of the order statistic, the rank statistic, and the derivations in a straightforward manner.

Consider the model (4.46) and define $u = g^{-1}(t)$, where $g \in G$, the group of strictly increasing and differentiable transformations of $(0, \infty)$ onto $(0, \infty)$. The conditional distribution of $U = g^{-1}(T)$ given Z has the hazard

$$\lambda_1(u)e^{Z'\beta},$$

where $\lambda_1(u) = \lambda_0(g(u))g'(u)$. Thus if the data were presented in the form $u_1, \ldots, u_n$ and $Z_1, \ldots, Z_n$ where $g(u_i) = t_i$, the inference problem about β would be the same provided that $\lambda_0(\cdot)$ [so $\lambda_1(\cdot)$] is completely unknown. The estimation problem for β is said to be invariant under the group G of transformations on t.

Consider now the effect of G on the sample space. The order statistic $O(t)$ can be mapped to any specified order statistic by an element of $g \in G$, while the rank statistic $r(t)$ is left unchanged by all $g \in G$. For example, if the transformation $u = t^2$ is applied to the sample above, we obtain $O(u) = [25, 144, 225, 289]$ and $r(u) = [1, 3, 4, 2] = r(t)$. Further, any specified order statistic can clearly be obtained for u by an appropriate choice of $g \in G$. Since the estimation problem for β is the same under any such transformation, and since the order statistic can be made arbitrary by such a transformation, only the ranks carry information about β when $\lambda_0(t)$ is completely unknown.

For inference about β, the marginal distribution of the ranks is available and the marginal likelihood (see Fraser, 1968; Kalbfleisch and Sprott, 1970) is proportional to the probability that the rank vector should be that observed. That is, the marginal likelihood is proportional to

$$P(r; \beta) = P\{r = [(1), \ldots, (n)]; \beta\}$$

$$= \int_{t_{(1)} < \cdots < t_{(n)}} \prod_1^n f[t_{(i)}; Z_{(i)}] \, dt_{(i)}$$

$$= \prod_{i=1}^n \frac{\exp[Z_{(i)}\beta]}{\sum_{\ell \in R(t_{(i)})} \exp(Z'_\ell \beta)}, \tag{4.47}$$

where, as before, $R(t_{(i)})$ is the set of labels attached to the individuals at risk just prior to $t_{(i)}$, so that $R(t_{(i)}) = [(i), (i+1), \ldots, (n)]$. This expression is identical to the partial likelihood for this case.

To handle censored data, some modification of this argument is required. If all items are put on test simultaneously and followed to the kth failure time (type II censoring), a marginal likelihood is again easily obtained. In this case, the group acts transitively on the censoring time and the invariant in the sample space is the first k rank variables $(1), \ldots, (k)$. The argument could be extended to progressive type II censoring patterns where items are withdrawn from test with each failure.

More general independent censoring cannot be handled directly by this approach since the censored model will not in general possess group invariance properties. We can note, however, that had the entire sample been observed, the rank statistic would be marginally sufficient for β. When a censored sample is obtained, only partial information is observed on the ranks. For example, if the observed survival times of four tested items were 114, 90[*], 63, 108[*], where the asterisks indicate censoring, the underlying rank statistic is known to be one of the following six possibilities:

$$[3, 2, 4, 1]; \quad [3, 4, 2, 1]; \quad [3, 2, 1, 4];$$
$$[3, 4, 1, 2]; \quad [3, 1, 2, 4]; \quad [3, 1, 4, 2].$$

To make an inference about β, the marginal probability that the rank statistic should be one of those possible can be used. The observed part of the marginally sufficient statistic r is generating the likelihood. Note that the exact time of censoring is ignored, but the invariance of the uncensored model suggests that the lengths of the intervals between successive failures is irrelevant for inference about β. Consequently, it would seem reasonable to suppose that the exact time of censoring, relative to adjacent uncensored times, should not contribute to the inference about β.

Suppose that k items labeled $(1), \ldots, (k)$ give rise to observed failure times $t_{(1)} < t_{(2)} < \cdots < t_{(k)}$ with corresponding covariates $Z_{(1)}, \ldots, Z_{(k)}$ and suppose further that m_i items with covariates $Z_{i1}, \ldots Z_{im_i}$ are censored in the ith interval $[t_{(i)}, t_{(i+1)})$, $i = 1, \ldots, k$, where $t_{(0)} = 0$ and $t_{(k+1)} = \infty$. The marginal likelihood of β is computed as the probability that the rank statistic should be one of those possible on the sample and is, therefore, the sum of a large number of terms like (4.47). The set of possible rank vectors can be characterized, however, by

$$t_{(1)} < \cdots < t_{(k)} \qquad t_{(i)} < t_{i1}, \ldots, t_{im_i} \quad (i = 0, 1, \ldots, k), \qquad (4.48)$$

where $t_{i1}, \ldots, t_{im_i}$ are the unobserved failure times associated with individuals censored in $[t_{(i)}, t_{(i+1)})$. Writing the event as (4.48) allows simple computation of the marginal likelihood since, given $t_{(i)}$, the event $t_{(i)} < t_{i1}, \ldots, t_{im_i}$ has the conditional probability

$$h(t_{(i)}) = \exp\left[-\sum_{j=1}^{m_i} \exp(Z'_{ij}\beta) \int_0^{t_{(i)}} \lambda_0(u)\, du \right], \qquad i = 0, 1, \ldots, k.$$

The marginal likelihood is then proportional to the probability of the event (4.48). This probability is

$$\int_{t_{(1)} < \cdots < t_{(k)}} \prod_1^k f(t_j; Z_{(j)}) h(t_{(j)})\, dt_{(j)} = \prod_{i=1}^k \frac{\exp(Z'_{(i)}\beta)}{\sum_{\ell \in R(t_{(i)})} \exp(Z'_l\beta)}, \qquad (4.49)$$

in exact agreement with the partial likelihood (4.14). It should be noted that (4.49) is not the probability of observing the event (4.48) in the censored experiment. This probability would depend on the censoring mechanism and in general also on $\lambda_0(t)$.

The expression (4.49) is the probability that, in the underlying uncensored version of the experiment, the event (4.48) would occur.

This same general idea of considering the set of all possible rank vectors consistent with the data leads also to a natural generalization of (4.49) to accommodate tied data, and this is given by (4.18), with the covariates $Z(t) = Z$ being time independent.

4.7.2 Efficiency of the Rank Analysis Under a Parametric Submodel

In this section we consider some aspects of the efficiency of the partial or marginal likelihood analysis of the relative risk model with fixed covariates $Z = Z(t)$ in comparison with parametric submodels. A more detailed discussion of asymptotic efficiency can be found in Section 5.8 following the development of asymptotic results for the partial likelihood. Some indication of finite sample relative efficiency can be obtained by comparing (expected) information matrices at fixed sample sizes. The information matrix gives the expected curvature of the likelihood function and hence an indication of the precision with which β is estimated. We make here a few comparisons of this type based on the parametric model in which $\lambda_0(t) = \lambda h_0(t)$, with $h_0(t)$ known. Following Kalbfleisch (1974), we compare the information matrix from the rank analysis with that based on a marginal likelihood analysis of this parametric model. Information calculations for the marginal and full likelihood analyses will be nearly the same.

First, define $t' = \int_0^t h_0(u)\,du$ so that the hazard function for t' is $\lambda e^{Z'\beta}$. This model for t' is invariant under the group of scale transformations and, in the absence of censoring, the variates $a_i = t'_i/t'_1, i = 2, \ldots, n$ form the maximum invariant in the sample space; their density function generates the marginal likelihood for β. For simplicity suppose that there is a scalar regressor variable with values $Z_1, \ldots, Z_n$, corresponding to the uncensored failure times $t_1, \ldots, t_n$. Without loss of generality we may take $Z_1 + \cdots + Z_n = 0$. The density function for $a = (a_2, \ldots, a_n)$ is

$$f_a(a; \beta) = (n-1)! \left(\sum_{i=1}^{n} a_i e^{\beta Z_i} \right)^{-n}, \qquad a_2, \ldots, a_n > 0,$$

where $a_1 = 1$. The probability distribution for the rank vector $r = [(1), \ldots, (n)]$ is, from (4.47),

$$f_r(r; \beta) = \prod_{i=1}^{n} \left(\sum_{j=i}^{n} e^{\beta Z_{(j)}} \right)^{-1}.$$

The information in the rank vector about β is

$$I_r(\beta) = -E\left[\frac{\partial^2}{\partial \beta^2} \log f_r(r; \beta) \right],$$

which at $\beta = 0$ reduces to

$$I(0) = \sum_{i=1}^{n} E_p \left[\frac{\sum_{j=i}^{n} \sum_{k=i}^{n} Z_{(j)} (Z_{(j)} - Z_{(k)})}{(n-i+1)^2} \right]$$

$$= \sum_{i=1}^{n} E_p(m_{2,i} - m_{1,i}^2),$$

where

$$m_{k,i} = (n-i+1)^{-1} \sum_{j=i}^{n} Z_{(j)}^k, \qquad k = 1, 2, \ldots.$$

Here E_p refers to the expectation over the permutation distribution on $\{(1), (2), \ldots, (n)\}$. It is easily verified that $E_p(m_{2,i}) = \mu_2$ and that

$$E_p(m_{1,i}^2) = \frac{\mu_2}{n(n-1)} \sum_{i=1}^{n} \frac{i-1}{n-i+1},$$

so that

$$I_r(0) = \frac{n\mu_2}{n-1} \sum_{i=1}^{n} \frac{n-i}{n-i+1},$$

where μ_k is the kth central moment of $Z_1, \ldots, Z_n$.

On the other hand, straightforward calculation verifies that the information on β contained in the variate a is

$$I_a(\beta) = -E \left[\frac{\partial^2}{\partial \beta^2} \log f_a(a; \beta) \right]$$

$$= \frac{n^2 \mu_2}{n+1}.$$

The relative efficiency at $\beta = 0$ for a sample of size n of the rank statistic compared to the statistic a is

$$R_n(0) = \frac{I_r(0)}{I_a(0)}$$

$$= \frac{n+1}{n(n-1)} \sum_{i=1}^{n} \frac{n-i}{n-i+1}. \tag{4.50}$$

Table 4.5   Relative Efficiency of Rank Analysis Versus Exponential at $\beta = 0$

n	2	3	5	7	10	15	20	40	60	100	∞
$R_n(0)$	0.75	0.78	0.82	0.84	0.89	0.91	0.94	0.95	0.97	0.99	1

As noted above, the asymptotic relative efficiency at $\beta = 0$ is

$$R(0) = \lim_{n \to \infty} R_n(0) = 1.$$

From (4.50) the relative efficiency for finite n can be evaluated, and this has a simple interpretation in terms of variance. For example, the case $n = 2$ gives $R_n(0) = \frac{3}{4} (\mu_2 \neq 0)$. This can be interpreted as the ratio of the asymptotic variances of the rank and parametric analyses in a twin study when the ith twin pair has its own failure rate λ_i and β is a regression parameter common to all twin pairs. Table 4.5 gives such relative efficiencies for several values of n, and the approach to full efficiency is readily seen to be rapid.

Consider now the case of two regression variables β_1, β_2, where β_1 is of interest, and suppose that the variable Z_2 takes on only a finite number of distinct values. It is easily seen that the asymptotic relative efficiency of the rank analysis for the estimation of β_1 is 1 at $\beta_1 = 0$ since the problem could be handled with full efficiency by considering the hazard for the jth possible value of Z_2 as $\lambda_{0j}(t)e^{\beta_1 Z_1}$, where the information regarding proportionality for the second variable is suppressed. Clearly, the small-sample efficiencies will be poorer; for example, if β_2 is very large and Z_2 takes only two values each with equal frequency, the small-sample relative efficiencies for the rank analysis will increase at about half the rate stated in Table 4.5. The relative efficency for $n = 20$ in this case would be about 0.89, compared to 0.94 with no auxiliary variable. This reduction in small-sample efficiency could conceivably be severe if Z_2 took many values and β_2 were reasonably large. This same line of reasoning suggests that no additional loss of asymptotic efficiency is incurred through unnecessary stratification, but that small-sample efficiencies based on q strata may approach asymptotic results at a rate as low as $1/q$ of that for unstratified analysis.

Further examination of the small-sample efficiencies by this approach could also allow the incorporation of censoring through consideration of type II or order statistic censoring in which individuals are removed at random from the risk set on the occurrence of failures. The group invariance properties are retained under this model, and information comparisons could be made.

4.8   SAMPLING FROM DISCRETE MODELS

Consider the discrete regression models discussed in Section 2.4.2 and inference about the regression parameter β and the baseline cumulative hazard function Λ_0

based on a right-censored (and possible left-truncated) sample. As before, the censoring and truncation is assumed to be independent.

As before, x represents a vector of fixed basic covariates and $Z(t)$ is a vector of derived, possibly time-dependent covariates whose elements are functions of x and t. Let Λ_0 be a discrete baseline cumulative hazard function with masses $\lambda_1, \lambda_2, \ldots, \lambda_k$ at the discrete times $a_1, a_2, \ldots, a_k$, where $0 < a_1 < a_2 < \cdots < a_k$, so that

$$\Lambda_0(t) = \sum_{j=1}^{k} \lambda_j I(a_j \le t).$$

Note that we have assumed that the number of possible mass points k included in the study is fixed so that the baseline hazard will be specified in terms of a finite number k of parameters. This allows straightforward asymptotic arguments to apply to the maximum likelihood methods.

Let $\Lambda(t, x)$ be the cumulative hazard function corresponding to covariate vector x. An examination of the three discrete models (2.21), (2.22), and (2.23) suggests the encompassing formulation

$$h[d\Lambda(t, x)] = h[d\Lambda_0(t)] + Z(t)'\beta, \tag{4.51}$$

where h is a monotone-increasing and twice-differentiable function mapping $[0, 1]$ into $[-\infty, \infty]$ with $h(0) = -\infty$. Note the choices:

1. $h(u) = \log[-\log(1 - u)]$ gives the grouped relative risk model (2.21).
2. $h(u) = \log u$ gives the discrete relative risk model (2.22).
3. $h(u) = \log[u/(1 - u)]$ gives the discrete logistic model (2.23).

Other discrete models could also be generated. For example, a discrete probit model is obtained by specifying $h(u) = \Phi^{-1}(u)$, where Φ^{-1} is the inverse standard normal cumulative distribution function (CDF). Note that if $d\Lambda(t, x) = \lambda(t, x) dt$ corresponds to a continuous model, then for each choice of h mentioned, the model (4.51) reduces to $\lambda(t, x) = \lambda_0(t) \exp[Z(t)'\beta]$. Thus, each of cases 1, 2, and 3 and even the discrete probit model can be viewed as a generalization of the continuous-time relative risk model (4.1) to discrete (and mixed) failure time variables.

It should be noted that (2) places some restrictions on β in order to satisfy the requirement that $d\Lambda(t, x) \le 1$ for all x and t. This relative risk model is useful for describing survival experience when $d\Lambda(a_j)$ is small for all j. This situation arises when a relatively large cohort is being observed over a relatively short period of time with many individuals censored. No restrictions on β are implied by model 1 or 3 above or by the discrete probit model.

Most commonly, discrete survival data arise when the survival time is subject to interval grouping, and model (1) arises through grouping the continuous-time relative risk model when the covariates $Z(t) = Z$ are time independent. In other instances, however, time may truly be discrete, as, for example, when T represents the number of attempts required to perform a certain task successfully.

Suppose that independent right-censored and/or left-truncated data are available from the discrete model (4.51). Let D_j represent the set of labels attached to individuals failing at a_j and R_j the set of labels attached to individuals censored at a_j or observed to survive past a_j. We suppose that an item censored at a_j contributes the information that its underlying survival time *exceeds* a_j but nothing further is known. In effect, censored observations at a_j are being supposed to follow failures at a_j. In grouping continuous data, this is equivalent to allowing censoring to occur only just prior to the end of an interval. It is possible, of course, that censoring may occur within an interval, and one way to handle this situation is to "reduce" the data by replacing all potential censoring times by potential censoring times at the immediately smaller partition point prior to applying the methods of this section.

A full maximum likelihood analysis of the model (4.51) is considered first in Section 4.8.1. In subsequent sections, we consider some special analyses available for the grouped relative risk and discrete logistic models.

4.8.1 Maximum Likelihood Estimation

Let $\gamma_j = h[d\Lambda_0(a_j)] = h(\lambda_j)$, $j = 1, \ldots, k$ and $\lambda_{k+1} = 1$. It follows that the model (4.51) can be rewritten as

$$d\Lambda(a_j, x) = g[\gamma_j + Z(t)'\beta], \tag{4.52}$$

where $g(u) = h^{-1}(u)$. Note that for cases 1, 2, and 3 above, $g(u)$ is given by $1 - \exp[-\exp(u)]$, $\exp(u)$, and $\exp(u)[1 + \exp(u)]^{-1}$, respectively. If the censoring is independent and occurs, as discussed above, at the end of grouping intervals, the log likelihood of $\gamma = (\gamma_1, \gamma_2, \ldots, \gamma_k)'$, $\beta = (\beta_1, \ldots, \beta_p)'$ can be written

$$\log L(\gamma, \beta) = \sum_{j=1}^{k} \left(\sum_{l \in D_j} \log g[\gamma_j + Z_l(t)'\beta] + \sum_{l \in R_j} \log \{1 - g[\gamma_j + Z_l(t)'\beta]\} \right),$$

$$\tag{4.53}$$

where the argument leading to (4.53) is basically the same as that leading to (1.12); the relationships (1.6) and (1.7) have been used to obtain the survivor and probability functions associated with (4.51).

The components of the score vector $c = c(\gamma, \beta)$ are

$$\frac{\partial \log L}{\partial \gamma_j} = \sum_{l \in D_j} \frac{g'_{jl}}{g_{jl}} - \sum_{l \in R_j} \frac{g'_{jl}}{1 - g_{jl}} \tag{4.54}$$

and

$$\frac{\partial \log L}{\partial \beta_u} = \sum_{j=1}^{k} \left[\sum_{l \in D_j} \frac{Z_{lu}(a_j) g'_{jl}}{g_{jl}} + \sum_{l \in R_j} \frac{Z_{lu}(a_j) g'_{jl}}{1 - g_{jl}} \right],$$

where $g_{jl} = g[\gamma_j + Z_l(a_j)'\beta]$, $g'_{jl} = g'[\gamma_j + Z_l(a_j)'\beta]$, and $1 \le j \le k, 1 \le u \le p$. The maximum likelihood estimator $(\hat{\gamma}, \hat{\beta})$ is a solution to

$$c(\gamma, \beta) = \left(\frac{\partial \log L}{\partial \gamma_1}, \ldots, \frac{\partial \log L}{\partial \gamma_k}, \frac{\partial \log L}{\partial \beta_1}, \ldots, \frac{\partial \log L}{\partial \beta_s} \right)' = 0.$$

Calculation of $(\hat{\gamma}, \hat{\beta})$ by a Newton–Raphson iteration requires second derivatives of $\log L$. The Fisher information observed can be written

$$H = \begin{pmatrix} H_{11} & H_{12} \\ H_{21} & H_{22} \end{pmatrix} = \begin{pmatrix} \dfrac{-\partial^2 \log L}{\partial\gamma\,\partial\gamma} & \dfrac{-\partial^2 \log L}{\partial\gamma\,\partial\beta} \\ \dfrac{-\partial^2 \log L}{\partial\beta\,\partial\gamma} & \dfrac{-\partial^2 \log L}{\partial\beta\,\partial\beta} \end{pmatrix},$$

where H_{11} is diagonal with jth element

$$\frac{-\partial^2 \log L}{\partial\gamma_j^2} = \sum_{l \in D_j} u_{jl} + \sum_{l \in R_j} v_{jl}.$$

The columns of H_{21} are

$$\frac{-\partial^2 \log L}{\partial\beta\,\partial\gamma_j} = \sum_{l \in D_j} z_{jl}u_{jl}Z_l(a_j) + \sum_{l \in R_j} v_{jl}Z_l(a_j), \qquad j = 1, \ldots, k,$$

and

$$H_{22} = \sum_{j=1}^{k} \left\{ \sum_{l \in D_j} [u_{jl}Z_l(a_j)^{\otimes 2}] + \sum_{l \in R_j} [v_{jl}Z_l(a_j)^{\otimes 2}] \right\},$$

where $u_{jl} = g''_{jl}/g_{jl} - (g'_{jl}/g_{jl})^2$, $v_{jl} = [g''_{jl}/(1-g_{jl})] - [g'_{jl}/(1-g_{jl})]^2$, and $g''_{jl} = g''[\gamma_j + Z_l(a_j)'\beta]$.

A Newton–Raphson iteration to compute $(\hat{\gamma}, \hat{\beta})$ involves updating current values (γ_0, β_0) to (γ_1, β_1) until convergence is reached using the formula

$$\begin{pmatrix} \gamma_1 \\ \beta_1 \end{pmatrix} = \begin{pmatrix} \gamma_0 \\ \beta_0 \end{pmatrix} + H_0^{-1}c_0,$$

where H_0 and c_0 represent H and c evaluated at (γ_0, β_0). Since $(k + p)$, the dimension of H, may be large, direct numerical inversion of H may be time consuming or inaccurate. The fact that the first $(k \times k)$ block of H is diagonal can be exploited, since

$$H^{-1} = \begin{pmatrix} H_{11} & H_{12} \\ H_{21} & H_{22} \end{pmatrix}^{-1} = \begin{pmatrix} H_{11}^{-1} + FJ^{-1}F' & \vdots & -FJ^{-1} \\ \cdots & & \cdots \\ -J^{-1}F' & \vdots & J^{-1} \end{pmatrix},$$

where $F = H_{11}^{-1}H_{12}$ and $J = H_{22} - H_{21}F$. Consequently, only a matrix J of dimension p need be inverted numerically.

A simple starting value is $\beta_0 = 0$ and $\gamma_0 = \hat{\gamma}(0)$, the maximum likelihood estimate at $\beta = 0$. Suppose that n_j individuals are at risk just prior to a_j (n_j is the total number of study subjects in $D_j \cup R_j$), of which d_j (the number of subjects in D_j) fail at a_j. The jth component of $\hat{\gamma}(0)$ is then

$$\hat{\gamma}_j(0) = g^{-1}(d_j n_j^{-1}) = h(d_j n_j^{-1}).$$

Instability in the Newton–Raphson procedure may occur if the numbers of failures in specific time intervals are small. Such situations would usually correspond to a rather fine grouping of failure times in which case estimation based on an approximate partial likelihood (4.19) or (4.20) may provide an attractive alternative. With course grouping or even with a relatively fine grouping and large sample sizes, many ties will occur in the failure time data and the methods of this section provide a useful estimation procedure in such circumstances. Further, with the choice $h(u) = \log[-\log(1 - u)]$, the model involves precisely the same relative risk parameter $\exp[Z(t)'\beta]$ as in the continuous model (4.1). Since the discrete model (4.51) or (4.52) involves only a finite number $(k + p)$ of parameters, asymptotic likelihood theory can be applied in a relatively straightforward way along the lines discussed for the parametric models in Chapter 3. This theory leads to an asymptotic normal distribution for $(\hat{\gamma}, \hat{\beta})$ with mean vector (γ, β) and variance matrix estimated by $\hat{H}^{-1}$, under some relatively mild restrictions on the $Z(t)$ vectors and the censoring.

Consider now estimation of the survivor function at a specified basic covariate vector x. The maximum likelihood estimator $\hat{F}(t; x)$ is a right-continuous step function with possible jumps only at the a_j's and with

$$\hat{F}(a_j; x) = \prod_{u=1}^{j} \{1 - g[\hat{\gamma}_u + Z(a_u)'\hat{\beta}]\}, \tag{4.55}$$

where $Z(t)$ is the derived covariate corresponding to x. Because of the fixed number of parameters, the asymptotic distribution of $\hat{F}$ can easily be determined. As with the Kaplan–Meier estimator, the range restrictions on $F(t; x)$ may make the approximation inaccurate in moderate sample sizes, especially when estimating $F(t; x)$ at small or large values of t. The adequacy of the approximations can often be improved by applying asymptotic results to $\hat{\psi} = \log[-\log \hat{F}(a_j; x)]$ rather than to

$\hat{F}(a_j; x)$ itself, since $\hat{\psi}$ is devoid of range restrictions. The asymptotic distribution of $\hat{\psi}$ is normal with mean $\psi = \log\left[-\log F(a_j; x)\right]$ and variance $\sigma_{\psi}^2 = w'H^{-1}w$, where $w' = (\partial\psi/\partial\gamma', \partial\psi/\partial\beta')$.

For the grouped relative risk model $g(u) = \log\left[-\log(1 - u)\right]$ in case 1, for example, w has components

$$\frac{\partial\psi_j}{\partial\gamma_u} = \frac{h_u}{\sum_1^j h_u} I(u \le j)$$

$$\frac{\partial\psi_j}{\partial\beta} = \frac{\sum_{u=1}^j Z(a_u)h_u}{\sum_{u=1}^j h_u},$$

where $h_u = \exp[\gamma_u + Z(a_u)'\beta]$. Note that at $\beta = 0$ the survival curve estimator (4.55) reduces to the Kaplan–Meier estimator

$$\hat{F}(a_j; x) = \prod_{u=1}^j \left(1 - \frac{d_u}{n_u}\right)$$

by virtue of $\hat{\gamma}(0)$. Also, the variance estimator $\hat{\sigma}_{\psi}^2$ reduces to the Greenwood estimator of the asymptotic variance.

The methods of this section with a grouped relative risk model are illustrated in Prentice and Gloeckler (1978) in the analysis of survival data on a large set (11,442) of breast cancer patients. Thompson (1977) provides an illustration for the discrete logistic model. It should be noted that these models can be fitted using software generally available for generalized linear models with family chosen to be binomial and a link function chosen to be complementary log-log, log, or logit for the three models specified earlier.

In the next two subsections, the discrete and continuous relative risk model and the discrete logistic model are considered. For each of these, it is possible to estimate the regression parameter β through special argument, rather like the use of the partial or marginal likelihood to estimate β in the continuous Cox model.

4.8.2 Estimating Equations in the Discrete and Continuous Cox Model

The discrete and continuous relative risk model is specified by the hazard relationship

$$d\Lambda(t, x) = \exp[Z(t)'\beta]\, d\Lambda_0(t) \tag{4.56}$$

in continuous, discrete, and mixed cases. We consider the discrete case, in which, as before,

$$d\Lambda_0(t) = \sum_j \lambda_j I(a_j \le t).$$

To estimate β in this model, we begin by considering the Breslow–Peto approximate partial likelihood (4.19). This likelihood gives rise to an estimating equation for β which can be written

$$U(\beta) = \sum_{j=1}^{k} U_j(\beta) = \sum_{j=1}^{k}[s_j - d_j \mathscr{E}(\beta, a_j)], \qquad (4.57)$$

where, as before, $\mathscr{E}(\beta, a_j) = \sum_{l \in R(a_j)} Z_l(a_j) p_l(\beta, a_j)$ and

$$p_l(\beta, a_j) = \frac{\exp[Z_l(a_j)'\beta]}{\sum_{i \in R(a_j)} \exp[Z_i(a_j)'\beta]}.$$

Let H_{a_j} specify the history of all failures and censorings in the period $(0, a_j)$ as well as the information that exactly d_j individuals fail at time a_j. By arguments similar to those used in Sections 4.2.1 and 4.2.2, and under the discrete model (4.56), it is easy to see that

$$E[U_j(\beta)|H_{a_j}] = 0,$$

so that $E[U_j(\beta)] = 0$ and (4.57) is an unbiased estimating equation for β. It can also be seen by arguments similar to those in Section 4.2.1 that the score components in (4.57) are uncorrelated. Thus, as with the partial likelihood estimating equation, under suitable regularity conditions, a central limit theorem for a standardized version of the score $U(\beta)$ might be expected to apply as the sample size becomes large.

As a consequence, we should expect that the estimator $\hat{\beta}$ that solves (4.57) is consistent for β, and further, that $n^{1/2}(\hat{\beta} - \beta)$ is asymptotically normal with mean 0 and covariance matrix that can be estimated using a standard sandwich type of estimator. Prentice and Kalbfleisch (2002) show that an unbiased estimate of the variance of $U(\beta)$ is given by

$$V_U(\beta) = I(\beta) - \sum_{j=1}^{k} \sum_{l \in R(t_j)} [Z_l(t_j) - \mathscr{E}(\beta, t_j)]^{\otimes 2} \exp[2Z_l(t_j)'\beta] \hat{\alpha}_0(\beta, t_j), \qquad (4.58)$$

where

$$\hat{\alpha}_0(\beta, t_j) = \frac{d_j(d_j - 1)}{\{\sum_{l \in R(t_j)} \exp[Z_l(t_j)'\beta]\}^2 - \sum_{l \in R(t_j)} \exp[2Z_l(t_j)'\beta]}$$

and

$$I(\beta) = \sum_{j=1}^{k} d_j \mathscr{V}(\beta, t_j)$$

is the observed information arising from the Breslow–Peto approximate partial likelihood (4.19). The quantities $V(\beta, t_j)$ and $\mathscr{E}(\beta, t_j)$ are defined as before. The asymptotic covariance of $\hat{\beta} - \beta$ is estimated with

$$I(\hat{\beta})^{-1} V_U(\hat{\beta}) I(\hat{\beta})^{-1}.$$

Prentice and Kalbfleisch (2002) present simulations that suggest that this variance estimate performs better than the naive estimate $I(\hat{\beta})^{-1}$ in yielding confidence intervals with more accurate coverage. The correction term here is relatively small unless the number of ties are substantial.

It should be noted that the approximate likelihood (4.19) that generates the estimating equation (4.57) is not a partial likelihood since the conditional probabilities that the specific items fail given H_{a_j} does not yield that likelihood. Nonetheless, under the model (4.56) the score function arising from it is an unbiased estimating equation with uncorrelated terms. This contrasts with the situation for the discrete logistic model in the next section, where conditioning on H_{a_j} generates a partial likelihood for the corresponding regression parameter.

4.8.3   Partial Likelihood for the Discrete Logistic Model

Suppose that a right-censored or left-truncated sample is available from the discrete logistic model

$$\text{logit}[d\Lambda(t, x)] = \text{logit}[d\Lambda_0(t)] + Z(t)'\beta. \tag{4.59}$$

Define H_{a_j} as in the last section and let $\mathscr{D}_j$ be the event that the specific individuals in D_j fail at a_j. Under the model (4.59), we find that

$$P[\mathscr{D}_j | H_{a_j}] = \frac{\exp[s_j(a_j)'\beta]}{\sum_{l \in R(a_j, d_j)} \exp[s_l(a_j)'\beta]}, \tag{4.60}$$

and a product over $j = 1, \dots, k$ yields the partial likelihood for β in (4.23) which can be used for inference. Here, the parameter β measures the effect on a logit scale. In the case, for example, of a two-sample problem where $Z(t) = Z$ simply specifies sample membership, the parameter β measures the log odds ratio of the failure rates in the two samples.

BIBLIOGRAPHIC NOTES

The relative risk model and partial likelihood analysis were proposed by Cox (1972). Cox (1975) formalized the partial likelihood argument further. The marginal likelihood derivation of Section 4.7.1, given by Kalbfleisch and Prentice (1973), generalized to censored data a result due to Savage (1957). Breslow

(1974) proposed the maximum likelihood approach discussed in Exercises 4.4, 4.5, and 4.6. Other maximum likelihood approaches were taken by Thompson and Godambe (1974), Jacobsen (1982,1984), and Johansen (1983). Early work on asymptotics based on empirical processes was done by Tsiatis (1978), and this has not been reviewed here. Reformulation of the relative risk model in terms of counting processes was an important development in the early 1980s and led to new methods of approaching asymptotics. These methods are reviewed in Chapter 5, and bibliographic notes are included there.

There have been many review papers on the Cox model. In addition, many books, for example, Cox and Oakes (1984), Lawless (1982), Klein and Moeschberger (1997), Andersen et al. (1993), and Fleming and Harrington (1991) all have extensive discussion of the Cox model and give many references. The last two approach the subject from a counting process perspective. Two related recent books are Therneau and Grambsch (2000), which provides a detailed discussion of model building and model testing for relative risk models and a detailed discussion of the use of certain available software packages for this purpose, and Hougaard (2000), which is focused on multivariate data but provides a review of the univariate relative risk model. A recent review article with special emphasis on material that has appeared in Biometrika is Oakes (2001).

The grouped relative risk model was proposed by Kalbfleisch and Prentice (1973), who derived the maximum likelihood estimate (4.37) from it. Cox (1972) made use of the discrete logistic model both in the analysis of tied failure times and as the basis for survivor function estimation. A recursive approach to computing the partial likelihood arising from the discrete logistic model is discussed in Howard (1972) and Gail et al (1981). The Nelson–Aalen estimator is equivalent to that proposed as approximate MLE by Breslow (1974), but it was derived in the context of counting process formulation of the relative risk model by Andersen and Gill (1982) and in the context of a Poisson model by Jacobsen (1983). The mixed discrete and continuous relative risk model (4.56) was proposed by Prentice and Kalbfleisch (2002), who noted its relationship to the Breslow approximate likelihood and obtained the variance correction in (4.58).

Inference based on interval censoring in the relative risk model with constant covariates has been considered by several authors. Finklestein (1986) develops a score test that is obtained by maximizing over the baseline survivor function. Huang and Wellner (1995) and Huang (1996) consider joint estimation of β and $F_0(\cdot)$ and show that the parameter estimates are asymptotically efficient, although the asymptotics applying to the baseline survivor function are order $n^{1/3}$ and numerical issues are substantial. Other approaches have attempted to implement a generalization of the marginal likelihood argument in Section 4.7 and Kalbfleisch and Prentice (1973) whereby the likelihood is generated by the probability that the underlying uncensored rank vector should be one in the class of rank vectors consistent with the data. The calculations are formidable, and various approximations and algorithms have been suggested. Monte Carlo versions of the expectation–maximization algorithm have been proposed by Wei and Tanner (1990) and Sinha et al. (1994). Satten (1996) considers an approximation using the Gibbs sampler, and Goggins et al. (1998) consider a Markov chain Monte Carlo method for sampling from the admissible rank vectors. The asymptotic properties of the estimates obtained by this approach are not known.

Small-sample efficiency properties of the estimate of β from the relative risk model have been considered by Kalbfleisch (1974), Kalbfleisch and McIntosh (1977), Efron (1977), Oakes (1977), and Kay (1979). Efron and Oakes give general expressions for asymptotic efficiency, and related results are given in Chapter 5. Crowley and Thomas (1975) considered efficiency of the log-rank statistic, and Andersen et al. (1993) and Fleming and Harrington (1991) also discuss efficiency.

EXERCISES AND COMPLEMENTS

4.1 Consider the partial likelihood equation (4.14) and show that the MLE $\hat{\beta}_l$ of a component β_l of β is ∞ (or $-\infty$) if and only if for each item that fails, the value of the corresponding component of $Z(t)$ is always the largest (or always the smallest) value in the risk set. That is, for item j that fails at t_j, $Z_{jl}(t_j) = \max_{i \in R(t_j)} Z_{il}(t_j)$, $j = 1, \dots, k$.

4.2 As in ordinary linear regression, the covariance and information matrices can be singular. At failure time t_j, construct the matrix A_j of dimension $p \times n_j$ whose columns are the vectors $Z_i(t_j), i \in R(t_j)$, and define the design matrix A obtained by concatenating the matrices $A_1, \dots, A_k$. Show that $I(\beta)$ is singular if and only if A is not of full (row) rank p.

4.3 Suppose that failure times are generated according to the relative risk model (4.1) but that observation begins on individual i at a time t_i^0 that may exceed 0. Show that a partial likelihood function can be developed in the manner of Section 4.2.2 and that it has the same form as (4.14), with $R(t_i)$ redefined to include only study subjects at risk and under observation t_i.

4.4 The Breslow derivation of the partial likelihood for β: Suppose that data are available from the relative risk model (4.10) as in Section 4.2.2. Consider approximating the underlying baseline hazard function as a step function

$$\lambda_0(t) = \lambda_j, \qquad t_{j-1} < t \leq t_j, \quad j = 1, 2, \dots, k + 1,$$

where the λ_j's are unknown positive constants, $t_0 = 0$ and $t_{k+1} = \infty$. Censorings that occur in the interval $[t_{j-1}, t_j)$ are shifted to the left endpoint of the interval. Suppose that the covariates $Z(t) = Z$ are fixed.

(a) Show that the log likelihood for λ, β where $\lambda = (\lambda_1, \dots, \lambda_k)$ is

$$l(\lambda, \beta) = \sum_{j=1}^{k} \left[\log \lambda_j + Z_j'\beta - \lambda_j(t_j - t_{j-1}) \sum_{l \in R(t_j)} \exp(Z_l'\beta) \right].$$

(b) Show that the profile or maximized likelihood for β obtained by maximizing this joint likelihood with respect to λ is proportional to the partial likelihood (4.14).

(c) Find the MLE of the baseline cumulative hazard function $\hat{\Lambda}_0(t)$. Note that this gives the Nelson–Aalen estimator at the failure times observed.

4.5 (*continuation*) Suppose that d_j failures occur at time $t_j, j = 1, \ldots, k$ and the data are as defined in Section 2.4.3, except that the covariates $Z(t) = Z$ are time independent. Repeat Exercise 4.4 and obtain the Breslow approximation (4.19) as the profile likelihood for β.

4.6 (*continuation*) Generalize the procedure in Exercise 4.5 to incorporate defined time-dependent covariates $Z(t)$. Under what conditions on the time-dependent covariates will this approach again yield (4.19) as the profile likelihood for β? Discuss how the time-dependent covariates might be approximated to satisfy these conditions.

4.7 A modified life-table estimate for the proportional hazards model: For data from the proportional hazards model with Z time independent, obtain a generalization of the estimate (1.17). Specifically, take the hazard function to be a step function

$$\lambda_0(t) = \lambda_j, \qquad t \in I_j = [b_0 + \cdots + b_{j-1}, b_0 + \cdots + b_j), \quad j = 1, \ldots, k,$$

where $b_0 = 0, b_k = \infty$ and $b_i > 0 (i = 1, \ldots, k - 1)$. Take $\beta = \hat{\beta}$ as estimated from the marginal or partial likelihood and show that the MLE of λ_j is $\hat{\lambda}_j = (d_j/S_j)$, where d_j is the number of failures in I_j and

$$S_j = b_j \sum_{\ell \in R_j} e^{Z_j'\beta} + \sum_{\ell \in D_j} (t_\ell - b_1 - \cdots - b_{j-1}) e^{Z_\ell'\beta},$$

where R_j is the risk set at $b_0 + \cdots + b_j - 0$ and D_j is the set of items failing in I_j. The estimate of the underlying survivor function $(Z = 0)$ is now given by $\hat{F}(t)$ in (1.17). (Holford, 1976)

4.8 (*continuation*) Compare the estimates in Example 4.7 with those obtained from (4.37) for the carcinogenesis data of Pike (Section 1.1.1) (see Figure 4.4).

4.9 (*continuation*) Generalize the modified life-table estimate to allow $Z = Z(t)$ to be a defined time-dependent covariate.

4.10 Construct an example in which the Nelson–Aalen estimator (4.38) has at least one increment $d\tilde{\Lambda}_0(t_j)$ that exceeds 1.

4.11 **(a)** Assuming that there are no ties in the data, use the method in Section 4.3 and expression (4.34) to obtain a maximized likelihood for β by maximizing over F_0. Write the result in the form

$$L_{\max}(\beta) = L(\beta)R(\beta),$$

where $L(\beta)$ is the marginal or partial likelihood for β, (4.14). Examine the form of $R(\beta)$ and show that for reasonably large data sets, the influence of $R(\beta)$ on the estimation of β is small.

(b) Compare the maximized likelihood and the marginal likelihood for the carcinogenesis data of Section 1.1.1.

4.12 Consider the variance formula V_U given in (4.58) for the mixed discrete and continuous Cox model. Show that this reduces to the hypergeometric variance for the logrank statistic in the $p + 1$ sample problem (see Sections 1.4 and 4.4).

4.13 **(a)** Consider the model of Section 4.2.4, in which the hazard for the carcinogenesis data is taken to be

$$\lambda_0(t) \exp[Z\beta_1 + Z \log t\beta_2]$$

with $Z = 0, 1$ a sample indicator. Construct a score function test of $\beta_2 = 0$ and compare the results with the test based on the partial likelihood estimate. Note that such score function tests involve considerably less computation than tests requiring remaximization of the partial likelihood.

(b) Generalize the score test in part (a) to yield a test of $\gamma = 0$ in the model

$$\lambda_0(t) \exp[Z'\beta + Z_1 g(t)\gamma],$$

where $Z = (Z_1, \ldots, Z_p)'$, $\beta = (\beta_1, \ldots, \beta_p)'$ and $g(t)$ is a given function of time. This provides one simple way to check for potential departures from the proportionality assumption in components of Z.

4.14 Let $(t_1, Z_1), \ldots, (t_n, Z_n)$ be independent observations from the exponential regression model with hazard

$$\lambda(t_i; Z_i) = \lambda \exp(Z_i'\beta), \qquad t_i > 0.$$

(a) Show that the estimation problem for β is invariant under the group G of scale transformations on the survival time t and thence that

$$a_i = \frac{t_i}{t_1}, \qquad i = 1, \ldots, n,$$

are jointly marginally sufficient for β.

(b) Obtain the marginal likelihood for β and compare the result with that obtained by maximizing the full likelihood over λ.

4.15 Suppose that n items are placed on test where the survival time t given Z follows the proportional hazards model with hazard

$$\lambda(t; Z) = \lambda_0(t) \exp(Z'\beta),$$

where Z is time independent. Consider a study design at which m_i items (chosen at random from those at risk) are censored immediately following the ith observed failure time, $i = 1, \ldots, r$, where $r + \sum m_i = n$. Show that for this design, the censored model is invariant under the group of strictly increasing differentiable transformations on t and that the marginal likelihood of β is proportional to (4.14).

CHAPTER 5

Counting Processes and Asymptotic Theory

5.1 INTRODUCTION

Convenient arguments leading to asymptotic results for many of the methods and approaches discussed in this book are based on counting processes and martingale limit theorems. In this chapter we outline some of these results. It has already been noted that notation based on counting processes is often convenient and is widely used in describing failure time models and methods. In this chapter we summarize some aspects of counting processes, martingales, and the associated asymptotic theory. We have not attempted to give a full treatment of this area, but rather, have outlined some of the main ideas and the nature of the theoretical results. This material rests on an extensive literature in probability theory, and more formal and complete accounts of applications to failure time data can be found in the excellent books by Andersen et al. (1993) and Fleming and Harrington (1991).

In the first three sections of the chapter we give the mathematical developments. Section 5.2 begins with a discussion of counting processes and sets the notation, with intuition being developed through a series of examples. In anticipation of developments to follow, basic covariates are allowed to include time-dependent stochastic covariates measured on each individual. More aspects of stochastic time-dependent covariates are discussed in Chapter 6. In Section 5.3 we give some basic developments of martingale theory and outline particular applications and results for discrete and continuous counting processes. In Section 5.4 we outline some of the asymptotic results using a version of Rebolledo's central limit theorem and an inequality (Lenglart) that is useful in establishing asymptotic results.

In the balance of the chapter we consider applications of the asymptotic results to the statistical methods developed previously. Specifically, the asymptotic results are applicable to situations where the data are subject to independent right censoring and/or left truncation (delayed entry). Thus, in Section 5.6, the Kaplan–Meier estimator and the logrank test are revisited and applications of the counting process and martingale results are described. In Section 5.7 we discuss the Cox model and asymptotics associated with the partial likelihood and the maximum partial likelihood estimator. Finally, in Section 5.8, the asymptotic results are used in parametric models and, in Section 5.9, to examine asymptotic efficiency using the partial likelihood analysis of the Cox model.

5.2 COUNTING PROCESSES AND INTENSITY FUNCTIONS

As noted earlier, a counting process $N = \{N(t), t \geq 0\}$ is a stochastic process with $N(0) = 0$ and whose value at time t counts the number of events that have occurred in the interval $(0, t]$. The sample paths of N are nondecreasing step functions that jump whenever an event (or events) occur. Any sample path (i.e., any realization) of N is right continuous with left-hand limits, and following standard terminology, we use the term *cadlag* from the French *continué à droit, limité à gauche*. For a counting process in continuous time, we assume that only jumps of size $+1$ can occur, and if there are several counting processes in continuous time, no two of them can jump at the same time. In discrete time, however, it is sometimes convenient to allow jumps of more than one at a given time, and to allow two processes to jump at the same time. As we have also done earlier, we use the notation $dN(t) = N(t^- + dt) - N(t^-)$ to indicate the number of events that occur in the interval $[t, t + dt)$, and $\Delta N(t) = N(t) - N(t^-)$ to indicate the number of events that occur at time t.

In this discussion, counting processes corresponding to failure time models are considered first. More general counting processes where individuals may experience more than one event, such as repeated infections or asthma attacks or equipment breakdowns, are considered later. Examples of such counting processes arise in Chapters 8, 9, and 10. There are also situations where individuals experience more than one type of event or failure. These are discussed here briefly and are dealt with in more detail in later chapters.

5.2.1 Failure Time Models

Suppose that n individuals having independent failure times are on study and let T_i be the time to failure of the ith, $i = 1, \ldots, n$. For each individual, there are covariates or covariate processes that are to be related to the rates at which events occur. Thus, we may have measurement at time $t = 0$ of a vector of (fixed) covariates x_i on the ith individual as described in Chapter 4. More generally, however, we consider a vector of basic covariates $x_i(t)$ for the ith individual that may vary over time. The components of $x_i(t)$ may include fixed covariates measured at time 0 as well as measurements of risk factors on the individual or on the environment that are evolving over time. Chapter 6 includes an examination of stochastic time-dependent covariates, and they are included here in anticipation of that discussion. Let $X_i(t) = \{x_i(u) : 0 \leq u < t\}$ specify the path or history of the covariate process up to time t^-.

As discussed earlier, hazard models can be used to describe aspects of the distribution of the time to failure. Specifically, let

$$d\Lambda_i(t) = d\Lambda[t; X_i(t)] = P\{T_i \in [t, t+dt)|X_i(t), T_i \geq t\}, \tag{5.1}$$

where in the continuous case,

$$d\Lambda_i(t) = \lambda_i(t)\, dt = \lambda[t; X_i(t)]\, dt. \tag{5.2}$$

In the discrete case with mass points at $a_1 < a_2 < \cdots$,

$$d\Lambda_i(t) = \begin{cases} \lambda_{il}, & t = a_l,\ l = 1, 2, \ldots \\ 0, & \text{otherwise}, \end{cases} \tag{5.3}$$

where $\lambda_{il} = P[T_i = a_l|X_i(a_l), T_i \geq a_l]$. Conversely, the cumulative hazard function is a right-continuous (cadlag) function,

$$\Lambda_i(t) = \begin{cases} \int_0^t \lambda_i(u)\, du, & T_i \text{ continuous} \\ \sum_{a_l \leq t} \lambda_{il}, & T_i \text{ discrete}. \end{cases} \tag{5.4}$$

In the case of fixed basic covariates, as discussed in earlier chapters, the cumulative hazard function is deterministic. However, in the case of stochastic covariates, where the elements of the basic covariates may involve time-dependent measurements on the individuals under study, $\Lambda_i(t) = \Lambda[t; X_i(t)]$ is a stochastic process. For interpretation of hazard models, this distinction does not matter greatly. In a broader context, however, stochastic covariates introduce some complications that are discussed in Chapter 6. On first reading of this chapter, little is lost through considering the basic covariates as fixed.

In preceding chapters, models for the hazard function were defined using $Z_i(t)$, a vector of derived or modeled covariates obtained as functions of the fixed basic covariates x and time t. When the basic covariates $x_i(t)$ are themselves time dependent, the derived covariates $Z_i(t)$ are defined as functions of t and the covariate paths $X_i(t) = \{x_i(u), 0 \leq u < t\}$. Once the covariates $Z(t)$ are defined, hazard models are specified as before. Thus, we might entertain a relative risk model of the form

$$d\Lambda_i(t) = \exp[Z_i(t)'\beta]\, d\Lambda_0(t).$$

Similarly, various parametric models could be hypothesized that relate the values of $Z(t)$ to the failure rate at time t. Sample paths for modeled regression variables Z_i are required to be left continuous.

These same models for the hazard function can also be specified with respect to an *underlying counting process* $\tilde{N}_i = \{\tilde{N}_i(t), 0 \leq t\}$ which counts events for the ith individual in the interval $(0, t]$, $i = 1, \ldots, n$. For failure time models, each individual experiences one event only and $\tilde{N}_i(t) = \mathbf{1}(T_i \le t)$ takes only values 0 or 1. The hazard model (5.1) can be written

$$d\Lambda_i(t) = P[d\tilde{N}_i(t) = 1 | X_i(t), \tilde{N}_i(t^-) = 0]. \tag{5.5}$$

As a matter of terminology, Λ_i is called the *cumulative intensity process* of the counting process $\tilde{N}_i$. The same continuous and discrete formulations (5.2) and (5.3) can now be described in terms of intensity models for $\tilde{N}_i$, and $\lambda_i(t)$, and λ_{il} are the corresponding intensity processes.

When individuals are subject to right censoring, not all events in the underlying processes $\tilde{N}_i$ are observed. Let $\{Y_i(t), t \ge 0\}$ be the *at-risk process* for the ith individual, $i = 1, \ldots, n$. Thus, the ith process is at risk of an observed event (i.e., uncensored and surviving at time t^-) if and only if $Y_i(t) = 1$. We assume that $Y_i(t)$ is a left continuous process. In the simplest case of random right censoring, for example, $Y_i(t) = \mathbf{1}(T_i \ge t, C_i \ge t)$ where C_i and T_i are the associated censoring and failure times. The *observed counting process* $N_i = \{N_i(t), t \ge 0\}$ counts the number of events on the ith individual that are observed to occur in the interval $(0, t]$. Thus, N_i will register a jump at time t if and only if $\tilde{N}_i$ has a jump at time t and $Y_i(t) = 1$. Equivalently, we may write $N_i(t) = \int_0^t Y_i(u) \, d\tilde{N}_i(u)$.

Consider specification of intensities for the observed counting processes N_i. For this purpose, we need to define an appropriate conditioning event and we do so by defining the history or *filtration*

$$\mathscr{F}_t = \sigma\{N_i(u), Y_i(u^+), X_i(u^+), i = 1, \ldots, n; \ 0 \le u \le t\}, \qquad t > 0, \tag{5.6}$$

where, for example, $Y_i(u^+) = \lim_{s \to u^+} Y_i(s)$. The notation $\sigma[\cdot]$ specifies the *sigma algebra of events* generated by the variables given in the brackets. For most practical situations, however, one can think of $\mathscr{F}_t$ as simply specifying the observed values of the variables in the brackets. [Note that we assume that $Y_i(0)$ and $x_i(0)$, $i = 1, \ldots, n$ are included in $\mathscr{F}_t$ for all $t \ge 0$.]

It is easy to see that $\{\mathscr{F}_t : t \ge 0\}$ is increasing ($\mathscr{F}_s \subseteq \mathscr{F}_t$, if $s \le t$), and right continuous so that

$$\mathscr{F}_t = \lim_{s \to t^+} \mathscr{F}_s$$

for all $t \ge 0$. This is assured by the right continuity of the sample paths $N_i(u)$, $Y_i(u^+)$, and $X_i(u^+)$, $i = 1, \ldots, n$ for $u > 0$ in (5.6). For each $t > 0$, let $\mathscr{F}_{t^-} = \sigma\{N_i(u), Y_i(u), X_i(u), i = 1, \ldots, n; 0 \le u < t\}$ denote the full history of the observed processes $N_i(u), Y_i(u)$, and $X_i(u), i = 1, \ldots, n$ up to but not including t. This is the information available to the experimenter or observer just prior to time t.

The intensities or rates for the processes N_i are defined with reference to the filtration $\mathscr{F}_t$. If the censoring process is independent, the intensity model for the counting process N_i is

$$P[dN_i(t) = 1 | \mathscr{F}_{t^-}] = Y_i(t) \, d\Lambda_i(t), \qquad i = 1, \ldots, n \tag{5.7}$$

for $t > 0$, where $\Lambda_i(t)$ is defined in (5.1). This relationship (5.7) is a characterization of independent censoring and specifies that the hazard rates of individuals under observation at time t are representative of the study population in terms of their hazard rates at time t.

In the continuous case, (5.7) can be written

$$P[dN_i(t) = 1|\mathscr{F}_{t^-}] = Y_i(t)\lambda_i(t)\,dt, \tag{5.8}$$

and it is assumed that no two individuals can fail at the same time. In the discrete case where $\Lambda_i(t)$ is given by (5.3), this reduces to

$$P[dN_i(a_l) = 1|\mathscr{F}_{a_l^-}] = Y_i(a_l)\lambda_{il}, \qquad l = 1, 2, \ldots \tag{5.9}$$

and $P[dN_i(t) = 1|\mathscr{F}_{t^-}] = 0$ elsewhere. It is usually assumed that individuals at risk at time t act independently of one another. In particular, for all $i \neq j$, it is assumed that

$$P[dN_i(a_l) = 1, \ dN_j(a_l) = 1|\mathscr{F}_{a_l^-}] = Y_i(a_l)Y_j(a_l)\lambda_{il}\lambda_{jl}, \qquad l = 1, 2, \ldots.$$

The definition (5.1) may seem unnecessarily restrictive in that the intensity for the ith individual can depend only on the basic covariates $x_i(\cdot)$ measured for that subject. In fact, however, $x_i(t)$ can include measurements on other individuals in the study, so the formulation is quite general in that respect. For example, one could include in $x_i(t)$ the values of the counting processes $N_j(t^-)$ for all $j \neq i$ and allow models for which the rate of failure of i depends on the total number of failures experienced earlier in the trial. Such a model might be appropriate, for example, if failure is due to a contagious disease in a confined population or group.

It is also important to note that the formulations above apply to experiments in which individuals are subject to delayed entry or left truncation. In this case, the at-risk process $Y_i(t)$ takes the value 1 when the individual comes under observation or enters the study. Individuals who experience failure prior to qualifying for entry to the study are completely unobserved. The condition (5.7) then characterizes independent left truncation or delayed entry as well as independent right censorship.

Counting processes have a close relationship to certain stochastic processes called martingales, and these are discussed in the next section. It is useful, however, to make reference to some of the links between counting processes and martingales since this provides a specific framework for the discussions to come.

Consider the continuous case (5.2) and, for any given i, define the process

$$M_i(t) = N_i(t) - \int_0^t Y_i(u)\lambda_i(u)\,du, \qquad t \geq 0.$$

Equivalently, we can define $M_i(t) = \int_0^t dM_i(u)$, where

$$dM_i(t) = dN_i(t) - Y_i(t)\lambda_i(t)\,dt.$$

It can be seen that $E[dM_i(t)|\mathscr{F}_{t-}] = 0$ for all t and that for all $s \leq t$,

$$E[M_i(t)|\mathscr{F}_s] = M_i(s).$$

A process that satisfies these (equivalent) conditions is a martingale. It is easy to deduce that $E[M_i(t)] = 0$ for all t and that the process $M_i(t)$ has uncorrelated increments. That is,

$$E\{[M_i(t) - M_i(s)]M_i(s)\} = 0$$

for all $0 < s < t$. Note that we have decomposed $N_i(t)$ into two processes as

$$N_i(t) = \int_0^t Y_i(u)\lambda_i(u)\,du + M_i(t). \tag{5.10}$$

The first term on the right of (5.10) is termed the *compensator* of the counting process N_i with respect to the filtration $\mathscr{F}_t$; the second term is the *counting process martingale* corresponding to $N_i(t)$. In terms of the differential increments of the process, (5.10) can equivalently be written as

$$dN_i(t) = Y_i(t)\lambda_i(t)\,dt + dM_i(t). \tag{5.11}$$

In the discrete case (5.9), the discrete-time martingale is

$$\begin{aligned} M_i(t) &= N_i(t) - \int Y_i(u)\,d\Lambda_i(u) \\ &= N_i(t) - \sum_{a_l \leq t} Y_i(a_l)\lambda_{il}, \end{aligned}$$

where the second term on the right side is the compensator of $N_i(t)$. In terms of the differential elements (or jumps) in the process, the decomposition in the discrete case is

$$dN_i(a_l) = Y_i(a_l)\lambda_{il} + dM_i(a_l). \tag{5.12}$$

One can think of the relationship (5.11) or (5.12) as writing the response $dN_i(t)$ as a conditional signal (the differential of the compensator) plus mean zero noise (the differential of the corresponding martingale). Note, however, that the compensator is in general a random process, although its differential and value at time t is fixed given $\mathscr{F}_{t-}$. The signal-plus-noise interpretation is thus conditional on $\mathscr{F}_{t-}$.

Before proceeding with a discussion of martingales in more general terms, we consider some examples.

Example 5.1. Suppose that n individuals having independent failure times are placed on trial at time 0 and that the continuous time to failure T_i of the ith item has hazard function $\lambda(t), i = 1, \ldots, n$. Suppose further that the ith item on trial is subject to a random (right) censoring time C_i which is distributed independently of T_i. Then $\tilde{N}_i(t) = I(T_i \leq t)$ and $N_i(t) = \mathbf{1}(T_i \leq t, \ C_i \geq T_i)$ count, respectively, the number of actual and observed failures in $(0, t]$ for the ith individual, and $Y_i(t) = \mathbf{1}(T_i \geq t, \ C_i \geq t)$. In this case, there are no covariates and $\mathscr{F}_{t^-} = \sigma\{N_i(u), Y_i(u^+),$ $i = 1, \ldots, n, 0 \leq u < t\}$ specifies the times of failure and or censoring up to but not including time t. It follows that

$$
\begin{aligned}
E[dN_i(t)|\mathscr{F}_{t^-}] &= P[dN_i(t) = 1|\mathscr{F}_{t^-}] \\
&= P[t \leq T_i < t + dt, \ C_i \geq t|\mathscr{F}_{t^-}] \\
&= P[t \leq T_i < t + dt, \ C_i \geq t|Y_i(t)] \\
&= Y_i(t)\lambda(t) \, dt.
\end{aligned}
\tag{5.13}
$$

The final equality clearly holds if $Y_i(t) = 0$. Now, $Y_i(t) = 1$ if and only if $T_i \geq t, \ C_i \geq t$, and the independence of T_i and C_i implies the result. The compensator for $N_i(t)$ is $\int_0^t Y_i(u)\lambda(u) \, du$, which is a random process. The corresponding counting process martingale with respect to the filtration or history $\mathscr{F}_t$ is

$$
M_i(t) = N_i(t) - \int_0^t Y_i(u)\lambda(u) \, du.
$$

Define the superposed counting process $N.(t) = \sum_{i=1}^n N_i(t), \ 0 \leq t$, which counts the number of failures across all individuals observed in the interval $(0, t]$. Since $P[dN.(t) = 1|\mathscr{F}_{t^-}] = E[dN.(t)|\mathscr{F}_{t^-}] = \sum E[dN_i(t)|\mathscr{F}_{t^-}]$, it follows that the intensity function of $N.(t)$, with reference to $\mathscr{F}_t$, is $\sum_{i=1}^n Y_i(t)\lambda(t)$. $\square$

Example 5.2. In the context of Example 5.1, suppose that T_i is discrete with cumulative hazard function $\Lambda(t) = \sum_{a_j \leq t} \lambda_j$. The superposed process $N.(t)$ may have jumps of more than 1 unit at the potential failure times $a_1, a_2, \ldots$. Again, $E[dN.(t)|\mathscr{F}_{t^-}] = \sum_{i=1}^n E[dN_i(t)|\mathscr{F}_{t^-}]$. Thus,

$$
E[dN.(a_l)|\mathscr{F}_{a_l^-}] = \sum_{i=1}^n Y_i(a_l)\lambda_l = Y.(a_l)\lambda_l.
$$

In fact, conditional on $\mathscr{F}_{a_l^-}$ the total number of failures at time $a_l, dN.(a_l)$, has a binomial distribution with index $Y.(a_l)$ and probability λ_l. $\square$

Example 5.3. Consider Example 5.1 again, but suppose that the censoring is type II, so that all items are put on trial at time 0 and followed until the time at which the kth failure occurs for some predetermined integer k. At this kth ordered failure time, $T_{(k)}$, all surviving items are censored. In this instance, $Y_i(t) = \mathbf{1}(T_{(k)} \geq t, T_i \geq t)$ and it can be seen that equations (5.13) hold without change. $\square$

Example 5.4. Suppose that n_i individuals are put on trial in the ith group, that all individuals have mutually independent failure times, and that the continuous-time hazard function for individuals in the ith group is $\alpha_i(t), i = 0, \ldots, p$. As before, each individual is subject to right censoring with a censoring time C_{ij} for the jth individual in the ith group distributed independently of the failure times. Let N_{ij} be the counting process and Y_{ij} be the at-risk process for the jth item in the ith group and let $\mathscr{F}_t = \sigma\{N_{ij}(s), Y_{ij}(s^+) : 0 \leq s \leq t; i = 0, \ldots, p; j = 1, \ldots, n_i\}$. Let $N_{i.}(t) = \sum_{j=1}^{n_i} N_{ij}(t)$ count the observed number of failures in the ith group. Then, $N(t) = [N_{0.}(t), \ldots, N_{p.}(t)]'$ is a multivariate counting process with respect to the filtration $\mathscr{F}_t$. The corresponding compensator is $E(t) = [E_0(t), \ldots, E_p(t)]'$, where

$$dE_i(t) = \sum_{j=1}^{n_i} Y_{ij}(t)\alpha_i(t)dt,$$

and $M(t) = N(t) - E(t)$ is a multivariate martingale with respect to the filtration $\mathscr{F}_t$. □

Example 5.5   Consider n independent individuals on trial subject to right censoring. Let $A_1, A_2, \ldots$ be independent, identically distributed (IID) positive random variables where A_i is the waiting time until some stress factor is applied to the ith individual. When the stress is applied, the failure or hazard rate changes from a baseline unstressed rate of $\lambda_0(t)$ to $\exp(\beta)\lambda_0(t)$. Let $x_i(t) = \mathbf{1}(A_i < t)$. In this case, $\mathscr{F}_t$ is defined in (5.6) and, under independent censoring,

$$P[dN_i(t) = 1|\mathscr{F}_{t^-}] = Y_i(t) \exp[x_i(t)\beta]\lambda_0(t)\, dt, \qquad t > 0.$$

The intensity at time t of the underlying process $\tilde{N}_i(t)$ is $\exp[x_i(t)\beta]\lambda_0(t)$, which is a random variable. Its value is fixed at time t once $x_i(t)$ [or $x_i(t^-)$] is known. □

5.2.2   More General Processes: Repeated Events

As noted earlier, there are some applications in which individuals may experience several events over time, and these same formulations extend in a natural way to those situations. Consider a trial with n subjects, and for the ith individual under study, let $\tilde{N}_i$ be the underlying process that counts the number of events that the ith individual experiences in the interval $(0, t]$. Let $x_i(t)$ and $X_i(t)$ be defined as before. In the continuous case, we assume that events can occur only one at a time, and the underlying intensity process is $\lambda_i(t)$, where

$$\lambda_i(t)\, dt = P[d\tilde{N}_i(t) = 1|X_i(t), \tilde{N}_i(u), 0 \leq u < t], \qquad t > 0.$$

Consider now the discrete case, where events occur only at times $a_1 < a_2 < \cdots$. If an individual can experience at most one event at a given time, the intensity process of $\tilde{N}_i$ is again given by (5.3), where $\lambda_{il} = P[d\tilde{N}_i(a_l) = 1|X_i(a_l), \tilde{N}_i(u), 0 \leq u < a_l]$.

In the discrete case, it is sometimes useful to allow an individual to experience more than one event at a given time and define the intensity process by

$$\lambda_i(a_l) = E[d\tilde{N}(a_l)|X_i(a_l), \tilde{N}_i(u), 0 \le u < a_l],$$

which specifies the (conditional) mean number of events that occurs at time $a_l, l = 1, 2, \ldots$. This reduces to the previous statements if at most one event can occur. To specify an intensity model completely in this case, however, one would also need to specify the distribution of the number of events; that is, we need to specify

$$P[d\tilde{N}_i(a_l) = j|X_i(a_l), \tilde{N}_i(u), 0 \le u < a_l], \qquad j = 1, 2, \ldots; \quad l = 1, 2, \ldots.$$

Continuous models could also be extended in this way. We encounter discrete counting processes with jumps of more than one unit, but they typically arise as sums or superpositions of processes with single jumps and, under independence, the distribution of the jump size can be deduced (see Example 5.2). Extensions that allow multiple jumps for an individual are discussed in Chapter 9.

As in univariate failure time models, repeated event data are often subject to right censoring, and the at-risk process $Y_i(t)$ is defined as before. The counting process observed is then $N_i(t) = \int_0^t Y_i(u) d\tilde{N}_i(u)$, and we define the filtration $\mathscr{F}_t$ as in (5.6). Thus, under independent right censoring and/or left truncation, the intensity function corresponding to $N_i(t)$ with reference to $\mathscr{F}_t$ is again given by (5.2) in the continuous case and (5.9) in the discrete case. Here again, the processes

$$M_i(t) = N_i(t) - \int_0^t Y_i(u) d\Lambda_i(u)$$

are martingales.

Example 5.6. Suppose that $\{\tilde{N}_i(t), t \ge 0\}$ is a nonhomogeneous Poisson process with intensity function $\alpha(t)$, $i = 1, \ldots, n$ independently, and that observations on the ith process are taken over the interval $[0, c_i]$, where $c_1 < c_2 < \cdots < c_n$ are fixed censoring times. Then $Y_i(t) = \mathbf{1}(t \le c_i)$, $N_i(t) = \min[\tilde{N}_i(t), \tilde{N}_i(c_i)], i = 1, \ldots, n$ and the filtration is $\mathscr{F}_t = \sigma\{N_i(u), Y_i(u^+), i = 1, \ldots, n, 0 \le u \le t\}$. It is easy to check that for each i, the process

$$N_i(t) - \int_0^t Y_i(u)\alpha(u) \, du$$

has mean zero and uncorrelated (actually independent) increments. The process $N.(t) = \sum N_i(t)$ counts the total number of observed events and

$$P[dN.(t) = 1|\mathscr{F}_{t-}] = \sum_{i=1}^n Y_i(t)\alpha(t) \, dt. \tag{5.14}$$

This is again a nonhomogeneous Poisson process with intensity function $(n - i + 1)\, \alpha(t)$ for $t \in (c_{i-1}, c_i]$, $i = 1, \ldots, n + 1$, where $c_0 = 0$ and $c_{n+1} = \infty$.

If the censoring times are random variables $C_1, \ldots, C_n$, it can be seen that (5.14) still holds but that the intensity process of N. is random. Between successive censoring times, the process is again a Poisson process, but the censoring times that constitute change points are random.   □

5.3 MARTINGALES

5.3.1 Introduction

Consider a probability model applying to processes over the interval $[0, \tau]$, where $\tau < \infty$ is a fixed value. Suppose that a filtration $\{\mathscr{F}_t, t \in [0, \tau]\}$ is given or has been defined. As in the special case of counting processes discussed above, $\mathscr{F}_t$ is a nondecreasing sequence of sigma fields so that ($\mathscr{F}_s \subseteq \mathscr{F}_t$ for all $0 \leq s \leq t \leq \tau$) and it is right continuous so that ($\mathscr{F}_s = \cap_{t>s}\mathscr{F}_t$ for all s). There is also a sigma field $\mathscr{F}_{t^-}$ which includes all events that are fixed just before time t and so represents the history of the process up to but not including time t; $\mathscr{F}_{t^-}$ represents the data available or known to the observer just prior to time t.

It is useful to define two technical terms.

1. A stochastic process $U = \{U(t), t \geq 0\}$ is said to be *adapted* to the filtration $\mathscr{F}_t$ if for each t, $U(t)$ is a function of (or is specified by) $\mathscr{F}_t$. In measure-theoretic terms, U is said to be adapted if $U(t)$ is $\mathscr{F}_t$ measurable for each $t \in [0, \tau]$. In less formal terms, this simply means that the value of $U(t)$ is fixed once $\mathscr{F}_t$ is given.

2. The stochastic process U is said to be *predictable* with respect to the filtration $\mathscr{F}_t$ if for each t, the value of $U(t)$ is a function of (or is specified by) $\mathscr{F}_{t^-}$. Again, in measure-theoretic terms, U is predictable if $U(t)$ is $\mathscr{F}_{t^-}$ measurable for all $t \in [0, \tau]$.

If U is adapted to $\mathscr{F}_t$ and has left-continuous sample paths, it easy to see that it is predictable. This is not, however, a necessary condition. For example, any deterministic function $U(t)$ is predictable, and we will also see important examples of right-continuous predictable processes. In Section 5.2 the processes $Y_i(t)$ and $X_i(t)$ are left continuous and, by the definition (5.6), they are obviously adapted to $\mathscr{F}_t$. They are, therefore, predictable with respect to the filtration $\mathscr{F}_t$.

A (real-valued) stochastic process $\{M(t), 0 \leq t \leq \tau\}$ is a martingale with respect to the filtration $\{\mathscr{F}_t\}$ if it is cadlag, adapted to $\mathscr{F}_t$, and satisfies the martingale property

$$E[M(t)|\mathscr{F}_s] = M(s) \qquad \text{for all} \quad s \leq t \leq \tau,$$

or equivalently,

$$E[dM(t)|\mathscr{F}_{t^-}] = 0 \qquad \text{for all} \quad t \in (0, \tau].$$

In essence, a martingale is a process that has no drift and whose increments are uncorrelated. We say that $M(t)$ is a mean zero martingale if $E[M(0)] = 0$, and hence $E[M(t)] = 0$ for all t. In what follows, we consider only martingales that have mean zero. The martingale $M(t)$ is said to be *square integrable* (or have finite variance) if $E[M^2(t)] = \text{var}[M(t)] < \infty$ for all $t \leq \tau$ or equivalently, if $E[M^2(\tau)] < \infty$.

The process $\{\bar{M}(t), 0 \leq t \leq \tau\}$ is a *submartingale* with respect to $\mathscr{F}_t$ if it is cadlag and adapted and satisfies

$$E[\bar{M}(t)|\mathscr{F}_s] \geq \bar{M}(s) \qquad \text{for all } s \leq t \leq \tau$$

or

$$E[d\bar{M}(t)|\mathscr{F}_{t^-}] \geq 0 \qquad \text{for all } t.$$

It is left as an exercise to show that a counting process $N(t)$ is a submartingale. As discussed later, another important example of a submartingale is provided by the process $M^2(t)$, where $M(t)$ is a square-integrable martingale.

The *Doob–Meyer decomposition theorem* establishes a decomposition of a submartingale. Specifically, let $\bar{M}$ be a nonnegative submartingale with $\bar{M}(0) = 0$. There exists an essentially unique decomposition of $\bar{M}$ into a (mean zero) martingale M and an increasing cadlag predictable process or *compensator*, C with $C(0) = 0$. That is, for all $t \in [0, \tau]$,

$$\bar{M}(t) = C(t) + M(t),$$

or equivalently,

$$d\bar{M}(t) = dC(t) + dM(t).$$

Conditioning on $\mathscr{F}_{t^-}$ and taking expectations gives the relationship

$$dC(t) = E[d\bar{M}(t)|\mathscr{F}_{t^-}],$$

which is useful in determining the compensator. Counting processes, as discussed in Section 5.2, provide our main example. See, for example, the decompositions given in (5.10) and (5.11).

The *predictable variation process* of a square-integrable martingale M is

$$\langle M \rangle(t) = \int_0^t \text{var}[dM(u)|\mathscr{F}_{u^-}]. \tag{5.15}$$

Equivalently, we can write

$$d\langle M \rangle(t) = \text{var}[dM(t)|\mathscr{F}_{t^-}].$$

In statistical terms, the primary role of the predictable variation process is that for given t, $\langle M \rangle(t)$ provides a systematic approach to estimating the variance of $M(t)$. To see this, note that $M^2(t)$ is a submartingale since

$$E[M^2(t)|\mathscr{F}_s] = E\{[M(t) - M(s)]^2 + M^2(s)|\mathscr{F}_s\}$$
$$\geq M^2(s) \qquad \text{for all } s \leq t,$$

where we have used the fact that $\mathscr{F}_s$ fixes the value of $M(s)$ so that $E[M^2(s)|\mathscr{F}_s] = M^2(s)$. Further, we find that

$$E[dM^2(t)|\mathscr{F}_{t^-}] = E[M^2(t^- + dt) - M^2(t^-)|\mathscr{F}_{t^-}]$$
$$= E\{[dM(t)]^2|\mathscr{F}_{t^-}\} = d\langle M \rangle(t).$$

It follows that the compensator of $M^2(t)$ is $\langle M \rangle(t)$, and from the Doob–Meyer decomposition theorem,

$$M^2(t) - \langle M \rangle(t)$$

is a mean zero martingale. Thus,

$$\text{var}[M(t)] = E[M^2(t)] = E[\langle M \rangle(t)] \tag{5.16}$$

and $\langle M \rangle(t)$ is an unbiased estimator of $\text{var}[M(t)]$. Usually, however, $\langle M \rangle(t)$ involves the parameters of the model, and in statistical problems the parameters in $\langle M \rangle(t)$ are replaced with estimates to obtain useful variance estimation formulas.

There is an alternative estimator of $\text{var}[M(t)]$ that in some problems is a function of observed quantities only. This is the *quadratic variation* or *optional variation process* $[M](t)$. In the types of martingales considered here, those based on discrete- or continuous-time counting processes,

$$[M](t) = \sum_{s \leq t}[\Delta M(s)]^2, \tag{5.17}$$

where $\Delta M(s) = M(s) - M(s^-)$ is the size of the jump in the process at s and the sum in (5.17) is over the points of discontinuity in M. It can be verified that $[M](t)$ also provides an unbiased estimator of $\text{var}[M(t)]$. We will typically use the predictable variation process rather than the optional variation process for purposes of variance estimation.

A result of considerable importance is the following. Suppose that $M(t)$ is a square-integrable martingale and $G(t)$ is a predictable process, both with respect to the filtration $\mathscr{F}_t$. Then, under quite general conditions on $G(t)$ [e.g., it is sufficient that $G(t)$ be bounded], the process $\{U(t), 0 \leq t \leq \tau\}$, where

$$U(t) = \int_0^t G(u)\,dM(u), \tag{5.18}$$

is a square-integrable martingale. This can be verified informally by noting that

$$E[dU(t)|\mathscr{F}_{t^-}] = E[G(t)\,dM(t)|\mathscr{F}_{t^-}] = G(t)E[dM(t)|\mathscr{F}_{t^-}] = 0,$$

where $G(t)$ can be taken outside the expectation since it is predictable and hence fixed by $\mathscr{F}_{t^-}$. The predictable variation process of $U(t)$ is

$$\langle U \rangle(t) = \int_0^t G^2(u)\,\mathrm{var}[dM(u)|\mathscr{F}_{u^-}] = \int_0^t G^2(u)\,d\langle M \rangle(u). \tag{5.19}$$

We have already used the idea of superposing martingales in Examples 2 and 4. We now consider this more formally. Suppose that M_i, $i = 1,\ldots,n$ are martingales with respect to the same filtration, $\mathscr{F}_t$. Then $M.(t) = \sum_{i=1}^n M_i(t)$ is a martingale with respect to $\mathscr{F}_t$. Its predictable variation process can be found, informally, through the argument:

$$\begin{aligned} d\langle M. \rangle(t) &= \mathrm{var}[dM.(t)|\mathscr{F}_{t^-}] \\ &= \sum \mathrm{var}[dM_i(t)|\mathscr{F}_{t^-}] + 2\sum_{i<j} \mathrm{cov}[dM_i(t), dM_j(t)|\mathscr{F}_{t^-}] \\ &= \sum d\langle M_i \rangle(t) + 2\sum_{i<j} d\langle M_i, M_j \rangle(t). \end{aligned} \tag{5.20}$$

Thus

$$\langle M. \rangle(t) = \sum \langle M_i \rangle(t) + 2\sum_{i<j} \langle M_i, M_j \rangle(t), \tag{5.21}$$

where $\langle M_i \rangle(t)$ is the predictable variation process of M_i and

$$\langle M_i, M_j \rangle(t) = \int_0^t \mathrm{cov}[dM_i(u), dM_j(u)|\mathscr{F}_{u^-}]$$

is called the *predictable covariation process* of M_i and M_j. If $\langle M_i, M_j \rangle(t) = 0$ for all t, M_i and M_j are said to be *orthogonal*. If the martingales M_i and M_j are orthogonal for all $i \neq j$,

$$\langle M. \rangle(t) = \sum_{i=1}^n \langle M_i \rangle(t). \tag{5.22}$$

It is interesting to note the similarity between formula (5.21) and the usual variance formula for the sum of random variables. This is, of course, no accident, given the interpretation in (5.20) in terms of the variances of the increments in $M.(t)$.

Finally, let M_i, $i = 1, \ldots, n$ be orthogonal square-integrable martingales and $G_i(t)$, $i = 1, \ldots, n$ be predictable processes with respect to the filtration $\mathscr{F}_t$. Combining the last two results, we see that

$$U.(t) = \sum U_i(t) = \sum \int_0^t G_i(t)\,dM_i(t)$$

is a martingale and its predictable variation process and optional variation processes are easily obtained. In particular, we find that

$$\langle U.\rangle(t) = \sum \int_0^t G_i^2(t)\,d\langle M_i\rangle(t)$$

with a similar expression for the optional variation process.

The notation used above may have, at first introduction, a rather foreign look to the statistical reader. In essence, however, $\langle M\rangle(t)$ and $[M](t)$ are for our purposes simply (unbiased) estimates of the variance of $M(t)$. It can also be noted that both of them may be functions of the parameters in the model, and typically, to obtain consistent estimators of the variance, we need to replace the parameters with consistent estimators. An alternative notation might use, for example, $V_U(t)$ for the predictable variation process of the martingale $U(t)$. The main advantage of the bracket notation $\langle U\rangle(t)$ is that it provides a simple mechanism for designating the particular process being considered, whereas the subscripts easily become cumbersome.

In the next two subsections, we give some martingale results for the particular cases of continuous- and discrete-time counting processes. These are among the main results used in this book.

5.3.2   Continuous-Time Counting Processes

Suppose that n individuals are on test and $\{N_i(t),\ 0 \le t \le \tau\}$ is a continuous-time counting process, as defined in Section 5.2, that counts the observed number of events in the interval $(0, t]$ on the ith subject. The filtration $\mathscr{F}_t$ is defined in (5.6) and the corresponding intensity process is given in (5.7). For each i, the process

$$M_i(t) = N_i(t) - \int_0^t Y_i(t)\lambda_i(t)\,dt$$

is a mean zero martingale with respect to the filtration $\mathscr{F}_t$. The corresponding predictable variation process (5.15) is

$$\langle M_i\rangle(t) = \int_0^t Y_i(u)\lambda_i(u)\,du \qquad\qquad (5.23)$$

162 COUNTING PROCESSES AND ASYMPTOTIC THEORY

since $\text{var}[dM_i(u)|\mathscr{F}_{u-}] = E[dN_i(u)|\mathscr{F}_{u-}] = Y_i(u)\lambda_i(u)\,du,\ 0 \le u \le \tau$. The optional variation process (5.17) is

$$[M_i](t) = N_i(t). \tag{5.24}$$

Both $\langle M_i\rangle(t)$ and $[M_i](t)$ provide unbiased estimates of $\text{var}[M_i(t)]$, but only the latter is a function of the data only.

If $G_i(t)$ is a predictable process with respect to $\mathscr{F}_t$, then $U_i(t) = \int_0^t G_i(u)\,dM_i(u)$ is a martingale with predictable variation process

$$\langle U_i\rangle(t) = \int_0^t G_i^2(u)Y_i(u)\lambda_i(u)\,du. \tag{5.25}$$

The optional variation process is

$$[U_i](t) = \int_0^t G^2(s)\,dN_i(s) = \sum_{s \le t} G^2(s)\,\Delta N_i(s). \tag{5.26}$$

It is assumed that no two of the continuous-time counting processes, $N_1(t),\ldots,N_n(t)$ can jump at the same time. As a consequence, for all $i \ne j$, and $t \in [0,\tau]$,

$$d\langle M_i,M_j\rangle(t) = \text{cov}[dM_i(t),dM_j(t)|\mathscr{F}_{t-}] = E[dN_i(t)\,dN_j(t)|\mathscr{F}_{t-}] = 0,$$

so that M_i and M_j are orthogonal. It then follows from (5.15) that the superposed martingale

$$M.(t) = \sum_{i=1}^n [N_i(t) - \int_0^t Y_i(u)\lambda_i(u)\,du]$$

has the predictable variation process

$$\langle M.\rangle(t) = \sum_{i=1}^n \int_0^t Y_i(u)\lambda_i(u)\,du. \tag{5.27}$$

The optional variation process is

$$[M.](t) = \sum N_i(t) = N.(t). \tag{5.28}$$

The corresponding results for the martingale $U.(t) = \sum_{i=1}^n \int_0^t G_i(t)\,dM_i(t)$, where for each i, $G_i(t)$ is a predictable process, are

$$\langle U.\rangle(t) = \sum_{i=1}^n \int_0^t G_i^2(u)Y_i(u)\lambda_i(u)\,du \tag{5.29}$$

and

$$[U.](t) = \sum_{i=1}^{n} \int_0^t G_i^2(u)\, dN_i(u). \tag{5.30}$$

5.3.3 Discrete-Time Counting Processes

Suppose again that $N_i(t)$ counts the number of observed failures (or events) for the ith individual in $(0, t]$. We consider the case in which events in the ith process occur only one at a time, so that jumps in the process N_i are of size $+1$ only. The process is, however, in discrete time, so that

$$P[dN_i(t) = 1|\mathcal{F}_{t-}] = E[dN_i(t)|\mathcal{F}_{t-}] = Y_i(t)\, d\Lambda_i(t),$$

where the underlying cumulative intensity function of the ith process is a step function with jumps at discrete time points $a_1 < a_2 < \cdots$ and

$$\Lambda_i(t) = \sum_{a_\ell \le t} \lambda_{il},$$

where the λ_{il} are the discrete hazard probabilities. The compensator of N_i is $\int_0^t Y_i(u)\, d\Lambda_i(u)$, which is a right-continuous step function with a jump of size λ_{il} at a_ℓ provided that the item is at risk at a_ℓ; that is, provided that $Y_i(a_\ell) = 1$. The corresponding martingale is

$$M_i(t) = N_i(t) - \int_0^t Y_i(u)\, d\Lambda_i(u)$$

and the predictable variation process is

$$\begin{aligned}
\langle M_i \rangle(t) &= \int_0^t Y_i(u)[1 - \Delta\Lambda_i(t)]\, d\Lambda_i(t) \\
&= \sum_{a_\ell \le t} Y_i(a_\ell)(1 - \lambda_{il})\lambda_{il}.
\end{aligned} \tag{5.31}$$

Note that the increment $d\langle M_i \rangle(t)$ at a_ℓ corresponds to $\mathrm{var}[dN_i(a_\ell)|\mathcal{F}_{t-}] = Y_i(a_\ell)$ $(1 - \lambda_{il})\lambda_{il}$, which if $Y_i(a_l) = 1$ is the variance of a Bernoulli random variable. Thus, as long as $Y_i(t) = 1$, the failure time process for the ith individual is being viewed sequentially as a sequence of Bernoulli trials.

If $G_i(t)$ is a predictable process, the predictable variation process of $U_i(t) = \int_0^t G_i(u)\, dM_i(u)$ is

$$\begin{aligned}
\langle U_i \rangle(t) &= \int_0^t G_i^2(u)[1 - \Delta\Lambda_i(u)]\, d\Lambda_i(u) \\
&= \sum_{a_\ell \le t} Y_i(a_\ell)G_i^2(a_\ell)(1 - \lambda_{il})\lambda_{il}.
\end{aligned} \tag{5.32}$$

which also follows immediately from thinking in terms of a sequence of Bernoulli trials.

The superposed process, $N.(t) = \sum_{i=1}^{n} N_i(t)$, may have jumps of more than 1 unit at the fixed mass points a_l. It is a submartingale, its compensator is the sum of the individual compensators, and thus $M.(t) = N.(t) - \sum_{i=1}^{n} \int_0^t Y_i(u)\, d\Lambda_i(u)$ is a martingale with predictable variation process

$$\langle M.\rangle(t) = \sum \langle M_i\rangle(t) + 2\sum_{i<j} \langle M_i, M_j\rangle(t). \tag{5.33}$$

If it is further assumed that at each time t, the individuals at risk are independent with respect to the failure mechanism operating at time t, then

$$d\langle M_i, M_j\rangle(t) = \mathrm{cov}[dN_i(t), dN_j(t)|\mathscr{F}_{t^-}] = E[dM_i(t)dM_j(t)|\mathscr{F}_{t^-}] = 0$$

for all t and $i \neq j$. In this case, the component martingales are orthogonal, and the predictable variation process for the martingale $U.(t) = \sum_{i=1}^{n} U_i(t)$ is

$$\langle U.\rangle(t) = \sum_{i=1}^{n} \int_0^t Y_i(u)G_i^2(u)[1 - \Delta\Lambda_i(u)]\, d\Lambda_i(u). \tag{5.34}$$

This expression (5.34) is sufficiently general to handle discrete, continuous, and mixed cases. In the discrete case, it reduces to

$$\langle U.\rangle(t) = \sum_{i=1}^{n} \sum_{a_\ell \leq t} Y_i(a_\ell)G_i^2(a_\ell)(1 - \lambda_{il})\lambda_{il},$$

where again the underlying Bernoulli structure can be seen.

5.4 VECTOR-VALUED MARTINGALES

Let $U(t) = [U_1(t), \ldots, U_k(t)]'$ be a vector-valued martingale with respect to the filtration $\mathscr{F}_t$. The corresponding predictable variation process $\langle U\rangle(t)$ is the $k \times k$ matrix-valued function

$$\langle U\rangle(t) = \int_0^t E[dU(t)\, dU(t)'|\mathscr{F}_{t^-}], \qquad t > 0,$$

the (i,j)th element of which is the predictable covariation process $\langle U_i, U_j\rangle(t)$.

It is worth mentioning one important way in which such vector processes arise. Suppose that $G_i(u) = [G_{1i}(u), \ldots, G_{ki}(u)]'$ is a $k \times 1$ vector of predictable processes with respect to the filtration $\mathscr{F}_t$. Suppose also that $M_1, \ldots, M_n$ are orthogonal mean zero martingales. Then

$$U(t) = \sum_{i=1}^{n} \int_0^t G_i(u) \, dM_i(u)$$

is a vector-valued martingale. It is left as an exercise to show that its predictable variation process is

$$\langle U \rangle(t) = \sum_{i=1}^{n} \int_0^t G_i(u) G_i(u)' \, d\langle M_i \rangle(u). \tag{5.35}$$

5.5 MARTINGALE CENTRAL LIMIT THEOREM

The central limit theorem due to Rebolledo (1980) is the main result useful for asymptotics in the applications in this book. This theorem drew together earlier work on central limit theorems for discrete martingales and gave a general version applicable to martingales arising from counting processes in discrete time, continuous time, or a mixture of the two. We write the theorem in terms of its application to martingales formed as stochastic integrals of basic processes arising, for example, from individuals or groups of individuals in a sample of size n, and consider the limit as $n \to \infty$.

Let $(M_1^{(n)}(t), \ldots, M_{r_n}^{(n)}(t))$, $t \in [0, \tau]$ be martingales with respect to a filtration $\mathscr{F}_t^{(n)}$ which may depend on n. In this, r_n is an integer which in some examples is a constant $r_n = r$ for all n; in other instances, r_n is increasing with n and in many applications, $r_n = n$. Let $G_{ji}^{(n)}(t)$, $j = 1, \ldots, k$; $i = 1, \ldots, r_n$ be predictable processes and let $U^{(n)} = (U_1^{(n)}, \ldots, U_k^{(n)})'$ be a vector of k martingales where

$$U_j^{(n)}(t) = \sum_{i=1}^{r_n} \int_0^t G_{ji}^{(n)}(s) \, dM_i^{(n)}(s)$$

for $j = 1, \ldots, k$. We consider conditions under which the distribution of $U^{(n)}$ approaches a normal limit as $n \to \infty$.

For given $\epsilon > 0$, we define a vector of related processes,

$$U_{\epsilon j}^{(n)}(t) = \sum_{i=1}^{r_n} \int_0^t G_{ji}^{(n)}(u) \mathbf{1}(|G_{ji}^{(n)}(u)| > \epsilon) \, dM_i^{(n)}(u), \tag{5.36}$$

$j = 1, \ldots, k$, $t \in [0, \tau]$, which registers all jumps of size ϵ or more in the original component processes $U_j^{(n)}$. Note that since the integrand of (5.36) is again a predictable process, $U_{\epsilon j}^{(n)}$ is a martingale with predictable variation process $\langle U_{\epsilon j}^{(n)} \rangle(t)$.

For a central limit theorem to apply to $U^{(n)}(t)$, we require essentially two things. The first is that the covariance of $U^{(n)}(t)$ approaches a limit as $n \to \infty$, and the second is that the processes $U_{\epsilon j}^{(n)}(t)$ approaches zero for all $\epsilon > 0$. The first condition requires that the G_{ji} functions must be appropriately standardized. The second condition is a Lindeberg-type condition which essentially guarantees that the influence of any single process is negligible in the limit.

Let $V(t)$ be a $k \times k$ positive semidefinite matrix on the interval $[0, \tau]$ where $V(0) = 0$ and $V(t) - V(s)$ is positive semidefinite for all s, t satisfying $0 \leq s \leq t \leq \tau$. Suppose further that $V(t)$ is right continuous. We are now in a position to state the main result:

Theorem 5.1 (Rebolledo's Theorem).   Let t be a fixed time in $[0, \tau]$ and consider the conditions:

(a) $\langle U^{(n)} \rangle(t) \overset{\mathscr{P}}{\to} V(t)$ as $n \to \infty$.

(b) $[U^{(n)}](t) \overset{\mathscr{P}}{\to} V(t)$ as $n \to \infty$.

(c) $\langle U_{\epsilon j}^{(n)} \rangle(t) \overset{\mathscr{P}}{\to} 0$ as $n \to \infty$, for all $j = 1, \ldots, k$ and $\epsilon > 0$.

Then either (a) or (b) together with (c) imply that $U^{(n)}(t) \overset{\mathscr{D}}{\to} N(0, V(t))$.

As stated, the theorem is directed toward the kind of result we typically wish to establish in this book, where interest centers on the convergence of some test statistic (e.g., a score statistic) or an estimator at a specified value of t; often, $t = \tau$, which corresponds to the totality of the data.

More general results are also available that involve the joint distribution of $[U^{(n)}(t_1), \ldots, U^{(n)}(t_n)]$ at given points $0 < t_1 < t_2 < \cdots < t_n \leq \tau$ or the convergence of the entire process over the specified interval. The following is a more general version of Rebolledo's theorem, which includes Theorem 5.1 as the simplest special case.

Theorem 5.2.   Extend conditions (a), (b), and (c) in Theorem 5.1 so that the convergence holds uniformly for all points $t \in K$ where $K \subseteq (0, \tau]$. If the resulting (c) together with either (a) or (b) holds and $t_1 < \cdots < t_r$ are r points in K, then

$$[U^{(n)}(t_1)', \ldots, U^{(n)}(t_r)']' \overset{\mathscr{D}}{\to} N(0, \Sigma), \qquad (5.37)$$

where Σ is a matrix of dimension $(kr) \times (kr)$ comprised of blocks of dimension $k \times k$. Thus

$$\Sigma = \begin{pmatrix} V(t_1) & V(t_1) & \cdots & V(t_1) \\ V(t_1) & V(t_2) & \cdots & V(t_2) \\ \vdots & \vdots & & \vdots \\ \vdots & \vdots & & \vdots \\ V(t_1) & V(t_2) & \cdots & V(t_r) \end{pmatrix}.$$

Further, if $K = (0, \tau]$, then $U^{(n)}$ converges weakly on K to a k-variate Gaussian martingale with covariance function $V(t)$. That is it converges to a Gaussian process with mean 0 and covariance function determined by

$$\text{cov}[U^{(\infty)}(s), U^{(\infty)}(t)] = V(s \wedge t).$$

Note that the weak convergence to the Gaussian process implies that all finite distributions converge as in (5.37). It also implies, however, that random variables, such as the supremum of the process $U^{(n)}$ on $(0, \tau]$, converge in distribution to the corresponding supremum of the limiting Gaussian process. A summary of weak convergence results can be found in Fleming and Harrington (1991, App. B).

Theorems 5.1 and 5.2 as stated apply to both discrete- and continuous-time martingales and so have quite a broad range of applicability. In the next two sections we discuss the use of this theorem to establish asymptotic results for many of the methods discussed previously.

A further result, which is of value in establishing asymptotic results, is *Lenglart's inequality* (Lenglart, 1977). Let $\bar{M}(t)$ be a submartingale on $[0, \tau]$ and suppose that its compensator $A(t)$ is nondecreasing. Then, for any $\eta > 0$ and $\delta > 0$,

$$P\left[\sup_{[0,\tau]} \bar{M}(t) > \eta\right] \leq \frac{\delta}{\eta} + P[A(\tau) > \delta]. \tag{5.38}$$

This states, in effect, that one can bound the probability of a large value of the submartingale $\bar{M}$ over the interval $[0, \tau]$ in terms of the probability of a large value of its compensator at the terminus τ of the interval.

One particular case of importance is $\bar{M}(t) = [M(t)]^2$, where $M(t)$ is a martingale. As noted earlier, this is a submartingale, and the corresponding compensator is the predictable variation process $\langle M \rangle(t)$. Thus, for any martingale $M(t)$ with nondecreasing $\langle M \rangle(t)$, (5.38) implies that

$$P\left[\sup_{[0,\tau]} |M(t)| > \sqrt{\eta}\right] \leq \frac{\delta}{\eta} + P[\langle M \rangle(\tau) > \delta]. \tag{5.39}$$

This result (5.39) is often applied when $X^{(n)}$ is a submartingale indexed by sample size n and $M^{(n)} = X^{(n)} - \tilde{X}^{(n)}$ is the corresponding martingale. If $\langle M^{(n)} \rangle(t)$ is nondecreasing in t for each n and $\sup_{[0,\tau]} |\langle M^{(n)} \rangle(t)| \xrightarrow{\mathscr{P}} 0$ as $n \rightarrow \infty$, it follows that $\sup_{[0,\tau]} |M^{(n)}(t)| \xrightarrow{\mathscr{P}} 0$. If, in addition, $\tilde{X}^{(n)}(\tau) \xrightarrow{\mathscr{P}} c$ for some constant c, then $X^{(n)}(\tau) \xrightarrow{\mathscr{P}} c$.

5.6 ASYMPTOTICS ASSOCIATED WITH CHAPTER 1

5.6.1 Nelson–Aalen and Kaplan–Meier Estimators

Suppose that n individuals having independent failure times are put on study at time 0 and data are collected over a finite interval $[0, \tau]$. We consider estimation of their common survivor or cumulative hazard functions, $F(t)$ or $\Lambda(t)$. For the purpose of asymptotic results, we assume that τ is fixed as the sample size $n \to \infty$. The processes $N_i, N., Y_i,$ and $Y.$ are defined as in Section 5.2 and $\mathscr{F}_t = \{N_i(u), Y_i(u^+), i = 1, \ldots, n, 0 \le t\}$. The censoring process is assumed independent so that

$$P[dN_i(t)|\mathscr{F}_{t-}] = P[dN_i(t)|Y_i(t)] = Y_i(t)\, d\Lambda(t)$$

for all $t \in [0, \tau]$ and $i = 1, \ldots, n$.

For convenience, define $J(u) = \mathbf{1}[Y.(u) > 0]$ and adopt the convention that $0/0$ is interpreted as 0. The Nelson–Aalen estimator of the cumulative hazard can then be written

$$\hat{\Lambda}(t) = \int_0^t \frac{J(u)}{Y.(u)}\, dN.(u), \qquad 0 \le t \le \tau. \tag{5.40}$$

The Kaplan–Meier estimator is

$$\hat{F}(t) = \mathscr{P}_0^t \left[1 - \frac{J(u)}{Y.(u)}\, dN.(u)\right] = \mathscr{P}_0^t [1 - d\hat{\Lambda}(u)] \tag{5.41}$$

for $0 \le t \le \tau$. If $t_{\max}$, the largest observed time, is a censoring time and less than τ, these estimators are still defined over the entire interval $[0, \tau]$ with $\hat{\Lambda}(t) = \hat{\Lambda}(t_{\max})$ and $\hat{F}(t) = \hat{F}(t_{\max})$ for $t_{\max} \le t \le \tau$. (We had before taken the estimators as undefined for $t > t_{\max}$.)

The counting process $N_i(t)$ has compensator $\int_0^t Y_i(u)\, d\Lambda(u)$. Thus, $M.(t) = N.(t) - \sum_{i=1}^n \int_0^t Y_i(u)\, d\Lambda(u)$ is a martingale with respect to the filtration $\mathscr{F}_t$. Since $J(t)[Y.(t)]^{-1}$ is a predictable process, it follows that

$$\hat{\Lambda}(t) - \Lambda^*(t) = \int_0^t \frac{J(u)}{Y.(u)}\, dM.(u),$$

where $\Lambda^*(t) = \int_0^t J(u)\, d\Lambda(u)$, is a mean 0 martingale. In considering asymptotic results, we will consider situations in which $\Lambda^*(t)$ approaches $\Lambda(t)$ for all t as $n \to \infty$.

A variance estimator of $\hat{\Lambda}(t) - \Lambda^*(t)$ can be obtained from its predictable variation process

$$\langle \hat{\Lambda} - \Lambda^* \rangle(t) = \int_0^t \left[\frac{J(u)}{Y.(u)}\right]^2 d\langle M. \rangle(u).$$

In the continuous case, $d\langle M. \rangle(u) = Y.(u)\, d\Lambda(u)$. Making this substitution and replacing $\Lambda(u)$ with $\hat{\Lambda}(u)$ gives a variance estimator appropriate for the continuous case:

$$\hat{V}(t) = \int_0^t \frac{J(u)}{[Y.(u)]^2}\, dN.(u). \tag{5.42}$$

If $\Lambda(t)$ is the cumulative hazard of a discrete failure time variable, then from (5.32), the corresponding predictable variation process is

$$V(t) = \langle \hat{\Lambda} - \Lambda^* \rangle(t) = \int_0^t \frac{J(u)}{[Y.(u)]^2} Y.(u)[1 - \Delta\Lambda(u)] \, d\Lambda(u).$$

Replacing $\Lambda(u)$ with its estimate gives the variance estimate

$$\hat{V}(t) = \int_0^t \frac{J(u)}{[Y.(u)]^2} \left[1 - \frac{\Delta N.(u)}{Y.(u)}\right] dN.(u), \tag{5.43}$$

which is also a valid estimate in the continuous case.

Large-sample properties of the Nelson–Aalen estimator can be shown to hold under relatively mild conditions and we give a brief outline here. Consider a sequence of experiments indexed by the sample size n and define corresponding processes $N.^{(n)}$, $Y.^{(n)}$, $J^{(n)}$, $\hat{\Lambda}^{(n)}$, $\Lambda^{*(n)}$, $V^{(n)}$, and $\hat{V}^{(n)}$ for each n. It is first established that the estimator is uniformly consistent on the interval $[0, \tau]$. For this purpose, it is sufficient to assume that $\inf_{t\in[0,\tau]} Y.^{(n)}(t) \xrightarrow{\mathscr{P}} \infty$ as $n \to \infty$. This simply guarantees that the number of individuals at risk at each time point becomes large. It is then easy to show that $\sup|\Lambda^*(t) - \Lambda(t)| \xrightarrow{\mathscr{P}} 0$ and it follows that

$$\sup_{s\in[0,\tau]} |\hat{\Lambda}^{(n)}(s) - \Lambda(s)| \xrightarrow{\mathscr{P}} 0.$$

With some additional assumptions, Theorem 5.1 can be applied. Suppose that $n^{-1}Y.^{(n)}(t)$ approaches a function $y(t)$ as n becomes large. More formally, it is assumed that $\sup_{t\in[0,\tau]}|n^{-1}Y.^{(n)} - y(t)| \xrightarrow{\mathscr{P}} 0$, where $y(t)$ satisfies

$$v(\tau) = \int_0^\tau \frac{1}{y(u)} [1 - \Delta\Lambda(u)] \, d\Lambda(u) < \infty,$$

which holds, for example, if $\inf_{t\in[0,\tau]} y(t) > 0$. This is sufficient to establish that for each $t \in [0, \tau]$:

(a) $nV^{(n)}(t) = n\int_0^t \frac{J^{(n)}(u)}{Y.^{(n)}(u)} [1 - \Delta\Lambda(u)] \, d\Lambda(u) \xrightarrow{\mathscr{P}} v(t).$

(b) For each $\epsilon > 0$,

$$n\int_0^t \frac{J^{(n)}(u)}{Y.^{(n)}(u)} I\left[\left| n^{1/2} \frac{J^{(n)}(u)}{Y.^{(n)}(u)} \right| > \epsilon \right] [1 - \Delta\Lambda(u)] \, d\Lambda(u) \xrightarrow{\mathscr{P}} 0.$$

These are the conditions needed for Theorem 5.1. It follows that for any fixed $t \in [0, \tau]$,

$$n^{1/2}[\hat{\Lambda}^{(n)}(t) - \Lambda(t)] \xrightarrow{\mathscr{D}} N(0, v(t)) \tag{5.44}$$

as $n \to \infty$. It also follows that $v(t)$ in (5.44) can be replaced by the consistent estimator $n\hat{V}^{(n)}(t)$ from (5.43). This establishes the result, less formally stated, that

$$\hat{\Lambda}(t) \sim N(\Lambda(t), \hat{V}(t)).$$

Stronger results are also available here. In particular, under the stated conditions, it can be shown that $n^{1/2}[\hat{\Lambda}(t) - \Lambda(t)]$ converges weakly on $[0, \tau]$ to a mean zero Gaussian process with covariance function $V(s \wedge t)$.

Asymptotic properties of the Kaplan–Meier estimator can also be obtained. It can be shown (see Exercise 5.8) that

$$\hat{G}(t) = \frac{\hat{F}(t)}{F^*(t)} - 1 = \int_0^t \frac{\hat{F}(s^-)J(s)}{F^*(s)Y.(s)} \, dM.(s), \tag{5.45}$$

where $F^*(t) = \mathscr{P}_0^t[1 - J(u)\,d\Lambda(u)] = \mathscr{P}_0^t[1 - d\Lambda^*(u)]$. Since the integrand on the right side of (5.45) is a predictable process, $\hat{G}(t)$ is a martingale. A variance estimate can again be obtained from the predictable variation process

$$V_G(t) = \langle \hat{G} \rangle(t) = \int_0^t \left[\frac{\hat{F}(s^-)J(s)}{F^*(s)Y.(s)}\right]^2 Y.(s)[1 - \Delta\Lambda(s)]\,d\Lambda(s).$$

Replacing $\Lambda(u)$ with its estimate, estimating $F^*(s)$ with $\hat{F}(s)$, and multiplying by $\hat{F}(t)^2$, we obtain the Greenwood estimate (1.14)

$$\hat{V}_F(t) = \text{vâr}[\hat{F}(t)] = [\hat{F}(t)]^2 \int_0^t \frac{1}{Y.(s)[Y.(s) - \Delta N.(s)]} \, dN.(s)$$

for the variance of $\hat{F}(t)$.

Uniform consistency of $\hat{F}(t)$ for $F(t)$ over the interval $[0, \tau]$ follows under the same conditions stated for the consistency of the Nelson–Aalen estimator. Under these conditions, Theorem 5.1 can be used to show that for any fixed $t \in [0, \tau]$,

$$[V_G(t)]^{-1/2}\,\hat{G}(t) = \frac{\hat{F}(t) - F^*(t)}{[V_G(t)]^{1/2}F^*(t)} \xrightarrow{\mathscr{D}} N(0, 1).$$

Since $\sup |F^*(t) - F(t)| \xrightarrow{\mathscr{P}} 0$ and $n\hat{V}_F(t) \xrightarrow{\mathscr{P}} n\,F(t)^2 V_G(t)$, it follows that

$$[\hat{V}_F(t)]^{-1/2}[\hat{F}(t) - F(t)] \xrightarrow{\mathscr{D}} N(0, 1)$$

for each fixed $t \in [0, \tau]$. This forms the basis for the approximate confidence intervals obtained in Section 1.4 for $F(t)$ or, using the delta method, for $\log[-\log F(t)]$. As with the Nelson–Aalen estimator, it can also be shown that the process $n^{1/2}[\hat{F}(t) - F(t)]$ converges weakly to a Gaussian process with covariance function estimated by $n\hat{V}_F(s \wedge t)$.

The discussion above has concentrated on the Greenwood estimate of variance, but there are other consistent estimates that might be used. For example, if there are no ties in the data, so that $N.(s)$ increases only in steps of size 1, it would be possible and perhaps natural to use the estimate of variance, analogous to (5.42), which arises from viewing $M.(s)$ as a martingale in continuous time. The Greenwood estimate, however, appears to do at least as well as this and other alternatives in simulations, even with continuous data, and has the advantage of providing a natural bridge between the discrete and continuous cases.

5.6.2 Log-Rank and Related Tests

Following Example 5.4 and Section 1.6, let $N_{ij}(t)$ and $Y_{ij}(t)$ be the counting process of observed failures and the at-risk process for the jth individual in the ith group, $j = 1, \ldots, n_i;\ i = 0, \ldots, p$. As before, let $N_{i.}(t) = \sum N_{ij}(t)$ record the number of observed failures in $(0, t]$ and $Y_{i.}(t) = \sum Y_{ij}(t)$ specify the number at risk at time t in the ith group. The filtration $\mathscr{F}_t = \{N_{ij}(u), Y_{ij}(u^+), j = 1, \ldots, n_i;\ i = 0, \ldots p;\ 0 \le u \le t\}$ is as defined in Example 5.4. Under independent censoring, the model can be written

$$P[dN_{ij}(t) = 1 | \mathscr{F}_{t-}] = Y_{ij}(t)\, d\Lambda_i(t) \qquad (5.46)$$

for all i, j and $t \in [0, \tau]$, where $\Lambda_i(t)$ is the continuous, discrete, or mixed underlying cumulative hazard that applies to individuals in the ith group.

As before, we restrict attention to the closed interval $[0, \tau]$ and examine the accumulated logrank statistic $w_i(\tau),\ i = 0, \ldots, p$ at time τ. From (1.21) and some algebra, it can be verified that for each $i = 0, 1, \ldots, p$,

$$
\begin{aligned}
w_i(\tau) &= \sum_{\ell=0}^{p} \int_0^\tau \left[\delta_{i\ell} - \frac{Y_{i.}(u)}{Y_{..}(u)} \right] dN_{\ell.}(u) \\
&= \sum_{\ell=0}^{p} \int_0^\tau \left[\delta_{i\ell} - \frac{Y_{i.}(u)}{Y_{..}(u)} \right] dM_{\ell.}(u),
\end{aligned} \qquad (5.47)
$$

where $\delta_{i\ell} = \mathbf{1}(i = \ell)$ and $M_{\ell.}(t) = N_{\ell.}(t) - \int_0^t Y_{\ell.}(u)\, d\Lambda(u)$ is a martingale under the null hypothesis $\Lambda_\ell = \Lambda,\ \ell = 0, \ldots, p$. Now, (5.47) expresses the logrank statistic as the integral of a predictable process with respect to a martingale. Under the null hypothesis, $\Lambda_l = \Lambda$, the predictable variation and covariation processes with Λ_i replaced by the common estimator $\hat{\Lambda}$ where $d\hat{\Lambda}(t) = \Delta\hat{\Lambda}(t) = [Y_{..}(t)]^{-1}\, dN_{..}(t)$ provide estimates of $\text{var}[w_i(\tau)]$ and $\text{cov}[w_i(\tau), w_\ell(\tau)]$. The result for $\text{var}[w_i(\tau)]$ is

$$W_{ii}(\tau) = \sum_{\ell=0}^{p} \int_0^\tau \left[\delta_{i\ell} - \frac{Y_{i.}(u)}{Y_{..}(u)} \right]^2 [1 - \Delta\hat{\Lambda}(u)]\, d\hat{\Lambda}(u),$$

which is applicable to discrete or continuous cases. With $\tau = \infty$, this reduces to

$$\sum_{j=1}^{k} n_{ij}(n_j - n_{ij})(n_j - d_j)d_j/n_j^3$$

in the notation of Section 1.5. The jth term in this expression differs from the hypergeometric expression in (1.19) by the factor $(n_j - 1)/n_j$. The covariance term arising here can also be seen to be that in (1.20) except for the factor $(n_j - 1)/n_j$. Since $\sum_{i=0}^{p} w_i(\tau) = 0$, let $w(\tau) = [w_1(\tau), \dots, w_p(\tau)]'$, and $W(\tau) = (W_{ij}(\tau))_{p \times p}$. The logrank statistic for testing equality of hazard rates over the interval $[0, \tau]$ across the $p + 1$ groups is

$$w(\tau)'[W(\tau)]^{-1}w(\tau). \tag{5.48}$$

Consider a sequence of experiments indexed by the total sample size $n = \sum n_i$. The conditions for Theorem 5.1 and the asymptotic normality of $w(\tau)$ follow if $n^{-1}W(\tau)$ converges in probability to a limiting nonsingular covariance matrix Σ as $n \to \infty$. The asymptotic normality of $n^{-1/2}w(\tau) \to N_p(0, \Sigma)$ then follows and the asymptotic χ_p^2 distribution of the quadratic form (5.48) is an immediate consequence.

Since the log-rank and weighted log-rank tests can be obtained as score tests based on the partial likelihood in the relative risk or Cox model, their asymptotic properties also follow directly from the results of Section 5.7.

5.7 ASYMPTOTIC RESULTS FOR THE COX MODEL

The asymptotic results for counting processes can also be applied to the partial likelihood results of Chapter 4 for right censored data from the Cox model. As before, let $N_i(t)$ be the right continuous counting process for the number of observed failures on $(0, t]$ and $Y_i(t)$ be the at-risk process for the ith individual. As discussed earlier, we extend the basic covariates to include measured time-dependent covariates $x_i(t)$. The processes $Y_i(t)$ is left continuous, and the filtration or history process is $\mathscr{F}_t = \{N_i(u), Y_i(u^+), X_i(u^+) : i = 1, \dots, n; 0 \le u \le t\}$, where $X_i(t) = \{x_i(u), 0 \le u < t\}$. Under independent censoring, the Cox model gives

$$P[dN_i(t) = 1 | \mathscr{F}_{t^-}] = Y_i(t)\lambda_i(t) = Y_i(t)\exp[Z_i(t)'\beta]\lambda_0(t)\,dt, \tag{5.49}$$

where $Z_i(t)$ is a vector of derived predictable covariates that are functions of $X_i(t)$ and t. Note that $\lambda_i(t) = \exp[Z_i(t)'\beta]\lambda_0(t)$ is the intensity or hazard function for the underlying uncensored counting process $\tilde{N}_i(t)$.

Consider the case of no ties, so that the partial likelihood is given by (4.14), and the corresponding score and observed information matrices are (4.15) and (4.17).

We consider the evolution of the partial likelihood or its score over a finite interval $[0, \tau]$. The score function based on data available up to a specified time $t \in [0, \tau]$ is

$$U(\beta, t) = \sum_{i=1}^{n} \int_0^t [Z_i(u) - \mathscr{E}(\beta, u)] \, dN_i(u), \qquad (5.50)$$

where, as before,

$$\mathscr{E}(\beta, u) = \sum_{\ell=1}^{n} Z_\ell(u) p_\ell(\beta, u)$$

and

$$p_\ell(\beta, u) = \frac{Y_\ell(u) \exp[Z_\ell(u)'\beta]}{\sum_{j=1}^{n} Y_j(u) \exp[Z_j(u)'\beta]}.$$

From (5.49), the compensator of $N_i(t)$ is

$$A_i(t) = \int_0^t Y_i(u) \lambda_0(u) \exp[Z_i(u)'\beta] \, du$$

and

$$M_i(t) = N_i(t) - A_i(t)$$

is a mean 0 martingale with respect to $\mathscr{F}_t$. Since $\int_0^t [Z_i(u) - \mathscr{E}(\beta, u)] \, dA_i(u) = 0$, it follows that (5.50) can be rewritten as

$$U(\beta, t) = \sum_{i=1}^{n} \int_0^t [Z_i(u) - \mathscr{E}(\beta, u)] \, dM_i(u). \qquad (5.51)$$

The ith term in this sum is a stochastic integral of a predictable vector process with respect to a martingale. Thus, U is itself a mean 0 vector-valued martingale with respect to $\mathscr{F}_t$. Since $M_i, \ldots, M_n$ are orthogonal martingales it follows from (5.35) that the corresponding predictable variation process is

$$\langle U(\beta) \rangle(t) = \sum_{i=1}^{n} \int_0^t nG_i(u)G_i(u)' \, \mathrm{var}[dM_i(u)|\mathscr{F}_{u^-}]$$

$$= \sum_{i=1}^{n} \int_0^t nG_i(u)G_i(u)' \, Y_i(u) \exp[Z_i(u)'\beta]\lambda_0(u) \, du, \qquad (5.52)$$

where $G_i(u) = n^{-1/2}[Z_i(u) - \mathscr{E}(\beta, u)]$.

The superposed counting process $N.(t) = \sum_{i=1}^{n} N_i(t)$ gives rise to the martingale

$$M.(t) = N.(t) - \int_0^t \sum_{i=1}^{n} Y_i(t) \exp[Z_i(t)'\beta]\lambda_0(u) \, du. \tag{5.53}$$

The Nelson–Aalen estimator is $\hat{\Lambda}_0(t) = \hat{\Lambda}_0(\hat{\beta}, t)$, where

$$d\hat{\Lambda}_0(\beta, t) = dN.(t) \bigg/ \sum_{i=1}^{n} Y_i(t) \exp[Z_i(t)'\beta] \tag{5.54}$$

arises as the moment estimate obtained by setting $M.(t) = 0$ in (5.53).

Finally, if $\lambda_0(u) \, du$ is replaced with $d\hat{\Lambda}_0(\beta, u)$ in (5.52), we obtain the observed information, $I(\beta, t) = -\partial U(\beta, t)/\partial\beta'$, on data up to time t. That is,

$$\begin{aligned}
I(\beta, t) &= \sum_{i=1}^{n} \int_0^t nG_i(u)G_i(u)' \, Y_i(u) \exp[Z_i(u)'\beta] \, d\hat{\Lambda}_0(\beta, u) \\
&= \int_0^t \mathcal{V}(\beta, u) dN.(u),
\end{aligned} \tag{5.55}$$

where

$$\mathcal{V}(\beta, u) = \sum_{i=1}^{n} [Z_i(u) - \mathcal{E}(\beta, u)]^{\otimes 2} p_i(\beta, u).$$

Let

$$\begin{aligned}
S^{(0)}(\beta, t) &= \sum_{i=1}^{n} Y_i(t) \exp[Z_i(t)'\beta], \\
S^{(1)}(\beta, t) &= \frac{\partial S^{(0)}(\beta, t)}{\partial\beta} = \sum_{i=1}^{n} Y_i(t)Z_i(t) \exp[Z_i(t)'\beta], \\
S^{(2)}(\beta, t) &= \frac{\partial^2 S^{(0)}(\beta, t)}{\partial\beta \, \partial\beta'} = \sum_{i=1}^{n} Y_i(t)Z_i(t)Z_i(t)' \exp[Z_i(t)'\beta].
\end{aligned} \tag{5.56}$$

Also, let $||A||$ be the largest absolute value of the entries in the scalar, vector, or matrix A; that is, for example, $||A|| = \max_{i,j}|a_{ij}|$ when A is a matrix. We denote the true value of β by β_0. The following conditions are adapted slightly from Andersen et al. (1993, p. 487).

C1. There exists an open neighborhood $\mathcal{B}$ of β_0 and functions $s^{(j)}(\beta, t), j = 0, 1, 2$ defined on $\mathcal{B} \times [0, \tau]$ which satisfy the following:

(a) $\sup_{\beta \in \mathscr{B}, t \in [0, \tau]} \| n^{-1} S^{(j)}(\beta, t) - s^{(j)}(\beta, t) \| \overset{\mathscr{P}}{\to} 0$ as $n \to \infty$.

(b) $s^{(0)}(\beta, t)$ is bounded away from 0 for $t \in [0, \tau]$. (This guarantees that the risk set becomes large at each t value and guarantees estimability of the baseline cumulative hazard function everywhere in $[0, \tau]$.)

(c) For $j = 0, 1, 2$, $s^{(j)}(\beta, t)$ is a continuous function of β uniformly in $t \in [0, \tau]$ and $s^{(1)} = \partial s^{(0)} / \partial \beta$ and $s^{(2)} = \partial^2 s^{(0)} / \partial \beta \, \partial \beta'$.

(d) $\Sigma(\beta, \tau) = \int_0^\tau v(\beta, u) s^{(0)}(\beta, u) \lambda_0(u) \, du$ is positive definite for all $\beta \in \mathscr{B}$, where $v(\beta, t) = s^{(2)}(\beta, t) / s^{(0)}(\beta, t) - e(\beta, t) e(\beta, t)'$ and $e(\beta, t) = s^{(1)}(\beta, t) / s^{(0)}(\beta, t)$. [Note that $e(\beta, t)$ and $v(\beta, t)$ are the probability limits of $\mathscr{E}(\beta, t)$ and $\mathscr{V}(\beta, t)$, respectively. Thus, the key condition (d) requires that an integral of a covariance matrix be positive definite.]

These conditions are sufficient to show that the partial likelihood estimator $\hat{\beta}$ obtained by maximizing the log partial likelihood

$$l(\beta, \tau) = \sum_{i=1}^n \int_0^\tau Z_i(u)' \beta \, dN_i(u) - \int_0^\tau \log[S^{(0)}(\beta, u)] \, dN.(u)$$

is consistent for β_0. In outline, the proof proceeds as follows. First, it is noted that the process

$$X(\beta, t) = n^{-1} [l(\beta, t) - l(\beta_0, t)]$$

$$= n^{-1} \sum_{i=1}^n \int_0^t \left\{ Z_i(u)(\beta - \beta_0) - \log\left[\frac{S^{(0)}(\beta, u)}{S^{(0)}(\beta_0, u)} \right] \right\} dN_i(u)$$

is a submartingale with compensator

$$\tilde{X}(\beta, t) = n^{-1} \int_0^t \left\{ S^{(1)}(\beta, u)'(\beta - \beta_0) - \log\left[\frac{S^{(0)}(\beta, u)}{S^{(0)}(\beta_0, u)} \right] S^{(0)}(\beta_0, u) \right\} \lambda_0(u) \, du.$$

$$(5.57)$$

Under condition C1, it can be seen that $\tilde{X}(\beta, \tau)$ converges to a function $f(\beta)$, $\beta \in \mathscr{B}$ obtained by replacing t with τ and $n^{-1} S^{(j)}$ with $s^{(j)}$ in (5.57). We find that $\partial f / \partial \beta |_{\beta = \beta_0} = 0$ and

$$-\partial^2 f(\beta) / \partial \beta \, \partial \beta' = \Sigma(\beta, \tau).$$

Further, it can be shown that the predictable variation process of $X(\beta, t) - \tilde{X}(\beta, t)$ converges to 0 on the interval $[0, \tau]$. Lenglart's inequality (5.39) implies that $X(\beta, \tau) \overset{\mathscr{P}}{\to} f(\beta)$ for $\beta \in \mathscr{B}$. It now follows from arguments based on convexity of $X(\beta, \tau)$ and $f(\beta)$ that $\hat{\beta} \overset{\mathscr{P}}{\to} \beta_0$.

To show the asymptotic normality of the score statistic or of $\hat{\beta}$, we require some condition on the covariates that limits their variability and individual influence. It is sufficient to assume that the covariates $Z_i(t)$ are bounded for all $t \in [0, \tau]$; in many ways, this is the natural condition to use since, in most applications, it is the occurrence of occasional extreme values of Z_i that casts doubt on the validity of the asymptotic normal approximations. A weaker condition which also implies the Lindeberg condition in Rebolledo's theorem, however, is sufficient for the asymptotic results.

C2.   There exists a $\delta > 0$ such that as $n \to \infty$,

$$n^{-1/2} \sup_{i,t} |Z_i(t)| \, Y_i(t) \, \mathbf{1}[Z_i(t)' \beta_0 > -\delta \, |Z_i(t)|] \xrightarrow{\mathscr{P}} 0.$$

This then gives the first main result:

Theorem 5.3.   If conditions C1 and C2 hold and $\int_0^\tau \lambda_0(u) \, du < \infty$, then as $n \to \infty$,

$$n^{-1/2} U(\beta_0, \tau) \xrightarrow{\mathscr{D}} N[0, \Sigma(\beta_0, \tau)], \tag{5.58}$$

$$n^{1/2} (\hat{\beta} - \beta_0) \xrightarrow{\mathscr{D}} N[0, \Sigma(\beta_0, \tau)^{-1}], \tag{5.59}$$

and

$$n^{-1} I(\hat{\beta}, \tau) \xrightarrow{\mathscr{P}} \Sigma(\beta_0, \tau). \tag{5.60}$$

Proof. The first step in the proof is to establish (5.58) using Theorem 5.1. The second step is to show that (5.60) is true using the consistency of $\hat{\beta}$ already shown. The final result (5.59) then follows from standard arguments involving a Taylor expansion of $U(\hat{\beta}, \tau)$ about β_0. $\square$

Proof of (5.58). Note that $G_i^{(n)}(u) = n^{-1/2}[Z_i(u) - \mathscr{E}(\beta_0, u)]$ is a vector of predictable processes and, from (5.51),

$$n^{-1/2} U(\beta_0, t) = \sum_{i=1}^n \int_0^t G_i^{(n)}(u) \, dM_i(u)$$

is a martingale with predictable covariation process

$$\langle n^{-1/2} U(\beta_0) \rangle(t) = \sum_{i=1}^n \int_0^t G_i^{(n)}(u)' G_i^{(n)}(u) Y_i(u) \exp[Z_i(u)' \beta_0] \lambda_0(u) \, du$$

$$= \int_0^t \mathscr{V}(\beta_0, u) S^{(0)}(\beta_0, u) \lambda_0(u) \, du.$$

Under condition C1,

$$\langle n^{-1/2} U(\beta_0) \rangle(t) \overset{\mathscr{P}}{\to} \int_0^t v(\beta_0, u) s^{(0)}(\beta_0, u) \lambda_0(u)\, du,$$

which establishes condition (a) of Theorem 5.1. It can also be shown that condition C2 (or the stronger condition of boundedness of the covariates) is sufficient to imply the Lindeberg condition (c) in Theorem 5.1. Together, these are sufficient to establish the result (5.58). □

Proof of (5.60). It can be shown that $n^{-1} I(\beta^*, \tau) \overset{\mathscr{P}}{\to} \Sigma(\beta_0, \tau)$ for any β^* that converges in probability to β_0. Since $\hat\beta \overset{\mathscr{P}}{\to} \beta_0$, the result follows. The proof is accomplished by adding and subtracting terms to the difference $n^{-1} I(\beta^*, \tau) - \Sigma(\beta_0, \tau)$ and showing that each term approaches zero. □

Proof of (5.59). The result for the maximum partial likelihood estimator follows by quite standard arguments. Using the consistency of $\hat\beta$, we obtain

$$0 = n^{-1/2} U(\hat\beta, \tau) = n^{-1/2} U(\beta_0, \tau) - [n^{1/2}(\hat\beta - \beta_0)]'[n^{-1} I(\beta^*, \tau)]$$

where β^* is between $\hat\beta$ and β_0. Since $n^{-1} I(\beta^*, \tau) \overset{\mathscr{P}}{\to} \Sigma(\beta_0, \tau)$ and (5.58) holds, the result (5.59) follows. □

Asymptotic results can also be obtained for the Nelson–Aalen estimator $\hat\Lambda_0(\hat\beta, t)$ from (5.54). The main result of use refers to the limiting distribution of the estimator at a specified time.

Theorem 5.4. If conditions C1 and C2 hold and $\int_0^\tau \lambda_0(u)\, du < \infty$, then for any given $t \in (0, \tau]$,

$$n^{1/2}[\hat\Lambda_0(\hat\beta, t) - \Lambda_0(t)] \overset{\mathscr{D}}{\to} N[0, b^2(t) + a(t)'\Sigma(\beta_0, \tau)^{-1} a(t)] \qquad (5.61)$$

as $n \to \infty$, where

$$b(t) = \int_0^t \frac{\lambda_0(u)\, du}{s^{(0)}(\beta_0, u)} \quad \text{and} \quad a(t) = \int_0^t e(\beta_0, u) \lambda_0(u)\, du.$$

The variance in (5.61) can be consistently estimated by

$$n[\hat b^2(t) + \hat a(t)' I(\hat\beta, \tau) \hat a(t)]$$

where

$$\hat b(t) = \int_0^t \frac{dN.(u)}{[S^{(0)}(\hat\beta, u)]^2} \quad \text{and} \quad \hat a(t) = \int_0^t \frac{\mathscr{E}(\hat\beta, u)\, dN.(u)}{S^{(0)}(\hat\beta, u)}.$$

More general results are again available which show that $n^{1/2}[\hat{\Lambda}_0(\hat{\beta}, t) - \Lambda_0(t)]$ converges weakly to a Gaussian process on the interval $[0, \tau]$ with covariance function given by $b^2(s \wedge t) - a(s)'\Sigma(\beta, \tau)^{-1}a(t)$.

For regression parameter estimation it is usual in relative risk and other regression modeling contexts to use the data at all time points where one or more study subjects are at risk without concern about specifying a follow-up interval $(0, \tau]$. For most purposes this is a reasonable approach to regression parameter estimation since the partial likelihood and related estimating functions are typically insensitive to contributions from time points where the risk sets are small, and with both right censoring and left truncation, data in either tail of the follow-up process can contribute usefully to the estimation. In fact, with right censoring only, the asymptotic results for $\hat{\beta}$ can be extended to hold for data on the entire interval $[0, \infty)$ by placing somewhat stronger conditions on the covariates (see, e.g., Arjas and Haara, 1988).

On the other hand, for the estimation of survivor functions or cumulative hazard functions at a time t, it is important that there be substantial follow-up over the entire interval $[0, t]$. More particularly, to make use of the weak convergence on the interval $[0, \tau]$, there should be follow-up over the entire interval. For random right censoring, for example, this will be true if τ is in the support of $T \wedge C$ or if $P[Y(\tau) = 1] > 0$. For left truncation we need $P[Y(0) = 1] > 0$. This may have implications for specification of the failure time variate and its time origin. For example, an epidemiologic cohort study on women 50 years of age or older could specify the failure time variate as age at failure (e.g., disease diagnosis) minus 50. If it should turn out that people really enter the study with appreciable frequency only at age 52 or more, the conditions necessary to estimate the survivor function would not hold. One could, however, estimate the conditional survivor function $F(t|T > 2)$, for example, and consider convergence over an interval $|2, \tau]$ for some suitably chosen τ. Alternatively, and more usually, the basic failure time variable in a cohort study is defined as time since entry into the cohort, with age accommodated through regression modeling or stratification. Not only does this specification avoid left truncation but also may tend to standardize the analysis of stochastic covariates since such covariate data are typically collected on a specified schedule as a function of time from cohort entry.

5.8 ASYMPTOTIC RESULTS FOR PARAMETRIC MODELS

The establishment of asymptotic results for parametric models was discussed in Chapter 3. It was noted there that provided that the censoring and/or truncation schemes are random and independent, asymptotic results can be derived by appealing to a central limit theorem for sums of independent random variables. Under certain regularity conditions, such a theorem can be applied to the total score. Asymptotic results for the MLE and likelihood ratio statistics can then be obtained by standard Taylor expansions of the score, again under standard regularity conditions. These arguments are quite general in nature and apply, for example, to independent random interval censoring and right truncation.

It is also possible, however, to obtain asymptotic results for the parametric models based on counting process and martingale formulations. These methods apply to the important cases of independent right censoring and left truncation. In one sense, they are more general than those based on independent random censoring since they apply to a larger class of right-censoring schemes. They are also less general, however, in that they do not cover other censoring schemes, such as interval censoring or right truncation, which also arise with some frequency. We give only a general outline of the arguments here without fully specifying regularity conditions for the asymptotic results to hold.

Suppose that right-censored and/or left-truncated data arise from a continuous parametric model and that $N_i(t)$, $Y_i(t)$, $\mathcal{F}_t$, and so on, are defined as in Section 5.7. Suppose that

$$P[dN_i(t) = 1|\mathcal{F}_{t-}] = Y_i(t)\lambda_i(t;\theta)\,dt, \qquad i = 1,\ldots,n,$$

where $\lambda_i(t;\theta) = \lambda(t;\theta,X_i(t))$ is the hazard for the ith sample and is assumed known up to a vector $\theta = (\theta_1,\ldots,\theta_p)'$ of unknown parameters. We suppose usual regularity conditions that $\lambda_i(t;\theta)$ is thrice differentiable in a neighborhood $\mathcal{N}$ of the true value θ_0, that θ_0 is an interior point of the parameter space, and that differentiation with respect to θ and integration with respect to t can be interchanged for all $\theta \in \mathcal{N}$.

The log-likelihood function arising from continuous failure time data on the interval $(0,\tau]$ can be written

$$l(\theta) = \sum_{i=1}^{n}\int_0^{\tau} \log\lambda_i(t;\theta)\,dN_i(t) - \sum_{i=1}^{n}\int_0^{\tau} Y_i(t)\lambda_i(t;\theta)\,dt.$$

The corresponding score process on data on $(0,t]$ is

$$U(\theta,t) = \sum_{i=1}^{n}\int_0^{t}\left[\frac{\partial}{\partial\theta}\log\lambda_i(u;\theta)\right]dN_i(u) - \sum_{i=1}^{n}\int_0^{t}Y_i(u)\left[\frac{\partial}{\partial\theta}\lambda_i(u;\theta)\right]du,$$

$0 < t < \tau$. This can also be written as

$$U(\theta,t) = \sum_{i=1}^{n}\int_0^{t}\left[\frac{\partial}{\partial\theta}\log\lambda_i(u;\theta)\right]dM_i(u),$$

where

$$M_i(t) = N_i(t) - \int_0^{t}Y_i(u)\lambda_i(u;\theta)\,du, \qquad i = 1,\ldots,n$$

are orthogonal martingales. Thus, the score process $U(\theta, t)$ is a martingale and its predictable variation process is

$$\langle U(\theta)\rangle(t) = \sum_{i=1}^{n} \int_0^t \left[\frac{\partial}{\partial \theta} \log \lambda_i(u; \theta)\right]^{\otimes 2} Y_i(u) \lambda_i(u; \theta)\, du.$$

The observed information matrix, $-\partial^2 l(\theta)/\partial\theta\,\partial\theta'$, is

$$I(\theta) = \sum_{i=1}^{n} \int_0^\tau \left[\frac{\partial}{\partial \theta} \log \lambda_i(u; \theta)\right]^{\otimes 2} dN_i(u)$$

$$- \sum_{i=1}^{n} \int_0^\tau \frac{\partial^2 \lambda_i(u; \theta)}{\partial\theta\,\partial\theta'} [\lambda_i(u; \theta)]^{-1}\, dM_i(u). \tag{5.62}$$

Since the second term in (5.62) is the integral of a predictable process with respect to a martingale and so has mean 0, it follows that

$$E[\langle U(\theta_0)\rangle(\tau)] = E[I(\theta_0)].$$

The central limit for the score function can now be applied with a set of relatively straightforward conditions. Thus we assume that, as $n \to \infty$, the following conditions hold:

1. $n^{-1}\langle U(\theta_0)\rangle(\tau) \xrightarrow{\mathscr{P}} \Sigma(\theta_0)$, where $\Sigma(\theta_0)$ is a positive definite $p \times p$ matrix.
2. For all j and $\epsilon > 0$,

$$n^{-1}\sum_{i=1}^{n} \int_0^\tau \left[\frac{\partial}{\partial \theta_j} \log \lambda_i(u; \theta_0)\right]^2 \mathbf{1}\left[\left|n^{1/2} \frac{\partial}{\partial \theta_j} \log \lambda_i(u; \theta_0)\right| > \epsilon\right] Y_i(u) \lambda_i(u; \theta) \xrightarrow{\mathscr{P}} 0.$$

These two conditions are sufficient to apply Theorem 5.1 to establish that

$$n^{-1/2} U(\theta_0, \tau) \xrightarrow{\mathscr{D}} N[0, \Sigma(\theta_0)]$$

as $n \to \infty$.

Additional conditions on the third derivatives of λ_i and $\log \lambda_i$ are sufficient to allow Taylor expansions to show the consistency of the MLE of θ arising from the maximum likelihood equation $U(\theta, \tau) = 0$ and to establish the asymptotic distributions of $\hat{\theta}$ and the likelihood ratio statistic. Thus, one obtains, for example, that

$$n^{1/2}(\hat{\theta} - \theta_0) \xrightarrow{\mathscr{D}} N[0, \Sigma(\theta_0)^{-1}]$$

as $n \to \infty$, and that $\Sigma(\theta_0)$ can be estimated consistently by $n^{-1}I(\hat{\theta})$.

It should be noted that in some instances, the normalization may be in terms of a different sequence of constants a_n which approaches ∞ as $n \to \infty$. In most instances, however, the $a_n = n^{1/2}$ standardization is appropriate.

5.9 EFFICIENCY OF THE COX MODEL ESTIMATOR

Consider now the efficiency of the estimator of β that maximizes the log partial likelihood function arising from the Cox model. We restrict attention to the continuous case in which tied failure times cannot occur. In this section the partial likelihood estimate is denoted $\hat{\beta}_c$ and initially, we restrict attention to the case considered in Chapter 4, in which the basic covariates x are time independent, so that $Z(t)$ is a vector of fixed covariates or deterministic functions of time.

Two important questions arise concerning the efficiency of $\hat{\beta}_c$. First, with the hazard function $\lambda_0(\cdot)$ unspecified, can $\hat{\beta}_c$ be improved upon in regard to (asymptotic) efficiency? That is, are there other procedures that would lead to more precise asymptotic estimation of β? Second, what is the relative efficiency of $\hat{\beta}_c$ with respect to that of a maximum likelihood estimator, $\hat{\beta}$, based on a special case in which $\lambda_0(\cdot)$ is specified up to certain unknown parameters?

Efron (1977) and, somewhat less explicitly, Oakes (1977) argue that the asymptotic variance matrix for $\hat{\beta}_c$ will be close to that for $\hat{\beta}$ provided the parametric family giving rise to $\hat{\beta}$ is reasonably rich. In essence, Efron argues that the asymptotic variance of $\hat{\beta}$ approaches that of $\hat{\beta}_c$ as the number of independent parameters in the parametric model increases. This leads to the conclusion that it is not possible to improve on the asymptotic accuracy of $\hat{\beta}_c$ without restrictive assumption on λ_0.

The second question concerns the efficiency of $\hat{\beta}_c$ relative to estimators from parametric models that may be highly specified. Some results in this case were given in Section 4.8 in assessing the efficiency of the Cox estimator compared to the exponential special case. In this section we consider more general results, due first to Oakes and Efron, which provide quite general efficiency expressions. Emphasis is placed on the interpretation of the efficiency expressions. The second question is addressed first, although the same approach leads to an answer to the first question also.

5.9.1 Comparisons with Parametric Submodels

The most extreme parametric specification of the $\lambda_0(\cdot)$ function in the Cox model is to suppose that it is known completely. Such a specification, however, involves an asymmetry in that the hazard function at $Z(t) = 0$ is known completely whereas at other $Z(t)$ values it is known only up to the factor $\exp[Z(t)'\beta]$. To avoid this situation we consider a model in which $\lambda_0(\cdot)$ is specified up to a scale parameter; that is,

$$\lambda_0(t) = \alpha h_0(t), \qquad (5.63)$$

where $h_0(t) > 0$ is a known function and α is an unknown positive constant.

We make extensive use of the notation from Sections 5.7 and 5.8, especially from (5.56) and condition C1. Suppose, as before, that follow-up extends over the interval $[0, \tau]$ and that conditions C1 and C2 for the asymptotic consistency and normality of $\hat{\beta}_c$ hold. The log-likelihood function under the parametric model (5.63) and independent censoring can be written

$$l_1 = \sum_{i=1}^{n} \int_0^{\tau} [\log \lambda_0(t) + Z_i(t)'\beta] dN_i(t) - \int_0^{\tau} S^{(0)}(\beta, t)\lambda_0(t)\, dt. \tag{5.64}$$

In order to investigate the efficiency of the regression estimator from the proportional hazards model, we compare the information content of (5.64) with the information from the log partial likelihood, given by

$$l_2 = \sum_{i=1}^{n} \int_0^{\tau} Z_i(u)'\beta\, dN_i(u) - \int_0^{\tau} \log[S^{(0)}(\beta, u)]\, dN.(u). \tag{5.65}$$

The observed information matrix from (5.64) has the (u, v) element

$$-\partial^2 l_1/\partial\beta_u\,\partial\beta_v = \sum_{i=1}^{n} \int Y_i(t)Z_{ui}(t)Z_{vi}(t)\exp[Z_i(t)'\beta]\lambda_0(t)\, dt, \tag{5.66}$$

where $u, v = 0, 1, \ldots, p$ and we have defined $\beta_0 = \log\alpha$ and $Z_{0i}(t) = 1$ for all i, t. The observed information matrix is then

$$I_1 = \begin{pmatrix} \int_0^{\tau} S^{(0)}\lambda_0(t)\, dt & \int_0^{\tau} S^{(1)'}\lambda_0(t)\, dt \\ \int_0^{\tau} S^{(1)}\lambda_0(t)\, dt & \int_0^{\tau} S^{(2)}\lambda_0(t)\, dt \end{pmatrix},$$

where $S^{(j)} = S^{(j)}(\beta, t)$, $j = 0, 1, 2$ is defined in (5.56). Recall that $S^{(0)}$ is a scalar, $S^{(1)}$ is a $p \times 1$ vector, and $S^{(2)}$ is a $p \times p$ matrix. We suppose that condition C1 holds, so that

$$n^{-1}I_1 \xrightarrow{\mathscr{P}} \mathscr{I} = \begin{pmatrix} \int_0^{\tau} s^{(0)}\lambda_0(t)\, dt & \int_0^{\tau} s^{(1)'}\lambda_0(t)\, dt \\ \int_0^{\tau} s^{(1)}\lambda_0(t)\, dt & \int_0^{\tau} s^{(2)}\lambda_0(t)\, dt \end{pmatrix}$$

represents the limiting average information per individual. If $(N_i(\cdot), Y_i(\cdot), Z_i(\cdot))$, $i = 1, \ldots, n$ are IID, $\mathscr{I}$ also represents the Fisher (or expected) information per observation. The marginal average information matrix for $\beta = (\beta_1, \ldots, \beta_p)'$, that is, the variance matrix for $\partial \log L/\partial\beta$ after maximizing out α, is then

$$\begin{aligned}
\mathbf{I}_1 &= \int s^{(2)}\lambda_0(t)\, dt - \left[\int s^{(0)}\lambda_0(t) dt\right]^{-1}\left[\int s^{(1)}\lambda_0(t)\, dt\right]^{\otimes 2} \\
&= \int v(\beta, t)s^{(0)}\lambda_0(t)\, dt + \int [e(\beta, t) - c]^{\otimes 2}s^{(0)}\lambda_0(t)\, dt,
\end{aligned} \tag{5.67}$$

where

$$c = \left[\int_0^\tau s^{(0)} \lambda_0(t) \, dt \right]^{-1} \int_0^\tau s^{(1)} \lambda_0(t) \, dt.$$

The information computation for the log partial likelihood (5.65) proceeds in a similar manner. The observed information matrix is

$$I_2 = E[-\partial^2 l_2 / \partial \beta \, \partial \beta']$$
$$= \int_0^\tau \mathscr{V}(\beta, t) \, dN.(t). \tag{5.68}$$

It follows that as $n \to \infty$,

$$n^{-1} I_2 \overset{\mathscr{P}}{\to} \mathbf{I}_2 = \int_0^\tau v(\beta, t) s^{(0)}(\beta, t) \lambda_0(t) \, dt, \tag{5.69}$$

which again can be interpreted as the average (in the limit) information per observation.

When the data are IID and a parametric analysis is used, it was noted that the limiting average information could also be interpreted as the Fisher information arising from each individual. In the Cox model, there is a small-sample correction and the average or expected information per observation actually varies with n. We noted this in the evaluation of small-sample efficiencies in Section 4.8. The expected information from the Cox likelihood is discussed in Exercise 5.5.

The partial likelihood estimator $\hat{\beta}_c$ will have full asymptotic efficiency if $\mathbf{I}_2 \mathbf{I}_1^{-1}$ converges to an identity matrix, or equivalently, if $(\mathbf{I}_1 - \mathbf{I}_2) \mathbf{I}_1^{-1}$ converges to a zero matrix. A comparison of (5.69) with (5.67) shows that full asymptotic efficiency for the estimation of β will be achieved if

$$e(\beta, t) = s^{(1)}(\beta, t) / s^{(0)}(\beta, t) = \text{constant}, \qquad t \in [0, \tau] \tag{5.70}$$

in which case the second term in (5.67) is zero. Recall that

$$\mathscr{E}(\beta, t) = \sum_{i=1}^n Z_i(t) p_i(t) \overset{\mathscr{P}}{\to} e(\beta, t).$$

Suppose now that $Z = Z(t)$ is time independent. If $\beta = 0$ and censorship does not depend on Z, all risk sets $R(t)$ of the same size are equally probable at time t and (5.70) is satisfied. The partial likelihood estimator is then asymptotically fully efficient. This also shows that the log-rank test for $\beta = 0$ will have full Pitman efficiency (see, e.g., Cox and Hinkley, 1974, p. 338, for definitions) under these circumstances. Censoring rates that depend on Z [e.g., differ among samples in the $(p+1)$-sample problem] give rise to a distribution of Z values with expectation that varies over time and thereby to some loss in efficiency since (5.70) is not satisfied. Crowley and Thomas (1975) give some calculations for the efficiency of the logrank test under random censorship but with differing censoring distributions in the samples. Even in situations with rather extreme differences in the censoring rates, the logrank has respectable efficiency, usually greater than 0.90.

Perhaps of more concern is the typical decline of the relative efficiency of $\hat{\beta}_c$ as β departs from zero. If, for example, $p = 1$, $\beta > 0$, and there is no censoring, then as t increases $R(t)$ with high probability contains more small values of Z than large and $e(\beta, t)$ decreases with increasing t. This dependence of efficiency on the size of β is a situation unfamiliar in ordinary linear regression.

As noted earlier,

$$\mathscr{E}(\beta, t) = \sum_{i=1}^{n} Z_i(t) p_i(t) \xrightarrow{\mathscr{P}} e(\beta, t),$$

and $\mathscr{E}(\beta, t)$ can be thought of as a weighted average of Z values in the risk set at time t. Oakes (1977) suggests plotting $e(t; \hat{\beta}_c)$ to obtain an empirical check on the efficiency of the partial likelihood analysis.

An explanation for the loss in efficiency of $\hat{\beta}_c$ when the average Z value over the risk set varies with time is that such variations introduce asymptotic correlations between the estimator of β and that of α. The parametric analysis can exploit such correlations, but the partial likelihood procedure cannot. As an extreme situation, consider a two-sample problem in which one sample is uncensored and the second is totally censored at time t_0. The contributions to the partial likelihood (5.65) at times after t_0 do not depend on β since all items in the risk set have the same covariate value. Parametric analyses, on the other hand, can utilize failures past t_0 to estimate the parameter $\lambda_0(\cdot)$ more precisely, and hence they yield more precise estimates of β. One might expect that if the parametric model were allowed to become more flexible, the information on failures beyond t_0 would be of less value in β estimation and the loss in efficiency in using (5.65) would be less severe. This line of thought is continued below, where the efficiency of the partial likelihood analysis is evaluated relative to more flexible parametric models. Before this, however, we consider in some detail a particular example of the type discussed above.

Example 5.7. Efron (1977) examined the limiting ratio $\mathbf{I}_2 \mathbf{I}_1^{-1}$ for the two-sample problem in the absence of censoring and $\tau = \infty$, where, without loss of generality, we may take $h_0(t) = 1$. Thus, the hazard for sample Z is $\alpha \exp(Z\beta)$, $Z = 0, 1$. Suppose that np and nq individuals are initially placed on test at time 0 in samples $Z = 1$ and $Z = 0$, respectively, and let $Y_{i.}(t)$ be the number of sample i individuals at risk at time $t, i = 0, 1$. Then,

$$S^{(0)}(\beta, t) = Y_{0.}(t) + Y_{1.}(t) \exp(\beta),$$
$$S^{(j)}(\beta, t) = Y_{1.}(t) \exp(\beta), \qquad j = 1, 2,$$

and

$$E[Y_{0.}(t)] = nq \, \exp(-\alpha t); \qquad E[Y_{1.}(t)] = np \, \exp(-\alpha t e^{\beta}).$$

After some calculation, with $u = \exp(\alpha t)$ and $\psi = \exp(\beta)$, we find from (5.69) that

$$\mathbf{I}_2 = pq\psi \int_0^1 \frac{du}{p\psi + qu^{1-\psi}}. \qquad (5.71)$$

Straightforward calculation gives the information per observation from the parametric analysis as $\mathbf{I}_1 = pq$, so that the asymptotic relative efficiency is

$$\text{eff}(\beta) = \psi \int_0^1 (p\psi + qu^{1-\psi})^{-1} du. \qquad (5.72)$$

The lower curves in Figure 5.1 correspond to this efficiency (5.72) for $p = 0.1$, 0.3, and 0.5. It can be seen that the efficiency of the partial likelihood exceeds 75% in

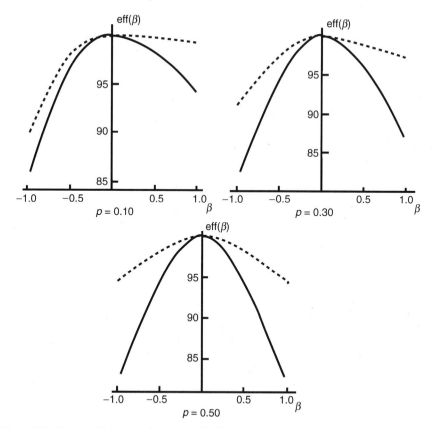

Figure 5.1 Percent efficiencies of the partial likelihood analysis in the two-sample problem against Weibull (dashed line) and exponential (solid line) true model.

most situations of interest, for example, when the failure rate ratio e^β for the two samples is between $\frac{1}{3}$ and 3. A Taylor expansion for the logarithm of (5.72) about $\beta = 0$ gives $\mathrm{eff}(\beta) \approx \exp(-\beta^2 pq)$, which provides a reasonable approximation for β not too large and p not too near 0 or 1. Kalbfleisch (1974) showed that with a single regressor variable the efficiency is approximately $\exp(-\beta^2 \mu_2)$, where μ_2 is the second central moment of the Z values.   □

Efron (1977) gives some numerical results indicating that efficiencies in the two-sample problem with censoring are in reasonable agreement with (5.72), at least for certain censoring patterns. Oakes (1977) also obtains the expression (5.72) with $p = q = \frac{1}{2}$ as a special case of an expression for the asymptotic efficiency of the rank analysis when censoring is exponential. Some of these results, and generalizations of them, are considered in Exercises 5.5, 5.6, and 5.7.

Suppose now that rather than specifying the $\lambda_0(\cdot)$ function up to a scale factor, we take

$$\lambda_0(t) = \exp[Z_{p+1}(t)\beta_{p+1} + \cdots + Z_{p+m}(t)\beta_{p+m}]h_0(t), \qquad (5.73)$$

where $Z_{p+1}(t), \ldots, Z_{p+m}(t)$ are specified functions of t [usually, with $Z_{p+1}(t) = 1$], and $h_0(t)$ is assumed completely known. The parametric likelihood can again be written as (5.64), and the notation has been chosen so that the (u, v) element in the observed information matrix $-\partial^2 l_1 / \partial \beta_u \partial \beta_v$, $u, v = 1, \ldots, p + m$, is again (5.66), where $\beta = (\beta_1, \ldots, \beta_p)'$ and $Z_l(t) = [Z_{1l}(t), \ldots, Z_{pl}(t)]$ as before, while $Z_{ul}(t) = Z_u(t)$ for $u = p + 1, \ldots, p + m$. Following the same steps as before, the information matrix can be written as the partitioned matrix

$$I_1 = -\frac{\partial^2 l_1}{\partial \beta \, \partial \beta'} = \begin{pmatrix} \int_0^\tau S^{(2)} \lambda_0(t) \, dt & \int_0^\tau S^{(1)} w' \lambda_0(t) \, dt \\ \int_0^\tau w S^{(1)'} \lambda_0(t) \, dt & \int_0^\tau S^{(0)} w w' \lambda_0(t) \, dt \end{pmatrix},$$

where the $S^{(j)} = S^{(j)}(\beta, t), j = 0, 1, 2$ are defined as before and

$$w = w(t) = [Z_{p+1}(t), \ldots, Z_{p+m}(t)]'.$$

The marginal information on $\beta = (\beta_1, \ldots, \beta_p)$, when $\beta_{p+1}, \ldots, \beta_{p+m}$ have been eliminated, can be written as (5.67), where c is replaced with the $p \times 1$ vector

$$c = \int_0^\tau s^{(1)} w' \lambda_0(t) \, dt \left[\int_0^\tau s^{(0)} w w' \lambda_0(t) \, dt \right]^{-1}. \qquad (5.74)$$

The asymptotic average information in the Cox partial likelihood is again denoted $\mathbf{I}_2$ and given by (5.68). The asymptotic relative efficiency of $\hat{\beta}_p$ is then given by $\mathbf{I}_2 \mathbf{I}_1^{-1}$, which depends on the size of the second term in (5.67). This term is

$$\int_0^\tau [e(\beta, t) - c][e(\beta, t) - c]' s^{(0)} \lambda_0 \, dt$$

with c given by (5.74). This is recognizable as the deviation from regression sum of squares for a weighted regression over the interval $[0, \tau]$ of $e(\beta, t) = s^{(1)}(\beta, t)/s^{(0)}(\beta, t)$ on $w(t) = [Z_{p+1}(t), \ldots, Z_{p+m}(t)]$ with weights given by $s^{(0)}(\beta, t)\lambda_0(t)$. This shows that $\hat{\beta}_c$ will be fully efficient with respect to $\hat{\beta}$ from (5.73) whenever the elements of $e(\beta, t)$ are in the space spanned by the functions $Z_{p+1}(t), \ldots, Z_{p+m}(t)$. The partial likelihood analysis will have close to full efficiency provided that the parametric modeling (5.73) is rich enough that a linear combination of $Z_{p+1}(t), \ldots, Z_{p+m}(t)$ closely approximates $e(\beta, t)$ or its finite sample counterpart $\mathscr{E}(\beta, t)$. Such closeness would occur, for example, if a parametric model were asserted in which $\lambda_0(\cdot)$ was taken to be arbitrary but constant within relatively narrow time intervals. Such a model would allow the mean function $e(\beta, t)$ to be well approximated by a linear function of the $Z_j(t), j = p + 1, \ldots, p + m$. This line of reasoning suggests that one cannot expect to find a β estimator that is asymptotically more efficient than $\hat{\beta}_c$ provided the $\lambda_0(t)$ function is unrestricted. Begun et al. (1983) establish the semiparametric efficiency of $\hat{\beta}_c$ using this basic idea.

Example 5.7 (continued). Consider again a two-sample problem in which the data arise from a Weibull regression model

$$\lambda(t; z) = \lambda_0(t) \exp(Z\beta), \qquad 0 < t < \infty,$$

where $Z = 0, 1$ and $\lambda_0(t) = \lambda\gamma(\lambda t)^{\gamma-1}, 0 < t < \infty$. We consider the efficiency of the partial likelihood analysis in the absence of censoring and $\tau = \infty$. This fits the framework above by choosing $h_0(t) = 1$ and $w(t) = (1, \log t)', 0 < t < \infty$, where $\beta_{p+1} = \log(\lambda\gamma)$, $\beta_{p+2} = \gamma - 1$, $p = 1$, and $m = 2$. It can be seen that the efficiency does not depend on λ, γ, so we take $\lambda = \gamma = 1$. It then follows that

$$\mathbf{I}_1 = \int_0^\infty s^{(2)}\lambda_0 \, dt - \int_0^\infty s^{(1)}w'\lambda_0 \, dt \left[\int_0^\infty s^{(0)}ww'\lambda_0 \, dt \right]^{-1} \int_0^\infty [s^{(1)}w']'\lambda_0 \, dt$$

$$= p - p^2(1 - a - \beta) \begin{pmatrix} 1 & -a - p\beta \\ -a - p\beta & b - 2pa\beta + p\beta^2 \end{pmatrix}^{-1} \begin{pmatrix} 1 \\ -a - \beta \end{pmatrix},$$

where p is the proportion in sample $z = 1$, $a = 0.5772$ is Euler's constant, and $b = (\pi^2/6) + a^2 = 1.9781$. The information in the partial likelihood is again given by (5.71) and the relative efficiency can now easily be computed.

Figure 5.1 gives a plot of the relative efficiency of the partial likelihood analysis compared with the exponential and Weibull special cases as a function of β for various values of p. It can be seen that the efficiency of the partial likelihood is substantially greater when the Weibull rather than the exponential is the alternative parametric analysis. For example, if $p = 0.5$, the efficiency is 93% or more for values of $\exp(\beta)$ between $\frac{1}{3}$ and 3, which covers most cases of interest. $\qquad\square$

5.10 PARTIAL LIKELIHOOD FILTRATION

Our analysis of the Cox model in Chapter 4 is based on the partial likelihood argument, and this also relates closely to the hypergeometric derivation of the logrank and related statistics in Section 1.6. However, in the asymptotic derivations and martingale versions of the relevant statistics discussed in this chapter, the conditioning is on $\mathscr{F}_{t-}$ from the filtration

$$\mathscr{F}_t = \sigma\{N_i(u), Y_i(u^+), X_i(u^+), \; i = 1,\ldots,n, \; 0 \le u \le t\},$$

which differs from the conditioning event in the partial likelihood of Section 4.2 in that it does not include information on $N.(t)$, the number of failures observed at time t. In this section we define a filtration that relates directly to the partial likelihood and the hypergeometric derivation of the logrank statistic.

The *partial likelihood filtration* is

$$\mathscr{F}_t^* = \sigma\{N_i(u), Y_i(u^+), X_i(u^+), \; j = 1,\ldots,n, \; 0 \le u \le t, \; K(t), R(t)\}, \qquad (5.75)$$

where $K(t) = \inf\{s : N.(t+s) > N.(t)\}$ and $R(t) = \Delta N.[K(t)]$ specify the time and the multiplicity of the next occurrence of an event. If there are no events after time t, we take $K(t) = \infty$ and $R(t) = 0$. It is easy to see that $K(t)$ and $R(t)$ are right-continuous processes so that $\mathscr{F}_t^*$ is right continuous. It is also easy to see that $N.(t)$ is a predictable process with respect to $\mathscr{F}_t^*$, in that the value of $N.(t)$ is specified by $\mathscr{F}_{t-}^*$.

In the continuous relative risk model, $\Delta N.(t) \le 1$, and under independent censoring, it can be seen that

$$P[dN_i(t) = 1|\mathscr{F}_{t-}^*] = p_i(\beta, t)dN.(t). \qquad (5.76)$$

The product of these probabilities across all times gives rise to the partial likelihood (4.14). The counting process $N_i(t)$ has compensator $\int_0^t p_i(\beta, u)\, dN.(u)$ with respect to the filtration $\mathscr{F}_t^*$, and the corresponding martingale is

$$M_i^*(t) = N_i(t) - \int_0^t p_i(\beta, u)\, dN.(u).$$

The score function process for the continuous-time Cox model can now be written

$$U(\beta, t) = \sum_{i=1}^{n} \int_0^t Y_i(u)Z_i(u)\, dM_i^*(u),$$

and the corresponding predictable variation process is

$$\langle U \rangle(t) = \sum_{i=1}^{n} \int_0^t Y_i(u)Z_i(u)Z_i(u)'\, d\langle M_i^* \rangle(t)$$

$$+ \sum_{i \ne j} \int_0^t Y_i(u)Y_j(u)Z_i(u)Z_j(u)'\, d\langle M_i^*, M_j^* \rangle(u). \qquad (5.77)$$

Substituting the expressions

$$\langle M_i^* \rangle(t) = \int_0^t [1 - p_i(\beta, u)\, \Delta N.(u)] p_i(\beta, u)\, dN.(u) \tag{5.78}$$

and

$$\langle M_i^*, M_j^* \rangle(t) = -\int_0^t p_i(\beta, u)\, \Delta N.(u) p_j(\beta, u)\, dN.(u) \tag{5.79}$$

into (5.77) yields the Fisher information process $I(\beta, t)$ given in (5.55). Asymptotic results for the score or for the partial likelihood estimator can now be deduced using arguments similar to those in Section 5.7.

Virtually identical arguments apply to the discrete logistic model (4.58). It can be verified that conditioning on the filtration $\mathscr{F}_t^*$ gives rise exactly to the partial likelihood (4.23) that was initially proposed by Cox for the discrete logistic model. The corresponding score statistic can be seen to be a martingale with respect to the new filtration, and as in the continuous Cox model, the predictable variation process of the score yields exactly the corresponding observed information process. In the $(p+1)$-sample problem, the score test for $\beta = 0$ arising from this partial likelihood is exactly the log-rank test that was used in Section 1.5.

BIBLIOGRAPHIC NOTES

The counting process formulation of a failure time process was first used for statistical purposes by Aalen (1975, 1978b) and subsequently developed more fully in the monograph by Gill (1980), who considered estimation of the cumulative hazard function and the asymptotic properties of log-rank type tests for the equality of survival distributions. Other early and influential work in demonstrating the utility of this formulation and its use in deriving asymptotic results can be found, for example, in Andersen and Gill (1982), Andersen et al. (1982), Harrington and Fleming (1982), and Andersen and Borgan (1985). The book by Fleming and Harrington (1991) gives a thorough treatment of failure time models and counting processes. The extensive and comprehensive book by Andersen et al. (1993) gives a complete treatment of asymptotic theory and many applications of counting processes in Markov-type models using multivariate counting processes. The treatment given here leans heavily on the Andersen et al. (1993) work, although it is less formal and aims only to give a general flavor of the mathematical treatment and results. We have not included a complete discussion of the asymptotic results for the Nelson–Aalen process nor of its asymptotic covariability with $\hat{\beta}$. Both of these topics are carefully developed in Andersen et al. (1993, pp. 503–506).

The martingale central limit theorem of Rebolledo (1979, 1980) followed discrete versions by various authors, including Billingsley (1961), Brown (1971), and McLeish (1974). Rebolledo's theorem together with the inequality by Lenglart

(1977) form the main tools for the asymptotic results. Andersen and Gill (1982) provided the first fully general asymptotic results from martingales to the Cox or relative risk model in an elegant paper that also applied to multiple failure types and multivariate counting processes. Prentice and Self (1983) considered applications to other relative risk forms. Tsiatis (1981) also considered asymptotics for the estimation of β and the cumulative hazard, but from an empirical process viewpoint. The martingale theorems are useful for the important cases of independent right censoring and left truncation. Other types of censoring, such as interval censoring, generally do not give rise to score processes that are martingales, and the methods discussed in this chapter are not available for developing asymptotic properties. Methods based on empirical processes as summarized, for example, in Bickel et al. (1993), Shorack and Wellner (1986), or van der Vaart and Wellner (1996) provide more generally applicable approaches. The very simple connection between counting processes and martingales, the simplicity of the limit theorems, and the availability of simple variance estimators make the martingale analysis very appealing when applicable.

Efficiency properties of the partial likelihood estimation of β were discussed by Efron (1977) and Oakes (1977). The derivations in Section 5.8 draw heavily on these two papers and especially on the ideas in Efron (1977). The idea of evaluating the efficiency with reference to increasing rich parametric families is a precursor to the idea of nonparametric information bounds and efficiency, as discussed, for example, in Begun et al. (1983) and Bickel et al. (1993, Chap. 5).

The partial likelihood filtration has not been discussed in the literature, although it does present a simple framework for the analysis of the Cox model and (weighted) log-rank tests.

EXERCISES AND COMPLEMENTS

5.1 Consider the signal plus noise specification (5.6) for the jumps in a discrete-time martingale. If $Y_i(a_l) = 1$, find the conditional distribution of $dM_i(a_l)$ given $\mathscr{F}_{a_l^-}$ and verify that $E[dM_i(a_l)|\mathscr{F}_{a_l^-}] = 0$.

5.2 Let $A_1, A_2, \ldots$ be independent uniform $(0, \tau)$ variates where τ is a known constant, and let $x_i(t) = \mathbf{1}(A_i < t)$, $i = 1, \ldots, n$. Let $N_i(t)$ count observed events for the ith individual and suppose that the ith individual is observed over the interval $(0, \tau_i]$, where the τ_i is a constant $0 < \tau_i \leq \tau$, $i = 1, \ldots, n$. Suppose that the intensity function of $N_i(t)$ with respect to the filtration (5.6) is

$$\lambda_i(t) = Y_i(t) \exp[x_i(t)\beta]\alpha, \qquad 0 < t < \tau,$$

where $i = 1, \ldots, n$ and $\alpha > 0$.

   (a) Obtain the intensity function for the counting process $N.(t) = \sum_{i=1}^{n} N_i(t)$ and define the corresponding martingale.

   (b) Find $E[N.(t)]$.

(c) Simulate 30 realizations of the process $[N.(t), 0 < t < 10]$ and the corresponding martingale when $\tau = 10$, $n = 10$, $\tau_i = i$, $i = 1, \ldots, 10$, and $\alpha = 1$. Plot the corresponding averages and observe the comparison with the theoretical mean for $N.(t)$ and the mean of the martingale.

5.3 Verify the result (5.35).

5.4 Explain why the information $\mathbf{I}_1$ for the Weibull special case of Example 5.7 does not depend on λ and γ.

5.5 Consider a two-sample problem in which the true hazard for sample Z is $\lambda \exp(Z\beta), Z = 0, 1$. Suppose that a proportion of p items are in sample 1 and $q = 1 - p$ are in sample 0. Let each sample be subject to censoring. The time to censoring is exponential with rate θ_i in sample $i = 0, 1$, where censoring and failure times are independent. Show that the limiting expected information per observation from the rank or partial likelihood analysis based on the proportional model is

$$\lim_{n \to \infty} \frac{I_\tau(\beta)}{n} = \int_0^\infty \left\{ \frac{\exp[t(\lambda + \theta_0)]}{q} + \frac{\exp[t(\lambda e^\beta + \theta_1)]}{pe^\beta} \right\}^{-1} \lambda \, dt.$$

Show also, for the parametric analysis, that the expected information per observation is

$$\frac{I_p(\beta)}{n} = \lambda e^\beta \left[\frac{e^\beta(\lambda + \theta_0)}{q} + \frac{\lambda e^\beta + \theta_1}{p} \right]^{-1}.$$

5.6 (*continuation*)

(a) Suppose that $\beta = 0$ and let $\gamma_i = \theta_i/\lambda, i = 0, 1$. Show that in this case the asymptotic efficiency of the rank analysis can be expressed as

$$\text{eff}(a) - \left(\frac{1}{q} + \frac{a}{p} \right) \int_0^\infty \left(\frac{e^s}{q} + \frac{e^{as}}{p} \right)^{-1} ds,$$

where $a = (1 + \gamma_1)/(1 + \gamma_0)$. Oakes (1977) gives some evaluations of this for $p = q = \frac{1}{2}$.

(b) The expression $\text{eff}(a)$ gives the Pitman efficiency of the log-rank test, although the integral can seldom be evaluated in closed form. Show, however, that if $p = q = \frac{1}{2}$ and $a = 2$, the efficiency is $3(1 - \log 2) = 0.92$. For the case $p = q = \frac{1}{2}$, evaluate this efficiency expression numerically for $0 < a < 4$. Note that for $a = 4$, the rate of censorship is more than four times as great in one sample as in the other.

5.7 Suppose that failure times arise in two populations from the same exponential distribution with failure rate λ. Suppose that the first sample $(Z = 0)$ is subject to no censoring, whereas the second $(Z = 1)$ is completely censored at t_0. Assuming that the data arise from a Weibull model, calculate the relative efficiency at $\beta = 0$ of the maximum partial likelihood estimator from the Cox model to that of the maximum likelihood estimator from the Weibull submodel as a function of λ and t_0. Compare these efficiency expressions with that based on the maximum likelihood estimator of β from a binary response model in which it is only noted whether or not failure time exceeds t_0.

5.8 Verify the relationship (5.45) in both the discrete and continuous cases.

5.9 Analogous to Section 5.8, suppose that independent left-truncated and/or right-censored data arise over the finite interval $[0, \tau]$ from a discrete parametric model. Thus, under independent censoring and truncation, we have

$$P[dN_i(t)|\mathscr{F}_{t-}] = Y_i(t)\, d\Lambda_i(t; \theta),$$

where $\Lambda_i(t; \theta) = \sum_{a_l \le t} \lambda_{il}(\theta)$ and $\lambda_{il}(\theta) = P[T_i = a_l|T_i \ge a_l, X_i(a_l)]$, $l = 1, 2, \ldots$, $i = 1, \ldots, n$.

(a) Show that the log-likelihood function can be written

$$l(\theta) = \sum_{i=1}^{n} \int_0^{\tau} \log \frac{\Delta\Lambda_i(t; \theta)}{1 - \Delta\Lambda_i(t; \theta)}\, dN_i(t)$$

$$+ \sum_{i=1}^{n} \int_0^{\tau} \log[1 - \Delta\Lambda_i(t; \theta)] Y_i(t) [\Delta\Lambda(t; \theta)]^{-1}\, d\Lambda_i(t; \theta).$$

(b) Show that the corresponding score process $U(\theta_0, t)$ can be written as a sum of stochastic integrals with respect to discrete-time martingales $M_i(t) = N_i(t) - \int_0^t Y_i(u)\, d\Lambda_i(u; \theta_0)$, $i = 1, \ldots, n$. That is, show that

$$U(\theta_0, t) = \sum_{i=1}^{n} \int_0^t \frac{\partial \Delta\Lambda_i(u; \theta_0)/\partial\theta_0}{\Delta\Lambda_i(u; \theta_0)[1 - \Delta\Lambda_i(u; \theta_0)]}\, dM_i(u).$$

(c) Assuming that the martingales in part (b) are orthogonal, find the predictable variation process $\langle U(\theta_0)\rangle(t)$. Show that $E[I(\theta_0)] = E[\langle U(\theta_0)\rangle(\tau)]$, where $I(\theta)$ is the observed information.

5.10 Verify that the predictable variation and covariation processes for the discrete-time martingales $M_i^*(t), i = 1, \ldots, n$ are as stated in (5.78) and (5.79) and thus verify that (5.77) reduces to $I(\beta, t)$.

5.11 Consider data arising from the discrete logistic model and the analysis using the partial likelihood filtration $\mathscr{F}_t^*$. Verify that the observed information arises as the predictable variation process of the corresponding score statistic martingale.

CHAPTER 6

Likelihood Construction and Further Results

6.1 INTRODUCTION

The construction of the likelihood for censored failure time data given in Chapter 3 assumes independence between the censoring or failure times of different individuals and so is valid for the special, although important case of a random censorship model. Also, notwithstanding some mention of more general covariates in Chapter 5, the basic covariates x that have been used in modeling to this point describe baseline $(t = 0)$ characteristics for each individual under study. In this chapter, the previous results are generalized to accommodate more general censoring schemes and covariates that vary with time. The construction of the likelihood with parametric models is considered in Section 6.2 for independent censoring schemes; time-dependent covariates are introduced in Section 6.3, and the likelihood construction is generalized to accommodate these. Partial likelihood, which formed the basis of the main analysis of the Cox or relative risk model in Chapter 4, is revisited in Section 6.4. The remainder of the chapter contains an example of time-dependent covariates in an analysis of the Stanford heart transplant data, a discussion of models that condition on only a part of the preceding covariate history, and some discussion of model checking and residuals.

6.2 LIKELIHOOD CONSTRUCTION IN PARAMETRIC MODELS

As in Section 3.2, consider n individuals to have been placed on test at time 0 and suppose that the risk of failure at time t is determined by the hazard $\lambda(t; x, \theta)$, where x is, as usual, a vector of fixed basic covariates measured in advance and θ is a vector of unknown parameters. The data for the ith individual are t_i, δ_i, x_i, where t_i is the failure time $(\delta_i = 1)$ or censored time $(\delta_i = 0)$. In what follows, we suppose the underlying failure times to be continuous. We also make use of the counting process notation and so let N_i and Y_i be, respectively, the right-continuous counting process of observed failures and the left-continuous at-risk process for the ith individual.

193

In Section 3.2, a derivation of the likelihood was given which assumed a random censorship model. It was noted there, however, that this model is not sufficiently general to encompass many censoring schemes (e.g., type II censoring) which are frequently used in some application areas. In this section it is shown that the likelihood (3.2) given by

$$L = \prod_{i=1}^{n} [f(t_i; x_i, \theta)^{\delta_i} F(t_i; x_i, \theta)^{1-\delta_i}] \tag{6.1}$$

$$= \prod_{i=1}^{n} \lambda(t_i; x_i, \theta)^{\delta_i} \exp\left[-\int_0^{\infty} \sum_{l=1}^{n} Y_l(u) \lambda(u; x_l, \theta) \, du \right] \tag{6.2}$$

is the appropriate likelihood under a broad class of censoring mechanisms.

For random censorship, the likelihood (6.1) or (6.2) can be derived by considering the independent contributions (t_i, δ_i, x_i), $i = 1, \ldots, n$. More generally, however, the likelihood is formed as the product of conditional contributions, of the entire study group, over successive infinitesimal time intervals. As in Chapter 5, let the history or filtration be

$$\mathcal{F}_t = \sigma\{N_i(u), Y_i(u^+), x_i, i = 1, \ldots, n, 0 \le u \le t\},$$

for all $t > 0$, so that $\mathcal{F}_{t^-}$ represents the complete history of the study up to but not including time t and $\mathcal{F}_0 = \{x_i, i = 1, \ldots, n\}$ represents the information available at time 0. The likelihood can be constructed as a product of the conditional terms

$$P[\mathcal{F}_{t^-+dt} \mid \mathcal{F}_{t^-}] = P[D_t(dt), C_t(dt) \mid \mathcal{F}_{t^-}]$$
$$= P[D_t(dt) \mid \mathcal{F}_{t^-}] P[C_t(dt) \mid \mathcal{F}_{t^-}, D_t(dt)], \tag{6.3}$$

where $D_t(dt) = \{l : dN_l(t) = 1\}$ is the set of labels associated with the individuals failing in $[t, t + dt)$ and $C_t(dt)$ is the set of labels associated with individuals censored in $[t, t + dt)$. It follows that

$$P[D_t(dt) \mid \mathcal{F}_{t^-}] = \prod_{l=1}^{n} \{ [\lambda(t; x_l, \theta) \, dt]^{dN_l(t)} [1 - \lambda(t; x_l; \theta) \, dt]^{Y_l(t) - dN_l(t)} \}, \tag{6.4}$$

where it has been assumed that:

1. Given $\mathcal{F}_{t^-}$, the failure mechanisms act independently over $[t, t + dt)$.
2. $P[dN_l(t) = 1 \mid \mathcal{F}_{t^-}] = Y_l(t) P[T_l \in [t, t + dt) \mid T_l \ge t, x_l]$. \tag{6.5}

We have essentially adopted the convention that censorings are presumed to follow failures in the interval $[t, t + dt)$. Such a convention is natural and necessary to deal with schemes such as type II censoring, where items are censored when certain failures occur. But more generally, we consistently break ties between censorings and failures by placing the failures first.

Assumption 2 above is the independent censoring assumption, and as discussed before, censoring mechanisms that satisfy it include type I and II censoring along with progressive type II censoring or indeed any censoring rule that when applied at time t, depends only on events that are in the history $\mathscr{F}_{t^-}$ and possibly on random mechanisms external to the failure process. Essentially, we require that conditional on x, the items withdrawn from risk at time t should be "representative" of the items at risk. In particular, items cannot be censored because they appear to be at unusually high or low risk of failure.

The total likelihood can now be written as a product integral

$$\mathscr{L} = P(\mathscr{F}_0)\mathscr{P}_0^\infty P(\mathscr{F}_{t+dt} \mid \mathscr{F}_{t^-})$$
$$= P(\mathscr{F}_0) \lim \prod_{i=0}^{m-1} P(\mathscr{F}_{\tau_i^- + \Delta\tau_i} \mid \mathscr{F}_{\tau_i^-}),$$

where $0 = \tau_0 < \cdots < \tau_m < \infty$, $\Delta\tau_i = \tau_i - \tau_{i-1}$, $\tau_m \to \infty$ as $m \to \infty$, and the limit is taken as $m \to \infty$ and $\Delta\tau_i \to 0$. Using (6.3) and assuming that $P(\mathscr{F}_0) = 1$ (or conditioning on $\mathscr{F}_0$), we have

$$\mathscr{L} = \mathscr{P}_0^\infty P[D_t(dt) \mid \mathscr{F}_{t^-}]\mathscr{P}_0^\infty P[C_t(dt) \mid \mathscr{F}_{t^-}, D_t(dt)].$$

The first factor on the right side arises from the failure information and, apart from differential elements, reduces to

$$L(\theta) = \prod_{i=1}^n \lambda(t_i; x_i, \theta)^{\delta_i} \mathscr{P}_0^\infty \prod_{l=1}^n [1 - \lambda(t; x_l, \theta)\, dt]^{Y_l(t) - dN_l(t)}$$
$$= \prod_{i=1}^n \lambda(t_i; x_i, \theta)^{\delta_i} \mathscr{P}_0^\infty [1 - \sum_{l=1}^n Y_l(t)\lambda(t; x_l, \theta)\, dt].$$

This reduces further to (6.2) and thence to (6.1).

The remaining factor in $\mathscr{L}$,

$$\mathscr{P}_0^\infty P[C_t(dt) \mid \mathscr{F}_{t^-}, D_t(dt)], \tag{6.6}$$

arises from the censoring contributions. If (6.6) depends on θ, the censoring mechanism is said to be *informative*, and otherwise, *noninformative*. If the censoring is informative, the likelihood (6.1) is not complete since it ignores the censoring contributions. Even in this case, however, if the censoring is independent, (6.1) has the interpretation of a partial likelihood [based on $\{D_t(dt)\}$ in the sequence $\{D_t(dt), C_t(dt)\}$ (see Section 4.2.1)] and so can still be used for inference, although there may be an associated loss in efficiency. This distinction between informative and noninformative censoring schemes is similar to the distinction made in stopping rules as discussed, for example, in Cox and Hinkley (1974, p. 40).

It is instructive to consider some specific censoring schemes in the construction above. For example, suppose that it is decided to censor all surviving items at the kth failure or a predetermined time τ_0, whichever occurs first. The censoring scheme is independent, and even though the marginal censoring probabilities clearly depend upon θ, it is noninformative, which can be seen by considering the contributions to (6.6). Given $\mathscr{F}_{t-}$ and $D_t(dt)$, this censoring scheme is deterministic and each contribution is unity. Realistic examples of informative but independent censoring schemes are hard to construct. An artificial example is provided by a hypothetical study in which the censoring time for each individual is determined as the failure time of a similar individual not included in the analysis. Censoring schemes that are not independent arise if items are censored selectively when they appear to be at high or low risk of failure compared to others at risk with the same covariate value x. For example, the censoring might be allowed to depend on a time-dependent covariate $x^*(t)$, which measures the health status of the individual at time t. If such censoring is in effect, the fact that an item is not censored will alter the conditional failure probabilities in (6.4) and violate assumption 2 in (6.5). Such censoring schemes can be made independent by including $x^*(t)$ in the model as discussed in the next section. But this can be at the expense of making other covariate (e.g., treatment) effects indecipherable.

6.3 TIME-DEPENDENT COVARIATES AND FURTHER REMARKS ON LIKELIHOOD CONSTRUCTION

In this section, conditions on the basic covariates x are relaxed to allow them to vary over time. Let $x_i(t)$ denote the covariate vector at time t for the ith individual under study, and let $X_i(t) = \{x_i(u); 0 \leq u < t\}$ give the covariate history up to time t. We assume that the data for the ith individual are $[t_i, \delta_i, X_i(t_i)]$, $i = 1, \dots, n$. Models are defined in terms of a vector of derived left continuous covariates $Z_i(t)$ whose elements are functions of $X_i(t)$ and t. We define the hazard function at time t as

$$\lambda[t; X(t)] \, dt = P\{T \in [t, t+dt)|X(t), T \geq t\}, \tag{6.7}$$

which conditions on the covariate path up to time t.

Time-dependent covariates fall into two broad general classes. A time-dependent covariate that satisfies the condition

$$P\{T \in [u, u+du)|X(u), T \geq u\} = P\{T \in [u, u+du)|X(t), T \geq u\} \tag{6.8}$$

for all u, t such that $0 < u \leq t$, is called *external*. A condition equivalent to (6.8) is

$$P[X(t)|X(u), T \geq u] = P[X(t)|X(u), T = u], \qquad 0 < u \leq t, \tag{6.9}$$

which formalizes the idea that whereas the covariate $x(\cdot)$ may influence the rate of failures over time, its future path up to any time $t > u$ is not affected by the occurrence of a failure at time u. A covariate that is not external is called *internal*. Internal covariates typically arise as time-dependent measurements taken on an individual study subject, the path of which is affected by the survival status. In many instances, an internal covariate has the property that it requires the survival of the individual for its existence, and its path thus carries direct information on the time to failure if failure is defined as the death of the individual.

6.3.1 External Covariates

One type of external covariate is the *fixed* covariate $x(t) = x$, whose value is measured in advance and fixed for the duration of study. Except for a brief discussion in Chapter 5, attention has been directed entirely to a vector of fixed basic covariates and derived covariates $Z(t)$ that were allowed to incorporate interactions with functions of time. A second type of external covariate is *defined* in that its total path up to any time t, $X_i(t)$ is determined in advance for each individual under study. One example would be a stress factor under control of the experimenter that is to be varied in a predetermined way; another example would be the age of an individual in a trial of long duration; interactions with time could also be viewed as defined covariates and incorporated into the basic covariates $x(t)$. A third type of external covariate is termed *ancillary*. A covariate of this sort is the output of a stochastic process that is external to the individual under study and whose probability laws do not involve the parameters in the failure time model under study. An example of such a covariate would be one that measures airborne pollution as a predictor for the frequency of asthma attacks. Ancillary covariates play the role of ancillary statistics for the failure time model. External time-dependent covariates also arise, for example, when the level of a stress factor to be applied to individuals at risk at time t is allowed to depend on the number and timing of previous failures in the trial.

For external covariates, we refer to (6.8) and the condition (6.9) and note that as a consequence, we can define a survivor function conditional on the covariate path,

$$F[t; X(t)] = P[T \geq t | X(t)].$$

It can be seen that the usual relationships between survivor and hazard functions hold since

$$F[t; X(t)] = \mathcal{P}_0^t \{1 - P[T \in [u, u + du)|X(t), T \geq u]\}$$

$$= \exp\left[-\int_0^t \lambda\{u : X(u)\} du\right].$$

As a simple example of a defined covariate, suppose that a voltage $\exp[x_i(t)]$, where $x_i(t) = \alpha_i \log t$ for some $\alpha_i > 0$, is to be applied at time t to the ith item in a test of insulation in electrical cable. Suppose further that the failure rate at time t for the ith item is the exponential regression model

$$\lambda[t; X_i(t), \beta] = \exp[\beta_0 + \beta_1 x_i(t)].$$

It is then easily seen that

$$\lambda[t; X_i(t), \beta] = \eta t^{\gamma_i - 1}$$

of the Weibull form, where $\eta = \exp(\beta_0)$ and $\gamma_i = \alpha_i \beta_1 + 1$. To some extent, such covariates offer nothing new but merely give a regression interpretation to certain parameters in the model. The log-likelihood function on data $[t_i, \delta_i, X_i(t_i)]$ for independent censoring is, from (6.1),

$$\sum \delta_i[\beta_0 + \beta_1 x_i(t_i)] - \int_0^\infty \sum_{l=1}^n Y_l(u) \exp[\beta_0 + \beta_1 x_l(u)] \, du, \qquad (6.10)$$

which may also be written as the product of Weibull densities and survivor functions. If the voltage applied at time t were randomly determined by some external process, essentially the same model would apply after conditioning and (6.10) would be the appropriate likelihood.

6.3.2 Internal Covariates

An internal covariate is typically the output of a stochastic process that is generated by the individual under study and in many instances is observed only as long as the individual survives and is uncensored. As a consequence, its observed value may carry information about the failure time. A simple example arises in a clinical trial where some measure of a patient's general condition is made at regular intervals and failure occurs when a patient dies. Suppose that at time t, values of 0 and 4 are assigned to $x(t)$ for dead and no clinical evidence of disease, respectively, and 3, 2, 1 represent intermediate levels of decreasing ability. A patient typically moves from one state to another over time and the hazard at time t depends markedly on $X(t)$. A second example arises in an immunotherapy trial in cancer. In such a trial, it may be of interest to examine the effect of an immunotherapeutic intervention on the risk of death given a current measure of immune status. In this case, the covariate $x(t)$ may be taken to specify immune status at time t, and models could be constructed to evaluate treatment effects while adjusting for immune status measurement or to allow treatment effects to depend on immune status.

It is clear that such covariates must be handled differently than ancillary or defined covariates since $X(t)$ can determine the failure information for the corresponding individual for times $u < t$. The hazard function is again defined by (6.8), which, at time t, conditions on the covariate process up to time t^-, but not further. For internal covariates, however, condition (6.9) does not hold, and as a consequence, this hazard is not related directly to a survivor function. Indeed, for the general condition covariate $x(t)$ described above,

$$P[T \geq t | X(t)] = 1,$$

provided that $x(t^-) \neq 0$.

With time-dependent covariates, whether internal or external, the formal construction of the likelihood proceeds in essentially the same way as in Section 6.2. The probability contribution of the interval $[t, t + dt)$ is written as

$$P(\mathcal{F}_{t^- + dt}|\mathcal{F}_{t^-}) = P[D_t(dt)|\mathcal{F}_{t^-}]P[\mathcal{X}(t + dt)|\mathcal{F}_{t^-}, D_t(dt)]$$
$$\times P[C_t(dt)|\mathcal{F}_{t^-}, D_t(dt), \mathcal{X}(t + dt)]. \qquad (6.11)$$

Here, $\mathcal{F}_{t^-} = \{N_i(u), X_i(u), Y_i(u) : i = 1, \ldots, n, 0 \leq u < t\}$ contains, in addition to the failure and censoring information, all information on the covariates up to time t, whereas $\mathcal{X}(t) = \{X_i(t), i = 1, \ldots, n\}$ contains only the covariate information. The product integral of the first factor on the right side of (6.11) gives, as before, the likelihood (6.2) under assumptions similar to (6.5). If the covariate is external, expression (6.1) also is valid, with F and f replaced with $F(t; X(t), \theta)$ and $f[t; X(t), \theta]$. When the covariate history $X(t)$ contains internal components, however, this expression is not useful or natural since the functions F and f do not have survivor and density function interpretations.

The other factors in (6.11) are

$$\mathcal{P}_0^\infty P[\mathcal{X}(t + dt)|\mathcal{F}_{t^-}, D_t(dt)] \qquad (6.12)$$

and

$$\mathcal{P}_0^\infty P[C_t(dt)|\mathcal{F}_{t^-}, D_t(dt), \mathcal{X}(t + dt)], \qquad (6.13)$$

which correspond to the contributions of the covariate and the censoring processes, respectively. Typically, neither of these involves the parameter θ of interest in the model $\lambda[t; X(t), \theta]$. If either (6.12) or (6.13) depends on θ, however, the likelihood (6.2) still has an interpretation as a partial likelihood based on $\{D_t(dt)\}$ in the sequence $\{D_t(dt), C_t(dt), \mathcal{X}(t + dt)\}$. It can still be used for inference but may be inefficient. It should be noted that the likelihood (6.2) gives information only on the instantaneous failure rate given $X(t)$. Inferences about the marginal distribution of T would require an integration over the distribution of $X(t)$ or a model for failure time in which $x(t)$ is suppressed.

In a randomized trial it is important to remember that an internal covariate process generally takes its values subsequent to the treatment assignment. As a consequence, such covariates may be *responsive* in that their values may be influenced by the treatment assignment. A conditional analysis gives information on instantaneous failure rates given the covariate values and the treatment assignment. If the effect of treatment is reflected predominantly in the covariate process, such an analysis will show little or no treatment differences. This can give useful information about the mechanism by which the treatment operates. Care must be exercised in interpretation, however, since treatment differences could be large despite the null results of this analysis. Suppose, for example, that the covariate for the general condition defined above is being used and that all individuals begin in the same state $[x(0) = 2$, say]. If one treatment decelerates the passage through the levels of $x(t)$, but the failure rate within each state is the same for both treatments, an analysis conditional on $x(t)$ shows no treatment difference. It is still possible, however, that the one treatment is greatly superior to the other, but this effect is predominantly through $x(t)$.

A censoring scheme that depends on the level of a time-dependent covariate $x(t)$ (e.g., general condition) is, as noted before, not independent if $x(t)$ is not included in the model. One way to circumvent this is to include $x(t)$ in the failure time model, but also to model and analyze $x(t)$ in relation to the treatment. Combining the fitted models gives estimates of the joint event of failure in $[t, t + dt)$ and $x(t)$. In principal, at least, this can be integrated to give estimates of the marginal hazard rates for each treatment group.

6.4   TIME DEPENDENCE IN THE RELATIVE RISK MODEL

6.4.1   Time-Dependent Covariates

In Section 4.2, the relative risk regression model was analyzed for fixed basic covariates x. It is straightforward to apply the same arguments to the relative risk model with time-dependent covariates. As before, $x(t)$ is a vector of time-dependent covariates, the elements of which may be internal or external, and $X(t) = \{x(u), 0 \leq u < t\}$ is the covariate history. The derived covariates $Z(t)$ are functions of $X(t)$ and t and we consider the hazard process

$$\lambda[t; X(t)]\, dt = P\{T \in [t, t + dt)|X(t), T \geq t\}$$
$$= \lambda_0(t) \exp[Z(t)'\beta]\, dt, \tag{6.14}$$

where Z has left-continuous sample paths.

As in Section 4.2, the sample consists of k failure times $t_1 < t_2 < \cdots < t_k$ with no ties, so that the remaining $n - k$ observations are right censored. The argument leading to a partial likelihood for β proceeds exactly as in Section 4.2, except that B_j now specifies the censoring *and covariate* information in the interval $[t_{(j-1)}, t_{(j)})$ plus the information that an individual fails at $t_{(j)}$, whereas A_j specifies the particular individual that fails. As before, the jth term in the partial likelihood is

$$L_j(\beta) = f(a_j|b^{(j)}, a^{(j-1)})$$
$$= \frac{\lambda[t_j; Z_j(t_j)]}{\sum_{l=1}^{n} Y_l(t_j)\lambda[t_j; Z_l(t_j)]}.$$

For the relative risk model (6.14), this gives rise to the same partial likelihood

$$L(\beta) = \prod_{j=1}^{k} \frac{\exp[Z_j(t_j)'\beta]}{\sum_{l=1}^{n} Y_l(t_j)\exp[Z_l(t_j)'\beta]}, \tag{6.15}$$

and generalizations and approximations to incorporate ties are also as before.

The partial likelihood (6.15) ignores the contributions to the likelihood (or partial likelihood) (6.2) that are made by the process $N.(t) = \sum_{i=1}^{n} N_i(t), t > 0$, which records the total number of failures up to time t. Thus the term in (6.11) that generates the likelihood (6.2) is being factored as

$$P[D_t(dt)|\mathscr{F}_{t-}] = P[D_t(dt)|\mathscr{F}_{t-}, dN.(t)]P[dN.(t)|\mathscr{F}_{t-}].$$

The product integral of the first term on the right side of this equation is the partial likelihood (6.15). The product integral of the second term gives the conditional contributions of the process $N.(t)$ to the likelihood (6.2). This second term is informative about β if a particular parametric form for $\lambda_0(\cdot)$ is specified. Without any restrictions on $\lambda_0(\cdot)$, however, the process observed $N.(t)$ can be accounted for by a hazard near 0 in the intervals containing no failures and a large contribution at the observed failure times. In this case, it would appear that little information about β is lost through ignoring the second term.

The estimation of the baseline cumulative hazard function $\Lambda_0(t) = \int_0^t \lambda_0(u)\,du$ can also be extended to more general time-dependent covariates without change. Thus, the arguments of Section 4.2. can be extended directly to obtain the Nelson–Aalen estimator $\hat{\Lambda}_0(t)$, where

$$d\hat{\Lambda}_0(t) = \frac{dN.(t)}{\sum_{i=1}^{n} Y_i(u)\exp[Z_i(t)'\beta]},$$

as before. Alternatively, the maximum likelihood estimator discussed in Section 4.2.4 can be generalized to this case. If the covariates are all external, the baseline survivor function $F_0(t) = \exp[-\int_0^t \lambda_0(u)\,du]$ can be estimated as in Section 4.2.4, but with internal time-dependent covariates, the function $F_0(t)$ has no simple interpretation, and summaries should be given in terms of the baseline cumulative hazard function.

Some uses of internal covariates are discussed in Chapter 8 in relation to the competing risks problem and the analysis of life-history event data.

6.4.2 Time-Dependent Strata

Time dependence can also enter the relative risk model through strata, and this facilitates many extensions of the model, as we shall see in the next section and more completely in Chapter 8. At any given time t, suppose that a study individual can be in any one of m distinct strata and that over time, individuals can move from one stratum to another. As before, let $X(t)$ be the individual's basic covariate history up to time t and $Z(t)$ be the derived covariate vector of dimension p. Denote the stratum occupied at time t by $J(t)$, which, like $Z(t)$, is a function of $X(t)$ with left-continuous sample paths. Consider the following model for the hazard rate:

$$\lambda[t; X(t)]\,dt = P[T \in [t, t+dt)|X(t), T \geq t]$$
$$= \lambda_{0j}(t)\exp[Z(t)'\beta] \qquad \text{if } J(t) = j, \tag{6.16}$$

where $j = 1, \ldots, m$ and $t > 0$. The special case of fixed strata was discussed in Section 4.4. Note that we have left a common regression parameter β over the different strata, which at first glance may suggest that the model requires the same relationship between elements of $Z(t)$ and the hazard in each stratum. However, this specification is without loss of any generality, since the components of $Z(t)$ may include interaction terms with stratum.

To apply this model, it is convenient to introduce a counting process notation. Let $N_i(t) = [N_{1i}(t), \ldots, N_{mi}(t)], t > 0$ be a multivariate counting process associated with the ith individual where the $N_{ji}(t)$ count the observed number of failures for individual i that occur in stratum j over the interval $(0, t], j = 1, \ldots, m, i = 1, \ldots, n$ and $t > 0$. Note that $N_{.i}(t) = \sum_{j=1}^{m} N_{ji}(t)$ is the usual counting process for individual i, and that $N_{j.}(t) = \sum_{i=1}^{n} N_{ji}(t)$ counts the total number of failures that occur in stratum j. The processes $N_{ji}(t)$ are assumed to have sample paths that are right continuous with left-hand limits (cadlag) and it is assumed that no two processes can jump simultaneously. As before, define an at-risk indicator for the ith individual as $Y_{.i}(t)$, which indicates that the individual is failure-free and uncensored at time t^-, and let $Y_{ji}(t) = \mathbf{1}[J_i(t) = j, Y_{.i}(t) = 1]$ be the stratum j at risk indicator for the ith individual, $i = 1, \ldots, n; j = 1, \ldots, m$. Let $Y_i(t) = [Y_{1i}(t), \ldots, Y_{mi}(t)]$. The history of filtration process is $\mathscr{F}_t = \{N_i(t), Y_i(t^+), X_i(t^+), i = 1, \ldots, n; 0 \le u \le t\}$. Under independent censoring,

$$P[dN_{ji}(t) = 1 \mid \mathscr{F}_{t^-}] = Y_{ji}(t) \exp[Z_i(t)'\beta]\lambda_{0j}(t)\, dt$$

for all $i, j, t,$ and $\mathscr{F}_{t^-}$.

A partial likelihood for β can be constructed, as for the case of fixed strata, by conditioning on the outcomes as they occur in each stratum separately and then combining across strata. Thus, terms in the partial likelihood from the jth stratum arise from expressions of the following form:

$$P[dN_{ji}(t) = 1 \mid \mathscr{F}_{t^-}, dN.(t)] = \frac{Y_{ji}(t) \exp[Z_i(t)'\beta]\lambda_{0j}(t)\, dt}{\sum_{l=1}^{n} Y_{jl}(t) \exp[Z_l(t)'\beta]\lambda_{0j}(t)\, dt},$$

from which $\lambda_{0j}(t)\, dt$ cancels. If there are no ties and $t_{j1} < t_{j2} < \cdots < t_{jk_j}$ are the k_j failure times in the jth stratum, $j = 1, \ldots, m$, the partial likelihood for β can be written

$$L(\beta) = \prod_{j=1}^{m} \left\{ \prod_{i=1}^{k_j} \frac{\exp[Z_i(t_{ji})'\beta]}{\sum_{l=1}^{n} Y_{jl}(t_{ji}) \exp[Z_l(t_{ji})'\beta]} \right\}.$$

The corresponding score function can be written

$$U(\beta) = \sum_{j=1}^{m} \int_0^{\infty} \sum_{i=1}^{n} [Z_i(t) - \mathscr{E}_j(\beta, t)]\, dN_{ji}(t), \qquad (6.17)$$

where

$$\mathscr{E}_j(\beta, t) = \sum_{l=1}^{n} Z_l(t) p_{jl}(\beta, t)$$

is the average of $Z(t)$ over the risk set in the jth stratum with weights

$$p_{jl}(t) = \frac{Y_{jl}(t) \exp[Z_l(t)'\beta]}{\sum_{u=1}^{n} Y_{ju}(t) \exp[Z_u(t)'\beta]}.$$

The corresponding observed information matrix is also just a sum of the information matrices arising from each stratum; thus,

$$I(\beta) = \sum_{j=1}^{m} \int_0^\infty \mathscr{V}_j(\beta, t)\, dN_{j\cdot}(t),$$

where $\mathscr{V}_j(\beta, t)$ is the variance of $Z(t)$ over the risk set in the jth stratum.

It can be verified that the score process can be written as a sum of integrals of predictable processes with respect to orthogonal martingales:

$$U(\beta, t) = \sum_{j=1}^{m} \int_0^t \sum_{i=1}^{n} [Z_i(u) - \mathscr{E}_j(\beta, u)]\, dM_{ji}(u), \qquad (6.18)$$

where

$$M_{ji}(t) = N_{ji}(t) - \int_0^t Y_{ji}(u)\lambda_{0j}(u) \exp[Z_i(u)'\beta]\, du$$

for $t > 0, j = 1, \ldots, m$ and $i = 1, \ldots, n$. Thus, $U(t)$ is itself a martingale whose predictable variation process is obtained easily from the results of Section 5.3. Asymptotic results for the MLE arising from the partial likelihood or partial likelihood score (6.17) can be obtained in a manner very similar to that used for the unstratified Cox model as discussed in Section 5.7. Andersen et al. (1993) derive the asymptotic results in this more general stratified context.

In Section 6.4.3 we look at one particular example where time-dependent strata and covariates arise naturally in the analysis. In Chapter 8 we consider competing risks and related topics involving Markov and modulated Markov processes where the notion of time-dependent strata is very valuable.

6.4.3 Example: Stanford Heart Transplant Data

An interesting data set that can be modeled using either time-dependent strata or covariates is the heart transplant data taken from Crowley and Hu (1977) and reproduced in Appendix A, data set IV. A brief description of these data is given in Section 1.2. From the time of admission to study until the time of death, a patient was eligible for a heart transplant. The time to transplant (the waiting time) is denoted by W and the time to death (from admission to study) by T. One issue of interest is a comparison of survival experience of transplanted and nontransplanted patients. Figure 6.1 gives a simple illustration of the process. An individual begins in state 1 (alive and nontransplanted) at admission to the program. From

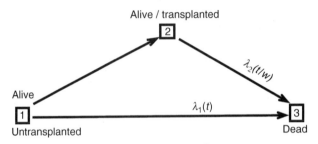

Figure 6.1 Compartment model for the heart transplant data.

there, the patient either moves next to state 2 (alive and transplanted) if a heart becomes available before death or directly to the death state, state 3.

Ignoring covariates for the moment, consider the hazard functions

$$\lambda_1(t)\,dt = P\{T \in [t, t+dt) | T \ge t, W \ge t\}$$

and for $t > w$,

$$\lambda_2(t|w)\,dt = P\{T \in [t, t+dt) | T \ge t, W = w\}.$$

Interest centers on the comparison of these two hazard functions. For these comparisons to be interpreted unambiguously, it is essential that there be no selection in the assignment of hearts to individuals; that is, the assignment must be made randomly to eligible individuals. The randomization may be restricted by taking account of covariates included in the model or of covariates not correlated with subsequent survival (in the absence of transplantation). The donor–recipient tissue-matching variables would presumably be variables used in the assignment, and it must be supposed that they fall into the latter category. Since assignments were not made at random, the possibility of selection exists, and this analysis is presented for illustrative purposes only.

One possible approach to the analysis of such data is to specify particular parametric forms for the hazards, $\lambda_1(t)$ and $\lambda_2(t|w)$. This approach could be extended to include covariates, and methods would follow the methods of Chapter 3 and likelihood constructions discussed in Section 6.2. Alternatively, one might adopt nonparametric models such as

$$\lambda_1(t) = \lambda_{01}(t); \quad \lambda_2(t|w) = \lambda_{02}(t), \qquad t > w, \tag{6.19}$$

where the hazard at each time $t > w$ may be affected by transplantation in an arbitrary way, but no further role is played by the waiting time w. It should be noted that in the nonparametric case, an assumption such as (6.19) is very useful if appropriate. Otherwise, $\lambda_2(t|w)$ will require separate estimation at each value of w. This could be handled, only with sufficient data, by grouping on or smoothing over the w variable. The model (6.19) is a simple stratified model and covariates can be incorporated as discussed in Section 6.4.2.

The basic covariates $x(t)$ can be taken to include baseline measurement taken at entry to the study, such as age, previous surgery, and year of admission, as well as donor–recipient matching data taken at the time of transplant. It also includes a time-dependent indicator of transplant status, $x_0(t) = \mathbf{1}(W < t)$. Let $Z(t)$ be a vector of p derived covariates and consider hazard models of the form

$$
\begin{aligned}
\lambda_1[t|X(t)] &= \lambda_{01}(t)\exp[Z(t)'\beta_1] \\
\lambda_2[t|w, X(t)] &= \lambda_{02}(t)\exp[Z(t)'\beta_2], \qquad t > w,
\end{aligned}
\tag{6.20}
$$

where β_1 and β_2 are vectors of regression parameters measuring, respectively, the effect of $Z(t)$ on the marginal and conditional hazards. Note that different covariates can be included in the two parts of (6.20). For example, if age is to appear in the first hazard but not the second, we merely take the coefficient in β_2 corresponding to age to be zero.

This model is a special case of those considered in Section 6.4.2, with transplant status corresponding to the (time-dependent) strata. In fact, (6.20) can be written in the form (6.16) by considering covariates $\{[1 - x_0(t)]Z(t)', x_0(t)Z(t)']$ and $\beta = (\beta_1', \beta_2')$. Thus, the arguments of that section lead to a partial likelihood for β_1 and β_2. For the ith patient, let $N_{1i}(t)$ and $N_{2i}(t)$ count the observed number of deaths in the interval $(0, t]$ without and with a transplant, $i = 1, \ldots, n$. Let $Y_{ji}(t)$ be the indicator that individual i is at risk and untransplanted ($j = 1$) or transplanted ($j = 2$) at time t. The log partial likelihood (if there are no ties) is then

$$
\log L(\beta_1, \beta_2) = \sum_{j=1}^{2}\sum_{i=1}^{n}\int_0^{\infty}\left(Z_i(t)'\beta_j - \log\left\{\sum_{l=1}^{n}Y_{jl}(t)\exp[Z_l(t)\beta_j]\right\}\right)dN_{ji}(t).
\tag{6.21}
$$

Note that (6.21) is a special case of the stratified partial likelihood of Section 6.4.2.

A model similar to (6.20) but that incorporates transplant status as a time-dependent covariate rather than as time-dependent strata provides a direct way to make comparisons between failure rates of the transplanted and nontransplanted groups. The hazard function is specified to be

$$
\lambda[t; X(t)] = \lambda_0(t)\exp[Z(t)'\beta_1 + x_0(t)\beta_0 + x_0(t)Z(t)'\beta_2],
\tag{6.22}
$$

a standard unstratified Cox model with time-dependent covariates. In this model, $\beta_0 + Z(t)'\beta_2$ measures the effect of transplantation at the covariate value $Z(t)$, and β_2 can be viewed as measuring the interactions between transplantation and the components of $Z(t)$. No absolute interpretation can be given β_0 since its meaning is affected by location changes in the covariates $Z(t)$. In the discussion above, the assumption is being made that a main effect for $Z_i(t)$ has been estimated in β_{1i}.

If a covariate is included in the time-dependent part and not in the constant part, the interpretation of its coefficient is unclear; its value is influenced both by a dependence of survival overall and by a dependence of posttransplant survival on that variable. Thus, such variables as waiting time or matching variables are difficult to interpret except as posttransplant prognostic factors; there is no comparable control group. It might, however be reasonable to assume that the matching variables are not in themselves correlated with subsequent survival experiences and would have an effect only when combined with transplantation.

The model (6.22) was fitted to the transplantation data. To remove ambiguity in the data, the following conventions were adopted: Ties between waiting times and failure (death) times were broken conservatively by placing the failure first except in the one case where the patient died on the day of transplant. As usual, ties between censored times and failure times were broken by placing failure times first.

Table 6.1 gives estimates of the regression coefficients and estimated standard errors for a number of models that were fitted to these data. As well as those reported here, a number of models using measures of mismatch were fitted, but none of these measures were found to correlate with subsequent survival. A remarkable feature of these data is the dependence of survival time on the time of acceptance to the study. The main effect of year of acceptance is significant at less than

Table 6.1   Regression Coefficients for Models of the Form (6.22) Fitted to the Heart Transplant Data [a]

Model	Main Effects, β_1 [b]			Transplant Status, β_0	Interactions, β_2 [b]		
	Z_1	Z_2	Z_3		Z_1	Z_2	Z_3
1	0.0138		−0.546	0.118	0.035		−0.291
	(0.018)		(0.611)		(0.027)		(0.758)
2		−0.265		−0.282		0.135	
		(0.105)				(0.140)	
3	0.0155	−0.274		−0.588	0.033	0.201	
	(0.017)	(0.105)			(0.028)	(1.42)	
4		−0.254	−0.236	−0.292		0.164	−0.550
		(0.107)	(0.628)			(0.141)	(0.775)
5	0.0150	−0.135	−0.420	0.077	0.027		−0.298
	(0.018)	(0.071)	(0.615)		(0.027)		(0.758)
6[c]	0.0152	−0.136	−0.621	0.048	0.027		
	(0.018)	(0.071)	(0.368)		(0.027)		

[a] Values in parentheses are estimated standard errors.
[b] Z_1 = age of acceptance − 48; Z_2 = year of acceptance − 1967; Z_3 = surgery (1 = yes, 0 = no).
[c] The estimated covariance matrix for model 6:

$$I^{-1} = 10^{-4} \begin{pmatrix} 3.06 & -0.41 & -2.30 & -19.39 & -3.03 \\ -0.41 & 50.27 & -50.99 & 14.12 & 2.85 \\ -2.30 & -50.99 & 1353.28 & -34.43 & 5.40 \\ -19.39 & 14.12 & -34.43 & 1037.97 & 15.07 \\ -3.03 & 2.85 & 5.40 & 15.07 & 7.37 \end{pmatrix}$$

the 5% level in all the models in which it is included. Unfortunately, the year of acceptance interaction with transplantation is also approaching significance but in the opposite direction. Together, these suggest that the overall quality of patient being admitted to the study was improving with time (possibly due to relaxation of admission requirements or to improving patient management) but the survival time of transplanted patients is not improving at the same rate. In fact, the sum of the two coefficients for year of acceptance would suggest a nearly constant survival pattern for the transplanted patients. It may be possible to explain such dependence on time of admission if there were some measure of general conditions of the patients placed in a study. An examination of such a variable could determine whether the selection for transplant moved in the direction of the poorer-risk patients as time progressed, while the overall general condition of patients improved.

If it is assumed that the interaction of year of acceptance and transplantation is zero, models can be fitted that involve only the main effect term for Z_2. From model 5 one finds that a test that the age interaction term is zero is nonsignificant ($p = 0.32$ on a two-tailed test). If the interaction is set to zero and only a main effect is estimated, this coefficient is also not significantly different from zero. The hypothesis that both the main effect β_{11} and the interaction β_{21} are zero yields a significance level of 0.09 and the hypothesis $\beta_{11} + \beta_{12} = 0$ is rejected at the 5% level. It would seem then that the effect of age is potentially complicated and a final model should include both β_{11} and β_{21}. For interpretation, it is important that models be hierarchical and it would be inappropriate to fit a model with only the β_{21} term and no main effect.

There is no evidence of an interaction between prior surgery and transplantation. In model 6 with just the main effect, $\hat{\beta}_{13} = -0.621$ with a standard error of 0.368 and a 10% significance level. There is perhaps some mild evidence of a better prognosis overall for patients who have had previous heart surgery, but this is based on only 16 cases with previous surgery.

This analysis suggests that transplantation may be beneficial for younger patients. The critical age is $48 + c$ with c estimated by $\hat{c}$, where $0.048 + 0.027\hat{c} = 0$ or $\hat{c} = -1.8$. Thus, it might be argued that for patients under 46 years of age, transplantation is beneficial, but for those over 46, there is no evidence of such a benefit as far as survival is concerned. Note, however, that Fieller's theorem could be used to place an approximate 95% confidence limit on c as the set of solutions to

$$\frac{|\, 0.048 + 0.027c \,|}{\sqrt{0.10397 + 0.003014c + 0.00737c^2}} \leq 1.96.$$

This gives the interval $(-\infty, \infty)$, which would suggest that c is estimated with considerable imprecision. This rather surprising result is due to the fact that the estimation of c is asymptotically equivalent to the estimation of the ratio of two normal means. The interval $(-\infty, \infty)$ arises since the data on the denominator mean are not significantly different from 0.

The conclusions above based on the model (6.22) are in close agreement with those reached when (6.20) is used. The difficulty with (6.20) is that there is no easily used measure of the difference between the transplanted and nontransplanted groups. Inferences must be based on the comparison of the estimated survivor functions.

6.5 NONNESTED CONDITIONING EVENTS

To this point, all hazard functions have been defined by conditioning on the entire past history $X(t) = \{x(u), 0 \leq u < t\}$ of the covariate process; see, for example, (6.7). In some instances, however, one may wish to specify models for the failure rates that use only a part of the preceding covariate history and, for example, that condition only on current (or perhaps recent) values of the covariate process. In this section we discuss some methods that can be used for inference in such models. Related ideas in the context of models for recurrent events are discussed in detail in Section 9.4.

We suppose that observations are collected on individuals over an interval $[0, \tau]$ and that the data $\{N_i(t), Y_i(t), X_i(t), 0 \leq t \leq \tau\}$ are independent and identically distributed, $i = 1, \ldots, n$. To set a specific framework for discussion, suppose that the covariate vector $x(t)$ has left continuous sample paths and consider models for the failure rate or hazard at time t that arise from conditioning only on the current value $x(t)$ of the basic (time-dependent) covariate vector. [It will be evident, however, that there is an immediate extension to conditioning on other incomplete specifications of $X(t)$.] Thus, let

$$\lambda[t; x(t)] \, dt = P\{T \in [t, t+dt) | T \geq t, x(t)\}, \qquad t > 0, \tag{6.23}$$

which can be compared with (6.7). Note, in fact, that

$$\lambda[t; x(t)] = E\{\lambda[t; X(t)] | x(t), T \geq t\},$$

so that (6.23) is the average failure rate that applies to individuals at risk at time t and whose covariate value at that time is $x(t)$. We assume that the censoring mechanism is such that

$$P[dN_i(t) = 1 | x_i(t), Y_i(t)] = Y_i(t)\lambda[t; x_i(t)] \, dt, \qquad i = 1, \ldots, n. \tag{6.24}$$

This is a condition of independent censoring with respect to the model (6.23). It is satisfied if the rate of censoring of items at risk at time t depends on $X(t)$ only through $x(t)$.

Various specific models could be considered for (6.23). It would be possible, for example, to consider parametric models for these average hazards, perhaps of the accelerated failure time type. We do not pursue that direction here, but rather, consider a relative risk model of the form

$$\lambda[t; x(t)] = \lambda_0(t) \exp[Z(t)'\beta], \tag{6.25}$$

where $Z(t)$ is a vector of (left-continuous) covariates whose elements are functions of t and $x(t)$.

At first glance, this does not seem much different from models like (6.7), and in fact, an argument rather like the partial likelihood can be applied to data from (6.25). Specifically, if there are no tied failure times, we can calculate

$$P[dN_i(t) = 1 \mid \mathscr{A}(t), dN.(t) = 1] = \frac{Y_i(t) \exp[Z_i(t)'\beta]}{\sum_{l=1}^{n} Y_l(t) \exp[Z_l(t)'\beta]}$$

at any failure time t. In this, the conditioning event is $\mathscr{A}(t) = \{x_i(t), Y_i(u), N_i(u), 0 \leq u < t\}$, and taking a product over the failure times gives rise to a product of terms identical to the usual partial likelihood. It is not a partial likelihood, however, since the conditioning events are not nested. Nonetheless, we could consider the usual score process

$$U(\beta; t) = \sum_{i=1}^{n} \int_0^t [Z_i(u) - \mathscr{E}(\beta, u)] \, dN_i(u)$$

$$= \sum_{i=1}^{n} \int_0^t [Z_i(u) - \mathscr{E}(\beta, u)] \, dM_i(u), \tag{6.26}$$

where $M_i(t) = N_i(t) - \int_0^t Y_i(u)\lambda_0(u) \exp[Z_i(u)'\beta] \, du$ and the $\mathscr{E}(\beta, t)$ are defined as before. Since the conditioning events $\mathscr{A}(t)$ are not nested across time, there is no filtration with respect to which $M_i(t)$ is a martingale. It can be seen, however, that $E[dM_i(t) \mid \mathscr{A}(t)] = 0$, and hence $E[M_i(t)] = 0$ for all $t \in [0, \tau]$ and $i = 1, \ldots, n$. But conditioning on $\mathscr{A}(t)$ does not fix the value of $M_i(s)$ at times $s < t$, and as a consequence, the increments in $M_i(t)$ will in general be correlated.

It follows that the total score has mean zero, so

$$U(\beta, \tau) = \sum_{i=1}^{n} U_i(\beta, \tau) = 0, \tag{6.27}$$

where $U_i(\beta, t) = \int_0^t [Z_i(u) - \mathscr{E}(\beta, u)] \, dM_i(u)$, is an unbiased estimating function. If the data arise from an IID model for (N_i, Z_i, Y_i) as discussed above, if $P[Y_i(\tau) > 0]$, and if the covariate process $Z_i(t)$ has bounded variation on the interval $[0, \tau]$, results from empirical process theory can be brought to bear on this problem. Arguments similar to those in Lin et al. (2000) and reviewed in Section 9.4 can be used to show that there is a consistent solution $\hat{\beta}$ to (6.27) and that as $n \to \infty$,

$$n^{-1/2} U(\beta_0, \tau) \xrightarrow{\mathscr{D}} N(0, \Sigma),$$

where Σ is the probability limit of $\text{var}[n^{-1/2}U\beta_0, \tau)]$. Further, Σ is consistently estimated by

$$\hat{\Sigma} = n^{-1} \sum_{i=1}^{n} \int_0^{\tau} \int_0^{\tau} [Z_i(u) - \mathscr{E}(\hat{\beta}, u)][Z_i(v) - \mathscr{E}(\hat{\beta}, v)]' d\hat{M}_i(u) \, d\hat{M}_i(v)$$

$$= n^{-1} \sum_{i=1}^{n} \hat{U}_i \hat{U}_i',$$

where $\hat{U}_i = \int_0^\tau [Z_i(u) - \mathscr{E}(\hat{\beta}, u)]' d\hat{M}_i(u)$, $\hat{M}_i(t) = N_i(t) - \int_0^t Y_i(u) \exp[Z_i(u)'\hat{\beta}] d\hat{\Lambda}_0(u)$, and $\hat{\Lambda}_0(t)$ is the Nelson–Aalen estimator. Under some additional regularity conditions for expansion of the score at β_0 about $\hat{\beta}$, it follows that $n^{1/2}(\hat{\beta} - \beta_0)$ is asymptotically normal and that its asymptotic variance can be estimated consistently with

$$\hat{V} = [n^{-1}I(\hat{\beta})]^{-1}\hat{\Sigma}[n^{-1}I(\hat{\beta})]^{-1}. \qquad (6.28)$$

In this, $I(\beta) = -\partial U(\beta, \tau)/\partial \beta'$ and (6.28) is of the sandwich form as arises from unbiased estimating equations. Note that $\hat{\Sigma}$ is the covariance matrix of the score residuals as defined in Section 6.6.

It might be argued that (6.28) is a more robust variance estimate for the time-dependent relative risk model. As we have seen before, $I(\hat{\beta})^{-1}$ provides an appropriate estimate of the variance of $\hat{\beta}$ if, given the entire history $X(t)$, the hazard at time t depends only on the current value of $x(t)$ in the way specified. The sandwich estimator in (6.28), however, is valid provided that the average dependence on $x(t)$ is correctly modeled even if given the entire covariate process, other aspects of $X(t)$ would affect the hazard at time t.

The heart transplant data provide an example. In the models considered in Table 6.1, the hazard function given $X(t)$ is assumed to depend only on the current transplant status and not further on the path of the transplant status process. In particular, there is no dependence on the time since transplant. Accordingly, the variance estimates in Table 6.1 are based on the information matrix only and are appropriate if this model is true. The sandwich estimators can also be obtained and would provide an accurate estimate of the variance provided that the average death rate is affected multiplicatively by the transplant status.

For model 6 in Table 6.1, the robust covariance matrix is

$$n\hat{V} = \begin{pmatrix} 0.000216 & -0.000015 & -0.000084 & -0.001250 & -0.000209 \\ -0.000015 & 0.005307 & -0.006888 & 0.003345 & 0.000243 \\ -0.000084 & -0.006888 & 0.128665 & -0.012486 & 0.000928 \\ -0.001250 & 0.003345 & -0.012486 & 0.100907 & 0.000476 \\ -0.000209 & 0.000243 & 0.000928 & 0.000476 & 0.000722 \end{pmatrix},$$

which differs only moderately from the uncorrected variance estimate in Table 6.1.

6.6 RESIDUALS AND MODEL CHECKING FOR THE COX MODEL

Consider again the Cox or relative risk model with hazard process defined in (6.7). We utilize the notation of Section 5.7, and examine ways of assessing the fit of the model. Residuals can be defined in various ways, and diagnostic plots can aid in the informal evaluation of the model or in discovering the nature of the failure variable's dependence on covariates that are not included in the model.

The Cox–Snell residuals were used in Section 4.5 to assess the fit of the fitted model to the head and neck clinical trial. For fixed covariates, these residuals are

$$\hat{r}_i = \exp(Z_i'\hat{\beta})\hat{\Lambda}_0(t_i), \qquad i = 1, \ldots, n \qquad (6.29)$$

where $\hat{\Lambda}_0$ is a cumulative hazard estimate. A piecewise constant estimate was used in the application of Section 4.5, although the Nelson–Aalen estimator is used more commonly. If the model is true, one might expect $(\hat{r}_i, \delta_i), i = 1, \ldots, n$ to be like a censored sample from a unit exponential distribution. As a consequence, plots of the Nelson–Aalen estimator based on these residuals, perhaps stratified by variables in the model, should approximate a straight line through the origin (see, e.g., Figures 4.10 through 4.13). With time-dependent covariates, the Cox-Snell residuals are given by

$$\hat{r}_i = \int_0^{t_i} Y_i(t) \exp[Z_i(t)'\hat{\beta}] \, d\hat{\Lambda}_0(t), \qquad i = 1, \ldots, n.$$

The *Cox-Snell residuals* were introduced by Kay (1977) and Crowley and Hu (1977). Their primary use has involved grouping covariate values and plotting estimates of cumulative hazards within strata so defined. These plots can give an overall view of fit, although the cause of departures from the expected linear form is not always easy to identify, as pointed out by Crowley and Storer (1983). In general, there is an advantage in having residuals that can easily be plotted against the observed levels of a given covariate so as to get a general view of fit with respect to that covariate or to suggest how it should be incorporated into the model. The martingale residuals discussed below are more satisfactory from this perspective.

The martingale associated with the ith individual under study is $M_i(t) = N_i(t) - \int_0^t Y_i(u) \exp[Z_i(u)'\beta] \, d\Lambda_0(u), \, t > 0, i = 1, \ldots, n$. As discussed in Section 5.7, these martingales are orthogonal and have expectation 0. It is then natural to consider the ith *residual processes*

$$\hat{M}_i(t) = \int_0^t dN_i(u) - Y_i(u) \exp[Z_i(u)'\hat{\beta}] \, d\hat{\Lambda}_0(u), \qquad t > 0 \qquad (6.30)$$

and the corresponding *martingale residual*

$$\hat{M}_i = \hat{M}_i(\infty) = \delta_i - \hat{r}_i. \qquad (6.31)$$

Analogous to the properties of the martingales, these residuals satisfy $\sum \hat{M}_i = 0$, and for large samples, $\mathrm{cov}(\hat{M}_i, \hat{M}_j) \approx 0$ for all $i \neq j$. As with residuals in the ordinary linear model, $\hat{M}_i$ can be plotted against covariates included in the model, or against covariates being considered for inclusion, to help identify the form of dependence. Note, from (6.31) that $1 - \hat{M}_i = \hat{r}_i + (1 - \delta_i)$, so the martingale residual is essentially the Cox–Snell residual except the censored residuals ($\delta_i = 0$) are increased by 1, the expected residual life in the unit exponential distribution. As this would suggest, censoring makes patterns in the raw martingale residuals difficult to identify, so a smoother is typically applied to the residual plots to make trends more easily seen. Such smoothers do help to discern trends but can also give the impression of too little variation. One needs to supplement the results of such plots and checks with more formal tests.

The score function arising from the Cox model can be written as a sum of integrals of predictable processes with respect to orthogonal martingales as

$$U(\beta) = \sum_{i=1}^{n} U_i(\beta) = \sum_{i=1}^{n} \int_0^\infty [Z_i(t) - \mathscr{E}(\beta, t)] \, dM_i(t).$$

The *score (vector) residuals* are

$$\hat{U}_i = \int_0^\infty [Z_i(t) - \mathscr{E}(\hat{\beta}, t)] \, d\hat{M}_i(t) \tag{6.32}$$

which arose before in the robust estimate of the variance of the score. It can be seen that

$$\hat{U}_i = \int_0^\infty [Z_i(t) - \mathscr{E}(\hat{\beta}, t)] \, dN_i(t) - \int_0^\infty [Z_i(t) - \mathscr{E}(\hat{\beta}, t)] Y_i(t) \exp[Z_i(t)'\hat{\beta}] \, d\hat{\Lambda}_0(t)$$

$$= \delta_i [Z_i(t_i) - \mathscr{E}(\hat{\beta}, t_i)] + \int_0^\infty [Z_i(t) - \mathscr{E}(\hat{\beta}, t)] p_i(\hat{\beta}, t) \, dN.(t). \tag{6.33}$$

The first term on the right side of (6.33) is the vector of *partial residuals* suggested by Schoenfeld (1982). For each individual i that fails, the partial residual compares its covariate value $Z_i(t_i)$ at the failure time with the estimated weighted average $\mathscr{E}(\hat{\beta}, t_i)$ over the risk set. It can easily be seen that the partial residuals sum to 0. The score residual U_i can be used to provide an approximation to the influence of the ith observation on the estimation of β. Specifically, it can be shows that

$$n^{1/2}(\hat{\beta} - \hat{\beta}^{(i)}) \approx U_i(\beta_0) I(\beta_0)^{-1}, \tag{6.34}$$

where $\hat{\beta}^{(i)}$ is the estimate of β with the ith observation omitted. Thus, a plot of an element of the appropriately standardized score residual can be used to identify observations with large leverage or influence on the corresponding parameter estimate, and the influence can also be related to the values of the covariate itself. The main advantage of using $\hat{U}_i I(\hat{\beta})^{-1}$ instead of $(\hat{\beta} - \hat{\beta}^{(i)})$ is purely computational, and in most instances, direct calculation of the latter is easily accomplished.

BIBLIOGRAPHIC NOTES

The likelihood construction for general independent censoring mechanisms is an extension of the work of Cox (1975), who generated the partial likelihood construction for censored data and who defined and discussed independent censoring. Other early general constructions were given by Efron (1977) and by Cornfield and Detre (1977), who considered the failure mechanism to be a nonhomogeneous Poisson process where at each time t the risk set $R(t)$ is taken as given. Kalbfleisch and MacKay (1978b) introduce the role of censoring explicitly as is done in Section 5.2. Andersen et al. (1993) also discuss likelihood construction and list various pertinent references, including Jacod (1975), Arjas (1989), and Gill and Johansen (1990).

Cox (1972, 1975) suggests the use of time-dependent covariates in proportional hazards regression models and gave the partial likelihood analysis. The categorization of internal and external time-dependent covariates is discussed by Prentice and Kalbfleisch (1979) and, with regard to the likelihood construction, by Kalbfleisch and MacKay (1978b). The particular application of time-dependent covariates to the heart transplant data was considered by Crowley and Hu (1977) based in part on a remark by Breslow (1975). The heart transplant data given here and more recent versions of them have been analyzed by many authors. A detailed account of the results of the clinical trial from which the data arose is given by Clark et al. (1971). Gail (1972) gives a critical view of their findings. The analyses most closely related to those considered in Section 6.4.3 are those by Crowley (1974a), Mantel and Byar (1974), Turnbull et al. (1974), and especially, Crowley and Hu (1977). Crowley (1974a) generalizes the log-rank test to accommodate such time-varying covariates as transplant status. Cox and Oakes (1984) also consider this study and examine hypotheses of an elevated postoperative risk by introducing additional time-dependent covariates.

Williams and Lagakos (1977) consider a random censorship model with no covariates and define a constant sum relationship between the failure and censoring mechanisms. They show that under this condition the likelihood (6.1) is appropriate. Kalbfleisch and MacKay (1979) show the constant-sum condition to be equivalent to the condition for independent censoring in Section 6.2. The notion of independent censoring is implicit in the work of Cox (1972, 1975) and is discussed by Andersen et al. (1993, Chap. 3) and Fleming and Harrington (1991, Chap. 1), who also give extensive references. Little and Rubin (1987), Heitjan and Rubin (1991), and Heitjan (1992) consider aspects of incomplete data. The latter two papers introduce the concept of coarsening at random, which relates closely to independent censoring, as well as more general censoring patterns, especially in the context of interval censoring.

The Cox–Snell residuals were introduced by Kay (1977) and discussed by Crowley and Hu (1977). Critical discussions can be found in Crowley and Storer (1983) and Lagakos (1981), who noted difficulties with interpretation and that the many parameters being fit can lead to residual hazard plots that resemble closely the unit exponential even if the fit is poor. Arjas (1988, 1989) and Arjas and Haara (1988) proposed alternative plotting schemes and proposed tests based on the Cox–Snell residuals. Schoenfeld (1982) and Andersen (1982) are other early references on partial and martingale residuals. Barlow and Prentice (1988) considered a broad class of residuals obtained by estimating $\int K_i(s)\, dM_i(s)$, where $K_i(s)$ is a predictable process. The score residuals are the special case $K_i(s) = Z_i(s) - \mathscr{E}(\beta_0, s)$.

The relationship (6.34) between the score residual $\hat{U}_i$ and the influence of the ith observation was noted by Cain and Lange (1984) and Reid and Crepeau (1985). The martingale residuals have a highly skewed distribution, and as a consequence, they are sometimes transformed into *deviance residuals* $D_i = \text{sign}(\hat{M}_i)\{-2[\hat{M}_i + \delta_i \log{(\delta_i - \hat{M}_i)}]\}^{1/2}$, $i = 1,\ldots,n$, which have a more symmetric distribution. Therneau et al. (1990) suggest that these have the advantage of detecting outliers in the left tail of the distribution. Andersen et al. (1993) give a brief summary of methods based on residuals, and Fleming and Harrington (1991) give a detailed treatment of the topic with many examples. More recent detailed discussion can be found in Therneau and Grambsch (2000) and Klein and Moeschberger (1997), who also give many worked examples. We have not discussed residuals and graphical methods in the context of parametric models; Escobar and Meeker (1992) give a summary of residual and graphical methods that can be applied with censored data from parametric models.

If an assumed relative risk model is false in an important way, the departures from it are typically smooth and often well described by model extensions to include simple interactions with time. Several authors have suggested tests of fit for the relative risk model through this approach, which in many ways is preferred. Some references are Breslow et al. (1984), O'Quigley and Pessione (1989), Moreau et al. (1985), and Lin (1991a). Therneau and Grambsch (2000) discuss these methods and relate many of them to score tests and to linear fits of residual plots. Some results based on score tests are outlined in Exercise 6.8.

EXERCISES AND COMPLEMENTS

6.1 Let (t_i, δ_i, x_i), $i = 1,\ldots,n$ be a censored sample on a discrete failure time variable with mass points $a_1, a_2, \ldots$ and cumulative hazard function

$$\Lambda(t_i; x_i, \theta) = \sum_{a_j \leq t} \lambda[a_j; Z_i(a_j), \theta]$$
$$= \sum_j \lambda_{ij}(\theta)\mathbf{1}(t \geq a_j),$$

where θ is an unknown parameter vector. Censorings tied with failure times are shifted an infinitesimal amount to the right so that ties are broken by ordering failure times first. Specify the condition for independent censoring and show that the likelihood, or partial likelihood, is $\prod_1^n L_i(\theta)$, where

$$L_i(\theta) = f(t_i; x_i, \theta)^{\delta_i} F(t_i; x_i, \theta)^{1-\delta_i},$$

and f and F are the probability function and the survivor function, respectively.

6.2 Let the failure times of two items to be tested be independent exponentials with failure rate λ. Suppose that a type II censoring scheme observation is terminated when the first failure occurs. The contribution to the score of the ith item is

$$U_i = \frac{\delta_i}{\lambda} - T, \qquad i = 1, 2,$$

where δ_i is 1 for a failure and 0 for a censored data point and T is the minimum of the two potential failure times. Show that U_1 and U_2 are not independent but that they are uncorrelated.

6.3 Consider a general parametric model with density and survivor function $f(t; x, \theta)$ and $F(t; x, \theta)$ that are differentiable in θ. Suppose for convenience that θ is a scalar parameter. Show that under independent censoring, the score contributions

$$U_i(\theta) = \delta_i \frac{\partial}{\partial \theta} \log f(t_i; x_i, \theta) + (1 - \delta_i) \frac{\partial}{\partial \theta} \log F(t_i; x_i, \theta), \qquad i = 1, \dots, n$$

are uncorrelated.

6.4 Suppose that n items are placed on test and that as soon as an item fails, it is immediately replaced with a new item so that at all times, n items are on test. Suppose further that testing terminates when the nth failure occurs. All items on test at time t are subjected to a temperature $x(t)$ and, given the whole covariate path, the failure rate is known to vary with temperature according to the relation

$$\lambda \exp[x(t)\beta].$$

Show that if $t_1 < \cdots < t_r$ are the times at which failures are observed, the likelihood of λ and β is proportional to

$$\lambda^r \exp\left\{ \beta \sum x(t_i) - n\lambda \int_0^{t_r} \exp[x(t)\beta] \, dt \right\}.$$

Construct a conditional test of the hypothesis $\beta = 0$.

6.5 Suppose that an individual is at risk of two different types of failure, each of which is certain to occur. Let T_1 and T_2 represent the continuous failure times and define

$$\lambda_i(t) = \lim_{\Delta t \to 0+} \frac{P(t \le T_i < t + \Delta t | T_1 \ge t, T_2 \ge t)}{\Delta t}, \qquad i = 1, 2,$$

$$\lambda_1(t_1 | t_2) = \lim_{\Delta t \to 0+} \frac{P(t_1 \le T_1 < t_1 + \Delta t | T_1 \ge t_1, T_2 = t_2)}{\Delta t}, \qquad t_1 > t_2$$

with a similar definition of $\lambda_2(t_2 | t_1), t_2 > t_1$. Show that the joint density function of T_1, T_2 is

$$f(t_1, t_2) = \lambda_2(t_2)\lambda_1(t_1 | t_2) \exp\left\{ -\int_0^{t_2} [\lambda_1(u) + \lambda_2(u)] \, du - \int_{t_2}^{t_1} \lambda_1(u | t_2) \, du \right\}$$

for $t_1 \geq t_2$ and a symmetric part for $t_1 < t_2$. Obtain this result by relating the conditional and marginal densities $f_1(t_1|t_2)$, $t_1 \geq t_2$, and $f_2(t_2)$ to the hazard functions given. Explain why this result would be anticipated from the product integral relationship between the hazard and survivor functions. (Cox, 1972)

6.6 (*continuation*)   Identify $T = T_1$ with failure time and $C = T_2$ with censoring time where censoring and failure times of different individuals are independent. Show that condition 2 for independent censoring in (6.5) is equivalent to

$$\lambda_1(t) = h_1(t), \qquad 0 < t < \infty,$$

where

$$h_1(t) = \lim_{\Delta t \to 0} \frac{P\{T \in [t, t + \Delta t)|T \geq t\}}{\Delta t}.$$

Thus, the condition requires that the information that an item is uncensored at time t cannot alter the instantaneous failure rate. This is a slightly weaker condition than full independence between T and C.

6.7 (*continuation*)   Williams and Lagakos (1977) give a constant-sum condition relating failure and censoring mechanisms and show that under this condition the usual likelihood obtained as a product of density functions and survivor functions is valid. Briefly, they define

$$a(u) = P(T < C|T = u), \qquad dB(u) = P[C \in (u, u + du)|T \geq u]$$

and impose the condition

$$a(u) + \int_0^u dB(v) = 1, \qquad u > 0.$$

Show that

$$a(u) = \lambda_1(u) \exp\left[-\int_0^u k(v)\,dv\right] \Big/ h(u)$$

and

$$dB(u) = \lambda_2(u) \exp\left[-\int_0^u k(v)\,dv\right] du,$$

where $k(v) = \lambda_1(v) + \lambda_2(v) - h(v)$. Under the assumption that $h(u)$ and $\lambda_1(u)$ are differentiable, verify that a model is a constant sum if and only if $h(u) = \lambda_1(u)$ for all $u > 0$.

6.8 Generalize the log-rank and weighted log-rank tests to allow time-dependent strata. Thus, suppose that individuals can move among $p + 1$ strata, and if the item is in stratum j at time t, it is subject to a failure rate of $\lambda_j(t), j =$

$0, \ldots, p, t > 0$ and consider a test of the hypothesis $\lambda_0(t) = \lambda_1(t) = \cdots = \lambda_p(t)$. Write the test statistic as in (5.47), making appropriate definitions of the variables therein. Apply the test to heart transplant data.

6.9 Consider a Cox model with hazard $\lambda_0(t) \exp[Z(t)'\beta]$, where $Z(t) = [Z_1(t), \ldots, Z_p(t)]'$. Suppose that the simple relationship with $Z_1(t)$ is being questioned and consider the expanded model

$$\lambda_0(t) \exp[Z(t)'\beta + Z_1(t)g(t)\gamma],$$

where $g(t)$ is of known form [e.g., $g(t) = t$ or $\log t$]. Develop a score test of the hypothesis $\gamma = 0$.

(a) Show that the test statistic is of the form

$$T(g) = \sum_{i=1}^{n} \int_0^\infty g(t)[Z_{1i}(t) - \mathcal{E}_1(\hat{\beta}, t)] \, dN_i(t),$$

where $\hat{\beta}$ is the MLE under the hypothesis and $\mathcal{E}_1(\beta, t)$ is the first entry of $\mathcal{E}(\beta, t)$. Note that the statistic $T(g)$ also arises in a linear regression of the partial residuals on $g(t)$ and that there is a simple interpretation of the weight function $g(t)$.

(b) Show that the asymptotic variance of $T(g)$ can be estimated with $W_{22} - W_{21} W_{11}^{-1} W_{12}$, where

$$W = \begin{pmatrix} W_{11} & W_{12} \\ W_{21} & W_{22} \end{pmatrix}$$

is the information matrix in the expanded model at $\hat{\beta}, \gamma = 0$. Give expressions for each W_{ij}.

(c) Apply this test to the carcinogenicity data of Table 1.1 with $g(t) = t$. (Therneau and Grambsch, 2000)

6.10 Verify that the condition (6.24) of independent censoring with respect to the model (6.23) is satisfied if the rate of censoring of items at risk at time t depends on $X(t)$ only through $x(t)$.

6.11 Using the model (6.19) for the heart transplant data, obtain nonparametric estimates of the cumulative hazard functions $\int_0^t \lambda_{01}(u) \, du$ and $\int_0^t \lambda_{02}(u) \, du$. Note the difficulties with estimating the latter function for small t.

CHAPTER 7

Rank Regression and the Accelerated Failure Time Model

7.1 INTRODUCTION

Linear rank tests have often been proposed as alternatives to parametric tests for comparison of two or more samples with uncensored data. Although the rank tests themselves are derived with certain alternatives in mind for which optimum parametric procedures exist, they generally possess greater robustness than the corresponding parametric tests and are generally less sensitive to outliers. In addition, for testing the null hypothesis, these tests typically involve only a small loss in efficiency compared to the parametric procedure when such a procedure is appropriate.

Censored data generalizations of some popular rank tests have been developed. One example of such a test has already been met in the form of the log-rank test or generalized Savage test. In addition, generalized forms of the Wilcoxon and Kruskal–Wallis tests (Gehan, 1965a; Breslow, 1970; Peto and Peto, 1972; Prentice, 1978) have frequently been used for survival comparisons in clinical trials. Such tests can be generated conveniently from the accelerated failure time model.

The accelerated failure time model specifies that the effect of a fixed covariate Z is to act multiplicatively on the failure time T or additively, on $Y = \log T$. A linear modeling gives

$$Y = \alpha + Z'\beta + \sigma e \tag{7.1}$$

with error density $f(e)$. As usual, $Z' = (Z_1, \ldots, Z_p)$ is a vector of dimension p and β is a corresponding vector of regression coefficients. A class of linear rank statistics and their right-censored counterparts are given in Sections 7.3 and 7.4 as score function tests based on data from the model (7.1). Methods of estimation of β based on these statistics are considered in Section 7.4. Other regression models that also rely on the ranks of censored response variables are considered in Section 7.5.

The accelerated failure time (AFT) model with unspecified error distribution can be considered as a semiparametric alternative to the relative risk or Cox models discussed in Chapters 4 and 6. Although censored data rank tests for testing $\beta = \beta_0$ under the AFT model are readily carried out, the corresponding estimation problem is more challenging. Specifically, the censored data rank tests are based on the ranks of the residuals $W = Y - Z'\beta_0$. When viewed as a function of β_0, these test statistics are step functions, and furthermore they are, in general, not monotone in β_0. This gives rise to the possibility of multiple solutions to estimating equations, defined by setting the censored data rank tests equal to zero. Also, the asymptotic variance of the regression parameter estimator involves the derivative of the hazard function for the error variate in (7.1) and hence may be difficult to estimate reliably. Despite the related numerical challenges, the AFT model as a semiparametric family has received considerable attention in the statistical literature over the past decade or so, and the developments are such that it can be considered a practical, although computationally intensive alternative to the Cox model for failure time regression. The AFT model postulates a direct relationship between failure time and covariates. This natural type of regression relationship led Sir David Cox (Reid, 1994) to remark that "accelerated life models are in many ways more appealing" than the proportional hazards model "because of their quite direct physical interpretation."

It is certainly desirable to have more than a single routinely available regression method for the analysis of failure time data. In particular, since the ability to check model assumptions may be limited, depending on sample size and other data characteristics, it can be desirable to examine whether key inferences concerning the regression parameter (e.g., treatment contrasts in a clinical trial) are consistent under relative risk and accelerated failure time modeling assumptions. Perhaps because much of the work to address numerical aspects of the application of AFT models is fairly recent, there are still rather few examples of its use in the literature.

7.2 LINEAR RANK TESTS

7.2.1 Introduction

Let $Y_1 = \log T_1, \ldots, Y_n = \log T_n$ be an uncensored sample of log failure times with corresponding covariates $Z_1, \ldots, Z_n$, where Z_i is a vector of time-independent (fixed) covariates for the ith individual. Let $Y_{(1)} < \cdots < Y_{(n)}$ be the order statistic with corresponding covariates $Z_{(1)}, \ldots, Z_{(n)}$. A linear rank statistic is one of the form

$$\mathbf{v} = \sum_1^n Z_{(i)} c_i \qquad (7.2)$$

where c_i is a score attached to the ith ordered sample value and we take $\sum c_i = 0$. It is convenient to think of the data as arising from the accelerated failure time model (7.1) and the scores are chosen, by methods discussed in Section 7.3, to be efficient for certain specifications of the error density $f(e)$. Consider first the null hypothesis

$\beta = 0$, under which the failure times are unrelated to the covariates and so are independent and identically distributed. Under this hypothesis, the mean and variance of $\mathbf{v}$ can be obtained by consideration of the permutation distribution of the rank labels $(1), \ldots, (n)$. Let E_p denote expectation over this permutation distribution. Then

$$E_p(\mathbf{v}) = \sum_{i=1}^{n} c_i E_p(Z_{(i)})$$

$$= \sum_{i=1}^{n} c_i \bar{Z} = 0,$$

where $\bar{Z} = \sum Z_i / n$. The covariance matrix is

$$V = E_p(\mathbf{vv'}) = \sum \sum c_i c_j E_p(Z_{(i)} Z'_{(j)})$$

$$= \sum_{i=1}^{n} c_i^2 E_p(Z_{(i)} Z'_{(i)}) + \sum_{i \neq j} c_i c_j E_p(Z_{(i)} Z'_{(j)}).$$

Now $\sum_{i \neq j} c_i c_j = -\sum c_i^2$ and $E_p(Z_{(i)} Z'_{(j)}) = [n(n-1)]^{-1}(n^2 \bar{Z} \bar{Z}' - \sum Z_i Z'_i)$, so that

$$V = (n-1)^{-1} \sum c_i^2 \tilde{Z}' \tilde{Z}, \tag{7.3}$$

where $\bar{Z} = n^{-1} \sum Z_i$ and $\tilde{Z}$ is the $n \times p$ matrix of regression vectors with columns standardized to add to zero.

The use of (7.2) in a significance test involves calculation of probabilities over the permutation distribution. Thus if $\mathbf{v} = \mathbf{v}_0$ is observed, an exact computation would involve summing the probabilities of all outcomes $\mathbf{v}$ for which $\mathbf{v}'V^{-1}\mathbf{v} \geq \mathbf{v}_0'V^{-1}\mathbf{v}_0$. If only one covariate is included, this procedure is equivalent to evaluating $P(|\mathbf{v}| \geq |\mathbf{v}_0|)$. Except in the simplest of problems, this computation is very laborious. Fortunately for most scoring procedures, an accurate approximation is obtained by comparing $\mathbf{v}'V^{-1}\mathbf{v}$ with the χ_p^2 distribution assuming that $\tilde{Z}'\tilde{Z}$ is nonsingular.

In the special case where $p + 1$ samples are being compared, Z is a vector of 0, 1 indicators for p of the samples so that $\mathbf{v}$ records the sum of the scores c_i for each of the p samples. In the simplest case, $p = 1$, (7.2) gives the sum of the scores for sample 1 and the variance, from (7.3), is $n_0. n_1. n^{-1}(n-1)^{-1} \sum c_i^2$, where $n_0.$ and $n_1.$ are the sample sizes. The log-rank or Savage exponential scores test is one example. It has previously been derived as a score function test in the proportional hazards model but also arises from the accelerated failure time model (7.1) when the error distribution is extreme value. In this case, (7.1) corresponds to a model for failure time with proportional hazards. The test statistic is of the form (7.2) with scores

$$c_i = \frac{1}{n} + \frac{1}{n-1} + \cdots + \frac{1}{n-i+1} - 1, \qquad i = 1, \ldots, n, \tag{7.4}$$

in which $c_i + 1$ is the expected value of the ith-order statistic in a sample of size n from the unit exponential. The covariance matrix is obtained by direct substitution in (7.3).

The Wilcoxon test provides another example of a linear rank test for which the scores are

$$c_i = 2i(n+1)^{-1} - 1, \qquad i = 1, \ldots, n. \tag{7.5}$$

This test is the optimum rank test if the error distribution in (7.1) is logistic. The particular case of this test that corresponds to the comparison of $p + 1$ samples is known as the Kruskal–Wallis test.

A conceptually simple generalization of these tests to the censored data situation can be given (Prentice, 1978). Suppose that $Y_{(1)} < Y_{(2)} < \cdots < Y_{(k)}$ are the observed log failure times and that $Y_{i1}, \ldots, Y_{im_i}$ are censored values in $[Y_{(i)}, Y_{(i+1)})$, $i = 0, \ldots, k$, where $Y_{(0)} = -\infty$ and $Y_{(k+1)} = \infty$. Consider linear rank statistics of the form

$$\sum_{i=1}^{k} \left(c_i Z_{(i)} + \sum_{j=1}^{m_j} C_i Z_{ij} \right), \tag{7.6}$$

where c_i and C_i are the scores for the uncensored and censored data points, respectively, and Z_{ij} is the covariate vector for the individual having censored log-failure time Y_{ij}. Note that all censored data points in the interval $[Y_{(i)}, Y_{(i+1)})$ receive the same score C_i.

For the exponential scores test, the appropriate scores are obtained from the score function at $\beta = 0$ in the proportional hazards model. We find that

$$c_i = \sum_{j=1}^{i} n_j^{-1} - 1, \qquad C_i = \sum_{j=1}^{i} n_j^{-1}, \tag{7.7}$$

where n_j is the number of items at risk at $t_{(j)}^-$. The test statistic (7.6) can then be written as

$$\mathbf{v} = \sum_{i=1}^{k} \left(n_i^{-1} \sum_{\ell \in R(t_{(i)})} Z_\ell - Z_{(i)} \right),$$

which is the negative of the score statistic from the proportional hazards model [see (4.25)]. The appropriate estimate of the variance can also be obtained as the variance estimate for the log-rank statistic as in Section 4.2.4 and can be written

$$V_0 = \sum_{i=1}^{k} n_i^{-1} \sum_{\ell \in R(t_{(i)})} (Z_\ell - \bar{Z}_i)(Z_\ell - \bar{Z}_i)', \tag{7.8}$$

where $\bar{Z}_i = n_i^{-1}\sum_{\ell \in R(t_{(i)})} Z_\ell$ is the average of the covariates of individuals at risk at $t_{(i)}^-$. Thus V_0 is the sum of the sample covariance matrices of the covariates of items at risk at each failure time.

The Wilcoxon test can be generalized by defining scores

$$c_i = 1 - 2\prod_{j=1}^{i} \frac{n_j}{n_j + 1}, \qquad C_i = 1 - \prod_{j=1}^{i} \frac{n_j}{n_j + 1}, \tag{7.9}$$

as discussed further in Section 7.3, where a corresponding variance estimator V_0 is given.

7.2.2 Illustration

Before presenting a development of rank tests of the form (7.6), we give an example of their use with censored and uncensored data.

An extremely simple illustration can be based on the data of Table 7.1. Data are given there for the log failure times, Y, for two samples and we consider testing the null hypothesis of no sample difference ($\beta = 0$ in the accelerated failure time model). Following the data, the next two lines in the table give the Wilcoxon and exponential ordered scores for each of the nine individuals. The last two lines in the table give the scores when two of the items are censored and can, for the moment, be ignored.

Let $Z = 0$ for observations in sample 1 and $Z = 1$ for sample 2. The Wilcoxon test statistic has value

$$\mathbf{v} = -0.2 + 0.4 + 0.6 + 0.8$$
$$= 1.6.$$

Table 7.1 Scores for the Comparison of Two Uncensored Samples

	Individual								
	1	2	3	4	5	6	7	8	9
Sample 1	2.1	4.7	6.8		7.9	8.6			
Sample 2				7.5			8.9	9.2	9.3
	All Items Uncensored								
Wilcoxon scores	−0.8	−0.6	−0.4	−0.2	0.0	0.2	0.4	0.6	0.8
Exponential scores	−0.889	−0.746	−0.621	−0.454	−0.254	−0.004	0.329	0.829	1.829
	Individuals 4,5 Censored								
Wilcoxon scores	−0.80	−0.60	−0.40	0.30	0.30	−0.12	0.16	0.44	0.72
Exponential scores	−0.889	−0.746	−0.621	0.379	0.379	−0.596	−0.266	0.334	1.334

Each of the 9! possibilities is equally likely under $\beta = 0$, so that the (two-sided) significance level of the hypothesis is the number of sequences for which $|\mathbf{v}| \geq 1.6$ divided by 9!. Because Wilcoxon scores are symmetric, this probability can be written

$$P(|\mathbf{v}| \geq 1.6) = 2\left(\frac{4!\,5!\,(4)}{9!}\right) = 0.063.$$

Sample sizes are too small to expect a normal approximation to be warranted. For comparative purposes, however, the Wilcoxon test variance is

$$V = (8)^{-1} \sum_{1}^{9} \left(\frac{2i}{10} - 1\right)^2 \sum_{1}^{9} (Z_i - \bar{Z})^2$$

$$= \tfrac{2}{3}.$$

Thus $1.6\left(\tfrac{2}{3}\right)^{-1/2} = 1.96$ is a realization of an "approximate" standard normal statistic under $\beta = 0$. The associated significance level of 0.05 is in surprisingly good agreement with the exact test.

Similarly, the exponential scores test statistic has value

$$\mathbf{v} = 0.454 + 0.329 + 0.829 + 1.829$$
$$= 2.53.$$

The associated exact significance level is $P(|\mathbf{v}| \geq 2.53) = 4(4!)(5!)(9!) = 0.032$, and again, in good agreement, the approximate standard normal statistic has value 1.93.

To illustrate the censored versions of the test, we suppose that items 4 and 5 are censored. The last two rows of Table 6.1 give the revised scores. The censored Wilcoxon statistic has the value

$$\mathbf{v} = 0.30 + 0.16 + 0.44 + 0.72 = 1.62$$

and the estimated variance, from Section 7.3, is $V_0 = 0.810$.

Although sample sizes are very small, for purposes of illustration one might compare the observed value $\mathbf{v}/\sqrt{V} = 1.80$ with the $N(0, 1)$ distribution to obtain an approximate significance level of 7%. The exponential ordered scores gives, in this cases,

$$\mathbf{v} = 0.379 - 0.266 + 0.344 + 1.334 = 1.781.$$

The estimated variance from (7.8) is $V_0 = 1.030$. Again, $\mathbf{v}/\sqrt{V_0} = 1.75$ is in close agreement with the Wilcoxon test.

7.2.3 Stratification

A useful extension of the rank test procedures can be provided when the study individuals are divided into strata on some auxiliary variable and comparisons are to be made on Z, the variable of primary interest. For example, patients in a clinical trial may be subdivided on age and then the treatments (given by Z) compared within each age group. For either the log-rank or the Wilcoxon test discussed above, we can let $\mathbf{v}_j$ be the test statistic, from (7.2) or (7.6), computed within the jth stratum, and let V_{0j} be the corresponding covariance matrix. The total score is then $\sum \mathbf{v}_j$ with covariance matrix estimated by $\sum V_{0j}$, and the stratified test statistic is

$$\left(\sum \mathbf{v}_j \right)' \left(\sum V_{0j} \right)^{-1} \left(\sum \mathbf{v}_j \right),$$

which again can be compared with χ_p^2. It has been assumed that $\sum V_{0j}$ is non-singular.

7.3 DEVELOPMENT AND PROPERTIES OF LINEAR RANK TESTS

7.3.1 Uncensored Data Rank Tests

Suppose that an uncensored sample of size n has been observed from the accelerated failure time model (7.1) with $f(e)$ of known form, and consider the problem of testing the hypothesis $\beta = \beta_0$. The methods in Section 7.2 were for the case $\beta = 0$. These are readily generalized to tests of a general $\beta = \beta_0$.

Let $W = Y - Z'\beta_0$ represent the "residuals" about the hypothesized value. From (7.1) we have that

$$W = \alpha + \sigma Z'\gamma + \sigma e,$$

where $\gamma = \sigma^{-1}(\beta - \beta_0)$ and the hypothesis $\beta = \beta_0$ is equivalent to the hypothesis $\gamma = 0$. Let $W_{(1)}, \ldots, < W_{(n)}$ represent ordered residuals with corresponding regression vectors $Z_{(1)}, \ldots, Z_{(n)}$, respectively. As in Section 4.7.1, the rank vector $r = r(w)$ is given by the corresponding labels $(1), \ldots, (n)$. That is,

$$r = [(1), (2), \ldots, (n)]$$

and the rank vector probability can be computed as

$$P(r) = \int \cdots \int_{\tau_{(1)} < \cdots < \tau_{(n)}} \prod_1^n f(\tau_{(i)} - Z_{(i)}\gamma) \, d\tau_{(i)}, \tag{7.10}$$

where $\tau_{(i)} = (W_{(i)} - \alpha)/\sigma$. This is the same calculation as was used in the marginal likelihood derivation in the proportional hazards model (in Section 4.7), and (7.10) might be viewed as a marginal likelihood of γ based on the residual ranks. Note that (7.10) is independent of α, σ, which is in contrast to the fully parametric likelihood for γ. A locally most powerful test of $\gamma = 0$, or equivalently, of $\beta = \beta_0$, can be based on the score statistic from (7.10); [see e.g., Puri and Sen (1971, p. 108) and Cox and Hinkley (1974, p. 188)]. Straightforward calculation gives

$$
\mathbf{v} = \frac{d \log P(r)}{d\gamma} \bigg|_{\gamma=0}
$$

$$
= \sum c_i Z_{(i)},
$$

a linear rank statistic, with

$$
c_i = \int \cdots \int_{\tau_{(1)} < \cdots < \tau_{(n)}} \left[\frac{-d \log f(\tau_{(i)})}{d\tau_{(i)}} \prod_{j=1}^{n} f(\tau_{(j)}) \, d\tau_{(j)} \right] \tag{7.11}
$$

$$
= n! \int \cdots \int_{u_1 < \cdots < u_n} \phi(u_i) \prod_1^n du_j \tag{7.12}
$$

$$
= E[\phi(u_i)], \tag{7.13}
$$

where $u_j = 1 - F(\tau_{(j)})$ is the jth-order statistic in a uniform $(0,1)$ sample of size n, and $\phi(u)(0 < u < 1)$ is given by

$$
\phi(u) = \phi(u, f) = \frac{-f'[F^{-1}(1 - u)]}{f[F^{-1}(1 - u)]},
$$

and, as usual, $F(\tau) = \int_\tau^\infty f(w) \, dw$. It is easy to see that the sum of scores is zero since

$$
\sum_1^n c_i = \sum_1^n E\left[\frac{-d \log f(\tau_{(i)})}{d\tau_{(i)}} \right]
$$

$$
= \sum_1^n E\left[\frac{-d \log f(\tau_i)}{d\tau_i} \right] = 0.
$$

The fact that u_i has expectation $i(n+1)^{-1}$ and variance $i(n-i+1)(n+1)^{-2}$ $(n+2)^{-1}$, $i = 1, \ldots, n$, leads to an asymptotically equivalent system of scores

$$
c_i = \phi[i(n+1)^{-1}]. \tag{7.14}
$$

Some interesting special cases of (7.13) and (7.14) are as follows: A logistic density $f(\tau) = e^\tau (1 + e^\tau)^{-2}$ gives Wilcoxon scores (7.5) for both (7.13) and (7.14). A standard normal density gives for (7.13) and (7.14), respectively, $c_i = E(\tau_{(i)})$, the normal scores test of Fisher and Yates (1938), and $c_i = G^{-1}[i/(n+1)^{-1}]$, the van der Waerden (1953) test, where G represents the standard normal distribution function. Similarly, an extreme value PDF, $f(\tau) = \exp(\tau - e^\tau)$, yields the exponential scores (7.4) from (7.13). The double exponential density $f(\tau) = 2^{-1}e^{-|\tau|}$ gives, from (7.14), the sign (median) scores, $c_i = \text{sign}[2i - (n+1)]$, with $c_i = 0$ if $2i = (n+1)$.

As illustrated in Section 7.2.2, the significance level based on the statistic **v** can be assessed in terms of its permutation distribution in that any of the $n!$ possible realizations of the rank vector for W are equally probable under $\beta = \beta_0$. As outlined there, an approximate significance test may be carried out upon noting that **v** has mean zero and variance matrix V given by (7.3). Under mild restrictions on the regression vectors (Hájek and Šidák, 1967, p. 159), **v** is asymptotically normal so that if $\beta = \beta_0$,

$$\mathbf{v}'V^{-1}\mathbf{v} \tag{7.15}$$

is asymptotically χ_p^2, where it has been assumed that V is nonsingular.

The rank tests considered above are asymptotically fully efficient (Pitman efficiency 1; see, e.g., Cox and Hinkley, 1974, p. 338) if the assumed score-generating distribution, f, and "actual" sample distribution, say f_0, agree up to location and scaling. More generally, the asymptotic relative efficiency is the square of the limiting correlation of **v** with the locally optimum test based on f_0. This efficiency may be written (Hájek and Šidák, 1967, p. 268)

$$\frac{\left[\int_0^1 \phi(u,f)\phi(u,f_0)\,du\right]^2}{\int_0^1 \phi^2(u,f)\,du \int_0^1 \phi^2(u,f_0)\,du}, \tag{7.16}$$

where it has been assumed that the Fisher information terms in the denominator are both finite. Under $f \neq f_0$, (7.16) typically indicates substantial improvement over the corresponding parametric test. For example, the normal scores test under mild conditions on f_0 (Puri and Sen, 1971, p. 118) has efficiency equal to or greater than that of the corresponding least squares procedure. Rank tests themselves differ somewhat in efficiency properties. For example, the Wilcoxon test has asymptotic efficiency 0.61 under Cauchy sampling, while the sign test has an even higher efficiency of 0.81. It is then important to consider the class of plausible sampling distributions in selecting a rank test. Birnbaum and Laska (1967) and Gastwirth (1970) give some consideration to the selection of robust rank tests when f_0 is restricted to an indexed family.

The approximate χ^2 statistic (7.15) can form the basis for more general inference on β. For example, an approximate confidence region for β is given by those β_0 values for which (7.15) does not exceed specified percentage points of the χ^2 distribution. In principle, exact significance levels could also be used to generate confidence regions, although the computation would usually be prohibitive. An estimator $\hat{\beta}$ may be defined as the β_0 value, or values, for which (7.15) is minimized. Jurečková (1969, 1971) has shown, under mild restrictions, that such a $\hat{\beta}$ is fully efficient if the assumed form for the score-generating density obtains, and that its efficiency, more generally, is the same as that of the corresponding test, as given in (7.16).

7.3.2 Censored Data Rank Tests

Consider now a censored sample from (7.1) with failure and censoring times independent given Z. Let $W_{(1)} < \cdots < W_{(k)}$ represent the distinct ordered residuals about the hypothesized model $\beta = \beta_0$, so that, as before, $W = Y - Z'\beta_0$. Further, let $W_{i1}, \ldots, W_{im_i}$ be right-censored residuals in the interval $[W_{(i)}, W_{(i+1)})$, $i = 0, \ldots, k$, where $W_{(0)} = -\infty$ and $W_{(k+1)} = \infty$. Also, let $Z_{(i)}$ and Z_{ij} represent the corresponding regression vectors. Note that it is being assumed for the moment that there are no ties among the uncensored residuals, as is implied by model (7.1).

To generalize the uncensored tests to this situation, it is necessary to consider generalizations of the rank vector to censored data. One possible extension (Peto, 1972a) takes the rank vector as the labels corresponding to the ordered sample of all censored and uncensored W values along with the associated censoring indicators. This statistic is the maximal invariant statistic under monotone increasing transformations on the W values observed, but its sampling distribution depends in a complicated way on the censoring mechanism (Crowley, 1974b). As a consequence, its use will usually not yield simple rank statistics for censored samples. An alternative approach, as discussed in Section 4.7 and by Kalbfleisch and Prentice (1973), views the rank vector of the underlying W values to be of primary interest. This rank statistic is observed only partially, owing to the censoring, and the censored data rank vector is taken to be the set of possible underlying rank vectors given the data. It should be noted that this definition of a censored rank vector does not utilize the ordering of censored W values between adjacent uncensored values. That is, no use is made of the ordering among $W_{i1}, \ldots, W_{im_i}$ in the rank vector probabilities, for any i.

The calculations follow very closely those of Section 4.7.1 that lead to the marginal likelihood for β in the proportional hazards model (Prentice, 1978). In evaluating the total probability that the uncensored rank vector should be one of those possible on the sample, we calculate first the probability of the event $W_{ij} \geq W_{(i)}$, $j = 1, \ldots, k$, given the uncensored residuals $W_{(1)} < \cdots < W_{(k)}$. This gives

$$\prod_{i=0}^{k} \prod_{j=1}^{m_i} F(\tau_{(i)} - Z_{ij}\gamma),$$

where, as previously, $F(\tau) = \int_{\tau}^{\infty} f(u)\, du$, $\gamma = \sigma^{-1}(\beta - \beta_0)$, and $\tau_{(i)} = (W_{(i)} - \alpha)\sigma^{-1}$, $i = 1, \cdots, k$. Note that the term in $i = 0$ can be dropped since $\tau_{(0)} = -\infty$ and empty products are interpreted as 1. The total accumulated probability of possible underlying rank vectors $\{r\}$ is then

$$P(\{r\}) = \int \cdots \int_{\tau_{(1)} < \cdots < \tau_{(k)}} \prod_{i=1}^{k} \left[f(\tau_{(i)} - Z_{(i)}\gamma) \prod_{j=1}^{m_i} F(\tau_{(i)} - Z_{ij}\gamma)\, d\tau_{(i)} \right]. \qquad (7.17)$$

Again, since the τ's enter (7.17) as dummy variables, $P(\{r\})$ is completely independent of α and σ. Note that at $\gamma = 0$ (i.e., $\beta = \beta_0$), (7.17) can be integrated directly, giving

$$P(\{r\})|_{\gamma=0} = \prod_1^k n_i^{-1},$$

where

$$n_i = \sum_{j=i}^k (1 + m_j)$$

is the number of individuals with W values known to be equal or greater than $W_{(i)}$.

As with uncensored data, a score test for $\beta = \beta_0$ may be based on (7.17) giving

$$\mathbf{v} = \frac{d \log P(\{r\})}{d\gamma}\bigg|_{\gamma=0}$$

$$= \sum_{i=1}^k [Z_{(i)}c_i + S_{(i)}C_i], \tag{7.18}$$

where $S_{(i)} = \sum_{j=1}^{m_i} Z_{ij}$, c_i is a score corresponding to $W_{(i)}$ and C_i is a score corresponding to each of $W_{i1}, \ldots, W_{im_i}$. This statistic is of the type discussed in Section 7.2. The scores are explicitly

$$c_i = \int \cdots \int_{\tau_{(1)} < \cdots < \tau_{(k)}} \frac{-d \log f(\tau_{(i)})}{d\tau_{(i)}} \prod_{j=1}^k [n_j F^{m_j}(\tau_{(j)}) f(\tau_{(j)}) \, d\tau_{(j)}]$$

and

$$C_i = \int \cdots \int_{\tau_{(1)} < \cdots < \tau_{(k)}} \frac{-d \log F(\tau_{(i)})}{d\tau_{(i)}} \prod_{j=1}^k [n_j F^{m_j}(\tau_{(j)}) f(\tau_{(j)}) \, d\tau_{(j)}].$$

As in (7.13), these scores can be expressed in terms of functions on (0,1). Set $u_j = 1 - F(\tau_{(j)})$, $j = 1, \ldots, k$, and define for $0 < u < 1$,

$$\phi(u) = \frac{-f'[F^{-1}(1 - u)]}{f[F^{-1}(1 - u)]}$$

$$\Phi(u) = \frac{f[F^{-1}(1 - u)]}{F[F^{-1}(1 - u)]} = (1 - u)^{-1} f[F^{-1}(1 - u)].$$

The scores can now be written

$$
c_i = \int \cdots \int_{u_1 < \cdots < u_k} \phi(u_i) \prod_1^k [n_j(1 - u_j)^{m_j} \, du_j]
$$

$$
C_i = \int \cdots \int_{u_1 < \cdots < u_k} \Phi(u_i) \prod_1^k [n_j(1 - u_j)^{m_j} \, du_j].
$$

(7.19)

The test statistic (7.18) is simply the average of uncensored test statistics over possible underlying rank vectors. This, along with the assumption that the censoring mechanism and the failure mechanism (7.1) are independent given Z, shows that the expectation of $\mathbf{v}$ is identically zero, under $\beta = \beta_0$, regardless of the actual sampling distribution of the error variate in (7.1). Note also that the sum of scores given by any realization of (7.18) is zero; that is,

$$
\sum_1^k (c_i + m_i C_i) = 0.
$$

(7.20)

This occurs since (7.20) is the average of sums of scores corresponding to possible underlying (uncensored) rank vectors, each of which, from Section 7.2, has value zero.

In order to list some specific scoring schemes, let

$$
J[g(u_i)] = \int \cdots \int_{u_1 < \cdots < u_k} g(u_i) \prod_{j=1}^k [n_j(1 - u_j)^{m_j} \, du_j],
$$

(7.21)

for an arbitrary function g. A simple calculation gives

$$
J[(1 - u_i)^l] = \prod_{j=1}^i \frac{n_j}{n_j + \ell}, \qquad l = 1, 2, \ldots
$$

(7.22)

A logistic score generating density $f(\tau) = e^\tau (1 + e^\tau)^{-2}$ gives $\phi(u) = 2u - 1$ and $\Phi(u) = u$, so that, from (7.22), the corresponding scores are those given by (7.9). This test was discussed in Section 7.2 and is a censored data generalization of the Wilcoxon test (Peto and Peto, 1972; Prentice, 1978).

An extreme value density $f(\tau) = \exp(\tau - e^\tau)$ yields $\phi(u) = -\log(1 - u) - 1$, $\Phi(u) = -\log(1 - u)$. Direct integration in (7.20) gives

$$
J[\log(1 - u_i)] = -\sum_{j=1}^i n_j^{-1},
$$

so that the corresponding scores are as given in (7.7) and discussed in Section 7.2.

A permutation approach is commonly used to calculate the variance of an uncensored linear rank statistic. This approach does not extend in a convenient way to arbitrarily censored data, because the expectation involved is a complicated function of the censoring mechanism. As in Chapters 3 and 4, however, the observed information matrix

$$
V_0 = \frac{-d^2 \log P(\{r\})}{d\gamma \, d\gamma'} \bigg|_{\gamma=0},
\tag{7.23}
$$

provides a variance estimator that is generally appropriate. After differentiation of (7.17), (7.23) can be written

$$
V_0 = \sum_{i=1}^{k} \left\{ Z_{(i)} Z'_{(i)} J[\psi_1(u_i)] + \sum_{j=1}^{m_i} Z_{ij} Z'_{ij} J[\psi_2(u_i)] \right\} - [J(bb') - \mathbf{v}\mathbf{v}'],
$$

where

$$
\psi_1(u) = \frac{-d^2 \log f(\tau)}{d\tau^2} \bigg|_{\tau = F^{-1}(1-u)},
$$

$$
\psi_2(u) = \frac{-d^2 \log F(\tau)}{d\tau^2} \bigg|_{\tau = F^{-1}(1-u)},
$$

and

$$
b = \sum_{i=1}^{k} [Z_{(i)} \phi(u_i) + S_{(i)} \Phi(u_i)].
$$

V_0 can be calculated explicitly for logistic and extreme value score-generating densities. A logistic density gives a test statistic (7.18) with scores (7.9) having variance estimator

$$
V_0 = \sum_{i=1}^{k} \left[a_i (1 - a_i^*) \left(2 Z_{(i)} Z'_{(i)} + \sum_{j=1}^{m_i} Z_{ij} Z'_{ij} \right) - (a_i^* - a_i) X_{(i)} \left(a_i X'_{(i)} + 2 \sum_{j=i+1}^{k} a_j X'_{(j)} \right) \right],
\tag{7.24}
$$

where

$$
a_i = \prod_{j=1}^{i} \frac{n_j}{n_j + 1}, \qquad a_i^* = \prod_{j=1}^{i} \frac{n_j + 1}{n_j + 2}, \qquad X_{(i)} = 2 Z_{(i)} + S_{(i)}, \qquad i = 1, \ldots, k.
$$

This expression was used in the previous illustration. An extreme minimum value distribution gives scores (7.7) and a test statistic with variance estimator (7.8).

7.3.3 Asymptotic Distribution Theory and Weighted Log-Rank Tests

The linear rank tests (7.18) can be recast as weighted log-rank tests of the form

$$\mathbf{v} = \sum_{i=1}^{k} Q_i (Z_{(i)} - \bar{Z}_{(i)}), \tag{7.25}$$

where $\bar{Z}_{(i)} = n_i^{-1} \sum_{\ell \in R(W_{(i)})} Z_\ell$ is the average of the Z values for individuals having $W \geq W_{(i)}$. The log-rank test is the special case $Q_i \equiv 1$, so that (7.25) is referred to as the class of weighted log-rank tests. The weighted log-rank form is convenient for studying asymptotic properties, and it is a broader class than (7.18).

To express (7.18) in the form (7.25), write

$$\mathbf{v} = \sum_{i=1}^{k} \left[c_i Z_{(i)} + C_i \left(n_i \bar{Z}_{(i)} - Z_{(i)} - n_{i+1} \bar{Z}_{(i+1)} \right) \right]$$

$$= \sum_{i=1}^{k} (c_i - C_i) Z_{(i)} + \sum_{i=1}^{k} n_i (C_i - C_{i-1}) \bar{Z}_{(i)},$$

where $C_0 = 0$ and $n_{k+1} = 0$. The coefficient of $\bar{Z}_{(i)}$ in this expression will equal $(c_i - C_i)$ if

$$n_i C_{i-1} = c_i + (n_i - 1) C_i,$$

corresponding to a *preservation of scores* at $W_{(i)}$ (Prentice and Marek, 1979). That such preservation holds generally can be shown by calculating C_i in (7.19) by integrating over $u_{i+1}, \ldots, u_k$, and applying integration by parts to the integral in u_i (Mehrotra et al., 1982; Andersen et al., 1982), giving

$$\mathbf{v} = \sum_{i=1}^{k} (c_i - C_i)(Z_{(i)} - \bar{Z}_{(i)}). \tag{7.26}$$

One can consider (7.25) for the broader class of weight functions with Q_i a fixed function of the risk set sizes $\{n_1, \ldots, n_i\}$ up to and including that at $W_{(i)}$. This includes, for example, the log-rank test and the generalized Wilcoxon test (7.9) characterized by $Q_i = - \prod_{j=1}^{i} n_j (n_j + 1)^{-1}$, as well as weight functions such as $Q_i = -n_i n^{-1}$, which characterizes the Gehan (1965a,b) generalized Wilcoxon test discussed further below. Such weights can be viewed as defining a step function Q that is constant between adjacent ordered censored or uncensored W values, with $Q(W_{(i)}) = Q_i$. As a function on the residual time scale, such Q processes will be predictable since risk set size is a left-continuous step function.

Asymptotic distribution theory for weighted log-rank tests can be developed in various ways. For example, the counting process methods of Section 5.6 for the comparison of several samples can be extended to regression tests that include a weight function. Perhaps more simply the asymptotic results for the Cox model can be adapted to this task by noting that (7.25) can be generated as a partial likelihood score test under a Cox model

$$\lambda(s, Z) = \lambda_0(s) \exp[Q(s)Z'(\beta - \beta_0)],$$

where $S = T \exp[-Z'(\beta - \beta_0)] = \exp[W(\beta_0)]$, and Q is in the class of predictable processes just mentioned. As such, a mean zero asymptotic normal distribution for $n^{-1/2}\mathbf{v}$ under $\beta = \beta_0$ can be asserted from the results of Section 5.7 for the score process at a zero value for the regression parameter. Explicitly, denote $Y_\ell(\beta_0, w) = 1$ if $W_\ell(\beta_0) \geq w$ and $Y_\ell(\beta_0, w) = 0$ otherwise. For the results of Section 5.7 to apply, we require that (7.25) be defined over a finite follow-up interval $[0, \tau]$ for the residuals $Y - Z'\beta_0$, where τ is such that the average number of at-risk individuals $n^{-1} \sum Y_\ell(\beta_0, \tau)$ converges in probability to a positive value. Condition C2 of Section 5.7 is satisfied without further restriction since the regression parameter is zero (under $\beta = \beta_0$) in the time-dependent Cox model above. Condition C1 of Section 5.7 in this setting simply requires that

$$S^{(j)}(\beta_0, w) = n^{-1} \sum_1^n Y_\ell(\beta_0, w) Z_\ell^{(j)} \tag{7.27}$$

converge in probability to fixed functions, uniformly over $[0, \tau]$, for $j = 0, 1, 2$, where $Z^{(0)} = 1, Z^{(1)} = Z, Z^{(2)} = ZZ'$, and requires the limiting variance matrix for $n^{-1/2}\mathbf{v}$ to be positive definite. Hence, $n^{-1/2}\mathbf{v}$ has a mean zero asymptotic normal distribution under these assumptions. Define

$$V = \sum_1^k Q_i^2 (Z_{(i)} - \bar{Z}_{(i)})(Z_{(i)} - \bar{Z}_{(i)})'. \tag{7.28}$$

Under condition C1 with $\beta = \beta_0$, $n^{-1}V$ converges in probability to a fixed matrix, and from Theorem 5.1 this fixed matrix is the asymptotic variance matrix for $n^{-1/2}\mathbf{v}$. Hence, $\mathbf{v}'V^{-1}\mathbf{v}$ has an asymptotic χ_p^2 distribution under these weak conditions.

Note that the weight process

$$Q(w) = - \prod_{j|W(j) \leq w} \frac{n_j}{n_j + 1} = -\hat{F}(w)$$

for the generalized Wilcoxon test (7.9) is a right-continuous predictable process that estimates the residual survivor function $F(w)$. In comparison, the weight process

$$Q(w) = n^{-1} \sum_{\ell=1}^n Y_\ell(\beta_0, w)$$

for the Gehan (1965a) generalized Wilcoxon test estimates a quantity that depends on the censoring. For example, under a random censorship model this weight function $Q(w)$ estimates the product of the failure and censoring residual survivor functions at $W = w$.

Cuzick (1985) considered approximate scores $c_i = \phi[\hat{F}(W_{(i)})]$ and $C_i = \Phi[\hat{F}(W_{(i)})]$ corresponding to (7.19), and showed that the resulting linear rank test was asymptotically equivalent to that using (7.19) provided that there exist $\alpha < \frac{1}{2}$ and $K < \infty$ such that

$$|s\phi'(s)| + |s^2\phi''(s)| \le Ks^{-\alpha},$$

where ϕ' and ϕ'' denote the first and second derivatives of ϕ. He also developed expressions for the efficiency of (7.18) for tests against local alternatives $\beta = \beta_0 + n^{-1/2}b$, for fixed vector b, giving expressions that generalize (7.16) to the correlation between certain censored data influence curves for the score-generating and actual error distributions in (7.1) in the special case where censoring is independent of Z. However, if censoring depends on Z, these tests are typically not fully efficient even if the actual error distribution is used to generate scores.

The finite interval condition used above to generate asymptotic results for (7.25) seems rather artificial, particularly for tests of $\beta \ne 0$. In the context of the more general problem of rank estimators for β under (7.1), Ying (1993) developed asymptotic results for weighted log-rank tests that avoid this finite interval condition, but at the expense of some stronger conditions to ensure tail stability for the estimating function. In particular, the regression variables were required to be uniformly bounded.

7.3.4 Discussion

An important aspect of rank tests is their invariance under monotone increasing transformations on the response variable. For example, if W were replaced by $\tilde{W} = h(W)$, where h is a strictly monotone increasing function, the rank vector, scores (7.25) would be unchanged. This feature is most easily appreciated in terms of a test of $\beta = 0$. Monotone transformations on Y itself, then, do not affect the test procedure, so that equivalent tests arise, for example, from taking Y to be normal or lognormal or indeed from taking any monotone function of Y to be normal. As a second example, the proportional hazards model of Chapter 4 can be thought of as arising from a model which specifies that some monotone increasing function of $Y = \log T$ has an extreme value error. For this reason, with uncensored data the log-rank scores are (locally) optimum within the entire proportional hazards class. See Kalbfleisch (1978b) for further discussion.

Since the Weibull model is a special case of both the proportional hazards and accelerated failure time models, we may compare the methods of Chapter 4 with those of this chapter by supposing the error density f in (7.1) to have the extreme-value form. It is convenient to consider the exponential special case $[\lambda_0(t) = \lambda]$ since β then has the same meaning in both models. Tests for $\beta = 0$ are equivalent from the two approaches, as just noted. More generally, however, the proportional hazards test of $\beta = \beta_0 \neq 0$ is based on the generalized rank vector $\{r(t)\}$ of the observed survival times t, whereas the log-rank methods of Section 7.3 are based on the generalized rank vector $\{r(w)\}$ based on centered values $W = \log T - Z'\beta_0$. As β_0 departs more radically from zero, the generalized rank vector $\{r(t)\}$ becomes increasingly predictable under the hypothesis and so less informative against local alternatives (values of β near β_0). If there is no censoring and a single covariate, Kalbfleisch (1974) shows that the proportional hazards test has efficiency approximately of the form $\exp(-\beta_0^2 \mu_2)$, where μ_2 is the variance of the finite population of Z values. In contrast, the log-rank test based on residuals about β_0 is fully efficient under these circumstances. It is important to note, however, that these are local efficiencies and the efficiency against distant alternatives may be somewhat different. The local efficiency comparison is somewhat unfair in that the proportional hazards test has the same efficiency when compared with any specific $\lambda_0(t)$ in the proportional hazards class, whereas the efficiency of this particular log-rank procedure changes if hazards are proportional but outside the Weibull class. Both procedures would be inefficient if model (7.1) were appropriate but the error density departs from the extreme value form. For the log-rank test based on $r(w)$, however, the reduction in efficiency is independent of β_0.

As mentioned above, Gehan (1965a) proposed a censored generalization of the two-sample Wilcoxon statistic that differs from (7.9). Gehan's statistic and the multiple sample generalization of Breslow (1970) have been widely used in the clinical trials setting. The censored and uncensored scores are, respectively,

$$c_i = i - n_i$$
$$C_i = i.$$

The scores are simply the number of residuals known to be smaller than the residual being scored minus the number known to be larger. In weighted log-rank form (7.25), this test is characterized by $q_i = c_i - C_i = -n_i$. As noted above, this statistic has asymptotic distribution theory covered by the previous presentation, but it has properties that depend not just on the failure time distribution under study, but also on the corresponding censoring distribution, giving rise to possible anomalies under heavy censorship (see Prentice and Marek, 1979, and Exercises 7.6 and 7.7). As discussed below, however, this particular weighted log-rank test is useful in addressing some numerical aspects of rank regression parameter estimation.

The rank tests discussed above have assumed that uncensored W values are distinct. Tied uncensored values may be handled in the manner suggested in Chapter 4. Specifically tied values at the same Z may be ordered arbitrarily because their actual ordering does not affect the statistic (7.25). More generally, if the fraction of tied values is small, tied values may be assigned the average of their possible scores. Perhaps more appropriate, but computationally difficult, the rank vector may be taken, as in Chapter 4, to consist of all possible underlying rank vectors that are consistent with the observed tied and censored data. The same idea allows the methods of this chapter to be extended to interval censored data.

7.3.5 Illustration

For illustration, we apply certain censored data rank tests to the carcinogenesis data of Chapter 1. The results may be compared with the parametric methods of Chapter 3 and the proportional hazards methods of Chapter 4. To begin, suppose that as in Chapter 3, $Y = \log(T - 100)$ adheres to a linear model of the form (7.1), where T represents time to carcinogenesis as given in Table 1.1. As previously, suppose that the scalar regression variable Z has value zero for group 1 and value 1 for group 2. A test of $\beta = \beta_0$ can be based on the generalized rank vector $r(w)$ of (7.6), where $W = Y - Z'\beta_0$. To test equality of the two survival curves, the hypothesis $\beta = 0$ is considered. In this case $W = Y$ and the generalized rank vector based on W coincides with that based on T itself, since $Y = \log(T - 100)$ is strictly monotone increasing. Monotonicity considerations also show the rank test for $\beta = 0$ to be independent of the "guarantee time," so that such a time need not be specified. The test (7.25) involves $n_j, j = 1, \ldots, k$, which under $\beta = 0$ are simply the number of rats alive just before the jth smallest (not necessarily distinct) failure time. Thus $k = 36, n_1 = 40, n_2 = 39, \ldots, n_{10} = 31, n_{11} = 29, \ldots, n_{36} = 2$. Note that since tied failure times occur only at the same Z value (same sample), they may be ordered arbitrarily. The generalized Wilcoxon scores (7.9) give a test statistic $\mathbf{v} = 2.887$ and a variance estimator $V_0 = 3.066$ using (7.19). This leads to an approximate χ_1^2 statistic (7.25) of value 2.72. In comparison, the log-rank scores (7.7) give $v = 4.584, V_0 = 7.653$ and an approximate χ_1^2 statistic of nearly identical value 2.75 as was obtained in Chapter 4 and compared there with other methods (e.g., maximum likelihood) in the proportional hazards model. From Chapter 3, parametric procedures using a Weibull regression model, that is, an extreme value error in model (7.1), yield $\hat{\beta} = 0.213$ and a χ_1^2 value of 3.90. The parametric procedure suggests somewhat stronger evidence for a difference between the survival curves than do the rank regression tests (0.05 versus 0.10 level of significance). The parametric method, however, is much more sensitive to the error modeling. As noted previously, for example, a parametric analysis using a log-normal model, that is, normal error in model (7.1) with $Y = \log(T - 100)$, gives $\hat{\beta} = 0.154$ and a χ_1^2 value of 1.32 for testing $\beta = 0$. Such a statistic is not suggestive of a survival difference between groups.

7.4 ESTIMATION IN THE ACCELERATED FAILURE TIME MODEL

7.4.1 Introduction

One can view the censored data rank test $\mathbf{v} = \mathbf{v}(\beta)$ in (7.25) as a function of β that has expectation zero at the true $\beta = \beta_0$. Hence, it is natural to consider estimating β by a value $\hat{\beta}$ such that $\mathbf{v}(\hat{\beta})$ is as close as possible to zero. For one-dimensional covariates, one can define $\hat{\beta}$ as a *zero crossing*, so that $\mathbf{v}(\hat{\beta}^-)$ and $\mathbf{v}(\hat{\beta}^+)$ have opposite signs. More generally, one can define $\hat{\beta}$ as a minimizer of $|\mathbf{v}'V^{-1}\mathbf{v}|$ or, perhaps more conveniently, as a minimizer of $\|\mathbf{v}\| = |\mathbf{v}'\mathbf{v}|$.

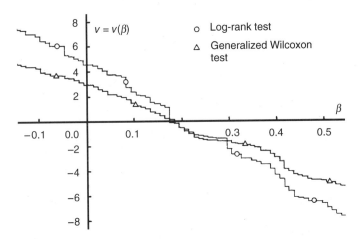

Figure 7.1 Values of certain rank statistics over plausible values of a two-sample regression coefficient β.

For example, the previous illustration can be continued to obtain rank-based estimators of β. Still with $Y = \log(T - 100)$, the statistic (7.25) was scanned over a fine grid of β_0 values. The test statistic is here a monotone decreasing function of β_0 for either generalized Wilcoxon (7.9) or log-rank (7.7) scores. These score functions are plotted in Fig. 7.1 and indicate approximate linearity over plausible values of β. An approximate 95% confidence interval for β is given by those β_0 values for which $\mathbf{v}'V^{-1}\mathbf{v}$ is less than 3.84, the upper 5% endpoint of a χ_1^2 distribution (Fig. 7.2). Using generalized Wilcoxon scores (7.9), one obtains $(-0.025, 0.414)$ as an approximate 95% interval for β, whereas log-rank scores give $(-0.038, 0.424)$. The two tests nearly coincide over plausible β values. The dashed line determines an approximate 95% confidence interval for β. The corresponding estimators $\hat{\beta}$ at which $\mathbf{v}'\mathbf{v}$ is minimized are 0.180 in each case, intermediate to those values from the two parametric analyses mentioned above.

7.4.2 Asymptotic Distribution Theory

Asymptotic distribution theory for $\hat{\beta}$ that minimizes $\|\mathbf{v}(\beta)\|$ proceeds by showing $\mathbf{v}(\beta)$ of (7.25) to be asymptotically linear in a neighborhood of the true β_0, whence the asymptotic normal distribution for $n^{-1/2}\mathbf{v}(\beta_0)$ induces an asymptotic normal distribution for $n^{1/2}(\hat{\beta} - \beta_0)$. Sufficient conditions for $\mathbf{v}(\beta)$ to be asymptotically linear were given by Tsiatis (1990). These were relaxed by Ying (1993), who was able to avoid a finite interval condition. The proofs of asymptotic linearity are somewhat tedious, owing to the fact that $\mathbf{v}(\beta)$ is a step function and need not be monotone in the elements of β. Specifically, the step function $\mathbf{v}(\beta)$ may change values whenever a change in β results in a change of ordering of any pair of censored or uncensored $W_i(\beta) = Y_i \wedge C_i - Z_i'\beta$ values, where C_i is the logarithm of the potential

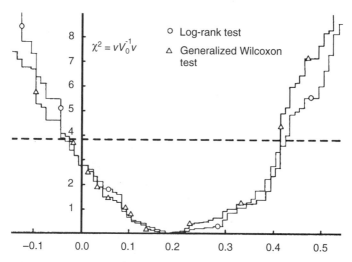

Figure 7.2 Values of approximate $\chi^2_{(1)}$ statistics corresponding to certain rank tests over plausible values of a two-sample regression coefficient β.

censoring time for the ith subject. Tsiatis tackled the linearity problem by bounding the number of such exchanges and the magnitude of jump in $\mathbf{v}(\beta)$ at any such exchange in a neighborhood of β_0. Ying applies a sequence of approximations to the processes that comprise (7.25). His sufficient conditions, under the usual independent failure time and independent censoring assumptions, can be written as follows:

1. The Z_i are uniformly bounded.
2. The error density f in (7.1) and its derivative f' are bounded, and f has finite Fisher information $\int_0^1 \phi^2(u, f)\, du < \infty$.
3. The random log-censoring times (C_i) have corresponding density g_i such that $g_i(t)$ is bounded uniformly for all (i, t).
4. The supremum over i of the expectation of $|e_i \wedge C_i|^\theta$ is finite for some $\theta > 0$.
5. The weight function $Q(\beta, w)$ has bounded variation and converges almost surely to a continuous function $q(\beta, w)$ in a neighborhood of β_0.

These conditions are enough to show that as in Section 7.3.3, $n^{-1/2}\mathbf{v}(\beta_0)$ converges to a mean zero normal distribution with variance matrix

$$\Sigma = n^{-1} \int_{-\infty}^{\infty} q(\beta_0, w)^2 [s^{(2)}(w) - s^{(1)}(w)s^{(1)}(w)'s^{(0)}(w)^{-1}]\lambda(w)\, dw,$$

where $s^{(j)}(w)$ is the probability limit of $S^{(j)}(\beta, w) = n^{-1}\sum_1^n Y_i(\beta, w)Z_i^{(j)}$ at $\beta = \beta_0$ for $j = 0, 1, 2$, where $Z^{(0)} = 1$, $Z^{(1)} = Z$, and $Z^{(2)} = ZZ'$, and where the risk indicator $Y_i(\beta, w)$ takes value 1 if $W_i(\beta) \geq w$ and value zero otherwise. Also in this expression, $\lambda(w) = f(w)/F(w)$ is the hazard function for the error variate in (7.1). Conditions 1 to 5 are also sufficient to show that $n^{1/2}(\hat{\beta} - \beta_0)$ converges to a mean zero normal distribution with variance $A^{-1} \sum A^{-1}$, where A is the slope matrix of the limiting linear form of $n^{-1/2}\mathbf{v}(\beta)$ in the vicinity of β_0. It can be written

$$A = n^{-1} \int_{-\infty}^{\infty} q(w, \beta_0)[s^{(2)}(w) - s^{(1)}(w)s^{(1)}(w)'s^{(0)}(w)^{-1}]\lambda'(w)\,dw,$$

where λ' denotes the derivative of λ. The matrix $\sum$ is readily estimated by $V(\hat{\beta})/n$, where $V(\hat{\beta})$ is (7.28) evaluated at $\beta = \hat{\beta}$, but because of the derivative λ', A is difficult to estimate.

Several proposals have been made for estimation of the variance of $n^{1/2}(\hat{\beta} - \beta_0)$. Tsiatis (1990) suggested nonparametric kernel density estimation of A, an approach that may require large-sample sizes to yield reliable estimators. Parzen et al. (1994) and Lin et al. (1998) propose a resampling technique wherein one obtains solutions $\beta_1^*, \beta_2^*, \ldots, \beta_m^*$, for some large m, to the estimating equation

$$\mathbf{v}(\beta_j^*) = d_j,$$

where $d_1, \ldots, d_m$ are independent observations from a multivariate normal distribution with mean zero and variance $V(\hat{\beta})$. Since $n^{1/2}(\hat{\beta}^* - \hat{\beta})$ generally has the same asymptotic distribution as $n^{1/2}(\hat{\beta} - \beta_0)$, one can use the sample variance of $\hat{\beta}^*$ to estimate the variance of $\hat{\beta}$, and the empirical distribution of $\hat{\beta}^*$ values can be used for estimating confidence intervals for the components of β. Chen and Jewell (2001) have considered a less computationally intensive estimator of the variance of $n^{1/2}(\hat{\beta} - \beta_0)$, attributed to Eugene Huang. This approach involves solving

$$\mathbf{v}(\tilde{\beta}_j) = b_j$$

for $\tilde{\beta}_j$, $j = 1, \ldots, p$, where $B = (b_1, \ldots, b_p)$ determines a decomposition $BB' = V(\hat{\beta})/n$ of the variance matrix for $n^{-1/2}\mathbf{v}(\beta_0)$. A consistent variance estimator for $n^{1/2}(\hat{\beta} - \beta_0)$ is then given by $(\tilde{\beta} - \hat{\beta})(\tilde{\beta} - \hat{\beta})'$. These variance estimators for $n^{1/2}(\hat{\beta} - \beta_0)$ evidently perform well with moderate-sized samples (e.g., Lin et al, 1998; Chen and Jewell, 2001).

The efficiency of $\hat{\beta}$ is the same as that of the test $\mathbf{v}(\beta_0)$ against local alternatives. It is maximized when A and $\sum$ are equal, as would be accomplished if the limiting weight function $q(w, \beta_0) = \lambda'(w)/\lambda(w)$, suggesting the possibility of semiparametric efficient estimation (Begun et al., 1983; Bickel et al., 1992) through an adaptive choice of weight function in (7.25). In fact, Lai and Ying (1991a) propose an adaptive construction of a semiparametric efficient estimator while allowing left truncation as well as right censorship. However, their estimator is expected to require large sample sizes to achieve such good efficiency. Note also that as for the test (7.25), the semiparametric efficient estimator $\hat{\beta}$ will typically be inefficient compared to the efficient parametric regression estimator based on a correct choice of error density in (7.1), except in the very special case where censoring is independent of Z.

7.4.3 Numerical Aspects of Rank Regression Parameter Estimation

Application of the asymptotic results summarized in Section 7.4.2 requires a reliable means of calculating $\hat{\beta}$ that minimizes $\|\mathbf{v}(\beta)\|$. As noted previously, this calculation problem is complicated by the fact that $\mathbf{v}(\beta)$ is a step function that need not be monotone. For example, the lack of monotonicity for the subclass of linear rank tests (7.18) essentially derives from a lack of monotonicity of the scores associated with the n ordered censored or uncensored residuals $W = Y \wedge C - Z'\beta$. Fygenson and Ritov (1994) have studied the monotonicity properties of the weighted log-rank statistic (7.25). They noted that the special case of Gehan's weights, where $Q_i = n_i n^{-1}$, does give a monotone estimating function, and they characterized the class of weight functions for which (7.25) is a monotone function of each component of β. The weights in this class are of the form $Q_i = n_i h_i$, where the sequence $h_1, h_2, \ldots$ satisfy $h_j \le h_{j-1} n_j n_{j-2}^{-1}$, with the Gehan special case given by $h_j \equiv 1$. Clearly, the log-rank (7.7) and generalized Wilcoxon (7.9) estimating functions are outside this class, and for a monotone estimating function, the weights in (7.25) generally need to depend on the censoring as well as the failure time pattern.

The estimating function with Gehan weights can be written

$$\mathbf{v}(\beta) = \sum_{i=1}^{n} \delta_i S^{(0)}(\beta, W_i)[Z_i - \bar{Z}(\beta, W_i)]$$

$$= \sum_{i=1}^{n} \sum_{j=1}^{n} \delta_i (Z_i - Z_j)\mathbf{1}[W_i(\beta) \ge W_j(\beta)], \qquad (7.29)$$

where δ_i is the censoring indicator for the ith individual and, once again, $\bar{Z}(\beta, w) = S^{(1)}(\beta, w)/S^{(0)}(\beta, w)$. Lin et al. (1998) note that (7.29) is the gradient of

$$\sum_{i=1}^{n} \sum_{j=1}^{n} \delta_i [W_i(\beta) - W_j(\beta)]^+, \qquad (7.30)$$

where $a^+ = a \vee 0$, so that the regression parameter estimate may be calculated by linear programming in which $\sum \sum \delta_i W_{ij}$ is minimized subject to $W_{ij} \ge 0$ and $W_{ij} \ge W_j(\beta) - W_i(\beta)$.

For other weight functions in (7.25), Lin and Geyer (1992) propose a simulated annealing approach to computing $\hat{\beta}$ when the dimension of β is too large to permit a direct grid search. Jin et al. (2002) have proposed a more convenient numerical procedure based on approximating (7.25) by

$$\mathbf{v}(\beta) = \sum_{i=1}^{n} \delta_i \frac{Q_i(\tilde{\beta}, \tilde{W}_i)}{S^{(0)}(\tilde{\beta}, \tilde{W}_i)} S^{(0)}(\beta, W_i)[Z_i - Z_i(\beta, W_i)], \qquad (7.31)$$

where $\tilde{\beta}$ is a consistent estimator of β, such as the Gehan estimator, and $\tilde{W}_i = Y_i - Z'\tilde{\beta}, i = 1, \ldots, n$. They note that (7.31) is monotone in β and that $\hat{\beta}$, which minimizes $\|\mathbf{v}(\beta)\|$ in (7.31), can again be obtained by linear programming. This leads to an iterative approach to finding $\hat{\beta}$ that minimizes $\|\mathbf{v}(\beta)\|$ in (7.25), in which $\tilde{\beta}$ is replaced by the updated solution to (7.31) at each step in the iteration.

7.4.4 Time-Dependent Covariates and Baseline Hazard Estimation

Under the accelerated failure time model (7.1),

$$T \exp(-Z'\beta) = \exp(\sigma e)$$

has a certain density, and a certain hazard function, say $\lambda_0(\cdot)$, that does not depend on Z. It follows that the hazard function for T, following a change in sign for β, can be written

$$\lambda(t, Z) = \lambda_0(te^{Z'\beta})e^{Z'\beta}. \tag{7.32}$$

It is natural to ask whether the semiparametric class (7.32) can be extended in a meaningful way to include time-dependent covariates, as allowance for such covariates is a key feature of relative risk (Cox) regression methods.

As in Chapter 6, we may consider a history of covariates $X(t) = \{x(u); 0 \le u < t\}$ of covariates up to time t, from which a modeled covariate p-vector $Z(t) = [Z_1(t), \ldots, Z_p(t)]'$ is derived. The notion that covariates accelerate or decelerate the rate at which a study subject progresses through the time axis can be adapted to time-dependent covariates by postulating a fixed distribution for $\int_0^T \exp[Z(u)'\beta] \, du$, in which case the hazard process for T can be written

$$\lambda[t; X(t)] = \lambda_0 \left\{ \int_0^t \exp[Z(u)'\beta] \, du \right\} \exp[Z(t)'\beta], \tag{7.33}$$

generalizing (7.32). This model was introduced by Cox and Oakes (1984, p. 66), without requiring an exponential form for the acceleration factor. It has been studied by Robins and Tsiatis (1992), who proposed rank estimators analogous to (7.25) for estimating β, and by Lin and Ying (1995), who provide corresponding asymptotic distribution theory. Briefly, if one defines $h_i(t, \beta) = \int_0^t \exp[Z_i(u)'\beta] \, du$, and $W_i(\beta) = h_i(T_i \wedge C_i, \beta), \tilde{Z}_i(\beta, t) = Z_i[h_i^{-1}(t, \beta)]$ and $\tilde{Y}_i(\beta, t) = I[W_i(\beta) \ge t]$, then one can define weighted log-rank tests as

$$\mathbf{v}(\beta) = \sum_{i=1}^n \delta_i Q_i(\beta, W_i) \left[\tilde{Z}_i(\beta, W_i) - \sum_1^n \tilde{Y}_j(\beta, W_j)\tilde{Z}_j(\beta, W_j) \Big/ \sum_1^n \tilde{Y}_j(\beta, W_j) \right] \tag{7.34}$$

and an estimator $\hat{\beta}$ as a minimizer of $\|\mathbf{v}(\beta)\|$. The asymptotic properties of $n^{-1/2}\mathbf{v}(\beta_0)$ under $\beta = \beta_0$, and the asymptotic properties $n^{1/2}(\hat{\beta} - \beta_0)$ and the numerical considerations in calculating $\hat{\beta}$, do not differ in any major way from the fixed covariate special case. See Lin and Ying (1995) for further detail on, and illustration of, the log-rank special case $Q_i(\beta, W_i) \equiv 1$.

With either fixed or time-dependent covariates, one can readily nonparametrically estimate the cumulative baseline hazard function Λ_0 given by $\Lambda_0(t) = \int_0^t \lambda_0(u)\, du$ by

$$\hat{\Lambda}_0(t, \hat{\beta}) = \int_0^t \sum_{i=1}^n d\tilde{N}_i(\hat{\beta}, u) \bigg/ \sum_{i=1}^n \tilde{Y}_i(\hat{\beta}, u),$$

where $\tilde{N}_i(\beta, t) = \delta_i I[W_i(\beta) \le t]$. An asymptotic Gaussian distribution for $n^{1/2}[\hat{\Lambda}_0(\cdot, \hat{\beta}) - \Lambda_0(\cdot)]$ can also be established over a finite follow-up interval. See Lin et al. (1998) for a closely related development in the context of recurrent events under an accelerated failure time model.

The results summarized in Sections 7.3 and 7.4, developed over the past couple of decades, now allow the AFT model to be considered as a practical semiparametric alternative to the relative risk (Cox) model. The efficiency properties of tests and estimators under the unweighted log-rank special case, for example, are likely to be good enough in most settings of practical interest.

7.5 SOME RELATED REGRESSION MODELS

A number of other semiparametric regression models have been proposed for the regression analysis of right-censored failure time data with fixed covariates. We will merely provide a list and brief commentary on a few of them here. If rather than $\log T$ in (7.1), one takes an unknown increasing function $h(T)$, one obtains

$$h(T) = Z'\beta + e. \tag{7.35}$$

This model is referred to as the *linear transformation model* if h is unspecified and the error density F is taken to be completely known (Kalbfleisch, 1978b; Dabrowska and Doksum, 1988). If f has extreme minimum form, then (7.35) has proportional hazards form, while (7.35) defines a proportional odds regression model if f is standard logistic. See Fine et al. (1998) for a recent contribution to the consistent estimation of β under this model using martingale methods. Also, (7.35) is invariant under the group of monotone transformations on T so that a marginal likelihood for β can be obtained. The integrals in this marginal likelihood, however, are generally intractable. Importantly, the model (7.35) suffers from a lack of ready interpretation for the regression parameter β, except in the extreme value and logistic error special cases.

Chen and Wang (2000) consider a hazard rate model

$$\lambda(t; Z) = \lambda_0[t \exp(Z'\beta)],$$

which they term the *accelerated hazards model*. This model also appears to lack a clear regression parameter interpretation, although it embraces a range of hazard ratio shapes. Recently, Chen and Jewell (2001) consider a class of hazard rate models

$$\lambda(t; Z) = \lambda_0[t \exp(Z'\beta_1)] \exp(Z'\beta_2)$$

which includes the proportional hazards model $(\beta_1 = 0)$, the accelerated failure time model $(\beta_1 = \beta_2)$, and the accelerated hazards model just mentioned $(\beta_2 = 0)$. This model has an identifiability problem in the Weibull special case $[\lambda_0(t) = \lambda_0 t^q]$. More generally, however, this model may have a useful role in discriminating between a proportional hazards and accelerated failure time models, given a specific linear form $Z'\beta$. Chen and Jewell provide a rank-based procedure for estimating β_1 and β_2 away from the Weibull submodel, along with corresponding asymptotic distribution theory and numerical considerations.

BIBLIOGRAPHIC NOTES

Introductory books dealing with uncensored rank tests include Lehmann (1975), Hájek (1969), Conover (1971), and Hollander and Wolfe (1973). More advanced works include Hájek and Sĭdák (1967) and Cox and Hinkley (1974, Chap. 4). Corresponding contributions to the use of residual ranks for estimating linear regression coefficients were made by Hodges and Lehmann (1963), Adichie (1967), and Jurečková (1969, 1971). See also Hogg (1974), Puri (1970), Puri and Sen (1971), Hettmansperger and McKean (1977), McKean and Hettmansperger (1978), and Hettmansperger (1984) for additional developments, review, and discussion of numerical aspects.

Rank tests with type II censored data were derived in a manner similar to that of Section 7.3.2 by Johnson and Mehrotra (1972). A rather general but somewhat ad hoc means of generating rank tests with censored data is given by Peto and Peto (1972) and Peto (1972a). The log-rank test (7.7) can be regarded as an outgrowth of a series of papers by Mantel and collaborators (Mantel and Haenszel, 1959; Mantel, 1963, 1966). Work on its properties has been carried out by Thomas (1969) and by Crowley and Thomas (1975). Crowley (1974b) discusses various likelihoods leading to the log-rank test. The more general approach considered here for ranks with censored data is that used by Kalbfleisch and Prentice (1973), which, in turn, amplified a remark of Peto (1972b). The generalized Wilcoxon test presented in (7.9) was given in Prentice (1978) and differs only slightly from that presented by Peto and Peto (1972). See also Peto et al. (1977), Tarone and Ware (1977), and Cuzick (1982) for related work and review.

Asymptotic distribution theory and properties of censored data rank tests were studied in the 1980s. Gill (1980) and Andersen et al. (1982) applied martingale convergence results to develop asymptotic distribution theory for such tests for the comparison of several samples. These authors, as well as Prentice and Marek (1979) and Mehrotra et al. (1982) noted the connection between linear rank tests of the form (7.19) and weighted log-rank tests (7.25). Harrington and Fleming (1982) proposed an interesting G^ρ class of censored data rank tests that include the log-rank and generalized Wilcoxon test (7.9) as special cases of a single parameter ρ. Leurgans (1983, 1984), Struthers (1984), and Cuzick (1985, 1988) considered further aspects of these tests, including efficiency aspects. Simulation studies to evaluate and compare small-sample properties of selected censored data rank tests are given in Lee et al. (1975), Peace and Flora (1978), Lininger et al. (1979), Latta (1981), and Kellerer and Chmelevsky (1983).

Generalizations of censored data rank tests to interval censored data under the AFT and proportional hazards models have been considered by Self and Grossman (1986), Satten (1996), and Sun (1996).

Nonparametric estimation under the accelerated failure time model with censored data has been considered by a number of investigators. Early work on this topic involved generalizations of least squares regression to censored data. Miller (1976) proposed an estimator that minimizes a weighted sum of squares of residuals, with weights determined by the Kaplan–Meier estimator of the residual survivor function. Buckley and James (1979) proposed an alternative estimator based on replacing censored observations by an estimate of their expectation and applying a least squares procedure. Lai and Ying (1991b) consider the Buckley–James estimator modified to control tail instability and develop asymptotic properties. Koul et al. (1981) provide an inverse censoring probability weighted modification of the least squares estimator.

Estimators based on censored data rank tests from (7.1) were proposed by Prentice (1978) and Gill (1980). Louis (1981) studied the log-rank test-based estimator in the two-sample problem and developed corresponding asymptotic theory, while Wei and Gail (1983) provide corresponding results for a broader class of censored data rank tests. Distribution theory and properties of parameter estimates based on weighted log-rank tests (7.25) in a more general regression context were derived by Tsiatis (1990), Ritov (1990), Wei et al. (1990), Lai and Ying (1991b), and in the greatest generality, by Ying (1993). Generalizations to include time-dependent covariates are described in Robins and Tsiatis (1992), and Lin and Ying (1995). Numerical aspects of fitting the accelerated failure time model and for estimating the variance matrix for the regression parameter are described in Lin and Geyer (1992), Fygenson and Ritov (1994), Parzen et al. (1994), Chen and Jewell (2001), and Jin et al. (2002).

Cox and Oakes (1984) propose a number of alternative regression models to the Cox model and accelerated models, beyond those mentioned in Section 7.5. The proportional odds regression model special case of (7.35) is sometimes recommended as providing a converging hazards alternative to the proportional hazards model. Estimation under a proportional odds model, or more general transformation model, has been considered by Bennett (1983), Pettitt (1982), Cuzick (1988), Dabrowska and Doksum (1988), Cheng et al. (1995), Wu (1995), Murphy et al. (1997), and Yang and Prentice (1999), among others.

EXERCISES AND COMPLEMENTS

7.1 Show explicitly that:

    (a) The expectation of (7.18) is zero under (1) censorship that is independent of Z, and (2) arbitrary independent censoring.

    (b) Write the permutation variance for (7.18) assuming censorship independent of Z. Compute this variance for both the generalized Wilcoxon (7.9) and the log-rank (7.7) scores for testing $\beta = 0$ with the carcinogenesis data (Table 1.1) as discussed in the illustration in Section 7.3.5. Compare these with the variance estimates based on the weighted log-rank (optional variation) estimator (7.28).

    (c) Show that (7.28) is an estimator of variance for (7.18) under general independent censoring.

7.2 Calculate explicitly the uncensored scores c_i (Section 7.3.2) based on a double exponential score generating density. Compare these with the sign (median) test scores.

7.3 Show that the uncensored rank test efficiency (7.16) is unchanged by a location or scale transformation on the data $Y_1, \ldots, Y_n$.

7.4 The censored data rank tests of Section 7.3.2 can be based on approximate quantile scores in the manner of (7.14) for the uncensored case. The scores can be specified as

$$c_i = \phi[\hat{F}(W_{(i)})]$$
$$C_i = \Phi[\hat{F}(W_{(i)})],$$

where $\hat{F}(W_{(i)}) = \prod_{j=1}^{i}[n_j/(n_{j+1})]$ (Prentice, 1978; Cuzick, 1985). Show that for the double exponential error density, this leads to a censored version of the sign test with scores

$$c_i = \begin{cases} \text{sign}\,[2\hat{F}(w_{(i)}) - 1], & \hat{F}(w_{(i)}) \neq 0.5 \\ 0, & \hat{F}(w_{(i)}) = 0.5 \end{cases}$$

$$C_i = \begin{cases} [1 - \hat{F}(W_{(i)})]/\hat{F}(w_{(i)}), & \hat{F}(w_{(i)}) \geq 0.5 \\ 1, & \hat{F}(w_{(i)}) < 0.5. \end{cases}$$

7.5 Derive conditions on the censoring mechanism under which the test statistic (7.18) based on the approximate scores of Problem 7.4 will be asymptotically equivalent to the same statistic with scores (7.19) (Cuzick, 1985).

7.6 Consider a two-sample problem with $Z = 0$ for sample 1 and $Z = 1$ for sample 2. In the absence of tied failure times, show that the log-rank statistic (7.7), the generalized Wilcoxon statistic (7.9), and the Gehan generalized Wilcoxon statistics can all be written in the form

$$\sum_{i=1}^{k} Q_i(Z_i - n_{1i}n_i^{-1}),$$

where i indexes the ordered failure times $t_1, \ldots, t_k$ in the combined sample, n_i is the size of the risk set just prior to t_i, n_{1i} is the size of the risk set in the second sample $(Z_i = 1)$ just prior to t_i, and the Q_i are weights given, respectively, by $Q_i = n_i$, $Q_i \equiv 1$, and $Q_i = F_i$ for the three statistics, where

$$F_i = \prod_{j=1}^{i} n_j(n_j + 1)^{-1}$$

is a survivor function estimator. Suggest a generalization for statistics in this form that will accommodate tied failure times. Present a variance estimator for these generalized statistics based on a partial likelihood argument and a hypergeometric distribution at each failure time.

7.7 (*continuation*) The Gehan generalized statistic has a weight function, $Q_i = n_i$, $i = 1, \ldots, k$, that depends on the censoring mechanism. Suppose that n study subjects are followed until the first failure time t_1, after which a random sample of m randomly selected subjects are followed until failure. Show that with equal initial sample sizes in the two groups, the standardized Gehan test statistic will approach either $+1$ or -1 (depending on which sample contains the initial failure) as $n \to \infty$ (with m fixed) and regardless of the relationship between the two failure time distributions being compared.

7.8 Show that the preservation of scores formula $n_iC_{i-1} = c_i + (n_i - 1)C_i$ of Section 7.3.3 holds for all $i = 1, \ldots, k$ by integrating C_i in (7.19) over $u_{i+1}, \ldots, u_k$ and by integrating by parts over u_i (Mehrotra et al., 1982).

7.9 Show that the weighted log-rank test (7.28) with Gehan scores $Q_i = n_i$ is monotone in each component of β (Fygenson and Ritov, 1994). Show that this weight function can be replaced by $Q_i(\beta) = h_i n_i(\beta)$ while retaining this monotonicity for fixed constants $h_1, \ldots, h_k$ that do not depend on β.

7.10 Show that the residual survivor function estimator $\hat{F}(w)$ in Problem 7.4 can be replaced by the Kaplan–Meier estimator $\prod_{j=1}^{i}[(n_j - 1)/n_j]$ or the Altshuler estimator $\exp(-\sum_{j=1}^{i} n_j^{-1})$ without changing the asymptotic distribution of the rank statistic under mild conditions. (Cuzick, 1985)

7.11 Apply the iteratively reweighted Gehan estimation procedure of Jin et al. (2001) (Section 7.4.3) to estimate β for the carcinogenesis data of Section 7.3.5. Apply both the resampling procedure of Parzen et al. (1994) and the procedure due to Eugene Huang described in Section 7.4.2 to estimate the variance of the log-rank-based estimator $\hat{\beta}$. Compare corresponding nominal 95% confidence intervals for β using these variance estimators to those given in Section 7.3.5.

CHAPTER 8

Competing Risks and Multistate Models

8.1 INTRODUCTION

In the preceding chapters we have presented methods for data analyses when there is a single, possibly censored failure time on each study subject. Failure data may be more general in two important respects. First, the failure on an individual may be one of several distinct failure types; such data are commonly referred to as *competing risks data*. Second, there may be more than one failure time on each study subject. Such multivariate failure times may correspond to repeated occurrences of some specific type of event or to the occurrence over time of events of several distinct types. In general, the life history of a typical individual under study may involve a variable number of failures, each with its own type or cause.

The hazard function or process is generalized in this chapter to permit the inclusion of competing risks and to certain life-history processes. In Chapters 9 and 10 we addresss additional important multivariate failure time topics. In Section 8.2, several distinct problems that have traditionally been discussed in the context of competing risks data are outlined and the methods of preceding chapters are generalized in order to study the relationship between covariates and certain *cause-* or *type*-specific hazard functions. This addresses in a natural way many issues of interest to investigators. Other questions concerning the relationship between the types of failure can be formulated and addressed, in some instances, through the use of internal time-dependent covariates that allow a deeper look at disease pathways. The modeling strategies discussed here relate specifically to directly estimable quantities. We argue that modeling through latent failure times and related multiple decrement models, so commonly used in the competing risks literature, is not to be recommended except in special types of applications where unobserved potential failure times can be given a clear meaning.

An introduction to the modeling and analysis of more general life-history processes is also considered briefly. In Section 8.3, hazard and counting process formulations are generalized to model the occurrence of several events that might occur in a study subject's follow-up. We consider stochastic processes with finite state space and consider Markov and semi-Markov models. Covariates are allowed to modulate the transition intensities in both types of models. We consider more specifically the use of the relative risk or Cox model in this more general setting.

8.2 COMPETING RISKS

8.2.1 Competing Risks Problems

Suppose that individuals under study can experience any one of m distinct failure types, and for each individual we observe, possibly subject to right censoring, the time to failure and the type of failure. For example, the failure types may correspond to death by several distinct causes or diagnoses, or we may be looking at the time to first breakdown of a complex mechanical system with the failure types corresponding to the failure of any one of m specified subsystems. Much of the literature on such competing risks data has presumed the existence of m *latent* or *potential* failure times for each individual. These hypothetical failure times correspond to the time to failure for each of the distinct types, and the data consist of the time and the type of the first failure. See, for example, Crowder (2001) for a recent review of this approach. In the presentation here, however, we consider primarily statistical models for observable quantities only and avoid reference to hypothetical and unobserved times to failure. The latent failure time approach is discussed in Section 8.2.5.

Suppose, as before, that each study subject has an underlying failure time T that may be subject to censoring, and a basic covariate vector x, or more generally, a covariate function $\{x(u) : u \geq 0)\}$. Suppose also that when failure occurs, it may be of any one of m distinct types or causes denoted by $J \in \{1, 2, \ldots, m\}$. For example, in the mouse radiation study mentioned in Section 1.1.1 (data listed in Appendix A, data set V), the failure types are death from thymic lymphoma, death from reticulum cell sarcoma, or death from other causes. In demographic mortality studies, failure types are often broad categories such as cancer, heart disease, or accidents. In industrial life testing a certain piece of equipment may develop one of several faults causing its failure.

Three distinct problems arise in the analysis of such data:

1. The estimation of the relationship between covariates and the rate of occurrence of failures of specific types.
2. Study of the interrelation between failure types under a specific set of study conditions.
3. The estimation of failure rates for certain types of failure given the *removal* of some or all other failure types.

For instance, problem 1 arose in the analysis of data from the University Group Diabetes Program (UGDP; e.g., Cornfield, 1971; Gilbert et al., 1975). The primary endpoint for treatment evaluation was patient survival time. Since the effects of treatment may differ markedly among causes of death, and since distinct causes of death may relate to different prognostic factors, an analysis of overall survival time may well be inadequate. In the UGDP study, significant treatment differences in respect to overall survival time were not found. An analysis of treatment differences with respect to the different types of failure led to the finding of primary interest and controversy in the study: namely, that one treatment, tolbutamide, appeared to give rise to a higher risk of cardiovascular death. Similarly, the Hoel data (Appendix A, data set V) may be used to study the effect of a germ-free environment on specific causes of death in irradiated male mice. Methods for the estimation of cause-specific failure rates are considered in Section 8.2.3.

Problem 2 is also of interest in a number of contexts. For example, Thomas et al. (1975a,b) summarized their experience with bone marrow transplantation in the treatment of acute leukemia. A little background is given here since this application serves to illustrate and clarify a number of points in relation to problems 2 and 3: End-stage leukemia patients in relapse are given ordinarily lethal doses of radiation and chemotherapy in an attempt to eradicate their leukemia. Following such conditioning and immunosuppression, bone marrow cells from an HLA (human lymphocyte antigen)-matched sibling donor are infused into the patient's bloodstream. These cells lodge in "spacings" in the patient's marrow, repopulate, and give rise to a new hemopoietic and immunologic system for the patient. Major causes of death relate to (1) recurrence of leukemia and (2) graft versus host disease (GVHD). The latter occurs as an immunologic reaction to the new marrow graft against the patient. GVHD is evident primarily through effects on the skin, liver, and gut.

A question of significant biological implication concerns the relationship between GVHD and relapse due to recurrent leukemia. For example, it may happen that a graft versus host (GVH) reaction of a certain degree of severity is useful in preventing recurrence. The immunologic reaction of the marrow graft on the host may even selectively kill residual or new leukemia cells. On the other hand, a severe GVH reaction may simply weaken the patient or be an indication of a poorly functioning marrow graft, which, in turn, may be associated with an increased risk of leukemic relapse. Knowledge of the interrelation between the two failure modes would be valuable, furthering understanding of the interaction between treatment and disease. It would also be useful in defining the desired degree of tissue similarity between marrow and patient and could be used to predict the usefulness of autologous marrow replacement following treatment in nonhematologic diseases. Note that this question is posed in terms of a patient group with well-defined eligibility criteria under a specific treatment program. The question does not involve, for example, inferences about recurrent leukemia rates if mortality associated with graft versus host disease could somehow be eliminated.

A second and more general application involves the relationship between failure and censoring due to withdrawal or removal from study. The methods of the preceding chapters all assume an independent censoring mechanism, and the estimators of regression coefficients and survivor functions may be seriously affected if, for example, study subjects are selectively censored when they appear to have become at a high risk of failure relative to other study subjects with comparable covariates at risk at the same follow-up time. In the formulation of this chapter we may regard censoring as an additional failure mode and attempt to study the relationship between time to censoring and the corresponding time to actual failure.

Several authors (e.g., Cox, 1959; Tsiatis, 1975; Peterson, 1975) have indicated that data of type (T, J) are not adequate to study the interrelation between failure types. These results extend directly to data (T, J, x) with x time independent. In Section 8.2.5 we discuss some approaches that utilize the relationship between the failure rates and measured time-dependent covariates $x(t)$ in order to study possible interrelationships among failure types. This offers a potentially useful way to address problem 2.

Problem 3 has often been regarded as *the problem* of competing risks. In order to venture solutions to problems of this type, one must assume that data under one set of study conditions in which m failure types are operative are relevant to a different set of study conditions in which only certain of the failure modes can occur. This is sometimes a reasonable assumption, with some applications involving, for example, pieces of equipment with quite separate and noninteracting subsystems that might fail. In most applications, however, it is clear that the failure rate function for a specific failure type may be affected in a variety of ways by *removal* of other failure types. In fact, *removal* itself may often be accomplished through a variety of mechanisms, and different removal mechanisms may have substantially different effects on the hazards associated with remaining failure types. For these reasons, problem 3 is primarily not a statistical problem. It is unrealistic to think that general statistical methods and formulas can be developed to estimate failure rates after some causes have been removed. Before reasonable methods can be proposed in any given setting, it is necessary to have a good deal of knowledge of the physical or biological mechanisms giving rise to the failures and failure types as well as a good understanding of the mechanism, and possible influences of the mechanism, by which some causes are being removed.

To illustrate difficulties associated with problem 3, consider again the marrow transplantation example. One can envisage two mechanisms that could potentially give rise to the removal of mortality associated with graft versus host disease. First, this could occur via use of a treatment that reduced the severity of the GVH reaction allowing the patient to survive through the acute GVHD time period (approximately 6 months) and be at subsequent risk only for the remaining types of failure. This mechanism would in no way alter the donor–recipient matching, the conditioning treatments prior to transplantation, or the marrow grafting procedure itself. On the other hand, it could substantially alter the relation between recurrent leukemia and the incidence rates of the other causes of failure. As a matter of fact, there are anti-rejection drugs [e.g., antithymocyte globulin (ATG)] that are effective in reducing the severity of the GVH reaction. Unfortunately, use of these drugs generally results in some suppression of the immune system, and this may result in increased risk of death from other causes, such as pneumonia. It is also not generally clear what effect such immunosuppressive drugs may have on the risk of recurrent leukemia.

This mechanism of removal would almost certainly change the rates of other types of failure and alter the environment under which the patient is at risk of leukemia recurrence.

A second conceivable mechanism to remove GVHD as a cause of death would involve a protocol change in the degree of genetic similarity (tissue matching) between donor and recipient before a patient undergoes transplantation. As an extreme special case, leukemia patients receiving marrow grafts from an identical twin donor do not experience graft versus host disease. Such a protocol change could give rise to a completely different immunologic response of the marrow graft against residual or new leukemia cells. Times to recurrent leukemia following such a change in matching criteria may not be closely related to recurrent leukemia times that would have been experienced under previous criteria.

8.2.2 Representation of Competing Risk Failure Rates

Suppose that failure time T is continuous and, as before, let $x(t), t \geq 0$ be a vector of possibly time-dependent covariates, $X(t) = \{x(u) : 0 \leq u < t\}$, and $Z(t)$ be a vector of p derived covariates which are defined as function of $X(t)$. As before, the overall failure rate or hazard function at time t is

$$\lambda[t; X(t)] = \lim_{h \to 0} h^{-1} P[t \leq T < t + h \mid T \geq t, X(t)].$$

To model competing risks, we consider a *type-specific* or cause-specific hazard function or process

$$\lambda_j[t; X(t)] = \lim_{h \to 0} h^{-1} P[t \leq T < t + h, J = j \mid T \geq t, X(t)]. \qquad (8.1)$$

for $j = 1, \ldots, m$ and $t > 0$. In words, $\lambda_j[t; X(t)]$ represents the instantaneous rate for failures of type j at time t given $X(t)$ and in the presence of all other failure types; that is, it specifies the rate of type j failures under study conditions. If only one of the failure types $(1, 2, \ldots, m)$ can occur, then

$$\lambda[t; X(t)] = \sum_{j=1}^{m} \lambda_j[t; X(t)]. \qquad (8.2)$$

In some applications, it may be natural to allow two or more types of failure to occur simultaneously. In the current framework, such coincidences may be accommodated by defining additional failure types.

Suppose now that all elements of $x(t)$ are fixed or external covariates (6.8) so that they have corresponding conditional survivor and density functions. It is convenient to denote by $X = \{x(u) : u \geq 0\}$ the entire path of the covariates in this case and simply condition on X so that we write $\lambda_j(t;X)$. The overall survivor function is

$$F(t;X) = P(T > t|X) = \exp\left[-\int_0^t \lambda(u;X)\,du\right]. \tag{8.3}$$

The (sub)density function for the time to a type j failure is

$$f_j(t:X) = \lim_{h\to 0} h^{-1} P(t \le T < t + h, J = j|X)$$
$$= \lambda_j(t;X)F(t;X), \qquad j = 1,\ldots,m, \tag{8.4}$$

and the density function of the time to failure is $f(t;X) = \sum_{j=1}^m f_j(t;X)$. From the characterization of external covariates [(6.8) and (6.9)], $F(t;X)$ and $f_j(t;X)$ depend only on X only through $X(t)$. When the covariates are of the fixed or external type, the likelihood on a sample subject to independent right censorship is written as a product of the survivor functions (8.3) for the censored data and the subdensities for the observed failure times. Thus (8.2) to (8.4) show that the likelihood function can be written entirely in terms of the type-specific hazard functions (8.1), as discussed further in the next section.

It is of interest to note some other functions and properties of the model that are sometimes useful. The *cumulative incidence function* for type j failures corresponding to the external covariate is

$$\bar{F}_j(t;X) = P(T \le t, J = j;X)$$
$$= \int_0^t f_j(u;X)\,du, \qquad t > 0, \tag{8.5}$$

for $j = 1,\ldots,m$. Note that

$$p_j = P(J = j) = \lim_{t\to\infty} \bar{F}_j(t;X), \qquad j = 1,\ldots,m \tag{8.6}$$

and $\sum_{j=1}^m p_j = 1$. It may also seem natural to consider the functions $F_j(t;X) = \exp[-\int_0^t \lambda_j(u;X)\,du], t > 0, j = 1,\ldots,m$, but these functions have no simple probability interpretation within the competing risks model, at least not without introducing strong additional assumptions.

Example 8.1. Suppose that $m = 2$ and that the covariate is a treatment indicator $x = 0, 1$. In the control group ($x = 0$), the hazard rates are $\lambda_1(t;x = 0) = 1$ and $\lambda_2(t;x = 0) = t^{1/2}$ for $t > 0$. Suppose further that the treatment affects the second type of specific failure rate by a multiplicative factor $\exp(\beta) = 0.5$ and does not alter the first rate. Thus $\lambda_1(t;x = 1) = 1$ and $\lambda_2(t;x = 1) = 0.5t^{1/2}$. Figure 8.1 gives a plot of the cumulative incidence functions (8.5) for the two failure types. Note that the treatment has an apparent effect on the incidence of type 1 failures, although this is as a result of the change in the type 2 rates. □

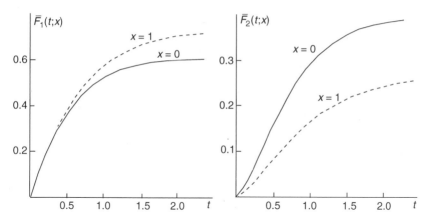

Figure 8.1 Cumulative incidence functions for Example 8.1.

These results can be extended to discrete and mixed models. Thus, again with fixed or external covariates, let $\Lambda_j(t; X)$ be the (left continuous) cumulative type j specific hazard function. Thus $d\Lambda_j(t; X) = \lambda_j(t; X)dt$ if t is a continuity point of $\Lambda_j(t; X)$ and

$$d\Lambda_j(t; X) = P(T = t, J = j \mid T \geq t; X)$$

if t is a jump point. Let $a_1 < a_2 < \cdots$ be the times of nonzero probability mass and let $d\Lambda_j(a_l; X) = \lambda_{jl}(X)$. As in the case of a single failure mode, the survivor function can be written

$$F(t; X) = \mathscr{P}_0^t[1 - d\Lambda(t; X)],$$

where $\Lambda(t; X) = \sum_{j=1}^m \Lambda_j(t; X)$, which reduces to (8.3) in the continuous case. In the discrete case, $F(t; X) = \prod_{a_l \leq t}[1 - \lambda_l(X)]$, where $\lambda_l(X) = \sum_{j=0}^m \lambda_{jl}(X)$. The general form of the jth subdensity function is

$$d\bar{F}_j(t; X) = d\Lambda_j(t; X)F(t^-; X),$$

which in the discrete case reduces to

$$d\bar{F}_j(a_l; X) = P(T = a_l, J = j; X) = \lambda_{jl}(X)\prod_{i=1}^{l-1}[1 - \lambda_i(X)], \qquad (8.7)$$

for $l = 1, 2, \ldots, j = 1, \ldots, m$, and an empty product is taken to be 1. With a single failure mode, the discrete failure time model can be viewed as a sequence of Bernoulli trials; in an analogous way, the discrete competing risks model corresponds to a sequence of multinomial trials. Thus, if the individual survives to time a_l^-, an $(m+1)$-sided die is rolled which shows face j with probability λ_{jl}, $j = 1, \ldots, m$ and face $m + 1$ with probability $1 - \lambda_l$. Face j corresponds to a type j failure at time a_l, whereas if the face $m + 1$ shows, the individual survives to the next potential failure time. The product integral again unifies the discrete and continuous cases and emphasizes this sequential interpretation.

8.2.3   Some Statistical Methods

Suppose that n study subjects from the continuous model give rise to a right-censored sample $[t_i, \delta_i, j_i, X_i(t_i)], i = 1, \ldots, n$, where t_i is the observed survival time, δ_i is the failure indicator, j_i is the failure type (which is unobserved and does not enter the likelihood if $\delta_i = 0$), and $X_i(t)$ is the covariate path for the ith individual. If the censoring is independent, the likelihood (or partial likelihood) is proportional to

$$L = \prod_{i=1}^{n} \left(\{\lambda_{j_i}[t_i; X_i(t_i)]\}^{\delta_i} \prod_{j=1}^{m} \exp\left\{ -\int_0^{t_i} \lambda_j[u; X_i(u)] \, du \right\} \right)$$

$$= \prod_{j=1}^{m} \left(\prod_{i=1}^{n} \{\lambda_j[t_i; X_i(t_i)]\}^{\delta_{ji}} \exp\left\{ -\int_0^{\infty} \sum_{i=1}^{n} Y_i(t) \lambda_j[t; X_i(t)] \, dt \right\} \right), \quad (8.8)$$

where $Y_i(t)$ is the at risk indicator and $\delta_{ji} = \mathbf{1}(j_i = j, \delta_i = 1)$ is the indicator of a type j failure for the ith subject. Arguments leading to this parallel those in Section 6.2. From (8.8), the likelihood factors into a separate component for each failure type $j = 1, \ldots, m$. Moreover, the likelihood factor involving a specific λ_j is precisely that which would be obtained by regarding all failures of types other than j as censored at the corresponding failure time. It follows that any of the methods of preceding chapters can be used for inference about the $\lambda_j[t; X(t)]$'s. In particular, a parametric model $\lambda_j[t; X(t), \theta_j]$ could be specified for the type-specific hazards (8.1). If there are no common parameters among failure types, θ_j can be estimated by maximum likelihood following exactly the methods outlined in Chapter 3.

It is also straightforward to generalize simple exploratory methods such as Kaplan–Meier and Nelson–Aalen estimators to competing risks data. Consider independent right censored data from a homogeneous (no covariates) competing risks model. We develop estimators of the cumulative "incidence" function $\bar{F}_j(t) = P(T \leq t, J = j) = \int_0^t f_j(u) \, du$ and (2) the cumulative hazard function $\Lambda_j(t) \, j = 1, \ldots, m$.

The nonparametric estimation technique of Section 1.3 is readily generalized to include competing risks as follows: Let $t_1 < t_2 < \cdots < t_k$ denote the k distinct failure times for all failure types combined. Suppose that failure type j occurs with multiplicity d_{ji} at time t_i, $i = 1, \ldots, k; j = 1, \ldots, m$. The likelihood function can be written

$$L = \prod_{i=1}^{k} \left(\prod_{j=1}^{m} \{[F_j(t_i^-) - F_j(t_i)] F(t_i^-)\}^{d_{ji}} \prod_{l=1}^{C_i} [F(t_{il})]^{c_{il}} \right), \quad (8.9)$$

where $t_{i1}, \ldots, t_{iC_i}$ denote the C_i censoring times in $[t_i, t_{i+1})$, where $t_{k+1} = \infty$ and c_{il} is the number of items censored at t_{il}. Following the arguments in Section 1.3 it can be seen that the nonparametric MLE places mass only at the observed failure times $1, \ldots, k$, so we seek a discrete model with discrete hazard component λ_{ji} for the jth failure type at time t_i. The partially maximized likelihood from (8.9) can now be rewritten using expressions for discrete models, to obtain

$$\hat{L} = \prod_{i=1}^{k} \left[\prod_{j=1}^{m} \lambda_{ji}^{d_{ji}} (1 - \lambda_i)^{n_i - d_i} \right], \tag{8.10}$$

where $d_i = \sum d_{ji}$ is the number of failures, n_i is the total number at risk, and $\lambda_i = \sum \lambda_{ji}$ is total discrete hazard at time t_i. Maximization of the multinomial likelihood (8.10) gives the MLE $\hat{\lambda}_{ji} = d_{ji}/n_i$. The cumulative hazard function is then estimated by $\hat{\Lambda}_j(t) = \sum_{i=1}^{k} \mathbf{1}(t_i \leq t) d_{ji}/n_i, t > 0$. This yields the Nelson–Aalen estimate of the total cumulative hazard and the Kaplan–Meier estimate of the overall survivor function $F(t)$.

The estimated cumulative incidence function is also discrete, and from (8.7) is given by

$$\hat{\tilde{F}}_j(t) = \sum_{\{i | t_i \leq t\}} d_{ji} n_i^{-1} \hat{F}(t_i^-) \qquad \text{for } j = 1, \ldots, m. \tag{8.11}$$

A plot of $\hat{\tilde{F}}_j(t)$ versus t gives estimates of the probability that a failure type j occurs before any specified time t within the range of observation. The cumulative hazard plot, $\hat{\Lambda}_j(t)$ versus t, provides supplementary information in that its "slope," over specific time intervals, estimates the average rate of occurrence of failures of type j.

Consider now a relative risk or Cox model for the cause-specific hazard functions

$$\lambda_j[t; X(t)] = \lambda_{0j}(t) \exp[Z(t)'\beta_j], \qquad j = 1, \ldots, m, \tag{8.12}$$

where $Z(t)$ is a vector of p derived covariates. Note that both the baseline hazards λ_{0j} and the regression coefficients β_j are permitted to vary arbitrarily over the m failure types and, as usual, Z is left continuous with right-hand limits.

Let $t_{j1} < \cdots < t_{jk_j}$ denote the k_j times of type j failures, $j = 1, \ldots, m$ and let Z_{ji} be the regression function for the individual that fails at t_{ji}. A partial likelihood can now be constructed by proceeding along the time axis and at each failure time conditioning on the previous history of failures and censorings and that, at time t_{ji}, a single type j failure occurs. The corresponding partial likelihood is

$$L(\beta_1, \ldots, \beta_m) = \prod_{j=1}^{m} \prod_{i=1}^{k_j} \frac{\exp[Z_{ji}(t_{ji})\beta_j]}{\sum_{l \in R(t_{ji})} \exp[Z_l(t_{ji})'\beta_j]}. \tag{8.13}$$

Estimation and comparison of the β_j's can be conducted by applying asymptotic likelihood techniques individually to the m factors in (8.13) exactly as in Chapters 4 to 6. Expression (8.13) accommodates tied failure times on different failure types, whereas a modification like that in (4.18), (4.19), or (4.20) would be required if tied failure times occur on the same failure type. The baseline cumulative hazard functions $\Lambda_{0j}(t) = \int_0^t \lambda_{0j}(u)\, du$ can be estimated using the methods of Chapters 4 and 6 upon inserting the maximum partial likelihood estimators $\hat{\beta}_1, \ldots, \hat{\beta}_m$ from (8.13). When the basic covariates are fixed or external, corresponding estimators of the cumulative incidence functions $\bar{F}_j(t; X)$ at any given X can be obtained in a natural way.

The cause-specific hazard functions can be modeled similarly using an accelerated failure time model of the form

$$\lambda_j(t; X(t)) = \lambda_{0j}\left\{ \int_0^t \exp[Z(u)'\beta_j]\, du \right\} \exp[Z(t)'\beta_j], \qquad j = 1, \ldots, m, \qquad (8.14)$$

where we have again assumed separate baseline and regression parameters for the m failure types. Because of the factorization (8.8), inferences about a particular β_j can be obtained using the parametric methods of Chapter 3 or the rank regression methods of Chapter 7 upon treating failure times of types other than j as censored.

If there are common parameters among the β_j's in (8.12), the semiparametric approaches of Chapters 4 and 6 would require joint maximization of the pertinent partial likelihood factors in (8.13). Similarly if (8.14) involves common β parameters, the related estimating equations in Section 7.4 would need to be simultaneously solved.

A specialization of (8.12) that, if applicable, would yield more efficient β_j estimators, is that given by the proportional risks model

$$\lambda_j[t; X(t)] = \lambda_0(t) \exp[\gamma_j + Z(t)'\beta_j] \qquad j = 1, \ldots, m, \qquad (8.15)$$

in which the cause-specific hazards are assumed to be proportional to each other (for uniqueness set $\gamma_1 = 0$). This proportionality assumption can be checked graphically using plots of $\log \hat{\Lambda}_{0j}(t)$, versus t, where the cumulative hazards are estimated from an analysis of model (8.12) without proportionality assumption. Such plots should be separated by approximately constant difference for various values of j if the model (8.15) holds. Estimation of the γ_j's and β_j's can be carried out using methods similar to those of Sections 4.2 and 4.3. A partial likelihood for these parameters can be obtained as the product, over successive failure times t_i, of conditional probabilities. At time t_i, the contribution is the conditional probability that i fails of cause j_i given the risk set at t_i and the information that an item fails of some unspecified cause at t_i. The partial likelihood can then be written

$$\prod_{i=1}^k \frac{\exp[\gamma_{j_i} + Z_i(t_i)'\beta_{j_i}]}{\sum_{j=1}^m \sum_{l=1}^n Y_l(t_i) \exp[\gamma_j + Z_l(t_i)'\beta_j]}. \qquad (8.16)$$

Again, an adjustment is needed to handle tied failure times.

The likelihood (8.16) can be analyzed using standard Cox likelihood software by defining a new sample with mn entries as follows. Each individual l in the sample is replicated m times and the jth replicate fails if $j_l = j$ and otherwise is censored at time t_l. The covariate vector $Z_{jl}^*(t)$ for the jth replicate of the lth individual has $m(p + 1)$ elements, all of which are zero except for the elements $(j - 1)(p + 1) +$ $1, \ldots, j(p + 1)$, which are $[1, Z_l(t)']$ for $l = 1, \ldots, n$ and $j = 1, \ldots, m$. The covariate vector is $\beta^* = (\gamma_1 = 0, \beta_1', \ldots, \gamma_m, \beta_m')'$.

Although it would often be more restrictive than is desirable, the proportional risk model (8.16) has some attractive properties. For instance, the probability that an individual with fixed covariate Z has failure type j is

$$P(J = j; Z) = \frac{\exp(\gamma_j + Z'\beta_j)}{\sum_{h=1}^{m} \exp(\gamma_h + Z'\beta_h)}, \qquad j = 1, \ldots, m, \qquad (8.17)$$

regardless of $\lambda_0(\cdot)$. It follows that T and J are statistically independent. Note that in the case of no regression variable the likelihood (8.16) reduces to the multinomial likelihood based on (8.17). The corresponding maximum likelihood estimators of the proportionality factors e^{γ_j}, subject to $\gamma_1 = 0$, are $\exp(\hat{\gamma}_j) = k_j/k_1, j = 2, \ldots, m$.

Example 8.2. Consider the mouse carcinogenicity data from Hoel (1972) discussed briefly in Section 1.1.1 and reproduced in Appendix A, data set V. Male mice were randomly assigned to a control or germ-free environment, subjected to 300 rad of radiation and then followed over time, the number of days until failure being recorded. There were three failure types, $j = 1, 2$, and 3, respectively, corresponding to the occurrence of thymic lymphoma, reticulum cell sarcoma, and death by some other cause prior to either type of cancer diagnosis. There were four unusually small observations (40, 42, 51, and 62) in the control group, which for purposes of illustration, we have eliminated from consideration in the analysis reported here. It is a useful exercise to repeat these analyses including those observations.

Figures 8.2 and 8.3 give cumulative hazard and cumulative incidence plots for the three failure types. It is evident that the differences between control and treated groups are marked for reticulum cell sarcoma and other causes. There seems relatively little difference between the groups with respect to rates or incidence of thymic lymphoma. Table 8.1 reports the results of a relative risk analysis utilizing the model (8.13) with the single treatment indicator $Z = 0$ for the control group and $Z = 1$ for the germ-free environment. This confirms the graphical analysis. It is evident that the germ-free environment has strong effects on the rate of occurrence of reticulum cell sarcoma and death by other causes. The result for reticulum cell sarcoma may possibly suggest some microbiological involvement in the occurrence of this cancer. There is no evidence of any dependence of thymic lymphoma rates on the germ-free environment. □

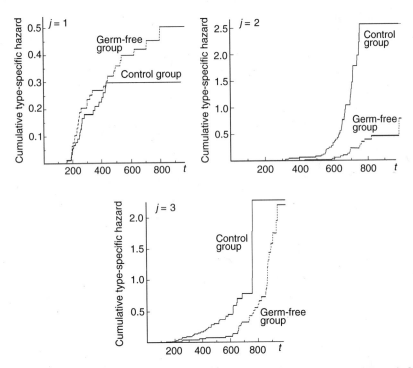

Figure 8.2   Estimates of the cumulative type-specific hazard functions for the data of Example 8.2.

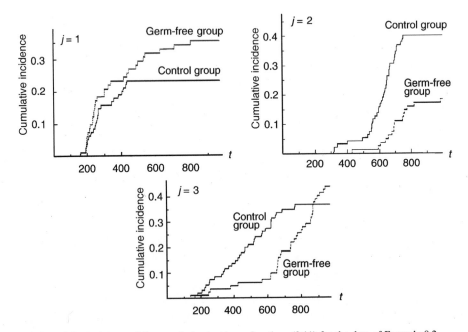

Figure 8.3   Estimates of the cumulative incidence functions (8.11) for the data of Example 8.2.

Table 8.1   Relative Risk Estimation for the Carcinogenicity Data of Example 8.2

Failure Type	$\hat{\beta}_j$	SE of $\hat{\beta}_j$	$\exp(\hat{\beta}_j)$	p Value
Thymic lymphoma	0.302	0.287	1.353	0.44
Reticulum cell sarcoma	-2.030	0.354	0.131	$\ll 0.01$
Other causes	-1.107	0.304	0.331	< 0.01

8.2.4   Identifiability and the Multiple Decrement Function

Much of the literature on competing risks approaches problems of types 1 through 3 in Section 8.2.1 by defining latent or conceptual failure times corresponding to each failure type. We consider continuous time and let $\bar{T}_1, \ldots \bar{T}_m$ denote these latent times. The actual failure time is defined to be

$$T = \min(\bar{T}_1, \ldots, \bar{T}_m),$$

and the corresponding failure type is J where $\bar{T}_J = T$. Problems of types 1 through 3 are then typically posed in terms of a joint distribution for $\bar{T}_1, \ldots \bar{T}_m$.

We restrict attention to a basic covariate vector x that is independent of t so that the derived covariates are either fixed or defined functions of time. (Since modeling here is done through a joint survivor function, it is not clear how to incorporate more general time-dependent covariates.) The multiple decrement function or joint survivor function is

$$Q(t_1, \ldots, t_m; x) = P(\bar{T}_1 > t_1, \ldots, \bar{T}_m > t_m; x). \tag{8.18}$$

This model gives a complete specification of the probability laws for the m variate failure time model. Thus, quantities introduced earlier can be expressed in terms of Q, such as the overall survivor function

$$F(t; x) = P(T > t; x) = Q(t, t, \ldots, t; x),$$

and the type-specific hazard functions

$$\lambda_j(t; x) = \lim_{\Delta t \to 0} \frac{P(t \le \tilde{T}_j < t + \Delta t \mid T \ge t; x)}{\Delta t}$$

$$= \frac{-\partial \log Q(t_1, \ldots, t_m; x)}{\partial t_j} \bigg|_{t_1 = \cdots t_m = t}, \qquad j = 1, \ldots, m. \tag{8.19}$$

Since the likelihood function (8.7) can be written entirely in terms of the $\lambda_j(t; x)$'s, it follows that functions of Q other than those given in (8.19) or functions thereof, such as the survivor function (8.3) or the cumulative incidence function (8.5), cannot be estimated without further assumptions. Such functions are said to be *non-identifiable* based on these data.

For example, the marginal distributions of the latent failure times are generally nonidentifiable. Let

$$Q_j(t_j; x) = P(\bar{T}_j > t_j) = Q(0, \ldots, 0, t_j, 0, \ldots, 0; x)$$

represent the marginal survivor function for $\bar{T}_j$ and let

$$h_j(t; x) = \frac{-d \log Q_j(t; x)}{dt} \tag{8.20}$$

be the corresponding hazard function. This is often called the *net hazard for type j failures* and at first introduction may appear to be of interest. However, because these quantities cannot be expressed as functions of (8.19), the marginal functions are nonidentifiable, and so, unlike the type-specific hazards, cannot be estimated without introducing additional model assumptions. One strong assumption asserts that $\bar{T}_1, \ldots, \bar{T}_m$ are statistically independent. That is,

$$Q(t_1, \ldots, t_m; x) = \prod_{j=1}^{m} Q_j(t_j; x). \tag{8.21}$$

It follows easily under (8.21) that

$$\lambda_j(t; x) = h_j(t; x), \qquad j = 1, \ldots, m, \quad t > 0, \tag{8.22}$$

and this assumption thus allows estimation of the marginal or net hazards. Actually, (8.21) is more restrictive than (8.22), in that for (8.22) to hold, we require only

$$Q(t, \ldots, t; x) = \prod_{j=1}^{m} Q_j(t; x),$$

a somewhat weaker condition than full independence.

The very restrictive assumption of independence, however, is wholly untestable. Independence requires the factorization (8.21) to hold, but both the marginal and joint survivor functions in this expression are nonidentifiable. Data of the type $(T, J; x)$ do not allow one, for example, to distinguish between an independent competing risk model and an infinitude of dependent models giving rise to the same type-specific hazards $\lambda_j, j = 1, \ldots, m$. Thus, even if a parametric model with dependent risks is assumed for Q, and all parameters are estimable within that model, it is impossible to distinguish between that assumed model and one with independent risks but the same type-specific hazard functions on the basis of data (t, J, x). The following example illustrates this point.

Example 8.3. Suppose it is assumed that $m = 2$ and

$$Q(t_1, t_2) = \exp\{1 - \alpha_1 t_1 - \alpha_2 t_2 - \exp[\alpha_{12}(\alpha_1 t_1 + \alpha_2 t_2)]\}, \qquad (8.23)$$

where $\alpha_1, \alpha_2 > 0$ and $\alpha_{12} > -1$. The parameter α_{12} measures the dependence between $\bar{T}_1$ and $\bar{T}_2$ within the model (8.23). The cause-specific hazard functions are

$$\lambda_j(t) = \alpha_j\{1 + \alpha_{12} \exp[\alpha_{12}(\alpha_1 + \alpha_2)t]\}, \qquad j = 1, 2. \qquad (8.24)$$

The likelihood function can be written in terms of the λ_j's and it can be seen that all three parameters are estimable. The likelihood obtained from (8.23), however, is identical to that arising from the independent risks model with cause-specific and net hazard functions given by (8.24) and corresponding joint survivor function

$$Q^*(t_1, t_2) = \exp\left[-\int_0^t \lambda_1(u)\, du - \int_0^t \lambda_2(u)\, du\right] \qquad (8.25)$$

It follows that the estimated value of α_{12} should not be taken as an indication of dependence unless there is clear external evidence to support the existence of the latent times and the specific model (8.23). To put it another way, the interpretation of $\hat{\alpha}_{12}$ as estimating the degree of dependence between $\bar{T}_1$ and $\bar{T}_2$ arises from a model assumption that cannot be checked from the data.

The marginal or net hazard functions for Q and Q^* are

$$h_j(t) = \alpha_j[1 + \alpha_{12} \exp(\alpha_j \alpha_{12} t)]$$

and (8.24), respectively. It is of interest to note that Q and Q^* give rise to proportional type-specific hazard functions with proportionality factor

$$\frac{\lambda_1(t)}{\lambda_2(t)} = \frac{\alpha_1}{\alpha_2},$$

which in the notation of Section 8.2.3 has maximum likelihood estimator k_1/k_2. □

The latent failure time formulation is also frequently used to study the association between specific failure types and fixed covariates (e.g., David and Moeschberger, 1978; Crowder, 2001). Such an approach involves the specification and estimation of a regression model $Q(t_1, \ldots, t_m; x)$. Usually, for tractability, the latent failure times are assumed independent or only the type-specific hazards are modeled. Otherwise, a parametric form must be imposed so that the multiple decrement function is estimable. Since only the type-specific hazards (8.19) enter the likelihood function and they are all that is needed to specify the joint distribution of (T, J), it is more straightforward to consider only these functions in the modeling.

8.2.5 Estimation of the Interrelation Between Failure Types

As indicated above, data of the type $(T, J; x)$ with x time-independent do not, without further assumptions, allow one to study the interrelation, or even test for independence, among competing failure modes. More comprehensive data are needed.

One possibility with human or animal data is to attempt to use the frequency with which multiple pathologic entities are present at failure (e.g., autopsy) to provide some information on the way such entities are related throughout the individual's lifetime. One difficulty, however, is that one may be observing only local phenomena just prior to failure, within which the presence of one pathologic entity substantially changes the risk of occurrence of others. Also, diagnostic procedures related to one pathologic entity may increase the probability of discovering certain other abnormalities. Similar statements could be made in relation to studies of equipment failure in which a number of different faults may be observed simultaneously at breakdown.

A second and more promising approach involves the use of time-dependent covariates. In some studies each, or at least some, failure types will have associated risk factors that can be measured regularly on the individuals over the course of the study. Suppose that there is such a risk factor available for the specific failure type j and that it can be used to define internal time-dependent covariates $\tilde{x}_j(t)$ that given other variables in the model are highly predictive of the rate of type j failures. Consider a model that relates $\tilde{x}_j(t)$ to the type-specific failure rate for another failure type, j' say. A test for no association between $\tilde{x}_j(t)$ and the failure rate for type j' failures could provide evidence for or against a hypothesis of no association between the two failure modes j and j'. In this view, two failure modes, j and j', are taken to be dependent if given baseline covariates included in the model, the same internal time-dependent covariate is predictive of both. This formulates the idea of interrelationships without requiring the definition of a multiple decrement function.

The examples mentioned in Section 8.2.1 can illustrate these ideas. Consider first the marrow transplantation example where individuals are at risk of death by two primary causes, either the recurrence of leukemia or the development of severe graft versus host disease (GVHD). Natural questions arise as to whether these two causes are associated. Regular measurements of the presence and severity of a graft versus host reaction on the skin, gut, and liver can be taken over the patients' posttransplantation course. These measurements provide rather direct information on the risk of a GVHD death, and the data may be used to define one or more time-dependent covariates. A test for no association between values of these covariates and the recurrent leukemia relapse rates then provides evidence for or against an association between GVHD and recurrent leukemia as causes of death. In addition, estimation of the coefficients for the time-dependent covariates may provide valuable information on characteristics of any suggested interrelation between failure types. As an illustration of this approach (also given in Prentice et al., 1978), a proportional hazards model of the form (8.12) was specified for recurrent leukemia relapse rates (failure $J = 1$). This model was then applied to

Table 8.2   Relative Risk Analysis of Recurrent Leukemia in Bone Marrow Transplant Data

Covariate	Coefficient β	Standard Error	Normal Deviate
GVHD risk indicator	-0.764	0.370	-2.06
Syngeneic (0) vs. Allogeneic (1)	0.054	0.340	0.16
Age in years/(10)	0.127	0.098	1.29

data on 135 marrow transplant recipients with the following covariates: (1) an indicator time-dependent covariate that takes value zero from the time of transplant to the date of diagnosis of acute GVHD and value 1 thereafter (this is an "internal" time-dependent covariate), (2) a fixed covariate giving the patient's age at transplant, and (3) an indicator variable indicating whether the marrow donor was a matched sibling (allogeneic transplant) or an identical twin (syngeneic transplant). Table 8.2 gives maximum likelihood estimates and corresponding standard errors for the regression coefficients.

This analysis indicates reduction in recurrent leukemia rates by an estimated factor $\exp(-0.764) = 0.47$ upon the onset of GVHD. Asymptotic likelihood theory applied to the GVHD indicator coefficient indicates that this reduction is significant at the 0.05 level. The analysis suggests a negative relationship between the two failure types. When the risk of GVHD mortality is high (i.e., when a GVH reaction has been diagnosed), the risk of leukemia relapse is evidently low. Several refinements of this analysis are possible. For example, more quantitative measures of the severity of GVHD could be utilized and additional prognostic factors for recurrent leukemia could be introduced.

A second illustration involves a test for independent censoring. Here a direct time-dependent measure of the risk of death, such as measurements of the patient's general condition or performance status over the course of the study, together with the initial performance status, could be related to the censoring rate. The test for a zero coefficient for the time-dependent covariate would test whether individuals are being selectively censored when they have relatively poorer or better prognosis than other study subjects with the same initial characteristics. If there were found to be such a dependence, this would indicate that the censoring mechanism is not independent in a model that did not include the time-dependent measurement. If the mechanism of dependence were well explained by the time-dependent performance status measurement, the censoring would be independent in regression models that appropriately included this variable. But this leads back to the earlier comments that the performance measure may be responsive and its time-dependent levels affected by treatment, so that its inclusion in the model may mask a treatment effect.

8.2.6   Failure Rate Estimation Following Cause Removal

As indicated in Section 8.2.1, detailed knowledge of the biological or physical mechanism giving rise to the various failure types as well as a knowledge of the mechanism of removal would usually be required to estimate failure rates for remaining types, given that others have been removed. An example of such knowledge would arise in circumstances where there is known to be a complete functional (biological, physical) independence between the organs or components giving rise to the various failure types. Assuming that the removal mechanism does not affect this independence, the type-specific hazard functions defined above are also the type-specific hazard functions given the removal of some or all of the other functionally independent failure modes. In these very special circumstances, the net hazard (all other failure types removed) and partial-crude hazard (some but not all other causes removed) are identical to type-specific hazards and so estimable on data of the type (T, J, X).

Chiang (1968, p. 246) suggests that these net or partial crude probability statements should generally be based on the type-specific hazard functions. This point of view, however, involves the very strong assumption that instantaneous failure rates for cause j under one set of conditions in which all m causes are operative are precisely the same as failure rates under a new set of conditions under which only some are operative. Other authors approach these problems in terms of latent failure times (Section 8.2.5). Upon hypothesizing latent failure times $\bar{T}_1, \ldots, \bar{T}_m$, one *assumption* concerning removal of a cause is that removal corresponds to marginalization. Thus, for example, if type 1 is removed, the new time to failure is $T = \min(\bar{T}_2, \ldots, \bar{T}_m)$. This leads to the marginal survivor function for the remaining latent failure times as the basis for failure rate estimation. In the case of all but a single cause being removed, for example, this suggests that the Q_j functions of Section 8.2.4 are relevant. However, these quantities are not identifiable from data of the type $(T, J; X)$. This approach involves strong untestable assumptions about the nature of the failure mechanism and how it operates under cause removal and yet does not lead to useful estimation techniques despite such assumptions. One needs to add untestable parametric or independence assumptions about the (often hypothetical) multiple decrement function. It is perhaps surprising that problems related to this approach (e.g., estimating net hazards in the first place) have received so much attention in the statistical literature.

As noted above, a statistical specification of failure rates given the removal of certain failure types will be sensible only in very special cases. In some specific applications, it may be possible to utilize existing data to reach sensible inferences. A detailed knowledge of the system giving rise to the failures and knowledge of the removal mechanism is required for such an extrapolation.

Consider again the bone marrow transplantation setting. GVHD is presumed to arise because of minor histocompatibility differences (genetic differences outside the major HLA loci) between the marrow donor and the patient. Patients receiving allogeneic marrow grafts who do not develop GVHD may be considered to have fortuitously experienced greater genetic similarity than those developing GVHD. The leukemia relapse hazard function for patients without GVHD may then provide a reasonable estimate of leukemia relapse rates under a mechanism that removes GVHD as a cause of death by obtaining greater tissue similarity between donor and recipient. The syngeneic (identical twin) transplants provide information on leukemia relapse rates under a limiting form of such a removal process in which minor histocompatibility differences are avoided and typically no GVHD arises.

8.2.7 Counting Processes and Asymptotic Results

Let $N_{jl}(t)$ be the right-continuous process that counts the number of observed type j failures on the lth study subject, and let $Y_l(t)$ be the left-continuous at-risk process. As above, let $X_l(t) = \{x_l(u) : 0 \leq u < t\}$, where $x_l(t)$ is the covariate process on the lth subject. We define the filtration or history as $\mathcal{F}_t = [N_{jl}(u), Y_l(u^+), X_l(u^+); j = 1, \ldots, m; l = 1, \ldots, n; 0 \leq u \leq t]$. Consider the continuous case where no two counting processes can jump simultaneously. Under independent censoring,

$$P[dN_{jl}(t) = 1 | \mathcal{F}_{t-}] = Y_l(t)\lambda_j[t; X_l(t)]\, dt$$

for $0 < t$ and all l, j. It follows that

$$M_{jl}(t) = N_{jl}(t) - \int_0^t Y_l(u)\lambda_j[u; X_l(u)]\, du, \qquad j = 1, \ldots, m, \quad l = 1, \ldots, n$$

are orthogonal martingales with respect to the filtration $\mathcal{F}_t$.

The derivation of the likelihood (8.8) follows closely the derivation in Section 6.2.1 and is constructed sequentially by tracking all failures, failure types, censorings, and evolving covariates as they occur in the study cohort.

The relative risk model (8.12) gives rise to the partial likelihood (8.13) for the regression parameters β_j, $j = 1, \ldots, m$. The score vector arising from (8.13) has jth vector component $U_j(\infty)$ corresponding to β_j, where

$$U_j(t) = \int_0^t \sum_{l=1}^n [Z_l(u) - \mathcal{E}(\beta_j, u)]\, dN_{jl}(u)$$

$$= \int_0^t \sum_{l=1}^n [Z_l(u) - \mathcal{E}(\beta_j, u)]\, dM_{jl}(u), \qquad j = 1, \ldots, m, \quad t > 0$$

and $\mathcal{E}(\beta, u)$ is the weighted average of $Z_l(u)$ over the risk set as before. Thus, the score vector is expressed as a stochastic integral of a predictable process with respect to a martingale and the score process $U(t) = [U_1(t)', \ldots, U_m(t)']$ is a martingale whose predictable variation process can be seen to be a block diagonal matrix. The asymptotic arguments and results of Section 5.5 apply directly to U_j and to $\hat{\beta}_j$ obtained by maximizing the jth term in the partial likelihood (8.13). These comments are assuming the usual case where there would be no common parameters among the β_j's. If there are common parameters, the corresponding score equations would correspond to sums across failure types, but the standard asymptotic results still hold.

Similarly, the Nelson–Aalen estimator of the baseline type-specific cumulative hazard function for the jth failure type can be written

$$\hat{\Lambda}_{0j}(t) = \int_0^t \frac{dN_{j\cdot}(u)}{\sum_{l=1}^n Y_l(u) \, \exp[Z_l(u)' \hat{\beta}_j]},$$

where it is presumed that the denominator sum is nonzero for all $u \in (0, t]$. This reduces to the estimator given in Section 8.2 in the case of no covariates. Asymptotic results again follow directly from the results of Section 5.5. In the case of external (or fixed) covariates, estimators of the baseline cumulative incidence functions (8.5) can also be obtained as

$$\hat{\bar{F}}_{0j}(t) = \int_0^t \exp\left[-\sum_{j'=1}^m \hat{\Lambda}_{0j'}(u) \right] d\hat{\Lambda}_{0j}(u).$$

These quite simple estimators agree to first order with estimators of Λ_{0j} and $\bar{F}_{0j}$ that could be obtained by maximizing the likelihood for given β_j as in Section 4.3.

8.3   LIFE-HISTORY PROCESSES

8.3.1   Multistate Models

The methods developed here extend in a natural and simple way to model and analyze situations in which study subjects move among a number $q > 1$ of discrete states over the course of the study. For example, in ordinary failure time analysis, $q = 2$ with states alive and dead (labeled 0 and 1 in Figure 8.4a). In the competing risks framework of Section 8.2, there are $q = m + 1$ states, with state 0 being alive and states $1, \ldots, m$ corresponding to the m distinct absorbing states (Figure 8.4b). An illness–death model arises when individuals begin in an initial healthy state (0) from which they may die (2) directly or, after some period of time, may experience an illness (state 1). From the illness state, they may recover or die (Figure 8.4c). If there is no possibility of recovery from the illness state, we obtain the progressive illness–death model (Figure 8.4d); the modeling of the heart transplant example in Section 6.4 provides an example of the latter. Much more complex models could also be considered. In an HIV/AIDS study, for example, individuals might begin in a healthy state (0), possibly experience HIV infection (state 1) and subsequent diagnosis of AIDS either with or without an opportunistic infection (states 2 and 3, respectively). Once diagnosed, individuals can experience repeated infections and recoveries. Death (state 4) may occur from any of the states (Figure 8.4e). Other examples are also easy to construct.

8.3.2   Markov and Modulated Markov Processes

Consider first a homogeneous population with no covariates. Let $A(t)$ be the state occupied at time t, $t \geq 0$. The simplest probability model that can be specified for

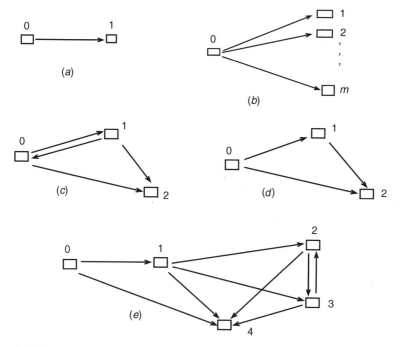

Figure 8.4 Compartment illustrations of multistate models for (*a*) failure time process; (*b*) competing risks; (*c*) illness–death model; (*d*) progressive illness death model; and (*e*) HIV, AIDS, and opportunistic infections.

$A(t)$ is the Markov process. If a randomly chosen individual is in state i at time t^-, the transition rate or intensity from i to j at time t is given by

$$d\Lambda_{ij}(t) = P[A(t^- + dt) = j \mid A(u), 0 \le u < t, A(t^-) = i]$$
$$= P[A(t^- + dt) = j \mid A(t^-) = i], \qquad t > 0, \qquad (8.26)$$

which holds for all $A(u)$, $0 \le u < t$ with $A(t^-) = i$, and $i, j \in \{0, 1, \ldots, q - 1\}$, $j \ne i$. The process is memoryless in that only the current state occupied is relevant in specifying the transition rates. The rates themselves are allowed to depend on the time t since the beginning of the study. It is convenient to define

$$d\Lambda_{ii}(t) = -\sum_{j \ne i} d\Lambda_{ij}(t)$$

so that the row sums of the matrix

$$d\Lambda(t) = \left[d\Lambda_{ij}(t) \right]_{q \times q}$$

are all 0.

In the discrete case, there exists a set of times $\{a_k: k = 1, 2, \ldots\}$, where $0 < a_1 < a_2 < \cdots$, at which transitions can occur and

$$\mathbf{P}_k = I + d\mathbf{\Lambda}(a_k) = (P[A(a_k) = j | A(a_k^-) = i])_{q \times q},$$

where I is the $q \times q$ identity matrix, is the usual one-step probability transition matrix of a nonhomogeneous Markov chain. Let $\mathbf{P}^{(r)} = (p_{ij}^{(r)})_{q \times q}$, where $p_{ij}^{(r)} = P[A(a_r) = j | A(0) = i]$ is the r-step transition probability, $r = 0, 1, \ldots$. It is well known that

$$\mathbf{P}^{(r)} = \prod_{k=1}^{r} \mathbf{P}_k = \mathbf{P}_1 \mathbf{P}_2 \ldots \mathbf{P}_r, \qquad r = 0, 1, \ldots, \qquad (8.27)$$

where an empty product is interpreted as I. It should be noted that the order of multiplication matters here since the $\mathbf{P}_k$ matrices will generally not commute. If the chain is homogeneous, so that $\mathbf{P}_k = \mathbf{P}$ for all $k = 1, 2, \ldots$, $\mathbf{P}^{(r)} = \mathbf{P}^r$.

In the continuous case, $d\Lambda_{ij}(t) = \lambda_{ij}(t)\, dt$ for all $i, j = 0, \ldots, q - 1$, so that $\lambda_{ij}(t), i \neq j$ is the continuous-time intensity function for i-to-j transitions, and $\lambda_{ii}(t) = -\sum_{j \neq i} \lambda_{ij}(t)$. In the homogeneous special case, $\lambda_{ij}(t) = \lambda_{ij}$ independent of t, and in this case the distribution of the sojourn time in the ith state is exponential with rate $-\lambda_{ii}, i = 0, \ldots, q - 1$. Let $P_{ij}(t) = P[A(t) = j | A(0) = i], i, j = 0, \ldots, q$, and let $\mathbf{P}(t) = [P_{ij}(t)]_{q \times q}$. Analogous to (8.27), we can write

$$\mathbf{P}(t) = \mathscr{P}_0^t[I + d\mathbf{\Lambda}(u)] = \mathscr{P}_0^t[I + \lambda(u)\, du] \qquad (8.28)$$

where $\lambda(u) = [\lambda_{ij}(u)]_{q \times q}$. This product integral is defined in the obvious way:

$$\mathscr{P}_0^t[I + d\mathbf{\Lambda}(u)] = \lim \prod_{i=1}^{M}[I + \mathbf{\Lambda}(u_i) - \mathbf{\Lambda}(u_{i-1})],$$

where $u_0 = 0 < u_1 < \cdots < u_M = t$ and the limit is taken as $M \to \infty$ and $\Delta u_i = u_i - u_{i-1} \to 0$.

Example 8.4. Let $q = 2$ and consider the failure time process of Figure 8.4a In this case,

$$\lambda(t) = \begin{pmatrix} -\lambda_{01}(t) & \lambda_{01}(t) \\ 0 & 0 \end{pmatrix},$$

and from (8.28) it is easy to verify that $P_{00}(t) = \mathscr{P}_0^t[1 - \lambda_{01}(t)\, du] = \exp[-\int_0^t \lambda_{01}(u)\, du]$ and $P_{01}(t) = 1 - P_{00}(t)$. The competing risks model of Figure 8.4.b corresponds to a transition intensity matrix with all entries 0 except those along the first row. It can be seen that $P_{00}(t) = \exp[-\sum_{j=1}^{m} \Lambda_{0j}(t)]$ is the overall survivor function, and $P_{0j}(t) = \int_0^t \lambda_{0j}(u) P_{00}(u) du = \bar{F}_j(t)$ is the jth cumulative incidence function, $j = 1, \ldots, m$. □

Estimation of the cumulative intensity functions $\Lambda_{ij}(t)$ and the related transition probabilities $P_{ij}(t)$ proceeds in a straightforward way. Consider a possibly right-censored and/or left-truncated sample of n individuals from the model (8.26). For $l = 1, \ldots, n$, let $N_{ijl}(t)$ be the right continuous process that counts the number of *observed* direct i-to-j transitions for lth individual, $i, j = 0, \ldots, q - 1, i \neq j$. Let $Y_l(t)$ be the corresponding at risk process and $1 Y_{il}(t) = \mathbf{1}[Y_l(t) = 1, A_l(t^-) = i]$ indicate the lth individual is in state i and under observation at time t^-, $i = 0, \ldots, q - 1$. Define the filtration or history process as

$$\mathcal{F}_t = [N_{ijl}(t), Y_l(u^+), 0 \leq u \leq t, l = 1, \ldots, n; i, j = 0, \ldots, q - 1]$$

and suppose that the censoring (and/or truncation) is independent so that

$$P[dN_{ijl}(t) = 1 | \mathcal{F}_{t^-}] = Y_{il}(t) \, d\Lambda_{ij}(t),$$

which must hold for all i, j, l and $t > 0$. In a manner completely analogous to earlier applications, the Nelson–Aalen estimator of $\Lambda_{ij}(t)$ is given by

$$d\hat{\Lambda}_{ij}(t) = dN_{ij\cdot}(t) [Y_{i\cdot}(t)]^{-1}$$

for all $i \neq j$. Let $\hat{\Lambda}_{ii}(t) = -\sum_{j \neq i} \hat{\Lambda}_{ij}(t)$ and let $\hat{\mathbf{\Lambda}}(t) = [\hat{\Lambda}_{ij}(t)]_{q \times q}$ be the matrix of cumulative intensity estimators. The corresponding estimate of the probability transition matrix is

$$\hat{\mathbf{P}}(t) = \mathcal{P}_0^t [I + d\hat{\mathbf{\Lambda}}(u)].$$

For example, in the failure time model of Example 8.3, we find that $\hat{P}_{00}(t)$ is the Kaplan–Meier estimate of the survivor function. In the competing risks model, also of Example 8.4, we find that $\hat{P}_{0j}(t) = \hat{\bar{F}}_j(t)$, the estimates of the jth cumulative incidence function, $j = 1, \ldots, m$ (8.11). A further simple example is provided by the progressive illness death model in Figure 8.4d, and estimation in this case is discussed in Exercise 8.6.

Suppose now that there is a vector of possibly time-dependent basic covariates $x(t)$, which, for convenience, we assume to include $A(t)$, and let $X(t) = \{x(u) : 0 \leq u < t\}$. Continuous-time modulated Markov models can be specified for the underlying (possibly random) intensity function

$$\lambda_{ijl}(t) = \lim_{h \to 0} h^{-1} P[A_l(t^- + h) = j | A_l(t^-) = i, X_l(u), 0 < u < t],$$

where the argument t on the left side indicates the basic time scale to be time since on study, as in (8.26). Parametric or semiparametric models for λ_{ijl} can be specified using any of the approaches discussed in previous chapters. For example, one could specify a parametric model depending on a vector of unknown parameters θ. Alternatively, a semiparametric relative risk model could be specified. In the latter case, a natural model to consider is

$$\lambda_{ijl}(t) = \lambda_{0ij}(t) \exp[Z_l(t)'\beta_{ij}], \tag{8.29}$$

for all i, j, l with $i \neq j$ and $t > 0$, where λ_{0ij} is an unknown baseline intensity function and the regression parameter vector is taken to be separate for each possible i, j transition. (As before, we have used the same Z_l for each ij, but this is without restriction since entries of β_{ij} can be set to 0.) In some instances, only one or two of the transitions i to j may be of interest, whereas in other instances, we may wish to model simultaneously all transition intensities.

Consider a possibly right-censored and/or left-truncated sample of n individuals from the model (8.29) and define the filtration or history process as

$$\mathscr{F}_t = [N_{ijl}(t), X_l(t^+), Y_l(u^+), 0 \leq u \leq t, l = 1, \ldots, n; i, j = 0, \ldots, q-1].$$

Suppose that the censoring (and/or truncation) is independent, so that

$$P[dN_{ijl}(t) = 1|\mathscr{F}_{t^-}] = Y_{il}(t)\lambda_{ijl}(t),$$

which must hold for all $i \neq j, l, \mathscr{F}_{t^-}$ and $t > 0$. Under a parametric model, the full log-likelihood on data over the interval $[0, \tau]$ is

$$\log L_M = \sum_{i \neq j} \left\{ \int_0^\tau \sum_{l=1}^n [\log \lambda_{ijl}(t; \theta) \, dN_{ijl}(t) - Y_{il}(t)\lambda_{ijl}(t; \theta) \, dt] \right\}, \tag{8.30}$$

which for each fixed $i, j (i \neq j)$ is of the same form as for a single sample, except that individuals come in and out of observation depending on the risk indicator $Y_{il}(t)$. This likelihood is again constructed by following the entire study cohort over time as discussed in Sections 6.2 and 6.3. In the parametric case, asymptotic properties for the score function and for the MLE $\hat{\theta}$ can again be deduced under regularity conditions either with independence assumptions across individuals or through martingale arguments as in Chapter 5.

The relative risk model (8.29) can be analyzed using partial likelihood arguments based on conditional probabilities of $dN_{ijl}(t), l = 1, \ldots, n$ given $[\mathscr{F}_{t^-}, dN_{ij.}(t), i, j \in [0, \ldots, q-1], i \neq j; t > 0]$. At each time $t \in [0, \tau]$ where a transition from i to j occurs, the contributing term to the partial likelihood is

$$P[dN_{ijl}(t) = 1|dN_{ij.}(t) = 1, \mathscr{F}_{t^-}] = \frac{Y_{il}(t) \exp[Z_l(t)'\beta_{ij}]}{\sum_{u=1}^n Y_{iu}(t) \exp[Z_u(t)'\beta_{ij}]}.$$

The log partial likelihood is then

$$\sum_{\text{all } i,j} \left[\int_0^\tau \sum_{l=1}^n Z_l(t)' \beta_{ij} \, dN_{ijl}(t) - \log\left[\sum_{l=1}^n Y_{il}(t) \, \exp[Z_l(t)' \beta_{ij}]\right] dN_{ij\cdot}(t) \right], \qquad (8.31)$$

and the (i,j)th term in this sum is the partial likelihood that arises from a model for time-dependent strata (see Section 6.4.2) but generalized to accommodate the competing risks or event types indexed by j.

Once estimates of β_{ij} are obtained through maximization of (8.31), it is possible to estimate the baseline cumulative incidence functions Λ_{0ij} using standard procedures, and corresponding estimates of baseline transition probability matrices can be obtained by adapting the methods discussed above for the homogeneous case. Asymptotics for parameter estimation and for estimation of the cumulative intensity functions follows from the general results on counting processes as discussed in Andersen et al. (1993) and outlined in Chapter 5.

In some instances, transitions to a specific state j may be of particular interest, such as, for example, when j corresponds to the death state and we wish to relate the intensity of type j events to the state currently occupied. In this instance it may be useful to consider, for that j, a model of the form

$$\lambda_{ijl}(t) = \lambda_{0j}(t) \exp[Z_l(t)' \beta_j] \qquad (8.32)$$

and incorporate in $Z_l(t)$ indicator variables $[Y_{0l}(t), \ldots, Y_{0,q-1}(t)]$ for the state currently occupied. This is analogous to the proportional risks model (8.15) discussed earlier in the context of competing risks. Partial likelihoods that are somewhat more efficient than (8.31) arise from (8.32) by conditioning on $\mathscr{F}_{t^-}, dN_{j\cdot}(t)$, and combining terms across the strata or states indexed by i.

The model (6.22) for the heart transplant data is of the form (8.32), where $q = 3$, the state $j = 2$ represents death, and the states 0 and 1 correspond, respectively, to "alive and untransplanted" and "alive and transplanted" (see Figure 8.4d). The vector $Z_l(t)$ contained terms involving the main effects of age, surgery, and year of acceptance, transplant status $Y_2(t)$, and interaction terms. In fact, the analyses of the heart transplant data outlined in Section 6.4 can be viewed as special cases of those discussed here.

The modulated Markov model uses time since on study as the basic time variable. In some instances, however, there may be a strong dependence on time since entry to a state. Such dependencies can be accommodated either by incorporating functions of sojourn time into the covariate part of the model (8.29), or by considering semi-Markov models as discussed below. In the former approach, aspects of the semi-Markov type model can be incorporated while maintaining the basic Markov structure, the straightforward analysis by partial likelihood, and asymptotics based on the martingale mehtods of Chapter 5. In the heart transplant data, for example, one might modulate the Markov rate for transplanted patients by allowing a regression dependence on some function of time since transplant.

In some instances it may be desirable to specify models that allow a separate set of baseline hazards for each new entry to state i. Thus, if state i is being occupied for the sth time at time t^-, the baseline hazard for i-to-j transitions could be taken to be $\lambda_{0ijs}(t)$. The partial likelihoods are now defined in terms of transitions within these new strata. The effects of repeated visits to state i on the intensities can also be ascertained by including s in the regression portion of (8.29).

8.3.3   Modulated Semi-Markov Models

In contrast, the transition intensities at time t in a semi-Markov or Markov-renewal model depend on the time $B(t) = \inf[s \mid A(t-s) \neq A(t^-)]$ since the individual entered the state occupied at time t^-. Thus, in the homogeneous case, the intensities for $i \neq j$ are given by

$$\lambda_{ij}(v) = \lim_{h \to 0} h^{-1} P[A(t^- + h) = j | A(t^-) = i, B(t) = v, A(u), 0 \leq u < t], \quad (8.33)$$

which are assumed to depend on $A(u), 0 \leq u < t$ only through i and v. The semi-Markov process is generally assumed to begin with entry to some specific state at time $t = 0$, or by specifying a different set of transition intensities for the first sojourn. It should be noted that in both this semi-Markov process and in the Markov models, once having entered a state i, say, an individual is subject to competing risks of different failure types corresponding to the states that might next be visited. As for the Markov model, it is possible to obtain nonparametric estimates of the cumulative intensities from (8.33).

The semi-Markov model can also be extended to incorporate covariates $x_l(t)$, and either parametric or semiparametric analyses may be performed. With a parametric model, it is assumed that the intensity of i-to-j transitions at time t for individual l is $\lambda_{ijl}(B_l(t); \theta)$, where $B_l(t)$ is the current sojourn time in the present state and θ is a vector of unknown parameters. The likelihood arising from a right-censored and/or left-truncated sample can now be written as

$$\log L_{SM} = \sum_{i \neq j} \sum_{l=1}^{n} \left\{ \int_0^\infty \log \lambda_{ijl}[B_l(t); \theta] \, dN_{ijl}(t) - Y_{il}(t) \lambda_{ijl}[B_l(t); \theta] \, dt \right\},$$

and again, for a fixed (i, j), this is of the standard form for a sample from a single hazard model.

The relative risk model and partial likelihood analysis can also be extended to modulated semi-Markov models. To do so, however, it is necessary to allow separate baseline hazards for each new entrance into state i. If at time t^-, individual l is in state i for the sth time, we let

$$\lambda_{ijl}(t) = \lambda_{0ijs}(v) \, \exp[Z_l(t) \beta_{ijs}],$$

where $v = B_l(t)$ again represents the current sojourn time. A partial likelihood can be obtained for β_{ijs} in a straightforward manner by analyzing the counting processes and at risk indicators as they unfold in the time scale v for the sth visit to state i. For this purpose, let $N_{ijsl}^*(v)$ count the observed number of transitions from i to j for individual l, and let $Y_{isl}^*(v)$ be the at-risk indicator for the lth individual at sojourn time v in the sth visit to state i. The log partial likelihood can be written

$$\log L = \sum_{l=1}^{n} \int_0^\infty Z_l(t_{isl})' \beta_{ijs} dN_{ijsl}^*(v) - \int_0^\infty \log\left\{ \sum_{u=1}^{n} Y_{isu}^* \exp[Z_u(t_{isu})' \beta_{ijs}] \right\} dN_{ijs\cdot}^*(v),$$

$$(8.34)$$

where $t_{isl} = a_{isl} + v$ and a_{isl} is the time at which individual l enters state i for the sth time.

In some instances, one may wish to consider a model of the form

$$\lambda_{ijl}(t) = \lambda_{0ij}(v) \exp[Z_l(t)\beta_{ij}],$$

where the baseline rates are taken to be the same for all s. In this case, however, there is no exact partial likelihood since one cannot define an appropriate filtration. The difficulty arises since individuals may enter the state more than once and so contribute more than once to the risk sets. One way to analyze this model would be to take a product over s of the stratified likelihood (8.34), with β_{ijs} replaced by β_{ij}. An alternative would be to proceed by analyzing the pseudo partial likelihood that arises by applying standard formula based to the sojourn time scale. Gill (1980) has investigated this approach and shown that the same asymptotic results apply as for the Markov models. We discuss related points in Chapter 9, where the analysis of renewal type processes is discussed.

BIBLIOGRAPHIC NOTES

The actuarial approach to competing risk problems in the absence of regressor variables has been discussed by many authors. Early contributions include Seal (1954), Cornfield (1957), Elveback (1958), Kimball (1958, 1969), Berkson and Elveback (1960), Chiang (1961; 1968, Chap. II; 1970). Pike (1970) and Hoel (1972) discuss models and statistical methods for the estimation of crude, partial crude, and net probabilities. An assumption of independence between latent failure times prevades most of this work, although Kimball (1958, 1969) considers a somewhat different but equally strong assumption. Chiang's work mostly involves the additional convenient assumption that cause-specific hazard functions are proportional within specified failure time intervals. As pointed out by Makeham (1874) and Cornfield (1957), the assumption that the elimination of certain causes of failure corresponds, nullifying the corresponding argument in the multiple decrement function Q, has no general validity even when the multiple decrement model is valid. Gail (1975) provided an excellent review of the foregoing literature. Aalen (1976, 1978a) considered some formal properties of nonparametric estimators for the multiple decrement model.

The fact that data of the form (T, J) do not allow one to discriminate between an independent risk model and an infinitude of dependent risk models has been pointed out in various contexts by Cox (1959; 1962b, p. 112), Tsiatis (1975), and Peterson (1975). Working within a multiple decrement model, Peterson (1976) provides bounds on the net survivor functions Q_j corresponding to given type-specific hazards, λ_j.

Crowder (2001) gives a rather lighthearted account of competing risks models and methods with many references. The main focus of that book is multiple decrement formulations. Parametric models are dealt with in detail with an extensive review of the literature and discussion of parametric identifiability issues relating to the question as to when the parameters of a parametric multiple decrement function can be fully estimated by competing risks data. Some of the key references are Marshall and Olkin (1967), Moeschberger and David (1971), Moeschberger (1974), Lagakos (1977), and David and Moeschberger (1978). Parametric identifiability is discussed, for example, by Anderson and Ghurye (1977), Basu and Ghosh (1978, 1980), and Arnold and Brockett (1983). The Marshall and Olkin bivariate models include a positive probability that the latent failure times are equal. The discussion of Section 8.2 is closely related to the presentation in Prentice et al. (1978). Estimation of cumulative incidence functions based on current status data for competing risks is considered in Jewell and Kalbfleisch (2002).

Issues of nonindependent censoring have been considered by several authors. Fisher and Kanarek (1974), for example, suppose that censored individuals possess a hazard rate that is proportional to that for the remainder of the sample. The effect of varying this proportionallity parameter on estimators of $F(t; x)$ can be readily investigated, although such a parameter is not estimable. In Section 8.2.5 and in Prentice et al. (1978), it is noted that internal time-dependent covariates can sometimes be used to test for independent censoring. More recently, such covariates have been used to obtain weighted estimates of the survivor function where the weights are based on models of the censoring distribution as it relates to internal time-dependent covariates (Robins and Rotnitzky, 1992; Murray and Tsiatis, 1996). Similar ideas have been used in Murray and Tsiatis (2001) and Robins and Finkelstein (2000) to develop tests for treatment effects that make adustments for dependent censoring. Other adjustments to marginal survival comparisons based on time to progression in illness–death models have been given by Gray (1994) and Finkelstein and Schoenfeld (1994). Satten and Datta (2000) give an elementary discussion of inverse probability of censoring weighted (IPCW) estimates.

Markov and semi-Markov models for multistate processes have been considered by several authors. Some early references are Cox (1973), Lagakos et al. (1978), Aalen (1978b), Prentice et al. (1981), and Andersen and Gill (1982). Aalen and Johansen (1978) obtained nonparametric estimates of the transition matrix for non-homogeneous Markov chains and Andersen et al. (1993) give an extensive discussion of nonparametric estimation of Markov transition rates and of probability transition matrices as outlined briefly in Section 8.3. They also develop and consider relative risk models and provide many references and some worked examples. The asymptotic results of Andersen et al. (1993) and Andersen and Gill (1982) are suffiently general to apply directly to the estimation of regression parameters and underlying cumulative transition rates in the model (8.28). Therneau and Grambsch (2000) also include a number of examples of Markov models and discuss implementation of some of the methods of Section 8.3 in S-plus and SAS.

EXERCISES AND COMPLEMENTS

8.1 A discrete competing risks model can be obtained by grouping a continuous time model. Consider the general continuous model with no covariates and type-specific hazard functions $\lambda_j(t), j = 1, \ldots, m$. Suppose that failure times are grouped into disjoint intervals $(0, a_1], (a_1, a_2], \ldots, (a_{r-1}, a_r = \infty]$ and let $\tilde{T} = i$ if $T \in (a_{i-1}, a_i]$, $i = 1, \ldots, k$. Find expressions for the type-specific hazard functions for the discrete failure time variable $\tilde{T}$.

8.2 Consider the discrete competing risks model as outlined in the material leading to equation (8.7) and let $\lambda_{jl}(X)$ be defined as in equation (8.7). A polychotomous logistic model offers one possible regression formulation in this case. Specifically, let

$$\lambda_{jl}(X) = \frac{\exp(\alpha_{0jl}) \, \exp[Z(a_l)'\beta_j]}{1 + \sum_{j'=1}^{m} \, \exp(\alpha_{0j'l}) \, \exp[Z(a_l)'\beta_{j'}]}, \qquad j = 1, \ldots, m, \quad l = 1, 2, \ldots,$$

(8.35)

where Z is a derived possibly time-dependent covariate of dimension p, β_j is the vector of regression parameters for the jth failure type, and $a_{0jl}, l = 1, 2, \ldots$ is a sequence of constants specifying the discrete baseline hazard for the jth failure type, $j = 1, \ldots, m$. This is a natural generalization of the discrete logistic model (2.23) to the competing risks framework. Generalize the partial likelihood argument in Section 4.8.3 to apply to the discrete polychotomous model. Note that at each time a_l we condition on the numbers $d_{jl}, j = 1, \ldots, m$ of observed type-specific failures. If there are no tied failure times, note that this reduces to the partial likelihood (8.11). Discuss how the parameters are to be interpreted.

8.3 A discrete and continuous Cox model for competing risks, analogous to that in Section 4.8.2, could be defined by considering hazard regression models of the form

$$d\Lambda_j[t; X(t)] = d\Lambda_{0j}(t) \exp[Z(t)'\beta], \qquad j = 1, \ldots, m, \quad t > 0 \qquad (8.36)$$

where $\Lambda_{0j}(t)$ is the baseline mixed, continuous, or discrete type-specific hazard function. Consider accommodating ties by introducing a Breslow-type approximation [see (4.19)] applied to the terms of the partial likelihood (8.13). Show that the score function arising from this partial likelihood yields an unbiased estimating equation under the model (8.36). Generalize the development to include variance estimators analogous to those in Section 4.8.2 to apply to parameter estimates in this competing risks model.

8.4 Write the model in (8.35) in a manner analogous to that in (8.36) so as to incorporate mixed, discrete and continuous models.

8.5 Give discrete versions of the proportional risks model (8.15) analogous to those in Exercises 8.2 and 8.3 and discuss their analysis.

8.6 Analyze the Stanford heart transplant data (Appendix A, data set IV) as a competing risks example with transplant and death coresponding to the two failure or event types. Relate the type-specific hazards to age, surgery, and year of acceptance. Interpret conclusions concerning year of acceptance in the context of the analysis in Chapter 6.

8.7 Consider a progressive illness–death model (see Figure 8.4d) and let $A(t)$ specify the state occupied at time t. Let

$$\lambda_j(t) = \lim_{h \to 0} h^{-1} P[A(t+h) = j | A(t^-) = 0], \qquad j = 1, 2$$

and

$$\lambda_{12}(t_2|t_1) = \lim_{h \to 0} h^{-1} P[A(t_2 + h) = 2 | A(t_2^-) = 1, W = t_1],$$

where W is the time of entry to illness state (state 1). Let $G(t) = P[A(t) = 1]$. If, for example, state 1 corresponds to remission and state 2 to progression of disease after therapy, $G(t)$ is the probability of being in remission at time t.

(a) Show that

$$G(t) = \int_0^t \lambda_1(x) \exp\left[-\int_0^x \lambda_1(u) + \lambda_2(u)\, du\right] \exp\left[-\int_x^t \lambda_{12}(y \mid x)\, dy\right] dx.$$

(b) Suppose that a Markov model holds so that $\lambda_{12}(y|x) = \tilde{\lambda}_{12}(y)$ and that independent right-censored data are available on n individuals. Adapting the counting process notation of Section 8.3.2 as necessary, obtain a nonparametric estimate of $G(t)$.

(c) Suppose that a semi-Markov model holds so that $\lambda_{12}(y \mid x) = \tilde{\lambda}_{12}(y - x)$. Find an estimate of $G(t)$ (Temkin, 1978).

8.8 (*continuation*) Apply the results of Exercise 8.7 to the Stanford heart transplant data. Specifically, estimate the probability that an individual has been transplanted and is surviving at time t after admission to the program under both the Markov and the semi-Markov assumptions.

8.9 Show that the sum over j of the cumulative incidence function estimates in (8.11) equals 1 minus the Kaplan–Meier estimate of the overall survivor function.

8.10 By incorporating interactions with time in the relative risk models for the mouse carcinogenicity data (Appendix A, data set V), test whether the proportional hazards model leading to the results in Table 8.1 is valid.

8.11 Consider the proportional risks model (8.15) and extend it to include a term $\psi_j g_j(t)$ into the regression portion of the jth type-specific hazard, $j = 1, \ldots, m$, where $\psi_1 = 0$ and $g_j(t)$ is a smooth function of time such as t or $\log t$. The proportional risks assumption corresponds to the hypothesis $\psi_j = 0$, $j = 2, \ldots, m$. Consider $m = 2$ and develop a score test of $\psi_2 = 0$.

8.12 Some authors (e.g., Crowder, 2001) have suggested that discrete competing risks models are naturally obtained by grouping continuous multiple decrement models, and in the discrete context, this leads naturally to competing risks models in which two or more failuretypes can occur simultaneously. Discuss whether and/or when this would be a sensible approach.

CHAPTER 9

Modeling and Analysis of Recurrent Event Data

9.1 INTRODUCTION

There are many application areas in which individual study subjects, or specific manufactured items, may experience multiple events or failures. In this chapter we consider the modeling and analysis of such recurrent event data. Examples of recurrent event data used in the statistical literature include bladder tumor recurrence times among patients in a randomized treatment trial (Byar, 1980; Lin et al., 1998), infection occurrence times among leukemia patients receiving bone marrow transplants (Prentice et al., 1981; Pepe and Cai, 1993), times to warranty claims for a particular automobile model (Kalbfleisch et al., 1991), times to valve seat replacement on diesel engines in a service fleet (Lawless and Nadeau, 1995), times to respiratory system exacerbations among cystic fibrosis patients (Therneau and Hamilton, 1997), and times to inpatient hospital admissions among intravenous drug users (Wang et al., 2001).

As these examples suggest, recurrent event data arise in diverse settings. Also, the goals of the recurrent event analysis can vary substantially among applications. For example, in some settings there may be interest in describing the relationship of recurrent event rates to the individual's preceding event history or the relationship to an evolving covariate history that may include treatment choices, or repair activities in an industrial setting, that are intended to reduce the risk of further events. In other settings, interest may focus on the overall rates of recurrent events in the study population and on the dependence of these rates on covariates. There are also intermediate settings in which, for example, one may be interested in event rate comparisons among individuals who have experienced a specified number of prior events without conditioning on the detailed history and timing of such events.

Statistical models and estimation procedures that are convenient for examining these types of associations have been discussed in the literature over the past couple of decades. The analysis of recurrent event data generally remains an active research area. Suppose that there is a single type of failure. Starting from a well-defined time origin (e.g., randomization into a controlled trial), one typically observes a point process $(T_1, T_2, \ldots)$ for a given study subject, where T_1 is the time to the individual's first failure, T_2 is the time (from the origin) to the individual's second failure, and so on, with follow-up continuing to a total follow-up time C that right-censors the point process. There may also be a baseline covariate vector x, or more generally a covariate process x, having history $X(t) = \{x(u), 0 \le u < t\}$ for the individual. Denote by $\tilde{N}(t)$ the number of failures on the individual by follow-up time t, and by $N(t)$ the corresponding observed number of failures in $(0, t]$. Note that $N(t)$ may be less than $\tilde{N}(t)$ because of censoring.

Recurrence rates that condition on the preceding failure and covariate histories for the individual constitute a logical starting point for modeling recurrent event data. Thus, for absolutely continuous event times, one can define the hazard or *intensity process* $\lambda(t)$ at follow-up time t, given the covariate history $X(t)$, by

$$\lambda(t)\, dt = P[d\tilde{N}(t) = 1 | \tilde{N}(u), 0 \le u < t, X(t)]. \tag{9.1}$$

This expression assumes that jumps in $\tilde{N}$ are of unit size only. However, recurrent event data sometimes include jumps of size greater than 1, as more than one event is recorded for an individual at a specific follow-up time. This typically occurs because event times from an underlying continuous process are grouped. It is also possible to consider continuous-time counting processes having jump sizes greater than 1, as, for example, in a queuing model that counts the arrival time of customers in continuous time, but where customers arrive in groups of various sizes.

One natural approach to the modeling of counting processes with jumps that may exceed 1 is to model the mean jump in $\tilde{N}$ across time. Hence, in this chapter, we consider models for the increments

$$d\Lambda(t) = E[d\tilde{N}(t) | \tilde{N}(u), 0 \le u < t, X(t)] \tag{9.2}$$

in the cumulative intensity process Λ. Note that (9.1) and (9.2) are equivalent to $\Lambda(t) = \int_0^t \lambda(u)\, du$ in the case of a continuous-time process having only unit jumps. In fact, whenever recurrence times are restricted to be absolutely continuous, we shall assume that $d\tilde{N}(t) \le 1$. We shall refer to (9.2) as the failure intensity at time t. Its interpretation as the expected number of events on an individual should be kept in mind if multiple events at specific time points may occur for a study subject.

Independent censorship requires that the condition $C \ge t$ can be added to the conditioning event without altering (9.2). This independent censorship assumption allows the censoring rate at time t to depend, possibly in a complicated fashion, on the preceding counting and covariate process histories, adding to its plausibility in specific settings. Also, as in the univariate failure time setting, the covariate process X could be defined to include "markers" that are associated with censoring rates to extend the applicability of an independent censoring assumption. It is convenient to let $Y(t) = \mathbf{1}(0 < t \le C)$. Independent censorship then requires that

$$E[dN(t) | N(u), Y(u); 0 \le u < t, X(t)] = Y(t)\, d\Lambda(t), \tag{9.3}$$

so that the failure intensity at time t for the observed point process N is equal to that (9.2) for the underlying process $\tilde{N}$ at all times at which the study subject is under follow-up (i.e., all $t \leq C$). Note that (9.3) also accommodates independent left truncation by setting $Y(u) = 0$ prior to active follow-up $[Y(u) = 1]$ on the point process but that $d\Lambda(t)$ may then involve the individual's counting process at times prior to active follow-up.

9.2   INTENSITY PROCESSES FOR RECURRENT EVENTS

Various models for recurrent event data may be considered, depending on the questions to be addressed in data analysis. These models are distinguished primarily by the choice of conditioning events. As was just noted, an overall intensity rate $d\Lambda(t)$ at time t conditions on the histories prior to t of counting (failure) and covariate processes. An increasing (as t increases) history that includes the preceding failure history for the individual is fundamental in the filtrations used in counting processes. It is the nested filtrations that generate the uncorrelated increments structure that produces simple variance formulas for tests and estimators. In many instances it may be challenging to model such conditional rates adequately, and frequently, principal interest resides in certain marginalized intensities that condition only on parts of the preceding histories.

The most studied marginal intensity rates drop the preceding failure history entirely from the conditioning event, and model

$$d\Lambda_m(t) = E[d\tilde{N}(t)|X(t)]. \qquad (9.4)$$

This rate arises as an expectation of the intensity rate (9.2) over the distribution of $[\tilde{N}(u); 0 \leq u < t]$ given $X(t)$. Identifiability of Λ_m typically requires that

$$E\{dN(t)|[Y(u); 0 \leq u < t], X(t)\} = Y(t) \, d\Lambda_m(t). \qquad (9.5)$$

Hence, use of (9.5) requires a stronger independent censoring condition than does use of (9.3), in that for (9.5) to hold, censoring rates are not allowed to depend on the individuals preceding failure history $[N(u); 0 \leq u < t]$. This excludes relatively common circumstances under which censoring rates at a given time tend to be higher among persons who have experienced a larger number of prior failures, or one or more "recent" failures. Furthermore, in view of the expectation leading to (9.4), covariates that are functions of $[\tilde{N}(u); 0 \leq u < t]$ need to be excluded from the conditioning event. In considering the covariate history $X(t)$ in (9.4), it is easiest to entertain fixed or external covariates; see also (6.8), for which

$$E\left[d\tilde{N}(u)|X(u)\right] = E\left[d\tilde{N}(u)|X(t)\right], \qquad \text{for all } t \geq u. \qquad (9.6)$$

Under this condition

$$E\big[\tilde{N}(t)|X(t)\big] = \int_0^t E[d\tilde{N}(u)|X(t)]$$

$$= \int_0^t E[d\tilde{N}(u)|X(u)]$$

$$= \Lambda_m(t),$$

so that $\Lambda_m(t)$ simply models the expected number of failures in $(0,t]$ as a function of $X(t)$, facilitating corresponding parameter interpretation. Technically, $X(t)$ in (9.4) could also include internal covariates, including covariates for which the association with failure rates is influenced by the corresponding $[\tilde{N}(u), 0 \le u < t]$. Careful attention to the interpretation of model parameters (e.g., regression parameters) is required for such covariates (see Exercise 9.5 for illustration), and careful consideration of the applicability of (9.6) is required if one wishes to interpret $\Lambda_m(t)$ as the mean number of failures in $(0,t]$.

Intensity-rate models that condition on some but not all of the preceding failure history also can be considered. For example, one could condition on the number of preceding failures $\tilde{N}(t^-)$ on the individual but not on other aspects of $[\tilde{N}(u); 0 \le u < t]$, giving intensity rates

$$d\Lambda_q(t) = E[d\tilde{N}(t)|\tilde{N}(t^-), X(t)], \tag{9.7}$$

where we have denoted $q = \tilde{N}(t^-)$. The natural independent censorship condition then requires that

$$E\{dN(t)|N(t^-), [Y(u); 0 \le u < t], X(t)\} = Y(t)\, d\Lambda_q(t). \tag{9.8}$$

This condition is weaker than that leading to (9.5) in that censoring rates at a given time can depend on the number of prior failures experienced by the individual but not on other aspects of the timing and spacing of prior failures. This independent censorship condition, although more stringent than that leading to (9.3), may be plausible in applications. Subsequent sections focus on modeling and estimation of Λ, Λ_m and on a class of intensities like Λ_q.

Wei et al. (1989) proposed a rather different approach to the modeling and analysis of absolutely continuous-recurrence-time data. They chose to model a hazard rate that can be written

$$d\Lambda_q^*(t) = P[d\tilde{N}(t) = 1, \tilde{N}(t^-) = q|X(t)], \tag{9.9}$$

for which distinct Cox models were specified for $q = 0, 1, \dots$. The joint event in the numerator of this probability tends to make the modeling and interpretation of parameters in Λ_q^* difficult. Hence we do not discuss the use of (9.9) for the analysis of recurrent events. However, the marginal hazard rate methods of Wei et al. (1989) are quite useful for the analysis of correlated failure times in other settings, as discussed in Chapter 10.

9.3 OVERALL INTENSITY PROCESS MODELING AND ESTIMATION

9.3.1 Cox-Type Intensity Models

Consider now the modeling of the overall intensity process (9.3) under independent censorship. Any of the classes of hazard rate models discussed in previous chapters can be considered for the hazard rates

$$d\Lambda(t) = E[d\tilde{N}(t)|\tilde{N}(u); 0 \le u < t, X(t)]$$

in (9.2). For example, one could specify a Cox-type model

$$d\Lambda(t) = d\Lambda_0(t) \exp\left[Z(t)'\beta\right], \tag{9.10}$$

where $Z(t)' = \left[Z_1(t), \ldots, Z_p(t)\right]$ is comprised of functions of $X(t)$ and $[N(u); 0 \le u < t]$ and product terms with t. Inference under (9.10) differs very little from that for univariate failure time data (see Chapters 4 and 6). In fact, Andersen and Gill's seminal (1982) paper on Cox model asymptotics was written to embrace absolutely continuous recurrent event data that adhere to (9.10). See also the discussion of modulated Markov processes in Section 8.3. The log-partial likelihood function, score statistic, and information matrix in stochastic integral notation are of the same form as for univariate failure time data, with integration extending over the entire risk period for each study subject. Specifically, these quantities can be written

$$\ell(\beta) = \log L(\beta) = \sum_{i=1}^{n} \int_0^\infty \{Z_i(t)'\beta - \log[S^{(0)}(\beta, t)]\} \, dN_i(t), \tag{9.11}$$

$$U(\beta) = \partial\ell(\beta)/\partial\beta = \sum_{i=1}^{n} \int_0^\infty [Z_i(t) - \mathscr{E}(\beta, t)] \, dN_i(t), \tag{9.12}$$

and

$$I(\beta) = -\partial^2\ell(\beta)/\partial\beta\,\partial\beta' = \sum_{i=1}^{n} \int_0^\infty \mathscr{V}(\beta, t) \, dN_i(t), \tag{9.13}$$

where, as before,

$$S^{(j)}(\beta, t) = \sum_{i=1}^{n} Y_i(t) Z_i(t)^{\otimes j} \exp\left[Z_i(t)'\beta\right], \qquad \text{for } j = 0, 1, 2,$$

$$\mathscr{E}(\beta, t) = S^{(1)}(\beta, t)/S^{(0)}(\beta, t),$$

and

$$\mathscr{V}(\beta, t) = S^{(2)}(\beta, t)/S^{(0)}(\beta, t) - \mathscr{E}(\beta, t)^{\otimes 2}.$$

A Newton–Raphson algorithm $\beta_{(\ell+1)} = \beta_{(\ell)} + I^{-1}(\beta_{(\ell)}) U(\beta_{(\ell)}), \ell = 0, 1, 2, \ldots$ can be used to calculate $\hat{\beta}$ that solves $U(\beta) = 0$. Typically, a starting value $\beta_{(0)} = 0$ will be adequate, and convergence will be achieved in just a few iterations. Testing and estimation can be based on the fact that under (9.10), $n^{-1/2}U(\beta)$ has an asymptotic normal distribution with mean zero and variance consistently estimated by $n^{-1}I(\beta)$, and on the fact that $n^{1/2}(\hat{\beta} - \beta)$ has an asymptotic normal distribution with mean zero and variance consistently estimated by $nI^{-1}(\hat{\beta})$ under mild conditions (see Chapter 5, and Andersen and Gill, 1982).

The same estimating function (9.12) applies and is unbiased if the underlying counting processes are allowed to have jumps of size greater than 1, and/or $d\Lambda_0$ is allowed to include mass points, but a more complex variance estimator for $U(\beta)$ is then typically required. A variance estimator $n\hat{\Sigma}$ appropriate to this situation is given below in (9.17), resulting in a sandwich variance estimator $I(\hat{\beta})^{-1}\hat{\Sigma}I(\hat{\beta})^{-1}$ for $\hat{\beta}$.

A simple special case of (9.10) arises if $Z(t)$ consists of fixed or external time-dependent covariates and interaction of such covariates with time only, in which case (9.10) is a nonhomogeneous Poisson process, modulated by covariates. Frequently, however, the intensities (9.10) will depend on the preceding failure history for the individual in a complex fashion, and $X(t)$ may include covariates that are responsive to failure events. The model (9.10) can be regarded as being flexible enough to be useful in these situations, although the modeling of the parametric factor, $\exp[Z(t)'\beta]$, may need to be quite complicated for an adequate representation of the data. A few special cases illustrate the plethora of intensity process modeling choices.

Beginning with (9.10) with $Z(t)$ defined in terms of fixed or external time-varying covariates, one could add to $Z(t)'\beta$ elements $\mathbf{1}[N(t^-) = 1]\gamma_1 + \mathbf{1}[N(t^-) = 2]\gamma_2 + \cdots$, thereby allowing the intensity to be altered by a multiplicative factor e^{γ_j} following the occurrence of the jth failure on the individual, compared to individuals at the same time who are without any failures. Additionally, one could supplement $Z(t)'\beta$ in (9.10) by

$$\mathbf{1}[N(t^-) = 1]Z(t)'\beta_1 + \mathbf{1}[N(t^-) = 2]Z(t)'\beta_2 + \cdots,$$

thereby allowing the relative risk parameter for the fixed or external covariates modeled in (9.10) to vary arbitrarily with number of prior failures. A further generalization would allow the shape of the baseline intensity to depend arbitrarily on the number of prior failures for the individual, giving

$$d\Lambda(t) = d\Lambda_{os}(t) \exp[Z(t)'\beta_s], \qquad (9.14)$$

where $s = N(t^-)$. This model simply stratifies the intensity rate $d\Lambda(t)$ on the number $s = 0, 1, \ldots$ of preceding failures for the study subject. It is readily applied by using a time-dependent stratified version of the score statistic and information matrix above (see Section 6.4).

Use of (9.14) may also require a complicated regression variable for an adequate description of the data. For example, the recent occurrence of a failure may, in some settings, have a profound effect on the subsequent recurrence rates. For example, the occurrence of a certain type of infection may trigger a response that implies a subsequent period of immunity, or the occurrence of a new tumor may engender a systemic therapeutic response that would reduce the probability of a further tumor for some period of time. In these circumstances the sojourn time, $v = t - T_{N(t^-)}$, since the immediately preceding failure time becomes influential, and an intensity rate model of the form

$$d\Lambda(t) = d\Lambda_{os}(v) \exp[Z(t)'\beta_s] \qquad (9.15)$$

may be parsimonious. A partial likelihood function is readily specified for $\beta_1, \beta_2, \ldots$ in (9.15), as are corresponding estimation procedures for β_s and Λ_{os} provided that $d\tilde{N}(t) \leq 1$. See Section 8.3.3 for a related discussion in the context of semi-Markov models.

Stratification choices other than $s = N(t^-)$ can also be considered in (9.14) and (9.15). For example, a finer stratification $s = 2N(t^-) + \mathbf{1}(t - T_{N(t^-)} > t_0)$ in (9.14) would allow the failure rate at t to depend arbitrarily on both the number of prior failures and on whether the sojourn time exceeds a prespecified value, t_0. Note that the stratification in (9.15) needs to be at least as fine as $s = N(t^-)$ for a partial likelihood argument to be used for the estimation of β_s, since then each study subject has at most a single at-risk time interval in each stratum.

9.3.2 Illustration

Prentice et al. (1981) consider an application of (9.14) and (9.15) to the analysis of selected types of infection among 76 bone marrow transplant recipients. The data set is rather small to expect asymptotic approximations to be accurate, but we will proceed to use such approximations for illustration purposes. Of particular interest is the association of chronic graft versus host disease (c-GVHD) and infection occurrence. The analysis was restricted to infections that occur after six months from marrow transplantation since c-GVHD diagnosis sometimes does not arise until the patient is a few months post transplant. In this data set 46 of the 76 patients have at least one, 29 have at least two, 20 have at least 3, 9 have at least 4, and 4 at least 5 post-six-month infections. Multiple infections on a patient at a particular follow-up time did not arise with this data set. Table 9.1, shows estimated regression coefficient estimates $\hat{\beta}_s$ and corresponding estimated standard errors under (9.14) and (9.15) for a c-GVHD indicator variable.

One sees that c-GVHD is associated with an estimated $\exp(1.076) = 2.94$-fold increase in the rate of first infection under either model. Corresponding relative risk estimates for the second to fifth infections under (9.14) appear to be somewhat variable. As mentioned previously, asymptotic distributional approximations may be crude with these very small sample sizes, especially in stratum 4, where there

Table 9.1 Relative Risk Analysis of Relationship of c-GVHD and Infection Incidence in Bone Marrow Transplant Recipients [a]

| Model | Stratum $[s = N(t^-)]$ Number | | | | | | All Strata $(\beta_0 = \cdots = \beta_4 = \beta)$ | |
| | 0 | 1 | 2 | 3 | 4 | | | |
	$\hat{\beta}_0$	$\hat{\beta}_1$	$\hat{\beta}_2$	$\hat{\beta}_3$	$\hat{\beta}_4$	$\log L(\hat{\beta})$	$\hat{\beta}$	$\log L(\hat{\beta})$
(9.14)	1.076	0.279	1.038	0.000	1.220	-293.06	0.796	-294.96
	(0.305)	(0.402)	(0.503)	(0.794)	(1.203)		(0.211)	
(9.15)	1.076	0.701	0.962	0.232	1.011	-350.78	0.891	-351.51
	(0.305)	(0.375)	(0.493)	(0.717)	(1.177)		(0.204)	

Source: Prentice et al. (1981).

[a] Estimated partial likelihood standard deviations shown in parentheses.

are but nine contributing patients. A likelihood ratio test for a common value $\beta_0 = \cdots = \beta_4 = \beta$ under model (9.14) takes value $2(294.96 - 293.06) = 3.80$, which is not extreme in relation to a χ_4^2 distribution.

Estimates of the regression coefficients $\{\beta_s\}$ under (9.15) also show little evidence of departure from a common value with a likelihood ratio statistic of value 1.46 for testing $\beta_0 = \cdots = \beta_4 = \beta$. Either model provides strong evidence against the global null hypothesis $\beta_0 = \cdots = \beta_4 = 0$, with approximate χ_5^2 score tests taking respective values 19.79 and 22.48 under (9.14) and (9.15).

Either model can be generalized through finer stratification or by adding additional terms to the regression vector. For example, (9.14) can be relaxed by defining a regression vector $Z(t) = [Z, Z_1(t)]$, where Z is again a c-GVHD indicator, and $Z_1(t) = Z \log[t - T_{N(t^-)}]$, where T_0 is defined to be six months from marrow transplantation. The c-GVHD relative risk in stratum s is then of the form $e^{\beta_{s0}}[t - T_{N(t^-)}]^{\beta_{s1}}$, which is monotone increasing, monotone decreasing, or constant as a function of the sojourn time $t - T_{N(t^-)}$, depending on whether β_{s1} is positive, negative, or zero. Application of this model with $\beta_{s1} \equiv \beta_1$ gives a maximum partial likelihood estimate of $\hat{\beta}_1 = -0.0087$ with corresponding standard deviation estimate of 0.199, providing little motivation for this generalization. One could similarly regress on time since marrow transplantation in the renewal process model (9.15).

In this setting, time (t) from bone marrow transplantation is a key factor in determining the patient's immune status, and perhaps it is not surprising that gap times appear to contribute little additional explanatory information. Regression coefficients seem to be more readily interpreted under (9.14) than under (9.15): Estimation under (9.14) is based on the comparisons of patients who are at the same length of time since transplantation and have the same number of post-six month infections. Regression estimation under (9.15) involves comparisons among patients who have a common number of post-six month infections and are at a common time duration since the most recent such infection. But these patients may be at quite different time durations from receipt of their marrow transplants. Note that the maximized log likelihoods under the two models cannot be compared directly, in view of the different compositions of the risk sets in the partial likelihood function.

It can be commented that models (9.10), (9.14), and (9.15) are readily applied using Cox model software for univariate failure times, provided that such programs accommodate time-dependent strata and covariates. Specifically, the sequence of failure times on a given study subject, including the final censored time, can be listed as if they were failure times on distinct subjects, with the time-dependent stratum feature being used to define the at-risk period for the "subject." Standard calculations then give estimators of β_s and of the cumulative baseline hazards Λ_{0s} and their estimated standard errors.

9.3.3 Other Intensity Models

Some quite elaborate intensity rate models have been proposed in the literature, often motivated by industrial reliability contexts. As a recent example, Peña and Hollander (2001) propose a model for absolutely continuous failure times given by

$$d\tilde{\Lambda}(t) = W\lambda_0[\zeta(t)]\rho[N(t^-)]\Psi[Z_i(t)'\beta],$$

where W is a multiplicative random effect, or frailty variate, the transformation ζ is a nonnegative function $[\zeta(0) = 0]$ that is monotone and differentiable between adjacent failure times on the individual with nonnegative derivative, ρ is a specified parametric function that allows the hazard rate to depend multiplicatively on the number of prior failures $[\rho(0) = 1]$, and Ψ is a nonnegative link function. Note that despite its generality, this model generally does not encompass the stratified models (9.14) and (9.15).

Although a detailed modeling of the intensity process may be of interest in some settings, in many other contexts interest will focus on such key quantities as covariate effects on expected event rates in the population. In these contexts it is advantageous to model marginalized intensities of the type mentioned in Section 9.2 and avoid modeling assumptions concerning other aspects of the failure rates. In fact, many of the proposals for recurrent event data analysis in recent years have focused on the modeling and analysis of the mean number of failures as a function of follow-up time t.

9.4 MEAN PROCESS MODELING AND ESTIMATION

9.4.1 Cox-Type Marginalized Rate Models

Consider again the marginalized intensity

$$d\Lambda_m(t) = E[d\tilde{N}(t)|X(t)]$$

and suppose that the censorship is such that (9.5) holds.

As noted previously, Λ_m arises as an expectation over the distribution of the preceding failure history $\{\tilde{N}(u); 0 \le u < t\}$ of the overall intensity rate, conditional on covariates. This marginal failure rate will be an attractive target for modeling and estimation when population aspects of failure rates and covariate effects are of interest in that it obviates the need to specifically model the dependence of failure rates on the prior failure history for the individual.

Suppose now that a Cox model

$$d\Lambda_m(t) = d\Lambda_0(t) \exp[Z(t)'\beta] \qquad (9.16)$$

is specified for the marginal intensities (9.4). It is worth considering the possible choices for elements of the modeled regression vector in (9.16) as compared to the overall intensity Cox model (9.10). First, (9.10) may include elements that are functions of $\{N(u); 0 \leq u < t\}$ such as $\mathbf{1}[N(t^-) = s], s = 0, 1, 2$, whereas such terms would be integrated out in developing Λ_m. Similarly, (9.10) could include terms such as $Z\mathbf{1}[N(t^-) = s]$, for $s = 0, 1, \ldots$, which would allow the multiplicative effect of a covariate Z (e.g., a treatment indicator) to vary with the number of prior failures. These types of regression variables, which depend on $\{N(u), 0 \leq u < t\}$ for their definition, would also need to be excluded from (9.16). As mentioned in Section 9.2, models for Λ_m may include fixed or external time-varying covariates or even internal time-varying covariates if one is careful about the interpretation of the regression coefficient. In general, the regression coefficient in (9.16) will have an averaged (over $\{\tilde{N}(u), 0 \leq u < t\}$) interpretation, whereas the corresponding coefficient in (9.10) has an interpretation that is conditional on the preceding failure time history. With fixed or external time-varying covariates or other covariates satisfying the condition (9.6), the expected number of failures by time t can be written, under (9.16), as

$$\Lambda_m(t) = \int_0^t \exp[Z(t)'\beta] \, d\Lambda_0(t).$$

In these circumstances, Λ_0 and β have an interpretation as parameters in a model for the mean number of failures (in the absence of censoring).

Estimation based on the Cox model (9.16) has been considered by Lawless and Nadeau (1995) and Lawless et al. (1997), with distribution theory developed under a discrete failure time model. Lin et al. (2000) presented a development for absolutely continuous event times which is outlined below.

The key idea for estimation of β and Λ_0 in (9.16) is that

$$dM_i(t) = dN_i(t) - Y_i(t) \exp[Z_i(t)'\beta] \, d\Lambda_0(t)$$

has expectation 0 under (9.16) at all values of t for each individual $i = 1, \ldots, n$ in the sample. Hence one can develop an unbiased estimating function

$$U(\beta) = \sum_{i=1}^{n} \int_0^{\infty} [Z_i(u) - \mathscr{E}(\beta, u)] \, dM_i(u)$$

$$= \sum_{i=1}^{n} \int_0^{\infty} [Z_i(u) - \mathscr{E}(\beta, u)] \, dN_i(u),$$

that is formally identical to (9.12), for the marginalized intensity Cox model (9.16). However, the contributions to $U(\beta)$ at distinct times are typically correlated since the conditioning events are not nested in time. Hence the processes $\{M_i; i = 1, \ldots, n\}$ are not martingales. Modern empirical process theory can, however, be used to produce the desired asymptotic distributional results, under independent and identically distributed assumptions on $(N_i, Y_i, Z_i), i = 1, \ldots, n$ and some additional conditions. Specifically, one can exploit the near independence of the contributions to $U(\beta)$ from each study subject to show that $n^{-1/2}U(\beta)$ converges in distribution to a mean zero Gaussian process with covariance matrix that is estimated consistently by the empirical estimator

$$\hat{\Sigma} = n^{-1} \sum_{i=1}^{n} \hat{U}_i \hat{U}_i', \qquad (9.17)$$

where $\hat{U}_i = \hat{U}_i(\infty)$ and $\hat{U}_i(t) = \int_0^t [Z_i(u) - \mathscr{E}(\hat{\beta}, u)] \, d\hat{M}_i(u)$, $\hat{\beta}$ solves $U(\beta) = 0$, $d\hat{M}_i(t) = dN_i(t) - Y_i(t) \exp[Z_i(t)'\hat{\beta}] \, d\hat{\Lambda}_0(t)$, and the estimator $\hat{\Lambda}_0$ of the cumulative baseline intensity is the Nelson–Aalen estimator

$$\hat{\Lambda}_0(t) = \int_0^t \sum_{i=1}^{n} dN_i(u) \Big/ \sum_{i=1}^{n} Y_i(u) \exp[Z_i(u)'\hat{\beta}]. \qquad (9.18)$$

These same methods can be used to show that under (9.16), expression (9.11) converges in probability to a concave function having maximum at the true β value yielding the consistency of $\hat{\beta}$ as an estimator for β, and a standard Taylor expansion, in conjunction with the asymptotic result above for $n^{-1/2}U(\beta)$ leads to an asymptotic mean zero normal distribution for $n^{1/2}(\hat{\beta} - \beta)$, with a variance that can be estimated consistently by the "sandwich" form $n^2 I(\hat{\beta})^{-1}\hat{\Sigma}I(\hat{\beta})^{-1}$, where $I(\hat{\beta})$ is given by (9.13).

Empirical process theory also allows one to show that $n^{1/2}(\hat{\Lambda}_0 - \Lambda_0)$ converges weakly to a mean zero Gaussian process with covariance function ζ such that $\zeta(s, t)$ can be estimated consistently by

$$\hat{\zeta}(s, t) = n^{-1} \sum_{1}^{n} \hat{\Psi}_i(s)\hat{\Psi}_i(t), \qquad (9.19)$$

where

$$\hat{\Psi}_i(t) = \int_0^t d\hat{M}_i(u) / S^{(0)}(\hat{\beta}, u) - \sum_{\ell=1}^{n} U_\ell'(t) I(\hat{\beta})^{-1} \int_0^t \hat{\mathscr{E}}(\hat{\beta}, u) S^{(0)}(\hat{\beta}, u)^{-1} \, dN_\ell(u).$$

This expression can be used to develop confidence intervals for $\Lambda_0(t)$ at specified values of t, or confidence bands for Λ_0 more generally.

We will not provide here a presentation of the empirical process convergence results that generate these asymptotic distributions. It can be commented, however, that these results are applicable broadly to the failure time methods described throughout this book, including the settings in which we have employed martingale methods, although under somewhat different conditions. Hence one can view the martingale methods as not necessary, although useful for obtaining convenient variance estimators. The conditions used by Lin et al. (2000) for the developments above are as follows: They assume a finite follow-up interval $[0, \tau]$ with simple right censorship such that $P(C > \tau) > 0$. They assume absolutely continuous failure times and suppose that (N_i, Y_i, Z_i) are independent and identically distributed, $i = 1, \ldots, n$. They also assume that $N_i(\tau)$ is bounded by a constant, that Z_i has bounded total variation, $i = 1, \ldots, n$, and finally that the information matrix $I(\hat{\beta})/n$ given in (9.13) converges in probability to a positive definite matrix as $n \to \infty$. These authors argue that the finite interval condition can be dropped, and their simulation studies do not suggest that bias is engendered by integrating over all times where one or more subjects are at risk.

9.4.2   Accelerated Failure Time Mean Models

For covariates satisfying (9.6), one can alternatively specify that

$$\Lambda_m(t) = \Lambda_0 \left\{ \int_0^t \exp\left[Z(u)'\beta\right] du \right\}, \qquad (9.20)$$

which has the interpretation that covariates accelerate or decelerate the expected number of failures in $[u, u + du)$ by the factor $\exp[Z(u)'\beta]$ for all $u > 0$. The corresponding failure intensity

$$d\Lambda_m(t) = \exp\left[Z(t)'\beta\right] d\Lambda_0 \left\{ \int_0^t \exp\left[Z(u)'\beta\right] du \right\} \qquad (9.21)$$

can be entertained more generally, but (9.6) is needed for Λ_0 and β to have an interpretation as mean model parameters. Estimation under (9.21) proceeds essentially as for univariate failure time data (Chapter 7), with integrals that define the estimating function and variance matrices extended to cover each individual's follow-up interval. As for the Cox-type models for Section 9.4.1, the contributions to the estimating function (defined below) at distinct failure times are typically correlated, and a corresponding empirical variance estimator is required.

   A brief description of the estimation procedure is as follows: Define transformed recurrence times $T_j^*(\beta) = \int_0^{T_j} \exp[Z(t)'\beta] \, dt$, corresponding to the point process $T_1, T_2, \ldots$ on a study subject. Let $C^*(\beta) = \int_0^C \exp[Z(t)'\beta] \, dt$ be the corresponding transformed censoring time. Now define the transformed counting and at risk processes $N^*(\cdot, \beta)$ and $Y^*(\cdot, \beta)$, so that $N^*(t, \beta)$ is the number of $T_j^*(\beta)$ values that are equal or less than t and so that $Y^*(t, \beta)$ is 1 if $t \leq C^*(\beta)$ and is zero otherwise. Also define the processes $M_i(\cdot, \beta), i = 1, \ldots, n$ by

$$dM_i(t, \beta) = dN_i^*(t, \beta) - Y_i^*(t, \beta) \, d\Lambda_0(t).$$

In view of (9.21), the M_i's have mean zero at the true β value. An estimating function for β can be defined by

$$U(\beta) = \sum_{i=1}^{n} \int_0^\infty W_i(t; \beta) \left[Z_i^*(t, \beta) - \bar{Z}(t, \beta) \right] dN_i^*(t, \beta),$$

where $Z_i^*(t, \beta) = Z_i(s)$ and s satisfies $t = \int_0^s \exp\left[Z(u)'\beta\right] du$, and $\bar{Z}(t, \beta) = \sum_1^n Y_i^*$ $(t, \beta) Z_i^*(t, \beta)/\sum_1^n Y_i^*(t, \beta)$, and $W_i(t, \beta)$ is a data-analyst-defined weight function. For example, $W_i(t; \beta) \equiv 1$ gives log-rank-type scores for estimating β. Substitution for N_i^* and simplification gives

$$U(\beta) = \sum_{i=1}^{n} \int_0^\infty W_i(t; \beta) \left[Z_i^*(t, \beta) - \bar{Z}(t, \beta) \right] dM_i(t, \beta),$$

so that $U(\beta)$ is an unbiased estimating function under (9.21). Under independent and identically distributed conditions on (N_i, Y_i, Z_i) and certain boundedness conditions on N_i, Z_i, W_i, and Λ_0, Lin et al. (1998) show $n^{-1/2} U(\beta)$ to converge to a normal distribution with mean zero and with variance matrix that can consistently be estimated by $n^{-1} \hat{\Sigma}$, where

$$\hat{\Sigma} = \sum_{i=1}^{n} \hat{U}_i(\hat{\beta}) \, \hat{U}_i(\hat{\beta})',$$

where

$$\hat{U}_i(\hat{\beta}) = \int_0^\infty \hat{W}_i(t, \hat{\beta}) \left[Z_i^*(t; \hat{\beta}) - \bar{Z}(t; \hat{\beta}) \right] d\hat{M}_i(t, \hat{\beta}),$$

$$d\hat{M}_i(t, \hat{\beta}) = dN_i^*(t, \hat{\beta}) - Y_i^*(t; \hat{\beta}) \, d\hat{\Lambda}_0(t, \beta),$$

$$d\hat{\Lambda}_0(t, \hat{\beta}) = \sum_{i=1}^{n} \int_0^t dN_i^*(u, \hat{\beta}) \Big/ \sum_{\ell=1}^{n} Y_\ell^*(u, \beta),$$

and where $\hat{\beta}$ minimizes $\| U(\beta) \|$. Also, $n^{1/2}(\hat{\beta} - \beta)$ was shown to converge to a mean zero Gaussian distribution. These authors also establish an asymptotic mean zero Gaussian distribution for $n^{1/2}(\hat{\Lambda}_0 - \Lambda_0)$ under the aforementioned conditions.

On the surface, the application of (9.21) seems deceptively similar to the use of a corresponding Cox model (9.16). However, as with the univariate failure time application of the accelerated failure time model, the calculation of $\hat{\beta}$ under (9.21) and the development of estimators of the variance of $\hat{\beta}$ and the covariance function for $n^{1/2}(\hat{\Lambda}_0 - \Lambda_0)$ are rather difficult. These difficulties arise since the score function $U(\beta)$ is a step function and the processes N^*, Y^*, and Z^* need to be recalculated at each β value. Calculation of β that minimizes the length of $U(\beta)$ generally requires some type of grid search, or use of simulated annealing (Lin and Geyer, 1992) if the dimension of $Z(t)$ is at all large. Perhaps better, the iteratively reweighted Gehan estimating function approach of Jin et al. (2001) evidently extends in a straightforward fashion to allow recurrent events. Furthermore, even though $U(\beta)$ is asymptotically linear in the vicinity of the true β, the coefficient matrix for the linear approximation involves derivatives of Λ_0, which are difficult to estimate nonparametrically.

Lin et al. (1998) recommend that the variance matrix of $n^{1/2}(\hat{\beta} - \beta)$ be estimated by a resampling technique. Specifically, a vector G is generated from a p-variate normal distribution having mean 0 and variance matrix $n\hat{\Sigma}$, and β^* is calculated as a solution to $U(\beta) = G$ using the numerical search procedures just described. The sample variance of β^* values under repeated sampling from this normal distribution then provides an estimate of the asymptotic variance matrix for $\hat{\beta}$, and approximate confidence interval for specific elements of β can also be obtained from the empirical distribution of β^* values. A resampling approach can also be used to estimate the asymptotic covariance function for $n^{1/2}(\hat{\Lambda}_0 - \Lambda_0)$. The variance estimation approach of Huang, mentioned in Section 7.4.2, could similarly be extended to accommodate recurrent events.

9.4.3 Illustration: Bladder Tumor Data

Byar (1980) discusses a randomized trial, conducted by the Veterans Administration Cooperative Urological Group, of patients with superficial bladder tumors. One question of interest concerned the effect of the treatment thiotepa on the frequency of tumor occurrences. The data set includes 48 patients assigned to placebo, in whom there were a total of 87 observed recurrence times, and 38 patients assigned to thiotepa in whom there were 45 recurrence times during trial follow-up, which averaged about 31 months. Tumors present at baseline were removed transurethrally prior to randomization. The possible dependence of the recurrence rate on the number and sizes of prerandomization tumors was also of interest. Table 9.2, shows data on recurrence and censoring times in months, along with treatment assignment, number of baseline tumors (truncated at 8), and diameter of the largest baseline tumor in centimeters. Not shown in the table are the number of tumors at each recurrence time. As with earlier analyses of these data, we define the point process N for each patient to jump by one at each of the recurrence times listed. An alternative and perhaps better analysis would allow N to jump by the number of tumors detected at each recurrence time.

Table 9.2 shows a number of tied recurrence times between patients, while individual study subjects are defined to have at most one recurrence at any follow-up time. Subsequent applications of the overall intensity model (9.10) use a Breslow approximation to accommodate these tied times, presumably contributing some slight conservatism in estimated standard errors for regression parameter estimates.

Table 9.3 presents some analyses of recurrence rates using the regression models described in this chapter. Fixed covariates are defined so that Z_1 takes value zero for

Table 9.2 Bladder Tumor Recurrence Data

Initial Tumors[a]		Censoring[b] Time	Recurrence Times[b] $T_1, T_2, \ldots$	Initial Tumors		Censoring Time	Recurrence Times $T_1, T_2, \ldots$
Number	Size			Number	Size		
Placebo Group							
1	1	0		1	5	30	2,17, 22
1	3	1		2	1	30	3, 6, 8, 12, 26
2	1	4		1	3	31	12, 15, 24
1	1	7		1	2	32	
5	1	10		2	1	34	
4	1	10	6	2	1	36	
1	1	14		3	1	36	29
1	1	18		1	2	37	
1	3	18	5	4	1	40	9, 17, 22, 24
1	1	18	12, 16	5	1	40	16, 19, 23, 29, 34, 40
3	3	23		1	2	41	
1	3	23	10, 15	1	1	43	3
1	1	23	3, 16, 23	2	6	43	6
3	1	23	3, 9, 21	2	1	44	3, 6, 9
2	3	24	7, 10, 16, 24	1	1	45	9, 11, 20, 26, 30
1	1	25	3, 15, 25	1	1	48	18
1	2	26		1	3	49	
8	1	26	1	3	1	51	35
1	4	26	2, 26	1	7	53	17
1	2	28	25	3	1	53	3, 15, 46, 51, 53
1	4	29		1	1	59	
1	2	29		3	2	61	2, 15, 24, 30, 34, 39, 43, 49, 52
4	1	29		1	3	64	5, 14, 19, 27, 41
1	6	30	28, 30	2	3	64	2, 8, 12, 13, 17, 21, 33, 49
Thiotepa Group							
1	3	1		8	3	36	26, 35
1	1	1		1	1	38	
8	1	5	5	1	1	39	22, 23, 27, 32
1	2	9		6	1	39	4, 16, 23, 27, 33, 36, 37
1	1	10		3	1	40	24, 26, 29, 40
1	1	13		3	2	41	
2	6	14	3	1	1	41	
5	3	17	1, 3, 5, 7, 10	1	1	43	1, 27
5	1	18		1	1	44	
1	3	18	17	6	1	44	2, 20, 23, 27, 38
5	1	19	2	1	2	45	
1	1	21	17, 19	1	4	46	2
1	1	22		1	4	46	
1	3	25		3	3	49	
1	5	25		1	1	50	
1	1	25		4	1	50	4, 24, 47
1	1	26	6, 12, 13	3	4	54	
1	1	27	6	2	1	54	38
2	1	29	2	1	3	59	

Source: Adapted from Andrews and Herzberg (1985, pp. 254–259).

[a] Initial number of tumors of 8 denotes 8 or more; Size denotes size of largest such tumor in centimeters.

[b] Censoring and recurrence times are measured in months.

Table 9.3 Regression Parameter Estimates of the Rate of Recurrence of Superficial Bladder Tumors under Various Models [a]

Regression Model	Treatment (0, placebo; 1, thiotepa)	No. Initial Tumors	Initial Tumor Size	No. Recurrences During Follow-up			
				1 vs. 0	2 vs. 0	3 vs. 0	4+
Overall intensity model (9.10)	−0.524 (0.187)	0.201 (0.044)	−0.040 (0.065)				
Overall intensity model (9.10)	−0.301 (0.195)	0.154 (0.049)	−0.001 (0.068)	0.557 (0.256)	1.610 (0.301)	1.319 (0.349)	1.498 (0.337)
Overall intensity model (9.14)	−0.324 (0.211)	0.118 (0.052)	−0.007 (0.072)				
Cox mean model (9.16)	−0.524 (0.262)	0.201 (0.064)	−0.041 (0.076)				
AFT mean [b] model (9.21)	−0.542 (0.312)	0.204 (0.066)	−0.038 (0.084)				

[a] Estimated standard errors in parentheses.
[b] Calculations from Lin et al. (1998) using log-rank weight function.

placebo and 1 for thiotepa treatment, Z_2 is the number of baseline tumors for the patient, and Z_3 is the size in centimeters of the largest baseline tumor. The first row of Table 9.3 assumes a Cox model (9.10) for the overall intensity, with Z_1, Z_2, and Z_3 as the modeled covariates. Under this model assumption the treatment indicator has a regression coefficient estimate of $\hat{\beta}_1 = -0.524$, the absolute value of which is 2.8 times its estimated standard error. Application of the asymptotic normal distributional approximation to $\hat{\beta}_1$ gives a significance level of 0.005 for testing $\beta_1 = 0$. This analysis also suggests that the number of tumors prior to trial enrollment is a strong predictive factor for failure recurrence, while the size of the largest such tumor does not appear to be influential.

As a model for the overall intensity (9.10), this analysis implicitly assumes that the failure intensity at a given follow-up time is independent of the preceding counting process history of failures for the individual. The second row of Table 9.3 tests this assumption by including time-dependent covariates $Z_4, \ldots, Z_7$ such that, for example, $Z_4(t) = 1$ if the patient has exactly one postrandomization recurrence time at time t [i.e., $N(t^-) = 1$] and $Z_4(t) = 0$, otherwise; while Z_5, Z_6, and Z_7 have corresponding definitions for values of 2, 3, and equal or greater than 4 for $N(t^-)$. This model provides evidence that the failure intensity depends strongly on the number of prior failures since randomization. For example, a (partial) likelihood ratio test for $\beta_4 = \cdots = \beta_7 = 0$ takes value $2(-504.411 + 523.289) = 37.76$, which is very extreme in relation to the χ_4^2 distribution. Under this more relaxed model the treatment parameter estimate is reduced to $\hat{\beta}_1 = -0.301$, so that thiotepa use is associated with an estimated relative risk of $\exp(-0.301) = 0.74$, as compared to an estimated relative risk $\exp(-0.524) = 0.59$ under the first analysis. The significance level for testing $\beta_1 = 0$ is increased to 0.12, suggesting that the data may not be able to clearly establish a benefit for thiotepa. On the other hand, the regression coefficient in the second analysis has a somewhat different interpretation than in the first, since it is most readily thought of as estimating the thiotepa log-relative risk at a given follow-up time among patients who have experienced the same number of postrandomization recurrences. This interpretation is more exact for the third analyses of Table 9.3, which uses the Cox model (9.14) for the overall intensity, with (time-dependent) stratification on the number of preceding failures $N(t^-)$. This analysis also yields an estimated significance level of 0.12 for testing $\beta_1 = 0$. Aside from issues of parameter interpretation each of these overall intensity models may be deficient in not adequately accommodating the dependence of failure intensity on the preceding counting process. For example, the length of time since the immediately preceding recurrence, or other aspects of the distribution and spacing of an individual's preceding recurrence times may influence the recurrence intensity.

Rather than proceeding with more detailed models for the overall intensity, one can instead apply mean models for the number of recurrences as a function of follow-up time t. The next analysis of Table 9.3 applies a mean model of Cox regression form (9.16) with regression variables Z_1, Z_2, and Z_3. As discussed above, such models average the overall intensity over the distribution of the preceding recurrence history $\{N(u), 0 \leq u < t\}$, with the result that contributions to the score statistic (9.12) may be correlated, requiring an empirical variance estimator (9.17). Hence, the fourth analysis of Table 9.3 has the same regression parameter estimates, as does the first analysis, but the latter model avoids the assumption that the failure rate is independent of the preceding history and the regression parameters have a population-averaged interpretation. Note that the empirical standard error estimators for these coefficients are considerably larger than those for the first analysis. For example, a test for $\beta_1 = 0$ under this population-averaged Cox model has asymptotic significance level of 0.05, providing some evidence of therapeutic benefit for thiotepa.

The next analysis of Table 9.3, obtained from Lin et al. (1998), fits a corresponding accelerated failure time mean model for the recurrence rates. Model (9.21) with fixed covariates $Z(t) = Z = (Z_1, Z_2, Z_3)'$ was applied using log-rank scores $[W(t, \beta) \equiv 1]$. It happens that the regression coefficients and empirical standard error estimates under the AFT and Cox mean models are very similar in this application. For example, a test for $\beta_1 = 0$ under the AFT model has an estimated significance level 0.08. This degree of agreement cannot be expected typically, as the two models overlap and have identical regression parameters only if $\Lambda_0(t)$ is proportional to t. For these data, $\hat{\Lambda}_0$ estimates under either (9.16) or (9.21) adhere closely to a straight line through the origin, as was noted by Lin et al. (2000).

The mean model analyses in Table 9.3, although avoiding strong assumptions about the dependence of recurrence intensities on the preceding counting process for the patient, require the recurrence intensities for the individuals under follow-up at each time t to be representative of that for the study population, given covariates. Here this precludes censoring rates at time t from depending on the preceding recurrence time history, at least insofar as such history affects recurrence intensities. To examine the reasonableness of this rather stringent independent censorship assumption it is instructive to consider regression analyses of the censoring rate, while including aspects of $\{N(u); 0 \leq u < t\}$ as part of the predictive information.

Table 9.4   Regression Parameters in Cox Model Analysis of Censoring Rates as a Function of the History of Bladder Cancer Recurrences and Other Factors [a]

Treatment (0, placebo; 1, thiotepa)	No. Initial Tumors	Initial Tumor Size	No. Prior Recurrences During Follow-up	Recurrence Within Past Month
0.013 (0.236)	0.027 (0.075)	−0.090 (0.081)	−0.045 (0.063)	
0.120 (0.225)	−0.162 (0.074)	−0.080 (0.082)		1.017 (0.375)

[a] Estimated standard errors in parentheses.

Table 9.4 shows some Cox model analyses of the censoring rates. The first row of Table 9.4 allow the censoring rate to depend on baseline covariates Z_1, Z_2, and Z_3 as well as on a time-dependent covariate that takes value $N(t)$ at time t, employing the usual convention that failures precede censoring at a specific time point. This analysis does not suggest any important dependence of censoring rates on the number of preceding recurrences during follow-up or on any of the three baseline variables. The censoring rate at time t, could, however, depend on $\{N(u); 0 \le u \le t\}$ in many ways other than through $N(t)$. For example, the second analysis of Table 9.4 includes a time-dependent covariate that indicates whether or not a recurrence had taken place within the past month [i.e., $N(t) \ne N(t^-)$], along with baseline factors. This analysis indicates that patients who have experienced a recent recurrence have a highly significant estimated $\exp(1.017) = 2.76$-fold increase in their censoring rates. Since patients censored on this basis are very likely at elevated risk for future recurrences (given baseline variates) the marginal intensity rate estimates can be expected to be biased, and treatment effect parameter estimates may also be biased. For example, if a treatment gave rise to comparatively early recurrences which caused tumor-prone patients to be censored early, the treatment may appear to be beneficial for artifical reasons. Under these circumstances it is necessary to base recurrence rates and treatment effect analyses on recurrence rate models that condition on $[N(u); 0 \le u < t]$ at time t, or on a sufficient subset of this history to render plausible a corresponding independent censorship assumption.

9.4.4   Discussion of Marginal Failure Rate Modeling

Averaging over some aspects of the conditioning event in the overall intensity (9.2) is a very useful device for the analysis of recurrent event data. Unnecessary assumptions can often be avoided, and parameter interpretation can be enhanced in many contexts. Just as averaging over $[\tilde{N}(u); 0 \le u < t]$ may be natural for some purposes, it may also be useful to average over some aspects of the covariate history $X(t) = [x(u); 0 \le u < t]$ prior to time t. For example, one may wish to examine the relationship of failure rate to current, or most recent, covariate measurements $x(t^-)$ without having to condition on the earlier history of the covariate process. A marginalized failure rate Λ_m at time t can then be defined by

$$d\Lambda_m(t) = E\{d\tilde{N}(t)|[\tilde{N}(u); 0 \le u < t], x(t^-)\}.$$

Compared to (9.2), these conditioning events no longer have the nested structure that allows martingale convergence results to be applied, but the empirical process results noted in Sections 9.4.2 and 9.4.3 would apply to Cox model or AFT model specifications for such failure intensities. Marginalizing over $\{\tilde{N}(u); 0 \leq u < t\}$ also would give a failure rate process

$$d\Lambda_m(t) = E[d\tilde{N}(t)|x(t^-)],$$

under an independent censoring assumption that would allow censoring rates to depend on $x(t^-)$ but not on the preceding counting process history or on other aspects of the preceding covariate history.

The requirement of a comparatively strong independent censoring assumption is one of the principal limitations of the use of these types of marginal failure rate models. In many applications there may be a greater chance of dropout from the study, or of other forms of censorship, among study subjects who have experienced frequent or recent events. If the events are life threatening, censoring due to death at time t will typically not be independent of $\{\tilde{N}(u); 0 \leq u < t\}$. Under these circumstances some augmentation in the conditioning event in the disease rate specification is needed in order that an independent censorship assumption be plausible.

Censoring due to lack of study subject survival is perhaps best accommodated by including $D \geq t$ in the conditioning event for the failure rate definition, where D denotes the lifetime for the study subject. The resulting recurrence rate function is then interpreted as a type-specific intensity, as in the competing risk modeling (8.1), with death constituting another failure type.

There are a couple of approaches to accommodate censoring that depends on the preceding recurrence history for the subject. Wang et al. (2001) have augmented the conditioning event in (9.4) by including a hypothetical random effect or frailty variate that is assumed to affect the failure intensity multiplicatively and to arise independently from a certain distribution (e.g., gamma distribution rescaled to have mean 1) across study subjects. An independent censoring assumption, leading to

$$E[dN(t)|X(t), [Y(u); 0 \leq u < t], W] = Y(t)\,d\tilde{\Lambda}_m(t),$$

where $d\tilde{\Lambda}_m(t) = E[d\tilde{N}(t)|X(t), W]$, may then be plausible, where W is the frailty variate. Wang and colleagues consider a Cox model specification

$$d\tilde{\Lambda}_m(t) = Wd\Lambda_0(t)\exp[Z(t)'\beta], \tag{9.22}$$

which leads to an interesting decomposition of the counting process intensity into size and shape components, with size reflecting the frequency and shape the pattern of recurrent events. Note, however, that any such augmentation of the conditioning event in (9.4) complicates the corresponding mean process model. For example, under (9.22) the failure intensity $d\{N(t)|X(t), [Y(u); 0 \leq u < t]\}$ can be written

$$Y(t)d\Lambda_0(t)\exp[Z(t)'\beta]E\{W|X(t), [Y(u); 0 \leq u < t]\}.$$

If the censoring rate depends on W, the expectation appearing in this intensity is complicated to the point that regression parameter interpretation will derive primarily from (9.22) and will be conditional on the postulated frailty variate.

A direct approach to augmenting this conditioning event in (9.4) would retain some aspect of the prior history of the preceding counting process in the marginal disease rate model. An obvious choice in many settings would be to include the number of preceding events $N(t^-)$ in the conditioning event at time t.

9.5 CONDITIONING ON ASPECTS OF THE COUNTING PROCESS HISTORY

9.5.1 Relative Risk and AFT Models

Consider the intensity specification (9.7),

$$dΛ(t) = E[d\tilde{N}(t)|\tilde{N}(t^-), X(t)], \qquad (9.23)$$

which conditions on $N(t^-)$ but not on other aspects of preceding recurrent event history. Additionally, the covariate history could be reduced to $x(t^-)$ or to some other subset of the preceding covariate history $X(t)$, and $D \geq t$ could be added to the conditioning event. An independent censoring assumption may be plausible for (9.23), giving

$$E\{dN(t)|N(t^-), [Y(u); 0 \leq u < t], X(t)\} = Y(t)\, dΛ(t).$$

A Cox model for (9.23) could stratify on $s = N(t^-)$, giving

$$dΛ(t) = dΛ_{0s}(t) \exp[Z(t)'β_s]. \qquad (9.24)$$

This model is formally identical to (9.14) for the overall intensity, but modeling assumptions concerning the relationship of $dΛ(t)$ to other aspects of the preceding failure history have been avoided. Pepe and Cai (1993) have proposed this type of model for the times to first and subsequent infections among bone marrow transplant recipients.

There are, of course, many other choices of conditioning event that can be entertained in an attempt to yield a plausible independent censorship condition while retaining a useful parameter interpretation. For example, one could consider the intensities

$$dΛ(t) = E[d\tilde{N}(t)|\tilde{N}(t^-), v, X(t)] \qquad (9.25)$$

that condition not only on the number of preceding events but also on the sojourn time $v = t - T_{\tilde{N}(t^-)}$ since the most recent event, with $T_0 = 0$. Cox-type models could be specified for (9.25) with the dependence of intensity on $N(t^-)$ and v accommodated through regression modeling or stratification.

The application of Cox-type models for such partially marginalized intensities is formally identical to the application of (9.10) or (9.14) except that, once again, an empirical estimate of the score statistic variance of the form (9.17) is required, as follows from the results of Lin et al. (2000). Further development of these models, including consideration of corresponding mean process models, and consideration of model checking procedures would seem to be well worthwhile.

Accelerated failure time models could also be considered for (9.23) or (9.25) or for intensity processes that condition on other aspects of $\{N(u); 0 \le u < t\}$ at time t. Asymptotic distributional results for such application will follow from the work of Lin et al. (1998) and once again will require variance estimators of sandwich form.

One additional note: Many of the models discussed in this chapter can be extended to allow censoring schemes more general than simple right censorship. A key requirement is that the aspects of $\{\tilde{N}(u), 0 \le u < t\}$ that are conditioned in the intensity process under consideration are available at times at which the study subject is at risk $[Y(t) = 1]$. This condition will be met more readily for marginalized intensities than for the overall intensity process.

9.5.2 Bladder Tumor Data Revisited

Table 9.5 shows some additional analyses of the bladder tumor data under Cox model specifications for (9.25) and for

$$d\Lambda(t) = E\{d\tilde{N}(t) | \tilde{N}(t^-), v, \mathbf{1}[\tilde{N}(t^-) \ne \tilde{N}(t^- - 1)], X(t)\}. \qquad (9.26)$$

The first analysis of Table 9.5 actually omits $\tilde{N}(t^-)$ from the conditioning event while including the gap time v as an element of the modeled covariate. This gives results somewhat different from the Cox mean model analysis of Table 9.3, with clear evidence for an association with gap time and a standardized test statistic of -1.77 for treatment. The second analysis of the table includes these same regression variables, along with time-dependent stratification on $\tilde{N}(t^-)$. This model provides an attempt to accommodate more fully the dependence of censorship on

Table 9.5   Further Analyses of Bladder Tumor Recurrence Data [a]

Regression Model	Treatment (0, placebo; 1, thiotepa)	Number Initial Tumors	Initial Tumor Size	Gap Time, v	Recurrence in Previous Month
Cox model for (9.25)	−0.362	0.154	−0.014	−0.063	
	(0.204)	(0.047)	(0.061)	(0.011)	
Cox model for (9.25) stratified on $\tilde{N}(t^-)$	−0.370	0.119	−0.012	−0.059	
	(0.177)	(0.047)	(0.061)	(0.025)	
Cox model for (9.26)	−0.345	0.161	−0.016	−0.076	−1.405
	(0.207)	(0.048)	(0.061)	(0.012)	(0.498)
Cox model for (9.26) stratified on $\tilde{N}(t^-)$	−0.346	0.122	−0.017	−0.082	−1.387
	(0.185)	(0.047)	(0.061)	(0.027)	(0.579)

[a] Estimated standard errors in parentheses.

the pattern of preceding recurrences, noted previously. The ratio of treatment parameter estimate to its sandwich-form standard deviation estimate is -2.09, providing evidence for a benefit from thiotepa.

The final two analyses of Table 9.5 correspond to the first two, except that an indicator variable for whether the patient was observed to have a recurrence within the previous month is added to the model. Note that there is evidence of lower recurrence risk among patients having a recurrence recorded within the previous month, and that evidence for a treatment effect is slightly weakened. Overall, (9.25) and (9.26) appear to provide good descriptions of these data. Whether or not there is stratification on $\tilde{N}(t)$, these analysis are consistent in suggesting a recurrence relative risk of about 0.7 associated with thiotepa, with a corresponding 95% confidence interval of about (0.5, 1.0).

BIBLIOGRAPHIC NOTES

There is a long history of point process modeling and estimation, with emphasis on Poisson processes and renewal processes (e.g., Cox and Lewis, 1966; Snyder, 1975; Cox and Isham, 1980; Andersen et al., 1993). Cox (1973) and Andersen and Gill (1982), respectively, discuss renewal and Poisson processes that are modulated by regression variables, and Andersen and Gill give a thorough account of the asymptotics for modulated Poisson processes using martingale methods. Prentice et al. (1981) consider these same types of regression models with modeling or stratification on the prior failure history for the individual. Special cases of these models were proposed earlier by Gail et al. (1980). Other authors (Aalen and Husebye, 1991; Lawless, 1987) have included random effects to extend the applicability of Poisson and renewal process assumptions.

The emphases on mean models for recurrent events dates back to Nelson (1988), with variance estimation for the mean number of failures given in Nelson (1995). Lawless and Nadeau (1995) extend these models to include a regression function that is multiplicative on the mean, and they give asymptotic results for the regression parameter and baseline mean function when failure times are measured in discrete units. Lin et al. (2000) provide corresponding distribution theory for Cox-type mean models with absolutely continuous failure times. Lin et al. (1998) consider an accelerated failure time form of mean model and develop corresponding asymptotic distribution theory. Ghosh and Lin (2000) include death in the conditioning event of mean models. Wang et al. (2001) introduce a multiplicative random effect into mean function Cox models for recurrent events, with the aim of relaxing the independent censoring assumption. There has also been some work on estimation of the distribution of the gap time between successive events under mean models (e.g., Wang and Chang, 1999; Lin et al., 1999). Lin et al. (2001) have extended the multiplicative mean modeling to a class of transformation models, with empirical process theory used to derive asymptotic results.

Pepe and Cai (1993) consider intensity models that condition on the number of preceding failures for the individual but not on other aspects of the preceding failure history. Note that the asymptotic distribution results developed by Lin et al. (2000) should be used in conjunction with Cox models for these intensities. This approach can be contrasted with that of Wei et al. (1989), who consider the rate of the joint event of failure at a specified time t and a specified value $N(t)$, without conditioning on any aspect of the prior failure history. These authors employ Cox models for these failure rates, while Lin and Wei (1992) consider accelerated failure time models for these rates. This form of modeling has been criticized in the context of recurrent events because a study subject is considered at risk for an sth failure $(s > 1)$ at time t even if the subject has yet to experience $s - 1$ failures. This criticism does not apply in the context of the correlated failure times discussed in Chapter 10, where this form of marginal modeling is discussed in greater detail.

EXERCISES AND COMPLEMENTS

9.1 Consider the intensity model (9.10) with a fixed or external time-dependent covariate satisfying (9.6). Show that in the absence of censoring the expected number of failures on an individual by time t is $\Lambda(t) = \int_0^t d\Lambda(u)$, so that (9.10) has a mean model interpretation. Also show that

$$P[\tilde{N}(t) = s | X(t)] = \Lambda(t)^s \exp[-\Lambda(t)]/s!, \qquad s = 0, 1, 2, \ldots.$$

9.2 Consider the mean model (9.16) with fixed or external time-dependent covariates and define $\hat{\Lambda}_0(t) = \int_0^t \sum_1^n dN_i(u)/\sum_1^n Y_i(u) \exp[Z_i(u)'\beta_0]$ with the convention that $0/0$ is zero. Show that the expectation of $\hat{\Lambda}_0(t) - \int_0^t I[Y.(u) > 0]\Lambda_0\,(du)$ is zero, and derive an expression for $\int_0^t \mathrm{var}[d\hat{\Lambda}_0(t) \,|\, X(t), Y(u), 0 \le u < t]\, dt$ assuming that $dN_i(t) \le 1$.

9.3 Consider the intensity model (9.10) with time-varying covariate $x(u) = x_1$ if $u < t^*$ and x_2 if $u \ge t^*$ for fixed t^* and as usual with $X(t) = [x(u), u < t]$. Consider an intensity model (Section 9.4.4)

$$d\Lambda_m(t) = E\{d\tilde{N}(t) | [\tilde{N}(u), 0 \le u < t], x(t^-)\}$$
$$= d\Lambda_0(t) \exp[Z(t)'\beta],$$

where $Z(t) = x(t^-)$, and with censoring that is independent given $[\tilde{N}(u), 0 \le u < t]$ and $x(t^-)$. Show that the estimating function $U(\beta)$ in (9.12) is unbiased. Discuss the circumstances under which the increments to $U(\beta)$ at distinct failure times are uncorrelated. Can you propose a variance estimator for $U(\beta_0)$, where β_0 is the true β value?

9.4 Consider the failure rate model

$$d\Lambda_m(t) = E[d\tilde{N}(t) | Z, W]$$
$$= W\, d\Lambda_0(t) \exp(Z'\beta),$$

where W is a frailty variate introduced to allow a plausible independent censoring assumption, and Z is a fixed covariate. Consider a follow-up time τ in the observation period for the point processes, so that $P(C > \tau|Z) > 0$. Derive a nonparametric estimator of the shape function

$$H(t) = \Lambda_0(t)/\Lambda_0(\tau)$$

in the interval $[0, \tau]$. Also show that the expectation of $N[CH(C)^{-1}]$ is $\Lambda_0(\tau)\exp[X'\beta]$, and hence derive estimating functions for Λ_0 and β under this model (Wang et al., 2001).

9.5 Consider an intensity rate model (9.2) of the form

$$d\Lambda(t) = d\Lambda_0(t)\exp\left[\beta_1\tilde{N}(t^-) + \beta_2 Z\right],$$

where the indicator variable Z takes value 1 for individuals receiving an active treatment, and value zero for control individuals. Suppose also that the number of failures $\tilde{N}(t)$ in treatment group Z is a Poisson variate with mean $\mu_Z t$ for $Z = 0, 1$.
Show that the marginal failure rates, $d\Lambda_m(t)$, in (9.4) satisfy

$$d\Lambda_m(t) = d\Lambda_0(t)e^{-\mu_Z t(1-e^{\beta_1})}e^{\beta_2 Z},$$

so that the failure rate ratio for the active $(Z = 1)$ versus control $(Z = 0)$ treatment groups at time t is

$$e^{-t(1-e^{\beta_1})(\mu_1-\mu_0)}e^{\beta_2}.$$

Comment on the special cases $\beta_1 = 0$ and $\beta_2 = 0$ concerning the interpretation of treatment effects and concerning the utility of modeling and estimation of Λ and Λ_m, respectively, for the elucidation of treatment effects.

CHAPTER 10

Analysis of Correlated Failure Time Data

10.1 INTRODUCTION

In Chapter 9 we considered the analysis of failure time data when individual study subjects may experience recurrent failures. This chapter focuses on the analysis of correlated or clustered failure time data. Such data may arise because study subjects having univariate failure times may be grouped in a manner that leads to dependencies within groups, or because individuals are simultaneously at risk for, and are followed to observe, failures of multiple types.

The former type of correlated failure time data arises in follow-up studies on family members in genetic epidemiology or in group randomized intervention trials. For example, Peterson et al. (2000) reported the results of a cigarette smoking prevention trial among 8388 students enrolled in any of 40 school districts in Washington State which were randomly assigned to intervention or control status. Beginning in the third grade, students in intervention schools were exposed to a social influences–based smoking prevention educational program, whereas those in control school districts received their usual health education instruction. Students were followed through two years beyond twelfth grade to observe the ages at smoking initiation or censoring. Such ages may be correlated within school districts, for example, on the basis of shared characteristics or exposures in particular geographic areas. A second example, where dependencies are substantial, is provided by the Australian twin study (Duffy et al., 1990). The strength of dependency between ages at appendectomy were compared between monozygotic and dizygotic twins to provide evidence concerning a genetic component to disease etiology. Some analyses of these data are described in Section 10.5.

Correlated failure time data on individuals also arise with some frequency. Such data are similar to the competing risk data discussed in Chapter 8, except that the observation period for an individual extends beyond the occurrence of a first type of event to allow the entire multivariate failure time vector to be observed, aside from censoring. For example, in the dietary modification component of the Women's

Health Initiative clinical trial (Women's Health Initiative Study Group, 1998), over 48,000 postmenopausal American women were randomized to either a low-fat eating pattern or to control status. Breast cancer and colorectal cancer incidence are the designated primary outcomes in this intervention trial, with coronary heart disease occurrence as a secondary outcome. Participating women continue to be followed after a breast cancer diagnosis, for example, for the occurrence of colorectal cancer, heart disease, and a range of other clinical outcomes. In contrast to the competing risk situation, data on these women can be used to examine dependencies among the occurrence times of various disease events among women who survive an initial disease event or events. In this context, most interest resides in the relationship between treatments or interventions and the risk of specific diseases. Such relationships can often be conveniently studied through regression models for the marginal hazard rates for each disease, as discussed in Section 10.2.

In such contexts there is usually some interest in the strength and nature of dependencies among the components of a multivariate failure time vector. In some settings, such as the twin study mentioned above, or other family studies in genetic epidemiology, elucidation of such dependencies may be a central objective. The presence of right censorship, however, complicates these assessments. Measures of pairwise dependency derive from nonparametric estimates of the bivariate survivor function. In Section 10.3 we consider some representations of the joint survivor function for two failure time variates, along with nonparametric survivor function estimators that arise by inserting certain hazard function estimators into these representations. In Section 10.4 we then discusses two possible choices of nonparametric summary measures of pairwise dependency. These methods are illustrated in Section 10.5 by application to Australian twin study data. The remainder of the chapter is more specialized. In Section 10.6 we overview the various approaches that have been considered toward identifying a nonparametric estimator of the bivariate survivor function having desirable properties. In Section 10.7 we consider survivor function representation and estimation in dimensions higher than 2.

10.2 REGRESSION MODELS FOR CORRELATED FAILURE TIME DATA

10.2.1 Multivariate Hazard Functions and Independent Censoring

Consider failure times $\tilde{T}_1, \ldots, \tilde{T}_m$ that may be correlated and suppose that failure time $\tilde{T}_i$ has an associated covariate vector x_i. The joint survivor function

$$F(t_1, \ldots, t_m; x) = P(\tilde{T}_1 > t_1, \ldots, \tilde{T}_m > t_m; x) \tag{10.1}$$

then characterizes the relationship between $x = (x_1', \ldots, x_m')'$ and $(\tilde{T}_1, \ldots, \tilde{T}_m)$. Hazard functions of orders $1, \ldots, m$ can be defined based on (10.1). Specifically, marginal (first-order) hazard functions are defined by

$$\Lambda_j(dt_j; x) = P(t_j \leq \tilde{T}_j < t_j + dt_j | \tilde{T}_j \geq t_j, x), \qquad j = 1, \ldots, m \tag{10.2}$$

while second-order hazard functions are defined by

$$\Lambda_{jk}(dt_j, dt_k; x) = P(t_j \leq \tilde{T}_j < t_j + dt_j, t_k \leq \tilde{T}_k < t_k + dt_k | \tilde{T}_j \geq t_j, \tilde{T}_k \geq t_k, x) \quad (10.3)$$

for $j \neq k, j, k \in 1, \ldots, m$. As elaborated in Section 10.7, the hazard functions of orders 1 up to m specify (10.1) completely so that regression modeling can focus on regression models for these hazard functions or on subsets of these. Note that we have changed the differential notation in this chapter only, writing $\Lambda_j(dt_j, x)$ rather than $d\Lambda_j(t_j, x)$, and $\Lambda_{jk}(dt_j, dt_k; x)$ rather than the more cumbersome and ambiguous $d^2\Lambda_{jk}(t_j, t_k; x)$.

More generally, a failure time variate $\tilde{T}_j$ may be accompanied by a covariate process x_j that evolves with follow-up. Define $X_j(t) = \{x_j(u): 0 \leq u < t\}$. Each of the hazard functions can be extended to include conditioning not only on baseline covariates for the set of failure times, but also the histories up to the hazard function arguments for the pertinent evolving covariates. Specifically, one can define marginal hazard functions by

$$\Lambda_j[dt_j, X_j(t_j)], \qquad j = 1, \ldots, m,$$

second-order hazard functions by

$$\Lambda_{jk}[dt_j, dt_k; X_j(t_j), X_k(t_k)],$$

and qth-order hazard functions by

$$\Lambda_{j_1 \ldots j_q}[dt_{j_1}, \ldots, dt_{j_q}; X_{j_1}(t_{j_1}), \ldots, X_{j_q}(t_{j_q})]$$

for all subsets $\{j_1, \ldots, j_q\}$ of $\{1, \ldots, m\}$, where each $X_j(t_j)$ includes all baseline covariate data for the m individuals, along with data prior to t_j for any covariates that evolve with $\tilde{T}_j$. Note that if $\tilde{T}_1, \ldots, \tilde{T}_m$ are failure time variates on a single individual the interpretation of second- and higher-order hazard rates may be difficult unless the covariates are restricted to be external (see Section 6.3.1).

Each of $\tilde{T}_1, \ldots, \tilde{T}_m$ may be subject to right censoring. Let $C_1, \ldots, C_m$ denote corresponding potential censoring times, let $T_j = \tilde{T}_j \wedge C_j$, and let $Y_j(t) = P(T_j > t)$, $j = 1, \ldots, m$. The censoring will be said to be independent if

$$P\{N_{j\ell}(dt_{j\ell}) = 1, \ \ell = 1, \ldots, q | Y_{j\ell}(u), 0 \leq u < t_{j\ell}, X_{j\ell}(t_{j\ell}), \ell = 1, \ldots, q\}$$

$$= \prod_{\ell=1}^{q} Y_{j\ell}(t_{j\ell}) \Lambda_{j_1 \ldots j_q}[dt_{j_1}, \ldots, dt_{j_q}; X_{j\ell}(t_{j\ell}), \ell = 1, \ldots, q]$$

$$(10.4)$$

for all $\{j_1, \ldots, j_q\} \subset \{1, \ldots, m\}$ and $t_{j_1}, \ldots, t_{j_q}$. Special cases of independent censoring include prespecified potential censoring times, and random censoring where

$(\tilde{T}_1, \ldots, \tilde{T}_m)$ and $(C_1, \ldots, C_m)$ are independent random variables, the latter having a joint survivor function $G(t_1, \ldots, t_m; x) = P(C_1 > t_1, \ldots, C_m > t_m; x)$, given a time-independent covariate x. Note that (10.4) is more restrictive than the independent censoring definition (9.3) for recurrent events in that the hazard rate in (9.3) is allowed to depend on the number and timing of preceding failures on the individual.

10.2.2 Regression on Marginal Hazard Functions

The same types of regression models considered in previous chapters for univariate hazard functions can be entertained for multivariate hazard functions up to a specified order. For example, one can consider Cox-type models

$$\Lambda_j[dt_j; X_j(t_j)] = \Lambda_{0j}(dt_j) \exp[Z_j(t_j)'\beta], \qquad j = 1, \ldots, m \qquad (10.5)$$

for the marginal hazard functions with $Z_j(t)$ a fixed-length vector formed from $X_j(t)$ and t, without specifying regression models for the second- or higher-order hazard functions. To estimate the relative risk parameter in (10.5) with absolutely continuous failure times one can make a working independence assumption concerning the m failure time variates, given corresponding covariate histories, and consider the likelihood function

$$L(\beta) = \prod_{j=1}^{m} \prod_{i=1}^{n} \left[\frac{\exp[Z_{ji}(t_{ji})'\beta]}{\sum_{\ell=1}^{n} Y_{j\ell}(t_{ji}) \exp[Z_{j\ell}(t_{ji})'\beta]} \right]^{\delta_{ji}}, \qquad (10.6)$$

where $(t_{1i}, \ldots, t_{mi})$ and $(\delta_{1i}, \ldots, \delta_{mi})$, $i = 1, \ldots, n$ denote the minimum of failure and censoring times and noncensoring indicators for n observations under independent censorship, and $Y_{j\ell}(t_{ji})$ is an at-risk indicator process for the jth component of the ℓth observation at time t_{ji}. Note that some components of $(\tilde{T}_1, \ldots, \tilde{T}_m)$ can be allowed to be missing by setting the corresponding at risk processes to be identically zero.

Expression (10.6) is not a partial likelihood function in general, unless the elements of $(\tilde{T}_1, \ldots, \tilde{T}_m)$ are independent given covariates and is sometimes referred to as a pseudolikelihood function. However, $\hat{\beta}$ that satisfies the pseudolikelihood score equation $\partial \log L(\beta)/\partial\beta = 0$ is generally consistent for β under independent and identically distributed conditions on the failure, at risk, and covariate processes, and $n^{1/2}(\hat{\beta} - \beta)$ generally has an asymptotic normal distribution with mean zero and with variance consistently estimated by $nI(\hat{\beta})^{-1}\hat{\Sigma}I(\hat{\beta})^{-1}$, where $I(\hat{\beta}) = -\partial^2 \log L(\hat{\beta})/\partial\hat{\beta}\partial\hat{\beta}'$ and $\hat{\Sigma}$ is an empirical estimator of the variance of $\partial \log L(\beta)/\partial\beta$ (Wei et al., 1989). Note that $\hat{\beta}$ and $I(\hat{\beta})$ can be obtained from standard Cox model software just by stretching the n m-dimensional data elements into a nm one-dimensional data elements, with the position $(1, 2, \ldots, m)$ in the original multivariate vector used to define m strata. Analogous to the score statistic variance estimator in previous chapters [e.g., (6.28) and (9.17)], one can write

$$\hat{\Sigma} = \hat{\Sigma}(\hat{\beta}) = \sum_{i=1}^{n} \hat{U}_{\cdot i}^{\otimes 2},$$

where $\hat{U}_{\cdot i} = \int_0^\infty \sum_{j=1}^m [Z_{ji}(t) - \mathscr{E}(\hat{\beta}, t)] \hat{M}_{ji}(dt)$, $\hat{M}_{ji}(dt) = N_{ji}(dt) - Y_{ji}(t) \exp[Z_{ji}(t)'$
$\beta] \hat{\Lambda}_{0j}(dt)$, $\hat{\Lambda}_{0j}(dt) = \sum_{i=1}^n N_{ji}(dt) S_j^{(0)}(\hat{\beta}, t)^{-1}$, $\mathscr{E}_j(\beta, t) = S_j^{(1)}(\beta, t) / S_j^{(0)}(\beta, t)$, and
where $S_j^{(q)}(\beta, t) = n^{-1} \sum_{\ell=1}^n Y_{j\ell}(t) Z_{j\ell}(t)^q \exp[Z_{j\ell}(t)'\beta]$ for $q = 0, 1$.

The estimating function for β can be written

$$\sum_{i=1}^n \int_0^\infty \sum_{j=1}^m [Z_{ji}(t) - \mathscr{E}_j(\beta, t)] N_{ji}(dt) = \sum_{i=1}^n \int_0^\infty \sum_{j=1}^m [Z_{ji}(t) - \mathscr{E}_j(\beta, t)] M_{ji}(dt).$$

In that the marginal martingales M_{ji}, $j = 1, \dots, m$ are typically correlated, one could consider introducing a weight matrix into this estimating function in an attempt to improve the efficiency of the regression parameter estimate. Doing so, however, may not be worthwhile unless dependencies among the M_{ji}'s are strong, particularly if there is heavy censorship (Cai and Prentice, 1995).

In some applications it may be appropriate to restrict the baseline hazard functions in (10.5) to be identical. A pseudolikelihood for β of the form (10.6) is then given by

$$L(\beta) = \prod_{j=1}^m \prod_{i=1}^n \left[\frac{\exp[Z_{ji}(t_{ji})'\beta]}{\sum_{k=1}^m \sum_{\ell=1}^n Y_{k\ell}(t_{ji}) \exp[Z_{k\ell}(t_{ji})'\beta]} \right]^{\delta_{ij}}.$$

Once again, $\hat{\beta}$ that solves $\partial \log L(\beta)/\partial \beta$ will generally be consistent for β and $n^{1/2}(\hat{\beta} - \beta)$ will have an asymptotic normal distribution with mean zero and sandwich form variance estimator (Lee et al., 1992). Under this model there is potential for more substantial efficiency gain through introducing a weight function into the estimating equation, especially if the cluster size (m) is large (Cai and Prentice, 1997).

These methods could be extended to include discrete and mixed discrete and continuous failure times by allowing the Λ_{0j} functions in (10.5) to include mass points. The same estimation procedures outlined above then continue to apply under independent and identically distributed assumptions on the counting, at risk and covariate processes for the n study subjects.

10.2.3 Regression on Marginal and Higher-Order Hazard Functions

Suppose now that one wishes to assess pairwise or higher-order dependencies among failure times $\tilde{T}_1, \dots, \tilde{T}_m$ while controlling for covariates. Some authors have addressed this problem by introducing a random effect, or frailty, variate for the cluster and by assuming the m failure times to be independent given the value of the frailty variate. For example, Nielsen et al. (1992) specify a hazard function for the jth component of the ith cluster by

$$\Lambda_j[dt_j; X_i(t_j), \alpha_i] = \alpha_i \Lambda_{0j}(dt_j) \exp[Z_j(t_j)'\beta], \tag{10.7}$$

for $j = 1, \dots, m$ and $i = 1, \dots, n$, where the unobserved frailties α_i are assumed to be independent and identically distributed across clusters and to arise from a gamma distribution that has been rescaled to have mean one and variance θ. An expectation-maximization algorithm was proposed for the estimation of (Λ_{0j}, β) in conjunction with likelihood profiling for the estimation of θ.

Although this frailty model approach has an appeal in that random effects have been introduced to allow for correlation in many other statistical modeling contexts, there are also some drawbacks. For example, the marginal hazard rates from (10.7) arise by taking an expectation over the distribution of α_i given $[T \geq t, X(t)]$; that is,

$$\Lambda_j[dt_j; X(t_j)] = \Lambda_{0j}(t_j) \exp[Z_j(t_j)'\beta] E[\alpha_i | T_j \geq t_j, X_j(t_j)],$$

and this expectation is typically a complicated function of $\{t_j, X_j(t_j), \Lambda_{0j}, \beta\}$. Hence the regression effects appear to have useful interpretation only conditional on the hypothetical frailty variable via (10.7). Also, (10.7) involves the strong assumption that the frailty variable affects the hazard function in a multiplicative fashion that is independent of time. As a result, the parameter θ characterizes all second- and higher-order dependencies among the failure times. This level of flexibility may be unacceptable if clusters are comprised of several different types of study subjects: for example, of family members having several different relationships to each other in an epidemiologic pedigree cohort study.

Hougaard (1986) has criticized frailty models in which the mixing distribution has a finite mean in that data on $\tilde{T}_1$ alone, for example, could then be used for estimation of the frailty parameter(s), which are intended to characterize dependencies among the failure times of cluster members. He proposed the use of positive stable frailty distributions that have an infinite mean and that preserve a Cox model form for the marginal hazard functions from (10.7) for time-independent modeled regression variables.

If $m = 2$, the Clayton (1978) model $F(t_1, t_2) = [F_1(t_1)^{-\theta} + F_2(t_2)^{-\theta} - 1]^{-1/\theta}$, $\theta \in [0, \infty)$ can be derived by integrating out a gamma-distributed frailty in (10.7) without covariates and reparameterizing. For a general cluster size m, this integration gives a joint survivor function

$$F(t_1, \ldots, t_m) = [F_1(t_1)^{-\theta} + \cdots + F_m(t_m)^{-\theta} - (m - 1)]^{-1/\theta}.$$

This model form suggests a regression extension

$$F(t_1, \ldots, t_m; x) = [F_1(t_1; x)^{-\theta} + \cdots + F_m(t_m; x)^{-\theta} - (m - 1)]^{-1/\theta}, \quad (10.8)$$

where, for example, Cox models of the form (10.7) could be specified to obtain the marginal survivor functions $F_j(t_j; x)$, $j = 1, \ldots, m$, where $x = (x_1, \ldots, x_m)$. Shih and Louis (1995) have proposed a two-stage procedure for estimation under this model with bivariate ($m = 2$) data. At the first stage, marginal Cox models are applied as in Section 10.2.2. The resulting parameter estimates $\{\Lambda_{0j}, \beta\}$ are then held fixed while θ is estimated using a likelihood-based estimating function. See also related work by Clayton and Cuzick (1985) and Glidden (2000).

Prentice and Hsu (1997) take a different approach to allowing pairwise dependencies among the failure times in a cluster while fitting Cox models for the marginal distributions. Specifically, using the theory of estimating equations for mean and variance parameters for correlated response data, they developed joint estimating equations for marginal hazard ratio parameters and for Clayton model pairwise dependence parameters. Relative to estimation under (10.8), this approach allows the possibility of different Clayton model parameter θ for each pair of failure times within a cluster, allows these parameters to depend on regression variables, and avoids assumptions concerning the form of higher-order hazard functions. The approach is still somewhat restrictive, however, in assuming each cross-ratio function

$$F(dt_j, dt_k; x)F(t_j^-, t_k^-; x)/[F(dt_j, t_k^-; x)F(t_j^-, dt_k; x)]$$

(e.g., Oakes, 1989) to be independent of (t_j, t_k) given covariates, $j, k = 1, \ldots, m$.

10.3 REPRESENTATION AND ESTIMATION OF THE BIVARIATE SURVIVOR FUNCTION

In Section 10.2.3 we described some proposed uses of semiparametric models for assessing dependencies among clustered failure times. Nonparametric assessments of such dependencies can derive from nonparametric estimators of the joint survivor function

$$F(t_1, t_2) = P(\tilde{T}_1 > t_1, \tilde{T}_2 > t_2), \tag{10.9}$$

excluding regression variables for the moment.

As we have seen in Section 1.4, the Kaplan–Meier estimator of a univariate survivor function can be developed by a nonparametric maximum likelihood estimation procedure (1.12), or by substituting the Nelson–Aalen estimator of the cumulative hazard estimator into a product integral representation (1.8) of the survivor function. Unfortunately, these approaches do not readily extend to two or higher dimensions. Specifically, a nonparametric maximum likelihood procedure for estimating (10.9) under independent censoring may have severe uniqueness problems. Also, the bivariate cumulative hazard function

$$\Lambda(t_1, t_2) = \int_0^{t_1} \int_0^{t_2} \Lambda(du_1, du_2) \tag{10.10}$$

does not determine the survivor function over a follow-up region $[0, t_1] \times [0, t_2]$. However, estimators of (10.10) in conjunction with estimators of the marginal cumulative hazard functions Λ_1 and Λ_2 do lead to convenient bivariate survivor function estimators having attractive properties.

In this section we consider a representation of (10.9) as a function of Λ, Λ_1, and Λ_2, and corresponding bivariate survivor function estimators that arise by inserting

Kaplan–Meier estimators of Λ_1 and Λ_2 along with some possible choices of nonparametric estimators of Λ in this representation. The resulting survivor function estimators are readily calculated and are adequate for most purposes. However, they are not in general nonparametric efficient, a topic that we return to in Section 10.6.

It is useful to extend our notation to allow $\tilde{T}_1$ and $\tilde{T}_2$ to have both discrete and continuous elements. Denote

$$F(dt_1, dt_2) = \begin{cases} F^{(11)}(t_1, t_2)\, dt_1\, dt_2 & \text{if } F \text{ is absolutely continuous in both} \\ & \text{components at } (t_1, t_2) \\ F^{(10)}(t_1, \Delta t_2)\, dt_1 & \text{if } F \text{ is continuous in its first, but not its} \\ & \text{second argument at } (t_1, t_2) \\ F^{(01)}(\Delta t_1, t_2)\, dt_2 & \text{if } F \text{ is continuous in its second, but not} \\ & \text{its first argument at } (t_1, t_2) \\ F(\Delta t_1, \Delta t_2) & \text{if } (t_1, t_2) \text{ is a mass point of } F \end{cases}$$

where $F^{(10)}$ denotes the partial derivative of F with respect to t_1, $F^{(11)}$ is the cross partial with respect to t_1 and t_2, and

$$F(\Delta t_1, \Delta t_2) = F(t_1^-, t_2^-) - F(t_1^-, t_2) - F(t_1, t_2^-) + F(t_1, t_2). \tag{10.11}$$

Hence F is given by the Stieltjes integral

$$F(t_1, t_2) = \int_{t_1}^{\infty} \int_{t_2}^{\infty} F(du_1, du_2)$$

for discrete, continuous, or mixed failure time variates. The bivariate hazard function is determined by

$$\Lambda(dt_1, dt_2) = F(dt_1, dt_2) / F(t_1^-, t_2^-). \tag{10.12}$$

As noted above, the survival probability $F(t_1, t_2)$ is not determined by $\Lambda(du_1, du_2)$ at times $u_1 \leq t_1$ and $u_2 \leq t_2$. Specifically, from (10.12),

$$\int_0^{t_1} \int_0^{t_2} F(du_1, du_2) = \int_0^{t_1} \int_0^{t_2} F(u_1^-, u_2^-)\Lambda(du_1, du_2),$$

which gives

$$F(t_1, t_2) = \Psi(t_1, t_2) + \int_0^{t_1} \int_0^{t_2} F(u_1^-, u_2^-)\Lambda(du_1, du_2), \tag{10.13}$$

where $\Psi(t_1, t_2) = [F_1(t_1) + F_2(t_2) - 1]$. Now (10.13) is an nonhomogeneous Volterra equation that has a unique Peano series solution that expresses $F(t_1, t_2)$ as a function of the hazard rates $\Lambda(du_1, du_2)$ for $u_1 \leq t_1$, and $u_2 \leq t_2$ and the marginal survivor functions F_1 and F_2, or equivalently, marginal hazard functions Λ_1 and Λ_2 at times $u_1 \leq t_1$ and $u_2 \leq t_2$.

In fact, it is not necessary to use this rather complex representation for F to obtain corresponding nonparametric estimators: Suppose that Kaplan–Meier estimators of F_1 and F_2 are specified so that positive mass is assigned only at grid points formed by uncensored T_1 and T_2 values within the risk region (where one or more pairs are being followed for failure on both components), and along half lines beyond the risk region. Equating $F(\Delta t_1, \Delta t_2)$ from (10.11) with $F(\Delta t_1, \Delta t_2) = F(t_1^-, t_2^-)\Lambda(\Delta t_1, \Delta t_2)$ from (10.12) at any such grid point gives

$$F(t_1, t_2) = F(t_1^-, t_2) + F(t_1, t_2^-) - F(t_1^-, t_2^-)[1 - \Lambda(\Delta t_1, \Delta t_2)]. \qquad (10.14)$$

This expression gives a simple recursive procedure for estimating $F(t_1, t_2)$ at all failure time grid points in the risk region starting with Kaplan–Meier estimates of $F(t_1, 0) = F_1(t_1)$ and $F(0, t_2) = F_2(t_2)$, given an estimate of the bivariate hazard rate $\Lambda(\Delta t_1, \Delta t_2)$ at such grid points. Such calculation specifies the bivariate survivor function estimator throughout the risk region, since the estimator is a step function on rectangles formed by the uncensored data grid points.

Survivor function estimators of this type that have been proposed all use Kaplan–Meier marginal survivor functions $\hat{F}_1$ and $\hat{F}_2$, in conjunction with various choices for an estimator $\hat{\Lambda}$ of the cumulative hazard function. The simplest such estimator, due to Peter Bickel (e.g., Dabrowska, 1988), inserts a simple empirical estimator for Λ, so that $\hat{\Lambda}(\Delta t_1, \Delta t_2)$ is simply the ratio of the number of double failures at (t_1, t_2) to the number of pairs at risk at that point; that is,

$$\hat{\Lambda}(\Delta t_1, \Delta t_2) = \#(T_1 = t_2, T_2 = t_2, \delta_1 = \delta_2 = 1)/\#(T_1 \geq t_1, T_2 \geq t_2). \quad (10.15)$$

The resulting survivor function estimator has reasonably good performance in moderate-sized samples, but is rather inefficient, due to a typical poor correspondence between $\hat{\Lambda}$ and the Kaplan–Meier marginals $\hat{F}_1$ and $\hat{F}_2$. Other estimators, due to Dabrowska (1988) and Prentice and Cai (1992), aim to improve this correspondence by estimating double failure hazard rates in a manner that acknowledges both the empirical rate (10.15) and the amount of mass assigned by the Kaplan–Meier marginals along $\tilde{T}_1 = t_1$ and along $\tilde{T}_2 = t_2$ that remains to be assigned at (t_1, t_2). These estimators were developed using certain product integral and Peano series representations for the ratio $F(t_1, t_2)/[F_1(t_1)F_2(t_2)]$. Using (10.13), however, they can be recast as arising from Kaplan–Meier marginal survivor functions along with certain double failure hazard function estimators: For the Prentice–Cai estimator the double failure hazard rate estimator is given by

$$\hat{\Lambda}(\Delta t_1, \Delta t_2) + \hat{L}_1(\Delta t_1, 0)[\hat{L}_2(t_1^-, \Delta t_2) - \hat{\Lambda}_2(t_1^-, \Delta t_2)]$$
$$+ \hat{L}_2(0, \Delta t_2)[\hat{L}_1(\Delta t_1, t_2^-) - \hat{\Lambda}_1(\Delta t_1, t_2^-)], \qquad (10.16)$$

where $\hat{L}_1(\Delta t_1, t_2^-) = -\hat{F}(\Delta t_1, t_2^-)/\hat{F}(t_1^-, t_2^-)$ and $\hat{\Lambda}_1(\Delta t_1, t_2^-) = \#\{T_1 = t_1, \delta_1 = 1, T_2 \geq t_2\}/\#\{T_1 \geq t_1, T_2 \geq t_2\}$ with corresponding specifications for $\hat{L}_2$ and $\hat{\Lambda}_2$.

The Dabrowska double failure hazard can be expressed in terms of these same quantities as

$$
\hat{L}_1(\Delta t_1, t_2^-)\hat{L}_2(t_1^-, \Delta t_2) + \frac{[1 - \hat{L}_1(\Delta t_1, t_2^-)][1 - \hat{L}_2(t_1^-, \Delta t_2)]}{[1 - \hat{\Lambda}_1(\Delta t_1, t_2^-)][1 - \hat{\Lambda}_2(t_1^-, \Delta t_2)]}[\hat{\Lambda}(\Delta t_1, \Delta t_2)
$$

$$
- \hat{\Lambda}_1(\Delta t_1, t_2^-)\hat{\Lambda}_2(t_1^-, \Delta t_2)].
\tag{10.17}
$$

We return to a discussion of the Bickel, Dabrowska, and Prentice–Cai estimators in Section 10.6. For now, however, it can be noted that all three survivor function estimators have been shown to be uniformly consistent and asymptotically Gaussian over a follow-up region $[0, \tau_1] \times [0, \tau_2]$ such that $P(T_1 \geq \tau_1, T_2 \geq \tau_2) > 0$ under random censorship, and bootstrap procedures (e.g., for variance estimation) have been shown to apply. A nice development of these properties is given by Gill et al. (1995).

10.4   PAIRWISE DEPENDENCY ESTIMATION

There are several standard nonparametric measures of the strength of dependency between pairs of random variables; for example, Spearman's rho, Kendall's tau, and the usual Pearson correlation are commonly used. The presence of right censoring implies that it may only be possible to assess dependency over a subset of the space of possible $(\tilde{T}_1, \tilde{T}_2)$ values. Accordingly, we consider the estimation of pairwise dependency over $[0, t_1] \times [0, t_2]$, where $P(T_1 \geq t_1, T_2 \geq t_2) > 0$.

One way to construct a dependency measure is to consider an appropriate weighted average over this region of a local dependency measure. A cross-ratio local dependency measure (e.g., Oakes, 1989) has a relative risk interpretation that is useful in many failure time data contexts. Specifically at $\tilde{T}_1 = s_1, \tilde{T}_2 = s_2$, define

$$
\begin{aligned}
\tilde{c}(s_1, s_2) &= F(ds_1, ds_2)F(s_1^-, s_2^-)/[F(ds_1, s_2^-)F(s_1^-, ds_2)] \\
&= \lambda_1(s_1|T_2 = s_2)/\lambda_1(s_1|T_2 \geq s_2) \\
&= \lambda_2(s_2|T_1 = s_1)/\lambda_2(s_2|T_1 \geq s_1).
\end{aligned}
$$

Hence, the cross ratio measures the relative increase in the hazard (λ_1) for $\tilde{T}_1$ when failure occurs at $\tilde{T}_2 = s_2$ compared to failure time $\tilde{T}_2$ at or beyond s_2, with a similar interpretation in regard to the hazard (λ_2) for $\tilde{T}_2$. For technical reasons it is easier to average the reciprocal $c(s_1, s_2) = \tilde{c}(s_1, s_2)^{-1}$ over $[0, t_1] \times [0, t_2]$. Fan et al. (2000) consider a dependency measure

$$
C(t_1, t_2) = \int_0^{t_1} \int_0^{t_2} c(s_1, s_2)F(ds_1, ds_2) \Big/ \int_0^{t_1} \int_0^{t_2} F(ds_1, ds_2),
\tag{10.18}
$$

which averages the reciprocal cross ratio with respect to the probability function for $(\tilde{T}_1, \tilde{T}_2)$, restricted to the region $[0, t_1] \times [0, t_2]$. It is important that the weighting function involve the failure time distribution only, not the censoring process, so that estimates of $C(t_1, t_2)$ from studies having the same failure distribution but different censorship conditions can be compared.

Kendall's concordance measure can be adapted in a similar fashion to censored data over a finite follow-up region. Kendall's tau is defined as

$$\tau = E[\text{sign}(\tilde{T}_{11} - \tilde{T}_{12})(\tilde{T}_{21} - \tilde{T}_{22})],$$

where $(\tilde{T}_{11}, \tilde{T}_{21})$ and $(\tilde{T}_{12}, \tilde{T}_{22})$ are independent variates having survivor function F. Hence τ assesses the degree of concordance of the elements of these variates. With censored data, such concordance can be determined only if the minima $\tilde{T}_{11} \wedge \tilde{T}_{21}$ and $\tilde{T}_{21} \wedge \tilde{T}_{22}$ are observed. Hence Oakes (1989) proposed

$$\tau(s_1, s_2) = E[\text{sign}(\tilde{T}_{11} - \tilde{T}_{12})(\tilde{T}_{21} - \tilde{T}_{22}) | \tilde{T}_{11} \wedge \tilde{T}_{21} = s_1, \tilde{T}_{21} \wedge \tilde{T}_{22} = s_2]$$

as a local concordance measure. In fact, the two local dependency measures are monotonically related by

$$\tau(s_1, s_2) = [1 - c(s_1, s_2)]/[1 + c(s_1, s_2)]$$

so that values of $c(s_1, s_2)$ of 1 (local independence), 0 (maximal positive dependence), and ∞ (maximal negative dependence) correspond to $\tau(s_1, s_2)$ values of $0, 1$ and -1 respectively.

A natural concordance measure over $[0, t_1] \times [0, t_2]$ is obtained by weighting $\tau(s_1, s_2)$ values according to the joint density

$$2F(s_1^-, s_2^-)F(ds_1, ds_2) + 2F(s_1^-, ds_2)F(ds_1, s_2^-)$$

of the component-wise minima $\tilde{T}_{11} \wedge \tilde{T}_{12}$ and $\tilde{T}_{21} \wedge \tilde{T}_{22}$. This gives the dependency measure

$$
\begin{aligned}
\mathscr{T}(t_1, t_2) &= E[\text{sign}(\tilde{T}_{11} - \tilde{T}_{12})(\tilde{T}_{21} - \tilde{T}_{22}) \mid \tilde{T}_{11} \wedge \tilde{T}_{21} \leq t_1, \tilde{T}_{21} \wedge \tilde{T}_{22} \leq t_2] \\
&= \frac{\int_0^{t_1} \int_0^{t_2} F(s_1^-, s_2^-)F(ds_1, ds_2) - \int_0^{t_1} \int_0^{t_2} F(s_1^-, ds_2)F(ds_1, s_2^-)}{\int_0^{t_1} \int_0^{t_2} F(s_1^-, s_2^-)F(ds_1, ds_2) + \int_0^{t_1} \int_0^{t_2} F(s_1^-, ds_2)F(ds_1, s_2^-)},
\end{aligned}
\tag{10.19}
$$

which takes values in $[-1, 1]$.

Nonparametric estimators of $\hat{C}(t_1, t_2)$, the average (reciprocal) cross ratio, and $\hat{\mathscr{T}}(t_1, t_2)$, the finite region concordance measure, are readily obtained by inserting a nonparametric estimator $\hat{F}$ for F everywhere. For example, the properties cited in Section 10.3 for the Dabrowska, Prentice–Cai, or Bickel estimators $\hat{F}$, along with the (compact) differentiability of the transformations (10.18) and (10.19), imply that

$\hat{C}(t_1, t_2)$ and $\hat{\mathcal{T}}(t_1, t_2)$ are consistent and asymptotically normal, with variances that can be estimated consistently using bootstrap procedures (Fan et al., 2000). Hence, $\hat{C}$ and $\hat{\mathcal{T}}$ provide nonparametric dependence measures of useful interpretation for testing and estimation on the nature of the dependency between $\tilde{T}_1$ and $\tilde{T}_2$.

10.5 ILLUSTRATION: AUSTRALIAN TWIN DATA

In this section we present some analyses of the ages at appendectomy for the members of twin pairs included in the Australian twin study (Duffy et al., 1990). These analyses were given in Prentice and Hsu (1997) and Fan et al. (2000), where further detail may be found. The Australian twin study was conducted, in part, to compare monozygotic (MZ) and dizygotic (DZ) twins with respect to the strength of dependency of disease risk between pair members, for various diseases. Twin pairs over the age of 17 were asked to provide information on the occurrence, and age at occurrence, of disease-related events, including the occurrence of vermiform appendectomy. Respondents not undergoing appendectomy prior to survey, or suspected of undergoing prophylactic appendectomy, give rise to right-censored times. In these analyses the study is regarded as a simple cohort study of twin pairs, without regard to the possible impact of such ascertainment criteria as continued survival and willingness to respond to the survey for both pair members.

As in Prentice and Hsu (1997), we consider data on 1953 twin pairs, comprised of 1218 MZ and 735 DZ pairs. Among MZ pairs there were 144 pairs in which both members, 304 pairs in which one member, and 770 in which neither pair member underwent appendectomy. The corresponding numbers for DZ twins were 63, 208, and 464, respectively.

Table 10.1 shows nonparametric estimates $\hat{C}$ of the average reciprocal cross-ratio measure (10.18) and estimates $\hat{\mathcal{T}}$ of the finite region concordance measure (10.19) at various ages (t_1, t_2) for the members of the twin pair. The Dabrowska estimator of the survivor function was used in these calculations, although very similar results would be expected if the Prentice–Cai or Bickel estimates were employed. Also shown in Table 10.1 are bootstrap estimates of the standard deviation of $\log \hat{C}$ and bootstrap estimate of the standard deviation of $\hat{\mathcal{T}}$ based on 200 bootstrap samples. These are given for $\log \hat{C}$, rather than for $\hat{C}$, since the distribution of $\log \hat{C}$ appeared to be better approximated by a normal distribution.

For example, for MZ twins an asymptotic 95% confidence interval for $C(10, 10)$ is given by $\exp[\log(0.13) \pm 1.96(0.36)] = (0.06, 0.26)$, indicating a very strong dependency between appendectomy risks of these twin pairs at ages up to 10 years. For DZ twins the average reciprocal cross-ratio estimate $\hat{C}(10, 10)$ is 0.24, somewhat closer to the independence value of 1.0, with a corresponding nominal 95% confidence interval of $(0.10, 0.57)$. The point estimators and confidence intervals for $C(30, 30)$ are 0.33 and $(0.27, 0.40)$ for MZ twins and 0.52 and $(0.39, 0.72)$ for DZ twins. Evidently, the dependencies are not as strong for either type of twin pair at larger ages. More specifically, over the age region $[20, 60] \times [20, 60]$ for the two twins the averaged reciprocal cross-ratio estimate is 0.76, with an

314 ANALYSIS OF CORRELATED FAILURE TIME DATA

Table 10.1 Average Appendectomy Relative Risk and Concordance Estimates at Selected Ages of the Members of Monozygotic (MZ) and Dizygotic (DZ) Twin Pairs in the Australian Twin Study (Duffy et al., 1990)

Age of First Twin[a]		Age of Second Twin[a]		
		10	30	50
Average Reciprocal Cross-Ratio Estimate[b]				
10	MZ	0.13(0.36)	0.24(0.22)	0.27(0.24)
	DZ	0.24(0.44)	0.63(0.47)	0.54(0.54)
30	MZ	0.32(0.26)	0.33(0.10)	0.36(0.10)
	DZ	0.34(0.27)	0.52(0.16)	0.61(0.17)
50	MZ	0.39(0.27)	0.39(0.11)	0.42(0.10)
	DZ	0.37(0.26)	0.58(0.17)	0.59(0.15)
Finite Region Concordance Estimate[c]				
10	MZ	0.78(0.083)	0.62(0.067)	0.59(0.075)
	DZ	0.61(0.150)	0.26(0.205)	0.30(0.203)
30	MZ	0.53(0.095)	0.51(0.031)	0.48(0.036)
	DZ	0.50(0.101)	0.32(0.071)	0.28(0.071)
50	MZ	0.47(0.107)	0.46(0.039)	0.43(0.038)
	DZ	0.47(0.107)	0.29(0.071)	0.28(0.063)

Source: Fan et al. (2000).

[a] The designations *first* and *second* are arbitrary within a twin pair.
[b] In parentheses are bootstrap estimates of the standard deviation of $\log \hat{C}$, based on 200 bootstrap samples.
[c] In parentheses are bootstrap estimates of the standard deviation of $\hat{\mathcal{T}}$, based on 200 bootstrap samples.

asymptotic 95% confidence interval of (0.45, 1.25) for MZ twins, and 0.77 with a confidence interval of (0.36, 1.65) for DZ twins, indicating that there is weak evidence, at best, for association in appendectomy risk when the members of the twin pair are beyond 20 years of age.

One can readily test the hypothesis of a common pairwise dependency between the appendectomy risks of MZ and DZ twins. For example, one can compare $C(30, 30)$ values by calculating the asymptotic normal test statistic $[\log (0.33) - \log (0.52)]/[(0.10)^2 + (0.16)^2]^{1/2} = -2.41$, providing evidence for stronger dependency between the incidence ages of MZ twins compared to DZ twins. Such analyses may suggest a genetic component to disease risk in view of the greater sharing of genetic information between MZ as compared to DZ twins, although differences in environmental factor sharing may also need to be considered.

Similar analyses arise using the finite region concordance summary measure. From the lower portion of Table 10.1 one sees that $\mathcal{T}(10, 10)$ has a point estimate of 0.78 and asymptotic 95% confidence interval of $0.78 \pm 1.96(0.083) = (0.62, 0.94)$, illustrating the strong dependency of risk for MZ twins. A test of equality of $\mathcal{T}(30, 30)$ values for MZ and DZ twins takes value 2.45, almost identical in absolute value to that based on $\hat{C}$.

The methods of Section 10.2.2 were applied to these data with year of birth as a covariate. The relative risk coefficient in (10.5) was estimated to be negative and highly significant, with little evidence of difference between MZ and DZ twins, indicating that appendectomy rates were reduced in more recent years. Simple estimating equation analyses using (10.6) indicate that cross-ratio estimates are not much altered by allowing marginal hazard rates to depend on year of birth in this application (Prentice and Hsu, 1997).

10.6 APPROACHES TO NONPARAMETRIC ESTIMATION OF THE BIVARIATE SURVIVOR FUNCTION

10.6.1 Introduction

In this section we give some more detail on the properties and limitations of the nonparametric bivariate survivor function estimators introduced in Section 10.3, and discuss estimators that derive from nonparametric maximum likelihood (NPML) considerations. Because there are few analytic results concerning the comparative efficiency of the estimators that have been proposed, we include some simulation results. As noted previously, these methods are somewhat specialized and technical, and this section and Section 10.7 can be omitted without affecting the readability of the remainder of the book.

10.6.2 Plug-in Estimators

In Section 10.3, bivariate survivor function estimators due to Bickel, Dabrowska, and Prentice and Cai were described. These estimators can be obtained by plugging marginal survivor function estimators and double failure hazard function estimators into the Peano series representation

$$F(t_1, t_2) = \Psi(t_1, t_2) + \int_0^{t_1} \int_0^{t_2} \Psi(u_1^-, u_2^-) \tilde{P}(u, t; \Lambda) \Lambda(du_1, du_2), \qquad (10.20)$$

that is the unique solution to the inhomogeneous Volterra equation (10.13), where the Peano series transformation $\tilde{P}$ is given by

$$\tilde{P}(u, t; \Lambda) = 1 + \sum_{m=1}^{\infty} \int_{A_m(u,t)} \prod_{j=1}^{m} \Lambda(du_{1j}, du_{2j})$$

and the region of integration $A_m(u, t)$ for the mth term in the summation is given by $(u_1 < u_{11} < \cdots < u_{1m} \le t_1; u_2 < u_{21} < \cdots < u_{2m} \le t_2)$. The survivor function is a complicated function of F_1 and F_2 [that appear in $\Psi(t_1, t_2) = F_1(t_1) + F_2(t_2) - 1$] and Λ, and considerable effort may be required to identify specifications that are mutually compatible in the sense that probability mass or densities are nonnegative throughout $[0, \infty) \times [0, \infty)$.

The Bickel estimator, given by Kaplan–Meier margins and (10.15), seems very natural at first glance in that it uses only empirical estimates of the first- and second-order hazard rates. However, the first- and second-order empirical hazard function estimators typically do not agree well.

Table 10.2 shows a simulated sample of size $n = 30$ with $\tilde{T}_1$ and $\tilde{T}_2$ independent unit exponential variates, subject to censoring by independent random variates C_1 and C_2 having exponential distributions with mean 0.5, with (C_1, C_2) independent of (T_1, T_2). Figure 10.1 displays these data in the quadrant $(\tilde{T}_1 \geq 0, \tilde{T}_2 \geq 0)$. For convenience of display, the uncensored failure times in each direction are equally spaced, and all censored observations are downward-shifted to the nearest uncensored observation. This shifting does not alter the survivor function estimator at failure time grid points in the risk region. Note, however, that the risk region outlined in Figure 10.1 is that for the shifted data and is slightly smaller than that for the original data. One can see that there are five uncensored observations ($\delta_1 = \delta_2 = 1$), denoted by dots; six singly censored observations ($\delta_1 = 1, \delta_2 = 0$), denoted by vertical arrows; seven singly censored observations ($\delta_1 = 0, \delta_2 = 1$), denoted by horizontal arrows; and 12 doubly censored observations ($\delta_1 = \delta_2 = 0$), denoted by diagonal arrows. These numbers can be compared to corresponding expected numbers of 3.33, 6.67, 6.67, and 13.33, respectively.

The upper part of Figure 10.2 shows the Bickel survivor function estimator at each of the grid points in the risk region. Note that the estimator is not a proper survivor function over this region. Its lack of monotonicity arises from negative mass assignments along half-lines beyond the risk region. The lower part of

Table 10.2   Simulated Bivariate Failure Time Sample of Size $n = 30$ [a]

T_1	δ_1	T_2	δ_2	T_1	δ_1	T_2	δ_2
0.456	0	0.419	0	0.473	1	0.220	1
0.462	1	0.172	0	0.117	1	0.792	1
0.364	0	0.116	1	0.391	0	0.054	1
0.275	0	0.277	1	0.552	0	0.765	0
0.066	1	0.311	0	0.657	0	0.180	0
0.012	0	0.099	0	0.015	0	0.759	0
0.484	0	0.148	1	0.174	1	0.297	1
0.267	0	0.572	1	0.004	0	1.682	0
0.948	1	0.269	0	0.111	0	0.271	0
0.510	0	0.028	0	0.315	1	0.164	0
0.028	0	0.044	0	0.297	1	0.171	0
0.079	0	0.286	1	0.113	1	0.257	1
0.143	0	1.247	1	0.170	1	0.252	1
0.974	0	0.472	0	0.219	0	0.088	0
0.083	0	0.066	0	0.355	1	0.226	0

[a] Failure times $\tilde{T}_1$ and $\tilde{T}_2$ arise from independent exponential distributions with mean 1, while censoring times C_1 and C_2 arise independently from independent exponential distributions with mean 0.5.

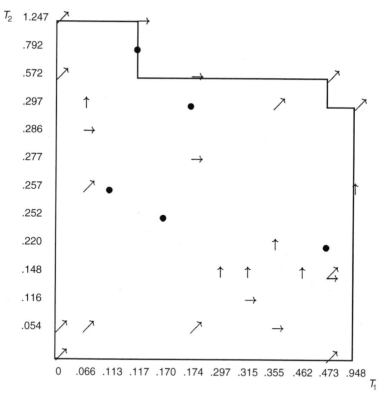

Figure 10.1 Display of data shown in Table 10.2. For convenience, uncensored observations are equally spaced for both failure time variates, and all censored values are shifted downward to the immediately preceding uncensored observation for that variate.

Figure 10.2 shows the Bickel estimator mass assignments within the risk region and along half-lines beyond the risk region.

The Dabrowska and Prentice–Cai estimators, given respectively by (10.17) and (10.16), in conjunction with Kaplan–Meier marginal survivor function estimators are nonparametric efficient under the complete independence of $\tilde{T}_1, \tilde{T}_2, C_1$, and C_2 (Gill et al., 1995). From a practical point of view this means that both estimators have nearly optimal asymptotic performance provided that the dependency between $\tilde{T}_1$ and $\tilde{T}_2$ is not too strong. Furthermore, simulation studies by various investigators indicate that the performance of both estimators is good in moderate-sized samples.

Since there are few available asymptotic efficiency results for comparison of the Bickel, Dabrowska, and Prentice–Cai estimators $\hat{F}_B$, $\hat{F}_D$, and $\hat{F}_{PC}$, we provide here a brief report on a simulation comparison. In addition, we include one of the earliest proposed bivariate survivor function estimators $\hat{F}_{CF}$, due to Campbell and Földes (1982), which is based on the factorization

$$F(t_1, t_2) = F_1(t_1)P(\tilde{T}_2 > t_2 \mid \tilde{T}_1 > t_1)$$

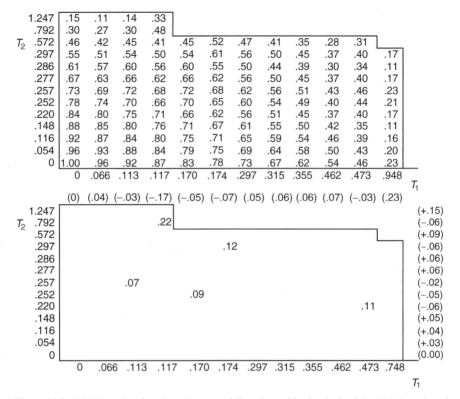

Figure 10.2 Bickel survivor function estimator at failure time grid points in the risk region (upper), and corresponding mass assignments in the risk region (lower) for data of Table 10.1. Mass assignments on half-lines beyond the risk region are shown in parentheses.

with Kaplan–Meier estimators plugged in for both factors, the latter based only on the observations for which $T_1 > t_1$. (An estimator interchanging the roles of $\tilde{T}_1$ and $\tilde{T}_2$ was also proposed by Campbell and Földes.) Sample sizes $n = 30$, 60, and 120 were considered with C_1 and C_2 independent exponential variates with mean 0.5 and with $\tilde{T}_1$ and $\tilde{T}_2$ unit exponential variates that are independent of (C_1, C_2) and that are either independent of each other or that arise from Clayton's (1978) bivariate survival model, for which

$$F(t_1, t_2) = [F_1(t_1)^{-\theta} + F_2(t_2)^{-\theta} - 1]^{-1/\theta}.$$

Here the parameter θ governs the strength of dependence between $\tilde{T}_1$ and $\tilde{T}_2$. The simulation uses $\theta = 4$, implying a fairly strong positive dependence between $\tilde{T}_1$ and $\tilde{T}_2$.

Table 10.3 shows sample means and sample standard deviations for $\hat{F}_{CF}, \hat{F}_B, \hat{F}_D$, and $\hat{F}_{PC}$ at selected percentiles of the distributions of the marginal distributions F_1 and F_2. Since all estimators use Kaplan–Meier marginal survivor functions,

Table 10.3 Simulation Summary Statistics for Selected Nonparametric Estimators of the Bivariate Survivor Function [a]

Sample Size, n	Estimator[b]	$(\tilde{T}_1, \tilde{T}_2)$ Percentiles			Negative Mass[c]	
		(70,70)	(70,55)	(55,55)	In	Out
		Independence Model				
	True $F(t_1, t_2)$	0.490	0.385	0.303		
30	$\hat{F}_{CF}$	0.495 (0.140)[d]	0.394 (0.178)	0.301 (0.209)	0.85	0.25
	$\hat{F}_B$	0.498 (0.120)	0.389 (0.136)	0.306 (0.161)	0	0.34
	$\hat{F}_D$	0.498 (0.122)	0.388 (0.141)	0.303 (0.167)	0.42	0.10
	$\hat{F}_{PC}$	0.499 (0.116)	0.393 (0.126)	0.323 (0.133)	0.21	0.11
60	$\hat{F}_{CF}$	0.491 (0.097)	0.390 (0.123)	0.310 (0.141)	1.11	0.30
	$\hat{F}_B$	0.490 (0.080)	0.389 (0.094)	0.308 (0.114)	0	0.44
	$\hat{F}_D$	0.490 (0.080)	0.390 (0.092)	0.307 (0.115)	0.58	0.12
	$\hat{F}_{PC}$	0.490 (0.079)	0.389 (0.089)	0.312 (0.095)	0.32	0.15
120	$\hat{F}_{CF}$	0.492 (0.066)	0.392 (0.081)	0.314 (0.088)	1.37	0.35
	$\hat{F}_B$	0.490 (0.056)	0.389 (0.067)	0.304 (0.077)	0	0.54
	$\hat{F}_D$	0.491 (0.055)	0.390 (0.064)	0.308 (0.070)	0.74	0.15
	$\hat{F}_{PC}$	0.491 (0.055)	0.389 (0.064)	0.307 (0.065)	0.43	0.18
		Clayton Model				
	True $F(t_1, t_2)$	0.608	0.516	0.468		
30	$\hat{F}_{CF}$	0.603 (0.131)	0.516 (0.172)	0.448 (0.206)	0.63	0.28
	$\hat{F}_B$	0.611 (0.113)	0.518 (0.133)	0.467 (0.164)	0	0.44
	$\hat{F}_D$	0.611 (0.114)	0.520 (0.138)	0.469 (0.179)	0.38	0.17
	$\hat{F}_{PC}$	0.607 (0.110)	0.512 (0.123)	0.462 (0.134)	0.20	0.15
60	$\hat{F}_{CF}$	0.606 (0.094)	0.519 (0.119)	0.468 (0.131)	0.85	0.31
	$\hat{F}_B$	0.608 (0.081)	0.519 (0.096)	0.468 (0.118)	0	0.54
	$\hat{F}_D$	0.609 (0.078)	0.521 (0.090)	0.474 (0.111)	0.54	0.20
	$\hat{F}_{PC}$	0.607 (0.077)	0.517 (0.088)	0.466 (0.093)	0.29	0.18
120	$\hat{F}_{CF}$	0.607 (0.062)	0.521 (0.076)	0.468 (0.086)	1.08	0.35
	$\hat{F}_B$	0.606 (0.053)	0.516 (0.063)	0.462 (0.079)	0	0.62
	$\hat{F}_D$	0.607 (0.052)	0.518 (0.059)	0.468 (0.068)	0.66	0.22
	$\hat{F}_{PC}$	0.606 (0.051)	0.516 (0.058)	0.463 (0.065)	0.38	0.21

[a] Based on 500 simulations at each sample size and sampling configuration.

[b] At $n = 30$, $\hat{F}$ values are based on 497, 480, and 425 runs with nonempty risk sets under the independence model and 499, 488, and 455 runs having nonempty risk sets under the Clayton model at the (70,70), (70,55), and (55,55) percentiles, respectively. Corresponding counts at $n = 60$ were 500, 497, and 468 under independence and 500, 499, and 490 under the Clayton model. At $n = 120$, the counts were 500, 500, and 494 under independence and 500, 500, and 498 under the Clayton model.

[c] The entries under "In" and "Out" are the average amount of negative mass assigned by the estimator inside the risk region and along half lines outside the risk region, respectively.

[d] Entries are the sample means of $\hat{F}$ values, with sample standard deviations in parentheses.

percentiles were selected that are away from the margins but not so far into the tails as to yield an unacceptable number of empty risk sets. For example, the point labeled (70,70) is given by (t_1, t_2), where $\hat{F}_i(t_i) = 0.7, i = 1, 2$. Note the near unbiasedness of each of the four estimators, even at sample size $n = 30$, where there are only 10 expected failures for each of T_1 and T_2. Also, from the sample standard deviations one can see that the Campbell–Földes estimator is considerably less efficient than the other estimators, while the Bickel estimator is somewhat less efficient than the Prentice–Cai estimator under both sampling configurations. The Dabrowska estimator also has sample standard deviations that are somewhat larger than those for the Prentice–Cai estimator under these sampling configurations, but the difference is small at the larger sample sizes. Table 10.3 also shows the average amount of negative mass assigned both within the risk region ("In") and along half-lines outside the risk region ("Out"). It seems evident that the propensity of these estimators to assign negative mass does not disappear as the sample size increases.

Despite the issues of negative mass assignments and lack of full efficiency away from the independence of $\tilde{T}_1$ and $\tilde{T}_2$, the Dabrowska and Prentice–Cai estimators will be adequate for most practical purposes. Some more specialized purposes motivate consideration of additional nonparametric estimators of the bivariate survivor function, including the desire to improve on the efficiency of marginal survivor function estimators by using the data on the other failure time to provide informative censorship information. For this purpose, estimators whose margins are not Kaplan–Meier estimators are needed.

10.6.3 Nonparametric Maximum Likelihood Estimators

In Chapter 1 it was noted that the Kaplan–Meier estimator of a univariate survivor function could be obtained by a nonparametric maximum likelihood argument. One can consider similarly a nonparametric maximum likelihood approach to estimation of the bivariate survivor function. The contribution to the likelihood for F from an observation $(T_1 = t_1, T_2 = t_2, \delta_1, \delta_2)$ is

$$F(dt_1, dt_2)^{\delta_1\delta_2} [-F(dt_1, t_2)]^{\delta_1(1-\delta_2)} [-F(t_1, dt_2)]^{(1-\delta_1)\delta_2} F(t_1, t_2)^{(1-\delta_1)(1-\delta_2)},$$

and the overall likelihood is the product of such terms over the sample of n observations. This likelihood can be maximized by placing all mass at grid points in the failure time risk region or along half-lines beyond the risk region, with the possibility of some additional mass within the intersection of the doubly censored observations. The resulting partially maximized likelihood function can be written

$$L = \prod_{i=0}^{I}\prod_{j=0}^{J}\left[p_{ij}^{n_{ij}^{11}} \left(p_{i\cdot} - \sum_{m=0}^{j} p_{im} \right)^{n_{ij}^{10}} \left(p_{\cdot j} - \sum_{\ell=0}^{i} p_{\ell j} \right)^{n_{ij}^{01}} \right.$$
$$\left. \times \left(1 - \sum_{\ell=0}^{i} p_{\ell\cdot} - \sum_{m=0}^{j} p_{\cdot m} + \sum_{\ell=0}^{i}\sum_{m=0}^{j} p_{\ell m} \right)^{n_{ij}^{00}} \right], \tag{10.21}$$

where $n_{ij}^{v_1 v_2} = \#(T_1 = t_{1i}, T_2 = t_{2j}, \delta_1 = v_1, \delta_2 = v_2)$ for $v_1 \in \{0,1\}$ and $v_2 \in \{0,1\}$ following a left or downward shifting of the censored T_1 and T_2 values to their nearest uncensored value, and where, $t_{11} < t_{12} < \cdots < t_{1I}$ and $t_{21} < t_{22} < \cdots < t_{2J}$ denote the ordered uncensored T_1 and T_2 values, respectively, and $t_{10} = t_{20} = 0$. Also in (10.21), $p_{ij} = F(\Delta t_{1i}, \Delta t_{2j})$ is the mass assignment at (t_{1i}, t_{2j}) in the risk region, while $p_{i\cdot} = -F(\Delta t_{1i}, 0)$ and $p_{\cdot j} = -F(0, \Delta t_{2j})$ are the marginal mass assignments along $\tilde{T}_1 = t_{1i}$ and along $\tilde{T}_2 = t_{2j}$ respectively. The log likelihood, $\log L$, is convex in its parameters (Campbell, 1981), but there may be regions interior to the risk region where the likelihood is flat, in which circumstance the NPMLE will not be unique.

This uniqueness problem is most easily illustrated in the special case of absolutely continuous failure times with $\tilde{T}_1$, but not $\tilde{T}_2$, subject to censoring, so that $n_{ij}^{10} \equiv n_{ij}^{00} \equiv 0$. Expression (10.21) then reduces to a multinomial likelihood that when maximized assigns mass n^{-1} to each uncensored observation $(n_{ij}^{11} = 1)$, and to the half-line formed by each singly censored observation $(n_{ij}^{01} = 1)$, but does not otherwise determine an estimator of F. Note that this lack of uniqueness may extend well inside the risk region, in contrast to the Kaplan–Meier univariate NPML estimator.

There has been quite a lot of work to repair the NPML procedure and circumvent these uniqueness problems. For example, some authors have added smoothness conditions to F so that mass assignments along rays formed by singly censored observations can borrow from neighboring data. Others have tried to characterize the subsets of the parameter space that receive unique mass assignments. The work of van der Laan (1996) is particularly noteworthy. Van der Laan began by truncating the $\tilde{T}_1$ data back to τ_1 and truncating $\tilde{T}_2$ back to τ_2, where once again $P(T_1 \geq \tau_1, T_2 \geq \tau_2) > 0$. This truncation does not alter survival probabilities in $[0, \tau_1) \times [0, \tau_2)$ and it allows one to eliminate $p_{i\cdot}$ and $p_{\cdot j}$ in (10.21) in favor of mass assignments along the boundary $\tilde{T}_1 = \tau_1$ or $\tilde{T}_2 = \tau_2$. In sufficiently large samples there will be one or more observations at risk at (τ_1, τ_2). Van der Laan then reduces the data somewhat by introducing a fixed partition in each direction and replacing all potential censoring times by their immediately smaller partition point. Mass is assigned in an NPML fashion within strips formed by the partition in either direction. Such assignment will become unique in sufficiently large samples. An expectation-maximization algorithm was proposed for calculation of this repaired NPML estimator, which was shown to be uniformly consistent, asymptotically Gaussian, and nonparametrically efficient for the reduced data over $[0, \tau_1) \times [0, \tau_2)$ under random censorship. Furthermore, van der Laan showed that an extension of this procedure could be used to develop an estimator having these same properties as well as nonparametric efficiency for the original data. To do so, he considered a nested sequence of partitions in which the maximum width of elements of the partition converged to zero at a slow rate as the sample size became large. Simulation studies in samples of moderate size (van der Laan, 1997) suggest that the repaired NPML procedure may yield marginal survivor function estimators of somewhat greater efficiency than Kaplan–Meier estimators under strong dependencies between $\tilde{T}_1$ and $\tilde{T}_2$, but that it is difficult to improve on the efficiency of the Dabrowska and Prentice–Cai estimators toward the tails of the failure time distribution.

Van der Laan's results are important in showing that asymptotically efficient nonparametric estimation of F is possible. Further development will be needed, however, before these estimators can be recommended for routine use. Specifically, the repaired NPML is only unique asymptotically, and moderate-sample size properties may depend importantly on the manner in which mass is distributed beyond the largest uncensored observation in a strip and on the choice of partition intervals more generally. Hence, additional work is needed to identify implementation plans that avoid bias while achieving good efficiency in samples of moderate size.

10.7 SURVIVOR FUNCTION ESTIMATION IN HIGHER DIMENSIONS

The representation of Section 10.3 can be extended to express a survivor function

$$F(t_1, \ldots, t_m) = P(\tilde{T}_1 > t_1, \ldots, \tilde{T}_m > t_m),$$

for $m \geq 1$ failure time variates $\tilde{T}_1, \ldots, \tilde{T}_m$ in terms of hazard functions of dimension $\leq m$. Denote by $F_{i_1, i_2 \ldots i_q}$ the survivor function with all arguments except $(i_1, i_2, \ldots, i_q)$ set to zero, so that, for example, $F_1(t_1) = F(t_1, 0, \ldots, 0)$, $F_{13}(t_1, t_3) = F(t_1, 0, t_3, 0, \ldots, 0)$, and $F_{12\ldots m}(t_1, \ldots, t_m) = F(t_1, \ldots, t_m)$. The m-dimensional hazard function $\Lambda = \Lambda_{1\ldots m}$ can be defined by

$$\Lambda(dt_1, dt_2, \ldots, dt_m) = (-1)^m F(dt_1, \ldots, dt_m) / F(t_1^-, \ldots, t_m^-), \qquad (10.22)$$

with a corresponding definition for all lower-dimensional hazards. From (10.22) one can write

$$\int_0^{t_1} \cdots \int_0^{t_m} F(ds_1, \ldots, ds_m) = (-1)^m \int_0^{t_1} \cdots \int_0^{t_m} F(s_1^-, \ldots, s_m^-) \Lambda(ds_1, \ldots, ds_m).$$

Integrating the left side of this expression gives

$$F(t_1, \ldots, t_m) = \Psi(s_1, \ldots, s_m) + (-1)^m \int_0^{t_1} \cdots \int_0^{t_m} F(s_1^-, \ldots, s_m^-) \Lambda(ds_1, \ldots, ds_m),$$

an nonhomogeneous Volterra equation with unique Peano series solution that expresses F in terms of Λ and the lower-dimensional marginal survivor functions, since

$$\Psi = (-1)^{m-1} + (-1)^{m-2} \sum_{i=1}^m F_i + (-1)^{m-3} \sum_{i \neq j} F_{ij} + \cdots$$

$$+ (-1)^0 \sum_{i_1 \neq i_2 \neq \cdots \neq i_{m-1}} F_{i_1 \cdots i_{m-1}}.$$

The same argument applies to each of the lower-dimensional survivor functions, showing, inductively, that F can be represented uniquely in terms of the hazard functions $\Lambda_{i_1,\ldots,i_q}$ for all nonzero subsets $(i_1,\ldots,i_q)$ of $\{1,2,\ldots,m\}$.

This representation makes clear that if Λ and all lower-dimensional hazard rates are unaffected by the censoring on $(\tilde{T}_1,\ldots,\tilde{T}_m)$, F will be identifiable from censored data $(T_i,\delta_i), i = 1,\ldots,m$, where $T_i = \tilde{T}_i \wedge C_i$ and where C_i is a potential censoring time. Hence, as in Section 10.2.1, one can define a censoring scheme to be independent if the m- and lower-dimensional hazards are not affected by the censoring.

A uniformly consistent, asymptotically Gaussian estimator $\hat{F}$ of F under independent censorship can be obtained over a region $[0,\tau_1] \times \cdots \times [0,\tau_m]$, where $P(T_1 \geq \tau_1,\ldots,T_m \geq \tau_m) > 0$, by inserting simple empirical estimators

$$\hat{\Lambda}_{i_1,\ldots,i_q}(dt_{i_1},\ldots,dt_{i_q}) = \#\{T_{i_1} = t_{i_1},\ldots,T_{i_q} = t_{i_q},\delta_{i_1} = 1,\ldots,\delta_{i_q} = 1\}/$$
$$\#\{T_{i_1} \geq t_{i_1},\ldots,T_{i_q} \geq t_{i_q}\}$$

for each such lower-dimensional hazard function. Alternatively, the Dabrowska and Prentice–Cai methods can be generalized to give higher-dimensional hazard rate estimators. With any of these estimators a recursive calculation can be used to generate a survivor function estimator from its corresponding m-dimensional hazard function estimator and lower-dimensional survivor function estimators. In particular, one can calculate

$$\hat{F}(t_1,\ldots,t_m) = \hat{F}(t_1^-,t_2,\ldots,t_m) + \cdots + \hat{F}(t_1,\ldots,t_{m-1},t_m^-)$$
$$- [\hat{F}(t_1^-,t_2^-,t_3,\ldots,t_m) + \cdots + \hat{F}(t_1,\ldots,t_{m-2},t_{m-1}^-,t_m^-)]$$
$$\cdots$$
$$+ (-1)^{m-2}[\hat{F}(t_1^-,\ldots,t_{m-1}^-,t_m) + \cdots + \hat{F}(t_1,t_2^-,\ldots,t_m^-)]$$
$$+ (-1)^{m-1}\hat{F}(t_1^-,\ldots,t_m^-)[1 - \hat{\Lambda}(dt_1,\ldots,dt_m)].$$

This expression allows the m-dimensional survivor function estimator to be readily calculated throughout the risk region from the $(m-1)$-dimensional marginal survivor functions and the m-dimensional hazard estimator. These joint survivor function estimators provide an avenue to the examination of dependencies among $(\tilde{T}_1,\ldots,\tilde{T}_m)$.

BIBLIOGRAPHIC NOTES

Additional general references for correlated failure time analyses include the book by Hougaard (2000) and Chapters 9 and 10 of Andersen et al. (1993). Chapter 10 of Andersen et al. provides a discussion on nonparametric estimation of a multivariate survivor function under independent censoring, including the Peano series estimators discussed in Sections 10.3 and 10.7. Hougaard's Chapter 14 provides further detail on characteristics of nonparametric maximum likelihood estimates of the bivariate survivor function, and he discusses and illustrates various graphical procedures for the display of bivariate failure time data.

Many authors beyond those cited above have considered aspects of nonparametric maximum likelihood estimation of the bivariate survivor function, including data grouping or smoothing to address NPML uniqueness problems. Examples include Campbell (1981), Campbell and Földes (1982), Hanley and Parnes (1983), Tsai et al. (1986), Pruitt (1991), van der Laan (1996, 1997), and Prentice (1999). Pruitt (1991) and van der Laan (1997) provide additional numerical comparisons among various nonparametric bivariate survivor function estimators. Gentleman and Vandal (2001, 2002) develop algorithms for identifying regions of the sample space for $(\tilde{T}_1, \tilde{T}_2)$ that are assigned unique mass by an NPML procedure, using graph-theoretic techniques.

The idea of estimating a bivariate survivor function nonparametrically by constructing estimators that weight observations according to the reciprocal of their estimated censoring probabilities was considered by Burke (1988). Jamie Robins and colleagues have extended this idea using estimating functions for nonparametric and semiparametric models to show how survivor function estimators can be defined that are efficient at a particular submodel (e.g., under independence, or under the Clayton model) (see Robins et al., 2000). Lin and Ying (1993a) and Tsai and Crowley (1998) discuss inverse censoring probability weighted estimators of the bivariate survivor function under univariate censoring $(C_1 = C_2 = C)$.

There is a substantial literature on the use of frailty models to induce multivariate failure time models. See Oakes (1989), Hougaard (2000, Chaps. 10–13), Andersen et al. (1993, Chap. 9), and Klein and Moeschberger (1997) for further discussion and illustration.

Additional material on the application of Cox regression models to marginal hazard functions in a correlated failure time context may be found in Lee et al. (1992) and Lin (1994). Some authors have focused specifically on tests of independence between pairs of failure time variates; see, for example, Pons and Turckheim (1991) and Hsu and Prentice (1996).

EXERCISES AND COMPLEMENTS

10.1 Calculate nonparametric estimators of the reciprocal cross-ratio and finite region concordance measures (10.18) and (10.19) for the data of Table 10.2 over the region $[0, 0.5] \times [0, 0.5]$, using the Bickel survivor function estimator shown in Figure 10.2. Calculate bootstrap variance estimators for these nonparametric estimates of pairwise dependency by resampling with replacement from the 30 observations of Table 10.2. Use 50- and 200-bootstrap samples. How different are the variance estimates between the smaller and larger number of bootstrap samples? Do the data of Table 10.2 suggest any dependency between $\tilde{T}_1$ and $\tilde{T}_2$ over this finite follow-up region?

10.2   Suppose that given the frailty variable α, $\tilde{T}_1$ and $\tilde{T}_2$ are independent continuous failure time variables with hazard functions $\alpha h_0(t)$ and $\alpha h_1(t)$, respectively. The corresponding conditional survivor functions are $H_0(t)^\alpha$ and $H_1(t)^\alpha$, where $H_j(t) = \exp[-\int_0^t h_j(u)\,du]$, $j = 1, 2$. Suppose also that α has a gamma distribution that is rescaled to have mean 1 and variance $\theta > 0$ (see Section 2.2.4).

(a) Give the density and moment generating function of α.

(b) Show that the joint (unconditional) survivor function of $\tilde{T}_1, \tilde{T}_2$ is the Clayton (1978) survivor function

$$F(t_1, t_2) = [F_1(t_1)^{-\theta} + F_2(t_2)^{-\theta} - 1]^{-1/\theta}$$

and express $F_j(t)$ as a function of $H_j(t)$. Show that $F(t_1, t_2)$ converges to $F(t_1)F(t_2)$ as $\theta \to 0$.

(c) Show that the Clayton survivor function can be extended to allow negative dependencies by setting

$$F(t_1, t_2) = \{[F_1(t_1)^{-\theta} + F_2(t_2)^{-\theta} - 1] \vee 0\}^{-1/\theta}.$$

(d) Show that this extended distribution approaches the upper Frechet bound of $F_1(t_1) \wedge F_2(t_2)$ for maximal positive dependency as $\theta \to \infty$, and approaches the lower Frechet bound of $[F_1(t_1) + F_2(t_2) - 1] \vee 0$ for maximal dependency as $\theta \to -1$.

(e) Show that the distribution of $(\tilde{T}_1, \tilde{T}_2)$ is absolutely continuous for $\theta > -0.5$.

(f) Generalize these developments to m failure time variates and a vector of fixed covariates, thereby giving a version of (10.8) that includes the possibility of negative dependencies. Over what range of θ values is the distribution of $(\tilde{T}_1, \ldots, \tilde{T}_m)$ absolutely continuous for $m > 2$?

10.3   Show that the extended Clayton model $F(t_1, t_2)$ in Exercise 10.2 gives a constant cross ratio

$$F(dt_1, dt_2)F(t_1^-, t_2^-)/[F(dt_1, t_2^-)F(t_1^-, dt_2)] = \theta + 1$$

for all (t_1, t_2), for $\theta > -0.5$.

10.4   Derive the Volterra integral equation (10.13) by integrating the left side of the preceding equation. Consider discrete failure time variables with $\tilde{T}_1$ taking values t_{11} and t_{12} and $\tilde{T}_2$ taking values t_{21} and t_{22}, with $t_{11} < t_{12}$ and $t_{21} < t_{22}$. Use (10.13) to express $F(t_{12}, t_{22})$ in terms of the marginal survivor function values $F_1(t_{11}), F_1(t_{12}), F_2(t_{21})$, and $F_2(t_{22})$ and the double failure hazards $\Lambda(dt_{11}, dt_{21}), \Lambda(dt_{11}, dt_{22}), \Lambda(dt_{12}, dt_{21})$, and $\Lambda(dt_{12}, dt_{22})$. Express the result in Peano series form [see (10.20)].

10.5 Calculate the Dabrowska and Prentice–Cai estimators for the data of Table 10.2 by using the recursive formula (10.14) with $\hat{\Lambda}$ set equal to (10.17) and (10.16), respectively, over the failure time grid points of Figure 10.1. Show that the row second from the top takes values 0.30, 0.27, 0.28, and 0.44 for the Dabrowska estimator and 0.30, 0.27, 0.28, and 0.37 for the Prentice–Cai estimator, close to the values shown in Figure 10.2 for the Bickel estimator. Also show that the corresponding estimates in the top row are 0.15, 0.00, 0.00, and 0.00 for the Dabrowska estimator, quite different from the corresponding Bickel estimator values and from the Prentice–Cai estimates of 0.15, 0.13, 0.13, and 0.17. Explain these zero values for the Dabrowska estimator and comment on whether this behavior could contribute to the relatively large sample variances at the smaller sample sizes in Table 10.3. Comment also on the extent to which the Dabrowska and Prentice–Cai estimators ameliorate the nonmonotonicity shown in Figure 10.2 for the Bickel estimator.

10.6 Consider a restricted version of an estimator $\hat{F}$ arising from the recursive calculation (10.14). Define a new step function survivor function estimator $\tilde{F}$ by setting $\tilde{F}(t_{1i}, t_{2j}) = [0 \vee \tilde{F}(t_{1i}^-, t_{2j}) + \tilde{F}(t_{1i}, t_{2j}^-) - \tilde{F}(t_{1i}^-, t_{2j}^-)] \vee \hat{F}(t_{1i}, t_{2j}) \wedge [\tilde{F}_{1i}(t_{1i}^-, t_{2j}) \wedge \tilde{F}(t_{1i}, t_{2j}^-)]$ at all grid points in the failure time region. By considering the 2×2 table defined by whether $\tilde{T}_1$ is equal or greater than t_{1i} and by whether $\tilde{T}_2$ is equal or greater than t_{2j} for pairs at risk at (t_{1i}, t_{2j}), show that $\tilde{F}(t_{1i}, t_{2j})$ avoids negative mass assignments both within the risk region and on half-lines outside the risk region. Compute the survivor function estimator over the points in the failure time risk region for the data of Table 10.2 using restricted versions of each of the Bickel, Dabrowska, and Prentice–Cai forms of $\tilde{F}$, and comment on how these compare with those shown in Figure 10.2 and with the calculations from Exercise 10.2. (*Comment*: To date, asymptotic properties have not been established for this restricted form of estimator.)

10.7 Consider a specialized censoring scheme, often referred to as *univariate censoring*, under which the potential censoring times are identical for the bivariate failure times T_1 and T_2. Set $C = C_1 = C_2$ and suppose that censoring is random with $P(C > t) = G(t)$. Define a new survivor function F^* as a step function given by

$$F^*(t_{1i}, t_{2j}) = \#(T_1 > t_{1i}, T_2 > t_{2j}) / [n\hat{G}(t_{ij})]$$

at failure time grid point (t_{1i}, t_{2j}), where $t_{ij} = t_{1i} \vee t_{2j}$ and $\hat{G}$ is the Kaplan–Meier estimator of G. By noting that (T_1, T_2) has survivor function FG, use empirical process theory and asymptotic results for $\hat{G}$ to show that F^* is a uniformly consistent estimator of F over a region $[0, \tau_1] \times [0, \tau_2]$ such that $F(\tau_1, \tau_2) G(\tau_1 \vee \tau_2) > 0$. Does this estimator assign negative mass? (Lin and Ying, 1993a)

10.8 Consider bivariate failure time data with random bivariate censoring and define a survivor function estimator F^* at failure time grid points by

$$F^*(t_{1i}, t_{2j}) = \#\left(T_1 > t_{1i}, T_2 > t_{2j}\right) / \left[n\hat{G}(t_1, t_2)\right],$$

where $\hat{G}$ is a strongly consistent estimator of the censoring survivor function $G(t_1, t_2) = P(C_1 > t_1, C_2 > t_2)$. Using empirical process theory, show that F^* is uniformly consistent for F over $[0, \tau_1] \times [0, \tau_2]$ such that $F(\tau_1, \tau_2)$ $G(t_1, t_2) > 0$ (Burke, 1988). Calculate F^* at grid points in the failure time risk region using the Bickel estimator of G for the data of Table 10.2 and compare with Figure 10.2. Does F^* assign negative mass?

10.9 Consider the partially maximized likelihood function (10.21) for the data of Table 10.1. Is the nonparametric maximum likelihood estimate uniquely determined for the data set? Identify the points, half-lines, or quadrants that are assigned positive mass by an NPML procedure.

CHAPTER 11

Additional Failure
Time Data Topics

11.1 INTRODUCTION

We conclude our discussion of failure time data analysis methods with brief presentations on some more specialized but important topics. These include the topics discussed in the final chapter of the first edition of this book, together with the additional topics of missing and mismeasured covariates and sequential testing.

In Section 11.2 we continue the presentation of Chapter 10 on the analysis of paired failure times by considering regression models that stratify on pair membership. Such models may be useful if the information on regression effects of interest resides wholly, or largely, within pairs. In Section 11.3 we consider failure time methods under a very specialized censoring scheme when all study subjects have the same potential censoring time, and make certain connections with binary response data analysis methods. In Section 11.4 we consider a topic important in epidemiologic and disease prevention clinical trials contexts, namely, failure time data analyses when covariate data are available only on individuals having uncensored failure times (the cases) and on a selected sub-sample of censored individuals (the controls). In subsequent sections we consider the related topics of failure time analysis when some covariate data are missing (Section 11.5) or are measured with error (Section 11.6). There has been considerable development on these important topics over the past couple of decades. In Section 11.7 we consider briefly group sequential testing in clinical trials using failure time endpoints, and related estimation problems. In Section 11.8 we provide an overview of Bayesian methods for failure time data analysis, especially in relation to the Cox regression model. Section 11.9 retains much of the discussion of some practical issues arising in the analysis of a mouse leukemia dataset from first edition while bringing in further analyses related to missing covariate data.

11.2 STRATIFIED BIVARIATE FAILURE TIME ANALYSIS

11.2.1 Background

Suppose that there is a single possibly censored failure time for each study subject, and a single failure type. As mentioned in Chapter 10, failure times may occur in pairs either naturally or by design. Natural pairing arises, for example, in twin studies and in studies in which two failure times are recorded on each individual or piece of equipment. In a matched-pair study, individuals sharing certain characteristics are assigned to a pair. Table 11.1 provides an additional example of naturally paired failure times. The survival times of closely and poorly HLA (human lymphocyte antigen) matched skin grafts on the same burned individual are given along with the percentage of body surface of a full-thickness burn.

Typically, certain primary regressor variables x (such as quality of the match in Table 11.1) are to be related to failure time, and other factors (such as characteristics of the individual receiving the graft) form the basis for pairing. The methods of Section 10.2 for examining regression effects on marginal hazard rates would often be appropriate for such data. In some specialized circumstances, the data pertinent to regression effects of interest will reside primarily within pairs. In such settings, stratified regression models, which make a rather comprehensive accommodation for pair effects in relation to failure time, may be considered. For example, in Table 11.1, information on the relationship between closely and poorly matched skin grafts and graft survival derives primarily from within-patient failure time comparisons. Similarly, the Diabetic Retinopathy Study Research Group (1981) conducted a clinical trial among 197 patients at high risk for loss of visual acuity, wherein one eye from each patient was selected for laser photocoagulation treatment while the other eye was observed without treatment. Each eye was then followed to observe the time from enrollment in the trial to the first occurrence of loss of visual acuity below a certain threshold, or right censorship. The laser treatment

Table 11.1 Days of Survival of Closely and Poorly Matched Skin Grafts on the Same Person [a]

Case number	4	5	7	8	9	10	11	12	13	15	16
Survival of close match graft	37	19	57^+ (57^+)	93	16	21–23	20	18	63 $(77,29)$	29	60^+
Survival of poor match graft	29	13	15	26	11	15–18	26	19–23	43	15 (18)	38–42
Amount of burn	30	20	25	45	20	18	35	25	50	30	30

Source: Batchelor and Hackett (1970).

[a] Amount of burn is percentage of body surface of full thickness of burn. A$^+$ indicates censoring, and the numbers in parentheses are survival times of other grafts on the same person.

effects in this study could also be estimated by within-patient contrasts. See Huster et al. (1989) and Prentice and Hsu (1997) for parametric and semiparametric marginal regression analyses of data from this study.

Estimation of the primary regression effects in this type of setting can also be carried out by allowing each failure time data pair to define a new stratum, leading to an analysis based on within-pair data. Such an analysis is valid even if matching factors can be modeled explicitly, but with some associated reduction in efficiency. The remainder of this section is devoted to models in which each pair includes a parameter or parameters specifically intended to describe failure time properties of the stratum defined by that pair.

Any of the parametric or partially parametric methods of preceding chapters can be considered for such data analysis. Cox model methods are considered in Section 11.2.2; log-linear (accelerated failure time) methods are considered in Section 11.2.3.

11.2.2 Cox Model Methods

Let $[t_{si}; X_{si}(t_{si})]$ represent the failure time $(\tilde{T}_{si} = t_{si})$ and regression variable history prior to t_{si} for the ith individual in the sth pair. Let $Z_{si}(u)' = [Z_{si1}(u), \ldots, Z_{sip}(u)]$ denote a corresponding modeled regression p-vector at follow-up time u.

Under a Cox model the hazard function for the sth pair may be written

$$d\Lambda[t, X(t), s] = \lambda_{0s}(t) \exp[Z(t)'\beta]. \tag{11.1}$$

A partial (also conditional) likelihood can be formed as the product over pairs of the conditional probability for the *pair rank* given the smallest failure time $t_{s(1)}$ in the sth pair. With continuous failure times the pair rank has only two possible values, depending on which member of the pair fails first. Denote

$$r_s = \begin{cases} 0, & t_{s1} < t_{s2} \\ 1, & t_{s2} < t_{s1} \end{cases}$$

and set $d_s = Z_{s2}[t_{s(1)}] - Z_{s1}[t_{s(1)}]$. The contribution to the conditional likelihood from pair s is then

$$P[r_s \mid t_{s(1)}; d_s] = \frac{\exp(r_s d_s'\beta)}{1 + \exp(d_s'\beta)}. \tag{11.2}$$

The conditional likelihood

$$\prod_{s \in S} \frac{\exp(r_s d_s'\beta)}{1 + \exp(d_s'\beta)}, \tag{11.3}$$

where S consists of all pairs for which r_s can be specified, is precisely a binary logistic likelihood. Standard asymptotic likelihood methods can be applied to (11.3). The score statistic is

$$U(\beta) = \sum_{s \in S} d_s \left[r_s - \frac{\exp(d_s'\beta)}{1 + \exp(d_s'\beta)} \right]$$

(11.4)

and the information matrix observed is

$$I(\beta) = \sum_{s \in S} d_s d_s' \frac{\exp(d_s'\beta)}{[1 + \exp(d_s'\beta)]^2}.$$

(11.5)

Further detail on estimation from (11.3) is given, for example, in Cox and Snell (1989).

Table 11.1 gives $r_s = 1$ for nine pairs and $r_s = 0$ for two pairs. A single indicator regressor variable for quality of the match (0, close match; 1, poor match) leads to $d_s \equiv 1$, all s, and a likelihood proportional to

$$\frac{\exp(9\beta)}{(1 + \exp\beta)^{11}},$$

so that $\hat{\beta} = \log\frac{9}{2} = 1.504$ with asymptotic variance $11/[(2)(9)] = 0.611$. This yields an approximate 95% confidence interval for β of $[1.504 \pm 1.96(0.611)^{1/2}] = (-0.028, 3.036)$. Correspondingly, an approximate interval for e^β, the ratio of hazard rates for poorly and well-matched grafts, is $(0.97, 20.83)$. In good agreement, an exact test for $\beta = 0$ based on the binomial parameter $p = e^\beta/(1 + e^\beta)$ yields a significance level of $2(1 + 11 + 55)/2^{11} = 0.065$. An additional regressor variable defined as the product of the indicator variable above and amount of burn can be used to test whether the importance of the HLA tissue typing depends on the severity of the burn.

11.2.3 Accelerated Failure Time Methods

We may also consider stratified log-linear failure time models for this type of paired data. Suppose that $Y_{si} = \log \tilde{T}_{si}$ follows a model

$$Y_{si} = \alpha_s + Z_{si}'\beta + \sigma_s v_{si},$$

(11.6)

where f_s is the density function for v_{si}, Z_{si} is time independent, and Y_{si} variates are independent for $s = 1, 2$, and $i = 1, \ldots, n$. Expression (11.6) is overparameterized, having more unrelated parameters than failure times. The assumption of common error distribution ($\sigma_s = \sigma$, $f_s = f$; all s) relieves this problem and leads to

$$D_s = Y_{s2} - Y_{s1} = d_s'\beta + \sigma w_s,$$

(11.7)

where the error w_s, corresponding to the pairwise difference in log failure times, has symmetric PDF.

$$g(w) = \int_{-\infty}^{\infty} f(v)f(w+v)\,dv. \tag{11.8}$$

For example, a Weibull density for $\tilde{T}_{si}$ leads to an extreme minimum density for v_{si}, which in turn gives a logistic density for w_s. Log-normal $\tilde{T}_{si}$'s, on the other hand, give a normal regression model for D_s.

Using D_s as the basic data eliminates the nuisance parameter α_s from the likelihood and avoids potential problems with consistency that can arise from (11.6). The model for D_s depends only on a fixed number of $p+1$ parameters, and with uncensored data the methods of Chapter 3 can be used for inference on β. These methods have somewhat greater efficiency than the partially parametric method of Section 11.2.2. For example, assuming a two-parameter Weibull model, the relative efficiency of the maximum likelihood estimator based on (11.3) to that based on (11.7) is 75% at $\beta = 0$ and less than 75% elsewhere. See also Table 4.5 and related discussion.

In the presence of censoring, however, a problem arises with pairs in which both members are censored. In this situation, basing inferences on the partial information available on D_s is not feasible since pairs with both times censored clearly can convey some information on β even though the range of possible values of D_s is unrestricted. Some authors have ignored such pairs in the analysis of the censored differences D_s even though there is potential for significant bias [see Holt and Prentice (1974) for further detail].

An alternative approach is to write the full likelihood for α_s, β, and σ using the distribution of Y_{s1} and Y_{s2}, but again, this yields the problem of many nuisance parameters. One approach to inference in this case is to suppose that $\alpha_1, \alpha_2, \ldots$ arise as IID random variables from some common parametric distribution and apply ordinary likelihood methods (e.g., Wild, 1983).

Following Chapter 7, semiparametric analysis from (11.6) may also be based on residual ranks for each pair. For simplicity, again assume a common error distribution ($\sigma_s = \sigma$, $f_s = f$). Now, in order to test $\beta = \beta_0$, form $w_{si} = Y_{si} - Z_{si}'\beta_0$ and let

$$r_s = r_s(\beta_0) = \begin{cases} -1, & w_{s1} < w_{s2} \\ 1, & w_{s2} \le w_{s1} \end{cases}$$

Then

$$P(r_s = -1; Z_{s1}, Z_{s2}) = \int_{-\infty}^{\infty} \int_{\tau_{s1}}^{\infty} \prod_{1}^{2} [f(\tau_{si} - Z_{si}'\gamma)\,d\tau_{si}],$$

where

$$\tau_{si} = \frac{w_{si} - \alpha_s}{\sigma}, \qquad \gamma = \frac{\beta - \beta_0}{\sigma},$$

and

$$P(r_s = 1; Z_{s1}, Z_{s2}) = 1 - P(r_s = -1; Z_{s1}, Z_{s2}).$$

The contribution from pair s to the score statistic for testing $\beta = \beta_0 (\gamma = 0)$ is simply

$$U_s = \frac{d \log P(r_s; Z_{s1}, Z_{s2})}{d\gamma}$$

$$= c r_s d_s,$$

where $d_s = Z_{s2} - Z_{s1}$, as before, and

$$c = 2 \int_{-\infty}^{\infty} \int_{\tau_1}^{\infty} \frac{-d \log f(\tau_1)}{d\tau_1} \prod_{1}^{2} [f(\tau_i) \, d\tau_i]$$

does not depend on s. Under $\beta = \beta_0$, r_s takes values -1 or $+1$, each with probability $\frac{1}{2}$. Hence, the variance of u_s is

$$V_s = c^2 d_s' d_s.$$

Let S denote the set of pairs for which $r_s = r_s(\beta_0)$ can be specified and set

$$U = \sum_S U_s, \quad V = \sum_S V_s.$$

A test for $\beta = \beta_0$ can then be based on an approximate χ^2 distribution (degrees of freedom the dimension of β, assuming V nonsingular) for

$$U'V^{-1}U = \sum_S d_s' r_s \left(\sum_S d_s d_s' \right)^{-1} \sum_S d_s r_s. \tag{11.9}$$

Note that (11.9) is independent of the choice of f.

The data of Table 11.1 lead to a $\chi^2_{(1)}$ statistic for testing the hypothesis of no association between HLA match and survival of value

$$U'V^{-1}U = (9 - 2)(11)^{-1}(9 - 2) = 4.45,$$

which is significant at the 0.05 level. Such a test is asymptotically valid regardless of the error distribution f.

The omission of pairs from (11.9) for which r_s cannot be specified (smaller w_{si} value is censored) does not bias the test since r_s takes value ± 1 with probability $\frac{1}{2}$

under $\beta = \beta_0$, under independent censorship. Interval estimation and partial tests for β can be based on (11.9), as discussed in Chapter 7. Note that unlike inference based on (11.3), the pairs utilized in testing $\beta = \beta_0$ via (11.9) depend on the hypothesized value β_0.

11.3   FIXED STUDY PERIOD SURVIVAL STUDIES

In some studies the only failure time data recorded are whether or not each individual survives some study period common to all individuals. Such studies can be viewed as special cases of classical quantal response bioassay. For example, Table 11.2 presents data on death or survival of insects after 5 hours of exposure to various levels of gaseous carbon disulfide.

Each of the parametric and semiparametric survival models of preceding chapters may be specialized to this type of data and yield binary response regression models. Consider first the log-linear models of Chapter 3. Log failure time Y is presumed to arise from a model

$$Y = \mu + Z'\gamma + \sigma v,$$

with error PDF $f(v)$ and fixed covariate Z. The probability that failure time t is less than a study period T_0 may be written

$$P(Z) = \int_{-\infty}^{\alpha + Z'\beta} F(v)\, dv,$$

Table 11.2   Mortality of Adult Beetle After 5 Hours' Exposure to Gaseous Carbon Disulfide

	Dosage ($\log_{10}$ CS$_2$ mg/L)							
	1.6907	1.7242	1.7552	1.7842	1.8113	1.8369	1.8610	1.8839
Insects	59	60	62	56	63	59	62	60
Killed	6	13	18	28	52	53	61	60
Probit fit	3.27	10.89	23.65	33.88	49.60	53.28	59.63	59.21
($\hat{\mu} = 1.771$,								
$\hat{\sigma} = 0.051$)								
Logit fit	3.45	9.84	22.45	33.89	50.10	53.29	59.22	58.74
($\hat{\mu} = 1.772$,								
$\hat{\sigma} = 0.029$)								
log $F(m_2 = 1)$	6.07	11.23	20.10	29.69	48.57	54.84	60.93	59.75
($\hat{\mu} = 1.818$,								
$\hat{\sigma} = 0.016$, $\hat{m}_1 = 2.79$)								

Source: Bliss (1935).

where $\alpha = (\log T_0 - \mu)/\sigma$ and $\beta = \gamma/\sigma$. Weibull, log-logistic, and log-normal models for T then yield the *multihit model*

$$P(Z) = 1 - \exp(-e^{\alpha + Z\beta}),$$

the logit model (11.2)

$$P(Z) = \frac{\exp(\alpha + Z'\beta)}{1 + \exp(\alpha + Z'\beta)},$$

and the probit model

$$P(Z) = \int_{-\infty}^{\alpha + Z'\beta} (2\pi)^{-1/2} \exp\left(-\frac{1}{2}v^2\right) dv,$$

respectively. The terminology *multihit*, *logit*, and *probit* usually apply to situations in which there is a scalar regression variable Z that gives the dosage of a toxic substance. The present notation includes the possibility that the regression variable Z may include more complicated functions of such a dosage variable and other covariates.

The proportional hazards model may also be used to induce a binary response model. A relative risk model, $\lambda_0(t)e^{Z'\beta}$, with fixed covariates gives once again the multihit model

$$P(Z) = 1 - \exp(-e^{\alpha + Z'\beta}),$$

where $\alpha = \alpha(T_0) = \log \int_0^{T_0} \lambda_0(u)\, du$.

The generalized F regression model (Chapters 2 and 3) offers additional special cases. For example, the subclass with $m_2 = 1$ gives

$$P(Z) = \left[\frac{\exp(\alpha + Z'\beta)}{1 + \exp(\alpha + Z'\beta)}\right]^{m_1},$$

whereas $m_1 = 1$ leads to

$$P(Z) = 1 - \left[\frac{\exp(-\alpha - Z'\beta)}{1 + \exp(-\alpha - Z'\beta)}\right]^{m_2}.$$

Model fitting and estimation from such binary response models will be considered very briefly. For individual i with regression vector Z_i, let

$$r_i = \begin{cases} 1, & t_i < T_0 \\ 0, & \text{otherwise.} \end{cases}$$

The likelihood is simply

$$\prod_1^n P(Z_i)^{r_i} Q(Z_i)^{1-r_i},$$

where $Q = 1 - P$. The score statistic in regard to a parameter $\theta = (\theta_1, \theta_2, \dots)$ that includes α, β, and possibly other shape parameters is given by

$$U = \sum_{i=1}^n [r_i P(Z_i)^{-1} - (1 - r_i) Q(Z_i)^{-1}] \frac{dP(Z_i)}{d\theta}, \qquad (11.10)$$

and the information matrix Σ is

$$\sum_{i=1}^n [P(Z_i) Q(Z_i)]^{-1} \left[\frac{\partial P(Z_i)}{\partial \theta} \right]^{\otimes 2}. \qquad (11.11)$$

Further,

$$\frac{dP(Z)}{d\alpha} = f(\alpha + Z'\beta), \qquad \frac{dP(z)}{d\beta} = Zf(\alpha + Z'\beta) \qquad (11.12)$$

so that model fitting with any specific f is typically straightforward if $P(Z)$ can be calculated conveniently. Score tests for log-normal (probit) and log-logistic (logit) failure time models relative to the generalized F regression model are also straightforward (Prentice, 1976b). Let $\theta = (\theta_1', \theta_2')'$, where $\theta_1 = (\alpha, \beta)$ and $\theta_2 = (m_1, m_2)$. Similarly, partition the score statistic (11.10) into $U = (U_1', U_2')'$ and covariance matrix (11.11) into

$$\Sigma = \begin{pmatrix} \Sigma_{11} & \Sigma_{12} \\ \Sigma_{21} & \Sigma_{22} \end{pmatrix}.$$

Consider now a score test for $\theta_2 = \theta_2^0$. A Newton–Raphson iteration based on (11.10)–(11.12) gives $\hat\alpha = \hat\alpha(\theta_2^0)$ and $\hat\beta = \hat\beta(\theta_2^0)$. Under $\theta_2 = \theta_2^0$, the asymptotic distribution of the test statistic

$$U_2'(\Sigma_{22} - \Sigma_{21}\Sigma_{11}^{-1}\Sigma_{12})^{-1} U_2 \qquad (11.13)$$

is χ_2^2, with all quantities in (11.13) evaluated at $\hat\alpha, \hat\beta$. For example, to test a log-logistic hypothesis, we need only note that

$$\frac{dP(Z)}{dm_j} = \frac{(-1)^j \log\{1 + \exp[(-1)^j(\alpha + Z'\beta)]\}}{1 + \exp[(-1)^j(\alpha + Z'\beta)]}, \qquad j = 1, 2.$$

With the data of Table 11.2 and $Z = \log CS_2$ per milligram per liter, (11.13) has value 7.95, which has significance level 0.019 on the basis of a χ_2^2 distribution. The log-normal model is most conveniently tested in terms of the parameters (q, p) of Section 3.8.2. Now take

$$\theta_2 = \binom{q}{p}$$

and note that

$$\frac{d \log P(Z)}{dq} = \frac{[(\alpha + Z'\beta)^2 + 2]\phi(\alpha + Z'\beta)}{6}$$

$$\frac{d \log P(Z)}{dp} = \frac{[(\alpha + Z'\beta)^3 + 3(\alpha + Z'\beta)]\phi(\alpha + Z'\beta)}{24},$$

where $\phi(w) = (2\pi)^{-1/2} \exp(-\frac{1}{2}w^2)$.

The statistic (11.13), with the data of Table 11.2, gives a χ_2^2 value of 7.40, which occurs at the 0.025 upper endpoint of the distribution. Table 11.2 also gives parameter estimates for the fit of probit and logit models as well as the generalized F model with $m_2 = 1$.

A final comment to complete this section: Logistic regression models for disease occurrence are often used for the analysis of epidemiologic cohort study data. Provided that the follow-up duration is not highly variable among study subjects and disease occurrence is rare, the resulting odds ratio parameter estimates and their standard error estimates can be expected to agree fairly closely with corresponding relative risk estimates and standard errors from a corresponding proportional hazards model (e.g., D'Agostino et al, 1990). Some authors have used binary logistic regression techniques even when follow-up times vary considerably among study subjects, but with follow-up time included as an additional component of the regression vector. Once again, one might expect substantial agreement between Cox model relative risk parameter estimates and logistic regression odds ratio parameter estimates for the other modeled covariates.

11.4 COHORT SAMPLING AND CASE–CONTROL STUDIES

Consider again a Cox regression model

$$\lambda[t, X(t)] = \lambda_0(t) \exp[Z(t)'\beta] \tag{11.14}$$

for a univariate failure time variate T having corresponding covariate history $X(t)$ up to follow-up time t, and modeled covariate p-vector $Z(t)$ at time t. The partial likelihood estimating function for β, based on the follow-up of a cohort of size n, can then be written [e.g., (5.50)] as

$$U(\beta) = \sum_{i=1}^{n} \int_{0}^{\infty} [Z_i(t) - \mathscr{E}(\beta, t)] \, dN_i(t), \qquad (11.15)$$

where

$$\mathscr{E}(\beta, t) = S^{(1)}(\beta, t) / S^{(0)}(\beta, t)$$

$$= \sum_{i=1}^{n} Y_i(t) Z_i(t) p_i(\beta, t)$$

and where

$$S^{(j)}(\beta, t) = \sum_{i=1}^{n} Y_i(t) Z_i(t)^{(j)} \exp[Z_i(t)'\beta] \qquad \text{for } j = 0, 1$$

and

$$p_i(\beta, t) = \exp[Z_i(t)'\beta] \left\{ \sum_{\ell=1}^{n} Y_\ell(t) \exp[Z_\ell(t)'\beta] \right\}^{-1}.$$

Note that $\mathscr{E}(\beta, t)$ is a weighted average of covariate values among cohort members at risk $[Y_i(t) = 1]$ at time t.

The key to the use of (11.15) for estimation of β is the zero expectation of the score function increments $\sum [Z_i(t) - \mathscr{E}(\beta, t)] \, dN_i(t)$ at the "true" β value at any follow-up time t. Since these increments are uncorrelated under (11.14), a simple variance estimator is available for the solution $\hat{\beta}$ to $U(\beta) = 0$.

In the context of large cohorts and infrequent failures, the assembly of covariate histories on all cohort members may be prohibitively expensive. For example, epidemiologic cohort studies and disease prevention clinical trials often include tens of thousands of study subjects, while only a few hundred subjects may have experienced the study disease (or diseases) by the time of data analysis. The covariates of interest in such settings may involve, for example, biochemical or genetic analysis of blood specimens or the extraction of detailed work histories from employer records in order to derive job exposure histories.

The weighted average $\mathscr{E}(\beta, t)$ can generally be replaced by a corresponding weighted average over the failing individual (the case) and a random sample of the other individuals at risk (the controls) at time t without affecting the zero expectation of the estimating function. More specifically, let $\tilde{R}(t)$ denote the case and time-matched controls at cohort follow-up time t. The estimating function

$$\hat{U}(\beta) = \sum_{i=1}^{n} \int_{0}^{\infty} [Z_i(t) - \hat{\mathscr{E}}(\beta, t)] \, dN_i(t), \qquad (11.16)$$

where

$$\hat{\mathscr{E}}(\beta,t) = \sum_{\ell \varepsilon \tilde{R}(t)} Y_\ell(t) Z_\ell(t) \tilde{p}_\ell(\beta,t),$$

with $\tilde{p}_\ell(\beta,t)$ defined below, will generally satisfy $E[\hat{U}(\beta)] = 0$. At failure time t

$$L_i(\beta,t) = P[dN_i(t) = 1|\tilde{R}(t), dN.(t) = 1, X_\ell(t), \ell \in \tilde{R}(t)]$$

$$= \exp[Z_i(t)'\beta] \Big/ \sum_{\ell \varepsilon \tilde{R}(t)} Y_\ell(t) \exp[Z_\ell(t)'\beta]$$

$$= \tilde{p}_i(\beta,t) \tag{11.17}$$

for $i \in \tilde{R}(t)$ provided only that the controls $[Y_\ell(t) = 1]$ are a random sample of the nonfailing cohort members at risk at time t. The expectation of zero then follows since the contribution to (11.16) at a failure time t is the score, $\partial \log L_i(\beta,t)/\partial \beta$, from (11.17). The contributions to (11.16) at distinct failure times are typically correlated, however, since the conditioning event in (11.17) excludes information [e.g., $\tilde{R}(u), u < t$] conditioned on at earlier failure times. A broad range of sampling procedures could be considered that will lead to a random sample of controls at each failure time. To date, most attention has been given to two specific sampling procedures: nested case–control sampling (Thomas, 1977; Prentice and Breslow, 1978), in which a random sample of controls of fixed size m, typically in the range 1 to 5, is selected for each failure time either without replacement or with replacement independently across time; and case–cohort sampling (Prentice, 1986a), in which a random subcohort provides the controls at each failure time, so that $\tilde{R}(t)$ is comprised of the failing individual and all subcohort members at risk $[Y(t) = 1]$ at time t.

Asymptotic distribution theory for $\hat{\beta}$ solving $\hat{U}(\beta) = 0$ and for a corresponding cumulative hazard function estimator under nested case–control sampling is given by Borgan et al. (1995); see also Goldstein and Langholz (1992). These authors augment the usual conditioning information $\mathscr{F}_{t^-} = \{N_i(u), Y_i(u), X_i(u); 0 \le u < t, i = 1, \ldots, n\}$ to $\mathscr{H}_{t^-} = \mathscr{F}_{t^-} \vee \{\tilde{R}(u), 0 \le u \le t\}$ and show that under mild conditions, the conditional probabilities (11.17) still obtain under $1, \ldots, m$ matched case–control sampling without replacement; that is,

$$p[dN_i(t) = 1|\mathscr{H}_{t^-}, dN.(t) = 1] = \tilde{p}_i(\beta,t). \tag{11.18}$$

In fact, (11.18) holds whether the controls at time t are selected without replacement, or with replacement independent of whether the same individual may have been selected earlier as a control. Arguments using counting processes and martingales apply to the partial likelihood score function (11.16) under nested case–control sampling, giving an asymptotic mean zero normal distribution for $\hat{\beta}$ with mean β and consistent variance estimator $\hat{I}(\hat{\beta})^{-1}$, where

$$\hat{I}(\beta) = -\partial \hat{U}(\beta)/\partial \beta' = \sum_{i=1}^{n} \int_0^{\infty} \mathcal{V}(\beta, t)\, dN_i(t)$$

and

$$\mathcal{V}(\beta, t) = \sum_{\ell \in \tilde{R}(t)} Y_\ell(t)[Z_\ell(t) - \hat{\mathscr{E}}(\beta, t)]^{\otimes 2} \tilde{p}_\ell(\beta, t)$$

and an asymptotic Gaussian distribution for a corresponding cumulative estimator, under mild conditions.

For many other sampling schemes, $P[dN_i(t) = 1 | \mathscr{H}_{t^-}, dN.(t) = 1]$ will not equal the right side of (11.18), but rather, the exponential terms will each be multiplied by a weighting factor (Borgan et al., 1995). For example, under case–cohort sampling, the matched set at a failure time t is comprised of the failing individual i at t along with all at-risk members of a fixed subcohort $\mathscr{C}$; that is, $\tilde{R}(t) = i \cup \mathscr{C}$. If the failing subject is outside the subcohort, then conditioning on $\{\tilde{R}(u), 0 \leq u \leq t\}$ will typically allow the failing individual to be identified, in which case $P[dN_i(t) | \mathscr{H}_{t^-}, dN.(t) = 1] = 1$ for the failing individual, and there would be no contribution to β estimation at that failure time. However, (11.17), which conditions on $\tilde{R}(u)$ only at $t = u$, holds for case–cohort sampling, and asymptotic distribution for $\hat{\beta}$ solving $\hat{U}(\beta) = 0$ can be developed (Self and Prentice, 1988) by writing

$$\hat{U}(\beta) = U(\beta) + \int_0^{\infty} [\mathscr{E}(\beta, t) - \hat{\mathscr{E}}(\beta, t)]\, dN_i(t). \tag{11.19}$$

The first term in (11.19) is the full-cohort score function, for which asymptotic results are available (e.g., Chapter 5). Finite population asymptotic results apply to the second term in (11.19), and the two terms turn out to be asymptotically uncorrelated. Hence, $\hat{\beta}$ generally has an asymptotic normal distribution with mean β. The corresponding variance matrix can be estimated consistently (Prentice, 1986a; Self and Prentice, 1988) by

$$\hat{I}(\hat{\beta})^{-1} \hat{V}(\hat{\beta}) \hat{I}(\hat{\beta})^{-1},$$

where the score function variance estimator $\hat{V}(\hat{\beta})$ is given by

$$\hat{V}(\hat{\beta}) = \hat{I}(\hat{\beta}) + 2\sum_{j=1}^{n} \delta_j \Delta_j \sum_{k|t_k < t_j} \delta_k \hat{v}_{kj}. \tag{11.20}$$

The second term in (11.20) estimates the inflation in score function variance due to the subsampling. In this term δ_j is the failure indicator for the jth study subject and $t_j = T_j \wedge C_j$ is the corresponding observed time; Δ_j is an indicator variable that takes value 1 if the jth subject is outside the subcohort and zero otherwise. Also, $\hat{v}_{kj}$

is an estimator of the covariance between the score contributions arising from the jth and kth failure times, given by

$$\hat{v}_{kj} = -\sum_{\ell \in \tilde{R}(t_j)} \frac{B_k + b_{jk} - b_{\ell k}}{R_k + r_{jk} - r_{\ell k}} \left(c_{\ell j} - \frac{B_j}{R_j} \right) r_{\ell j} R_j^{-1},$$

where $r_{\ell m} = Y_\ell(t_m) \exp[Z_\ell(t_m)'\hat{\beta}]$, $b_{\ell m} = Y_\ell(t_m)Z_\ell(t_m)\exp[Z_\ell(t_m)'\hat{\beta}]$, $c_{\ell m} = Y_\ell(t_m)$ $Z_\ell(t_m)$ for all $\ell, m \in \{1,\ldots,n\}$ and R_m and B_m are sums over $\ell \in \tilde{R}(t_m)$ of $r_{\ell m}$ and $b_{\ell m}$ values, respectively. Wacholder et al. (1989) and Barlow (1994) develop other variance estimators for $\hat{\beta}$ under case–cohort sampling.

The nested case–control and case–cohort regression estimators just described each has its advantages. Either can lead to major cost savings relative to full-cohort approaches. The availability of the case–cohort subcohort $\mathscr{C}$ may be useful for study monitoring and can provide a comparison group for each of the multiple study diseases that may be of interest in a cohort study. For example, Sorensen and Andersen (2000) generalize case–cohort sampling methods to competing risk data. The fact that a subcohort member serves in the comparison group for all cases in that individual's risk period, whereas a matched control does so only at the failure time (or times) where the control is specifically selected, can also lead to a modest efficiency advantage for the case–cohort estimator (e.g., Self and Prentice, 1988) in some circumstances. However, the nested case–control estimator has the advantage of aligning the cases and controls in a very direct manner, and this can lead to noteworthy efficiency advantages in the presence of highly variable follow-up periods among study subjects (e.g., Langholz and Thomas, 1990). The efficiency of the nested case–control estimator compared to the full-cohort estimator is about $m(m+1)^{-1}$ at $\beta = 0$ and varies somewhat as β departs from zero, depending in part on the covariate distribution (e.g., Whittemore, 1981; Breslow et al., 1983). Wacholder et al. (1989) provide simple procedures for estimating power and efficiency for the case–cohort estimator.

Modifications of either estimator can be considered toward ensuring good properties. For example, in an assembled cohort, one could generalize the case–cohort procedure by stratifying the cohort on the at-risk period of study subjects and comparing failures to subcohort members within strata to ensure a good correspondence between cases and comparison individuals. Also, alternative estimators can be considered to enhance efficiency under nested case–control sampling in an assembled cohort. For example, Samuelsen (1997) considers an estimator $\hat{\beta}$ that again solves

$$\sum_{i=1}^{n} \int_0^\infty \frac{\eta_i}{\pi_i} [Z_i(t) - \hat{\mathscr{E}}(\beta, t)] \, dN_i(t) = 0 \tag{11.21}$$

but with

$$\hat{\mathscr{E}}(\beta, t) = \sum_{i=1}^{n} \frac{\eta_i}{\pi_i} Y_i(t)Z_i(t) \exp[Z_i(t)'\beta] \bigg/ \sum_{i=1}^{n} \frac{\eta_i}{\pi_i} Y_i(t) \exp[Z_i(t)'\beta]. \tag{11.22}$$

In this expression, $\eta_i = 1$ for cases and all sampled controls, and $\eta_i = 0$ otherwise; π_i is the overall sampling probability, given the failure and at-risk data on the cohort, for the ith subject. This sampling probability is given by $\pi_i = 1$ if the ith individual is a case $(\delta_i = 1)$ and

$$\pi_i = 1 - \prod_{j=1}^{n}\left[1 - m\frac{Y_i(t_j)}{Y.(t_j) - 1}\right]^{\delta_j}$$

is the probability that the ith censored individual $(\delta_i = 0)$ is selected as a control under matched case–control sampling with replacement, and $t_j = T_j \wedge C_j$ is the time observed for the jth cohort member. This estimation procedure essentially constructs estimators of the full-cohort $S^{(1)}(\beta, t)$ and $S^{(0)}(\beta, t)$ terms in (11.15) in forming $\hat{\mathscr{E}}(t, \beta)$. The estimation procedure thereby uses a selected control as a part of the comparison group for all cases in its risk period and allows the same set of controls to provide a comparison group for several types of failure. Samuelsen (1997) presented simulation results indicating that estimation based on (11.21) and (11.22) may improve on the efficiency of the usual estimator based on (11.16) and (11.17) under nested case–control sampling, but a complete account of the corresponding asymptotic distribution theory has yet to be given for this estimator. Chen and Lo (1999) consider a closely related estimation procedure under case–cohort sampling.

It may happen that some covariates are available routinely on the entire cohort, whereas others await expensive analysis of stored materials. Under these circumstances a cohort sampling method may allow the sampling rates to depend on the available covariate value, so-called *two-stage sampling* (e.g., Breslow and Cain, 1988). Borgan et al. (2000) consider exposure stratified case–cohort designs and corresponding relative risk parameter estimates.

The cohort sampling procedures considered in this section can be viewed as special missing covariate procedures, with covariates missing by design. Some rather general approaches to accommodating missing covariates in the Cox model are discussed in the next section. These approaches provide some additional relative risk estimators and provide potential for efficiency improvement over the estimators discussed in this section, particularly if some elements of the modeled covariate are available on all cohort members.

There is a strong connection between the cohort sampling procedures discussed here and the classical case–control study with its associated odds ratio estimators. A classical case–control study may begin by defining a cohort to be comprised of all persons meeting certain eligibility criteria in a geographic area covered by a disease register. The study may then involve the collection of survey materials and biological specimens on all cases identified by the register in a defined case accession period and from a matched control group who are still without disease at the end of the accession period. Variations exist according to whether controls are time matched to cases or are individually or frequency matched to cases on other characteristics. Provided that controls (and cases) can be drawn randomly from the study cohort and that the occurrence of disease, or knowledge of disease occurrence, does not affect the materials collected or covariate measurements taken, the classical case–control study can be a very efficient alternative to a cohort study, especially for studies of rare diseases. Estimates of relative risk under a time-matched case–control study, or odds ratios under a fixed study period control selection (see Section 11.3), do not require that a cohort roster be available, although corresponding cumulative hazard or disease probability estimates do. Of course, the ability to avoid selection bias and recall bias are major considerations in deciding between a cohort study and a classical population-based case–control study.

11.5 MISSING COVARIATE DATA

Consider again the Cox model (11.14) and suppose that the modeled covariate is partitioned as $Z(t) = [Z_1(t)', Z_2(t)']'$, where Z_1 may be missing for some study subjects, by design or by happenstance, while Z_2 is always available. In this section we provide a brief account of various approaches to the estimation of $\beta = (\beta_1', \beta_2')'$ and of their associated assumptions and properties. Available research largely confines attention to the fixed covariate special case $Z(t) = Z = (Z_1', Z_2')$, as we do in this section and the next.

The hazard rate at time t for an individual having Z_1 missing is (e.g., Prentice, 1982)

$$\lambda_0(t)E\{\exp[Z(t)'\beta]\}, \tag{11.23}$$

where the expectation is conditional on available components of the covariate history, including Z_2, and on $T \geq t$. The latter conditioning implies that this expectation will generally be a complicated function of β and Λ_0.

Consider a random censorship model where $\tilde{T}$ is the underlying failure time, C is the potential censoring time, and $\tilde{T}$ and C are independent given Z. Let $T = \tilde{T} \wedge C$ and $\delta = \mathbf{1}(\tilde{T} \leq C)$. If the probability that Z_1 is missing may depend on Z but not on (T, δ), the set of study subjects having Z_1 values available will constitute a random sample of the full cohort risk set at each failure time. In this situation, consistent estimation of β can proceed by applying standard partial likelihood procedures with $Y_\ell(t)$ replaced by $Y_\ell(t)\eta_\ell$, where $\eta_\ell = 1$ if $Z_{2\ell}(t)$ is observed and $\eta_\ell = 0$ otherwise, for all $\ell = 1, \dots, n$. This "complete case" analysis can be quite inefficient if there are appreciable missing values, and will generally be biased if the missingness probabilities depend on (T, δ).

Several authors have proposed simply calculated estimators of β to improve on the complete case estimator. For discrete covariates, and in the context of Z_2 being a surrogate variable for Z_1, Zhou and Pepe (1995) propose that β estimation proceed by replacing the expectation in (11.23) if Z_1 is missing, by a nonparametric estimate formed from the set of Z values for individuals at each failure time, and by maximizing a resulting estimated partial likelihood function. The maximum estimated partial likelihood estimator allows the missingness rate $\eta_\ell(t)$ to depend on Z_2 but not on (Z_1, T, δ). Lin and Ying (1993b) propose an approximate partial likelihood procedure with estimating function $\hat{U}(\beta) = [\hat{U}_1(\beta)', \hat{U}_2(\beta)']'$, where

$$\hat{U}_1(\beta) = \sum_{i=1}^{n} \int_0^{\infty} \eta_i(t)[Z_{1i} - \hat{\mathscr{E}}_1(\beta, t)] \, dN_i(t),$$

$$\hat{U}_2(\beta) = \sum_{i=1}^{n} \int_0^{\infty} [Z_{2i} - \hat{\mathscr{E}}_2(\beta, t)] \, dN_i(t),$$

and where $\hat{\mathscr{E}}(\beta, t) = [\hat{\mathscr{E}}_1(\beta, t), \hat{\mathscr{E}}_2(\beta, t)]'$ is of standard form but based only on study subjects having Z_1 available. This estimator can, but need not, improve on the efficiency of the complete case estimator, but it requires a "missing completely at random" (MCAR) assumption (e.g., Little and Rubin, 1987) for its validity; that is, the missingness probability for η is not allowed to depend on any aspect of (T, δ, Z).

More comprehensive estimators of β (and Λ_0) generally require further modeling assumptions, either concerning the probability distribution for the covariates, or concerning the missingness probabilities, or both. Inverse sampling probability weighted (IPW) estimators using (11.21) and (11.22) for β estimation will typically be consistent under "missing at random" (MAR) assumptions, wherein the missingness probability can depend arbitrarily on (T, δ, Z_2), provided that the missingness model $\pi = P[\eta = 1 | T, \delta, Z_2]$ is known or can be estimated consistently. However, these estimators can be quite inefficient if there is considerable missing data, in that only individuals having complete data $(\eta = 1)$ contribute to the summation (11.21).

Toward more efficient parameter estimation Chen and Little (1999) consider a nonparametric maximum likelihood (NPML) approach, whereas Wang and Chen (2001), building on results given in Robins et al. (1994), propose an augmented inverse probability weighted (AIPW) estimation procedure. Both methods involve expectations over the distribution of missing components of the covariate vector and hence require a model for the covariates to be specified.

The NPML procedure involves maximizing the full likelihood for the failure and covariate data available. This likelihood function can be written

$$L(\beta, \Lambda, \theta) = \prod_{i=1}^{n} \left(\lambda_0(t_i)\{e^{Z_i'\beta}\exp[-\Lambda_0(t_i)e^{Z_i'\beta}]f(Z_i, \theta)\}^{\eta_i} \right.$$
$$\left. \times \left\{ \int e^{Z_i'\beta}\exp[-\Lambda_0(t_i)e^{Z_i'\beta}]f(Z_i, \theta) \, dZ_i \right\}^{1-\eta_i} \right), \qquad (11.24)$$

where the integral in the second factor is over Z_{1i} given Z_{2i}. Also in (11.24), f denotes the probability density for Z, which is allowed to depend on a fixed-length parameter vector θ. The maximization proceeds by approximating Λ_0 by a step function that is constant between adjacent pairs of uncensored failure times, leading to a profile likelihood for β and θ to which standard likelihood formulas apply under mild conditions, as was shown by Chen and Little building on the work of

Murphy et al. (1997). Chen and Little describe an expectation-maximization algorithm to obtain the NPML parameter estimates and estimates of their covariance matrices.

Conditions under which the asymptotic results for the NPML procedure apply include identifiability and smoothness properties for f, a positive probability that Z_1 is observed at all (T, δ, Z) values, and a positive probability for an uncensored failure at all (T, Z). The integral in (11.24) and its related likelihood derivatives implies that numerical integration may be required to implement this procedure, although discrete covariates provide an important exception. Chen and Little also allow more complex missing data patterns than that considered here.

An important point concerns the independent censoring assumption that is needed for the NPML estimation procedure. The likelihood contribution from the ith subject to (11.24) obtains only if the potential censoring time C_i is independent of the failure time T_i given the observed covariate. Hence the censoring model for the ith subject is not allowed to depend on the value of missing Z_{1i}'s, and a stronger than usual independent censorship assumption attends the NPML procedure.

The AIPW estimator of Wang and Chen (2001) avoids this stronger censorship condition through inverse probability weighting. Their estimator of β solves an estimating equation of the form

$$\sum_{i=1}^{n}\left\{ \eta_i \pi_i^{-1} \int_0^\infty [Z_i(t) - \hat{\mathcal{E}}(\beta, t)]\, dN_i(t) + A_i(\beta) \right\} = 0, \qquad (11.25)$$

where the augmentation term is given by

$$A_i(\beta) = (1 - \eta_i \pi_i^{-1})\left\{ \int_0^\infty E[Z_i dN_i(u)|T_i, \delta_i, Z_{2i}] - \hat{\mathcal{E}}(\beta, u) E[dN_i(u)|T_i, \delta_i, Z_{2i}] \right\}$$

and where

$$\hat{\mathcal{E}}(\beta, t) = \hat{S}^{(1)}(\beta, t)/\hat{S}^{(0)}(\beta, t)$$

with

$$\hat{S}^{(j)}(\beta, t) = \sum_{\ell=1}^{n} \eta_\ell \pi_\ell^{-1} Y_\ell(t) Z_\ell^j \exp(Z_\ell'\beta) + (1 - \eta_\ell \pi_\ell^{-1}) Y_\ell(t) E[Z_\ell^j \exp(Z_\ell'\beta)|T_i, \delta_i, Z_{2i}]$$

for $j = 0, 1$. Upon inserting the missingness probabilities $\pi_\ell, \ell = 1, \ldots, n$, or consistent estimates thereof, this estimator can be calculated using an EM procedure based on (11.25),

$$d\Lambda_0(t) = \sum_{1}^{n} dN_i(t)/\hat{S}^{(0)}(\beta, t),$$

and a likelihood-based estimating function for θ that parameterizes the distribution of Z_1 given (T, δ, Z_2). A sandwich-type variance estimator can be calculated for the AIPW estimator. The AIPW estimator has a nice robustness property: It will generally provide a consistent asymptotically normal estimator of β under a MAR assumption and some other conditions (Robins et al., 1994, App. B), even if the missingness model is not specified correctly, provided that the distribution of Z_1 given (T, δ, Z_2) is specified correctly. This follows since (11.25) comprises zero mean and asymptotically independent contributions from each study subject. On the other hand, if the missingness probabilities are specified correctly, (11.25) has mean zero and consistent β estimation will generally hold, even if the covariate distribution is misspecified. The latter robustness feature would seem valuable in the context of nested case–control sampling, for example, where the π_i's are a known function of failure time and at-risk data. In that setting, it may be possible to choose a normal distribution, or other distribution of convenience, for Z_1 with little deterioration in the properties of β estimators.

Simulation studies (Chen and Little, 1999; Wang and Chen, 2001) with Z_1 and Z_2 binary suggest good efficiency properties for the NPML and AIPW estimators relative to the other estimators mentioned above. The efficiency of the NPML estimator seems to be particularly good, although the bias that arises if censoring depends on Z_1 was noticeable in the Wang and Chen simulations. An illustration of the various estimators is given in Section 11.9.

11.6 MISMEASURED COVARIATE DATA

Consider again the Cox model (11.14) with $Z(t) = Z = (Z_1', Z_2')'$, where Z_1 may be missing. Suppose also that a covariate vector W is always available, where W is a poorly measured version of Z_1. The covariate W is said to be a *surrogate* for Z_1 if $\lambda(t; Z_1, Z_2, W)$ does not depend on W, and hence is given by the right side of (11.14). Estimation of $\beta = (\beta_1', \beta_2')'$ based on these data can proceed using any of the approaches of Section 11.5 simply by extending Z_2 to $(Z_2', W')'$ in (11.14) and restricting the regression coefficients corresponding to W to be zero. In fact, the surrogacy assumption can be checked by testing for zero values of these regression coefficients. If W is highly informative concerning the missing Z_1 values, the efficiency of estimates of $\beta = (\beta_1', \beta_2', 0')'$ may be enhanced substantially by inclusion of the surrogate data. The subset of the sample where Z_1 is observed is sometimes referred to as a *validation subsample*, whereas the remainder of the cohort, having only (T, δ, Z_2, W) available, constitutes the nonvalidation sample. The estimated partial likelihood of Zhou and Pepe (1995) was aimed at the situation of a surrogate variable and a validation sample. It constitutes a convenient method if covariates are discrete and missingness probabilities depend only on (Z_2, W). Zhou and Wang (2000) extended this method to continuous covariates using kernel estimation, although the method may not be practical if the dimension of (Z_2, W) is high.

An alternative simple procedure, referred to as *regression calibration* (e.g., Carroll et al., 1995), approximates the expectation in (11.23) by $\exp[E(Z_1|Z_2, W)'\beta_1 + Z_2'\beta_2]$ and estimates $E(Z_1|Z_2, W)$, typically using a simple least squares procedure, based on the validation subsample. This method is also applicable if missingness depends only on (Z_2, W). Because of the relative risk approximation, the resulting regression parameter estimates typically have some asymptotic bias. Wang et al. (1997) develop the asymptotic theory for this estimator, along with a suitable variance estimator. In extensive simulations, they showed the bias to be surprisingly modest in situations of practical interest. The bias can be substantial, however, for large β values, depending on the censorship pattern. The more complicated NPML and AIPW estimation procedures described in Section 11.5 can be applied if missingness rates depend on (T, δ) in addition to (Z_2, W), and the AIPW procedure allows the missingness rate to depend on Z_1 as well.

It quite often happens that measurements of Z_1 are unavailable for the entire sampled cohort; that is, there is no validation subsample. In these circumstances it is necessary to make error model assumptions to connect the covariate measurement W to the "true," but missing, covariate Z_1. Often, a classical measurement model assumption

$$W = Z_1 + \varepsilon \qquad (11.26)$$

is made, where the additive error ε is assumed to be independent of (Z_1, Z_2) and of the corresponding failure and censoring times, and ε is assumed to have mean zero and a variance σ^2. Repeat measurements $W_1, W_2, \ldots$ on some study subjects are needed to estimate σ^2 or other aspects of the error distribution. The error variates corresponding to multiple measurements on a study subject are usually assumed to be independent, as are the error variates across study subjects. The measurements $(W_1, W_2, \ldots)$ meeting these conditions are said to constitute a *covariate reliability sample*.

Before proceeding, it is important to note that (11.26) and its attendant assumptions may be oversimplified or inappropriate in many important circumstances. For example, consider an epidemiologic study of nutrient consumption in relation to disease risk. One might expect consumption estimates from repeated self-reports of dietary intakes to include systematic bias (e.g., errors having different means depending on such characteristics as body mass, age, and ethnicity) as well as positive within-person correlations. In such settings it is crucial to identify biomarkers or other objective measures that plausibly adhere to (11.26). Such objective measures are probably too expensive to be practical in the entire cohort in an epidemiologic study, so that a more comprehensive measurement model may be needed that assumes (11.26) for the objective measure on a subset, along with a more flexible model for the self-report data. In effect, the objective measures data can then be used to calibrate the self-report data on the entire cohort.

Assuming a reliability sample adhering to (11.26) to be available, Xie et al. (2001) adapt the regression calibration approach and extend it by improving the approximation to the relative risk (11.24) to $\exp[E(Z_1|Z_2, W_1, W_2, \ldots; T \geq t)'\beta_1 + Z_2'\beta_2]$. This yields a recalibration within each risk set prior to applying a partial likelihood estimation procedure. More specifically, simple variance component arguments lead to estimates of the mean and covariance of $(Z_1, \bar{W}, Z_2)$ at each failure time, where $\bar{W}$ is the average of W values available for an individual. A joint normality assumption for $(Z_1, \bar{W}, Z_2)$ then leads to an estimator of $E(Z_1 | \bar{W}, Z_2, T \geq t)$ for use at time t in the partial likelihood function. Asymptotic distribution theory and a variance estimator were provided by Xie et al. for the ordinary and risk set regression calibration estimators under reliability sampling. Both estimators typically incorporate some asymptotic bias, but the recalibration within risk sets extends the set of configurations where the bias will be negligible.

If only a reliability sample adhering to (11.26) is available, consistent estimation of Cox model parameters is possible using a corrected score function approach. This approach (e.g., Nakamura, 1992) involves replacing the terms in the (standardized) score function (11.15) by consistent estimates based on the reliability sample. Some corrected score proposals require distributional assumptions on Z_1 or ε to hold, but recent work by Huang and Wang (2000) avoids distributional assumptions in either the true covariate or the error variate for consistent estimation of β, assuming that two or more W values are available for each individual. Briefly, consider $n^{-1}U(\beta)$ from (11.15) with time-independent covariates. This function can be estimated consistently by

$$n^{-1} \sum_{i=1}^{n} \int_0^{\infty} \left[\binom{\bar{W}_i}{Z_{2i}} - \frac{\hat{S}^{(1)}(\beta, t)}{\hat{S}^{(0)}(\beta, t)} \right] dN_i(t), \qquad (11.27)$$

where

$$\hat{S}^{(j)}(\beta, t) = \sum_{i=1}^{n} Y_i(t) \mathscr{A}_i \left\{ \binom{\tilde{W}_{1i}}{Z_{2i}} \exp \left[\binom{\tilde{W}_{2i}}{Z_{2i}}' \beta \right] \right\}, \qquad j = 0, 1$$

and the operator $\mathscr{A}_i$ is a summation over all distinct pairs of $\tilde{W}_{1i}$ and $\tilde{W}_{2i}$ selected from the set $\{W_{1i}, W_{2i}, \ldots\}$ of W values available on the ith study subject. The independence of the error terms in (11.26) for these W values implies that $n^{-1}\hat{S}^{(j)}(\beta, t)$ estimates the corresponding $n^{-1}S_j(\beta, t)$ aside from the factor $E(e^{\varepsilon\beta_1})$, which cancels out of the ratio in (11.27). This factor needs to be estimated to obtain a cumulative hazard estimator, requiring a further assumption. For example, an assumption of symmetry of the error distribution is sufficient for this purpose.

The corrected score regression parameter estimator of Huang and Wang (2000) performed well in simulations reported by these authors. The lack of monotonicity of the estimating function (11.27), however, presents some numerical challenges that have yet to be fully addressed.

11.7 SEQUENTIAL TESTING WITH FAILURE TIME ENDPOINTS

It is increasingly common for failure time data from controlled clinical trials to be monitored on an ongoing basis for potential early stoppage. Early stoppage may occur if key hypotheses have been answered or if serious participant safety concerns arise. In fact, many trials have a data and safety monitoring board (DSMB), external to the trial organization, that meets periodically to review trial data. It is the responsibility of such a board to assess the appropriateness of trial continuation in view of emerging trial data and any pertinent data from outside the trial. In the context of a therapeutic trial, there is typically a designated primary outcome, such as death or disease recurrence. A two-armed trial then typically compares a test treatment to a standard treatment in respect to the times to primary outcome events. In addition to monitoring the primary outcome, the DSMB may examine data on a number of other outcomes, including biomarkers or symptoms, that may be affected by one or both of the regimens under test.

For logistical reasons, data analyses by treatment arm and meetings of the DSMB typically take place on a prescribed schedule in chronological time, such as once or twice per year. Each data analysis, or inspection, offers an opportunity for early stoppage if, for example, the accumulated data allow rejection of the primary outcome null hypothesis of no difference between treatments. The question then arises as to how the significance level (e.g., $\alpha = 0.05$) for rejecting the null hypothesis given that it is true can be preserved in the presence of the multiple sequential tests. Proposed sequential testing procedures rely mostly on log-rank or weighted log-rank tests at each inspection time. A set of critical values needs to be chosen for each of the test statistics in sequence, so that the type I error rate is equal to the prespecified α value for the group sequential test. Several proposals have been made for determining the sequence of critical values to be compared to the absolute values $|v_1|, \ldots, |v_K|$ of standardized (weighted) log-rank tests at prespecified K inspection times.

Under the assumption that equal increments in information present at each time point, Pocock (1977) proposes a common critical value c, so that the null hypothesis is rejected if $|v_j| \geq c$ at any inspection time. In some contexts this choice of *stopping rule* may expend too much of the significance probability (α) at early inspection times. Hence, O'Brien and Fleming (1979) proposed that the null hypothesis be rejected after a group sequential test j if $|v_j| \geq c(K/j)^{1/2}$, with c again chosen to yield the desired α level. These stopping boundaries are quite extreme at early inspection times but tend to require a smaller expected sample size to achieve a given power compared to the Pocock boundary, depending somewhat on the size and shape of the hazard ratio between two treatment groups. Haybittle (1971) and Peto et al. (1976) propose the simple but appealing approach of rejecting the null hypothesis if $|v_j| \geq 3$ for any of $j = 1, \ldots, K - 1$, or if $|v_K| \geq c$, where c is selected to yield the desired α level. Wang and Tsiatis (1987) propose a family of tests that include the Pocock and O'Brien–Fleming tests as special cases.

The asymptotic distribution of the score processes under a Cox model provides a context for determining the value of c that determines the critical values for each of the group sequential test procedures above. In a typical clinical trial the inspection times occur at designated points in chronological time, but the time axis of most importance for assessing treatment effects is the study subject's time from randomization (or enrollment) in the clinical trial. The multistate models considered in

Section 8.3 are convenient for describing this testing problem. Specifically, one can consider a study subject in state zero from the (chronological) time $(t = 0)$ at which the trial is launched until randomization into the trial (state 1), with the comparison of rates of occurrence of the primary outcome (transition from state 1 to state 2) being of principal interest. As in Section 8.3.3, one can specify a modulated semi-Markov model

$$\lambda_{12}(t) = \lambda_{120}(v)\exp[Z(t)\beta] \tag{11.28}$$

for this transition intensity, where v is the time from the individual's trial enrollment. Also, the modeled covariate may be a fixed treatment indicator ($Z = 1$ for test and $Z = 0$ for standard) if the intensities are assumed to be proportional by treatment group, or $Z(t)$ may be comprised of defined time-dependent covariates according to a prescribed form for the hazard ratio. In either case, a partial likelihood procedure is available for inference on β, under independent left-truncation and censoring assumptions, and a test for $\beta = 0$ gives a log-rank $[Z(t) \equiv Z]$ or weighted log-rank test at each inspection time. Denote by $U_j = U_j(0)$ the partial likelihood score test for $\beta = 0$ from (11.28), and by $I_j = I_j(0)$ the corresponding observed information at inspection time j, for $j = 1, \ldots, K$. The partial likelihood development and related martingale representation of the score function shows the score increments $U_1, U_2 - U_1, \ldots, U_K - U_{K-1}$ to be uncorrelated under the null hypothesis $\beta = 0$. Asymptotically, these increments are mean zero normal random variables with variances $I_1, I_2 - I_1, \ldots, I_K - I_{K-1}$. Sequential test design often assumes these information increments to be equal, giving a simple expression for test size (α) and a simple calculation of critical values for the standardized tests under the sequential procedures above. Moreover, for small $\beta \neq 0$ and a simple proportional hazards model (11.28), each U_j is approximately normally distributed with mean βI_j and variance I_j (Exercise 11.5), with the U_j's having uncorrelated increments. Hence, if the information increments are assumed to be equal, the score increments are approximately independent normal variates having mean βI and variance I, allowing one simply to calculate the approximate power of the sequential test under local alternatives. These size and power calculations are fairly insensitive to moderate departures from common information increments. Critical value determination and power projections under other assumptions about information increments or more complex relative risk models can be carried out by computer simulation using a joint normal distribution for the score tests at the K time points. There are also adaptive testing procedures where one specifies the null hypothesis rejection probability to be expended at each inspection time and chooses a critical value for $|v_j|$ in a manner that acknowledges data accumulated prior to the jth inspection time (Slud and Wei, 1982; Lan and DeMets, 1989). This error spending approach does not restrict the information increments to be equal or approximately equal.

Estimation of β at the conclusion of the clinical trial may also need to acknowledge the sequential monitoring procedure. If the trial proceeds to planned termination, and most of the significance probability is retained for the test at planned trial termination, as is typically the case, for example, with the O'Brien–Fleming testing procedure, the maximum partial likelihood estimate $\hat{\beta}$ can be expected to exhibit little bias. On the other hand, if the trial stops early on the basis of rejecting the null hypothesis, $\hat{\beta}$ will tend to be biased away from zero, perhaps substantially so. This happens because the test statistic will tend to have a value toward the extremes of its distribution when the stopping boundary is first crossed. Various approaches have been proposed for correcting such bias. Whitehead (1986) proposed a numerical method for estimating this bias under certain triangular stopping boundaries, and Emerson and Fleming (1990) propose the use of the conditional expectation of $\hat{\beta}$ at the first inspection time conditional on the time and standardized test statistic at trial termination as an approximate minimum variance unbiased estimate of β. The latter estimator effectively avoids bias but has somewhat larger mean-squared error than Whitehead's corrected estimator in simulation studies. More recently, Wang and Leung (1997) have proposed the use of parametric bootstrap methods to reduce bias following a sequential test. Note that if β in (11.28) is of the form $(\beta_1', \beta_2')'$, where monitoring is based on tests for $\beta_1 = 0$, partial likelihood estimates of β_2 will also tend to be biased if $\hat{\beta}_1$ and $\hat{\beta}_2$ are correlated.

Some clinical trials have several failure time endpoints that are of interest in evaluating treatment effects. This is particularly the case for disease prevention trials among healthy persons if the treatment or intervention under test has the potential to affect the risk of several diseases, beneficially or adversely. Trial monitoring may be based on the values of standardized test statistics for several failure time endpoints, and testing and estimation procedures need to acknowledge the multivariate aspect of the failure time data. For example, the ongoing Women's Health Initiative Clinical Trial (Women's Health Initiative Study Group, 1998) is evaluating the use of certain hormone replacement therapy (HRT) preparations for their potential to reduce the risk of coronary heart disease, within the overall context of testing whether overall health benefits exceed health risks. The trial involves the randomization of over 27,000 postmenopausal women to either active HRT or placebo, with a planned average 8.5 years of follow-up. Biannual trial monitoring involves weighted log-rank tests at inspection times over a 10-year period for the primary outcome designated, coronary heart disease, and also for an anticipated adverse effect on breast cancer incidence. In addition, a *global index*, defined to be the time to first occurrence of any one of several serious diseases (including death from other causes) that are plausibly affected by the HRT treatment, is formally monitored. Trial monitoring guidelines specify consideration of early stoppage for benefit if the primary outcome test is significant in favor of the treatment and the global index is supportive, and early stoppage for harm if the breast cancer test statistic is significant in the direction of harm, or any of the elements of the global index are significant in the direction of harm following a Bonferroni correction, and the global index does not show benefit (e.g., Freedman et al., 1996). O'Brien–Fleming stopping boundaries and trial power projections under a sequence of design assumptions were estimated using computer simulations. Lin (1991b) provides methodology for sequential testing based on marginal weighted log-rank tests for a multivariate response.

11.8 BAYESIAN ANALYSIS OF THE PROPORTIONAL HAZARDS MODEL

In this section we outline briefly a Bayesian analysis of failure time data arising from the Cox model. This requires a discussion of some methods of nonparametric Bayesian inference in Section 11.8.1. In Sections 11.8.2 and 11.8.3 we consider an analysis of the Cox model using a gamma process prior, and in Section 11.8.4, some related discrete models. Some of the relevant literature is reviewed briefly in the Bibliographic Notes at the end of the chapter.

11.8.1 Dirichlet, Gamma, and Other Neutral to the Right Processes

We consider here specific application of some nonparametric Bayesian procedures to survival distributions. This derives from work of Ferguson (1973, 1974), Doksum (1974), Susarla and Van Ryzin (1976), Kalbfleisch (1978a), Ferguson and Phadia (1979), Wild and Kalbfleisch (1981), and Hjort (1990), among others.

Suppose that the survivor function F_0 for a failure time variate T is itself the realization of a stochastic process (the prior distribution) to be defined. We allow F_0 to be discrete, continuous, or mixed and note that, in general, F_0 is related to the corresponding cumulative hazard function Λ_0 through the product integral

$$F_0(t) = P(T > t|F_0) = \mathscr{P}_0^t[1 - d\Lambda_0(u)]. \tag{11.29}$$

In the continuous case, this reduces to $\exp[-\Lambda_0(t)]$. To describe a class of prior distributions for F_0, we consider a partition of $(0, \infty)$ into a finite number k of disjoint intervals

$$(a_0 = 0, a_1](a_1, a_2], \ldots, (a_{k-1}, a_k = \infty)$$

and define the (random) hazard contribution in the ith interval as

$$q_i = P\{T \in (a_{i-1}, a_i] | T > a_{i-1}, F_0\} \tag{11.30}$$

if $P(T > a_{i-1}|F_0) > 0$, and $q_i = 1$, $i = 1, \ldots, k$ otherwise. It follows that

$$-\log F_0(a_i) = \sum_{j=1}^{i} -\log(1 - q_j) = \sum_{j=1}^{i} r_j, \qquad i = 1, \ldots, k, \tag{11.31}$$

where $r_j = -\log(1 - q_j)$.

A probability distribution can be specified on the space of survivor functions $[F_0(\cdot)]$ by specifying the distributions of $q_1, \ldots, q_k$ for every possible partition $(a_{i-1}, a_i], i = 1, \ldots, k$ for all integer $k > 0$ (Doksum, 1974). By (11.31), this will also define all finite-dimensional distributions of F_0. The simplest construction assigns independent prior probability densities for $q_1, \ldots, q_k$ subject to consistency conditions. The resulting processes are called *tail-free* or *neutral to the right* by Doksum. From (11.31) it is clear that by this construction, $-\log F_0(t)$ is a nondecreasing independent increments process. Accordingly, we specify independent priors for the r_i's (or for the q_i's) subject to the Kolmorogorov consistency condition (e.g., Hjort, 1990) that the distribution of $r_i + r_{i-1}$ must be the same as would be obtained by direct application of the rules that define the prior to the combined interval $[a_{i-1}, a_{i+1})$. We define two particular processes of this type.

The Dirichlet process, introduced by Ferguson (1973), has been used widely as a random probability measure in a Bayesian context. Ferguson defined the process in very general probability spaces by giving the finite-dimensional distributions of a random probability measure on the space. In the context above, however, where the observable variate T has a natural ordering, it is convenient to think of the process as unfolding sequentially in time, and specifying the distribution through increments in $-\log F_0$. In this context, the Dirichlet process prior is defined in terms of two quantities: a positive real number c, and a completely specified (right continuous) survivor function $F^*(\cdot)$. Referring to (11.30), we suppose that for a given partition $(a_{i-1}, a_i]$, $i = 1, \ldots, k$, $q_1, \ldots, q_k$ are independent with

$$q_i \sim \text{beta}[(\gamma_{i-1} - \gamma_i), \gamma_i], \qquad i = 1, \ldots, k,$$

where $\gamma_i = cF^*(a_i)$, $i = 0, \ldots, k$ and $\text{beta}(a, b)$ denotes a beta distribution with density function

$$\frac{\Gamma(a+b)}{\Gamma(a)\Gamma(b)} x^{a-1}(1-x)^{b-1}, \qquad 0 < x < 1,$$

where $a, b \geq 0$. The following conventions are used: $\text{beta}(0, b)$ with $b > 0$ is the distribution with unit mass at 0, and $\text{beta}(a, 0)$ with $a \geq 0$ is the distribution with unit mass at 1. The Kolmogorov consistency condition is satisfied since the distribution of the hazard $q_c = 1 - (1 - q_i)(1 - q_{i+1})$ corresponding to the combined interval $[a_{i-1}, a_i + 1)$ is

$$q_c \sim \text{beta}(\gamma_{i-1} - \gamma_{i+1}, \gamma_{i+1}), \tag{11.32}$$

which is obtained from the joint distribution of q_i, q_{i+1} as defined above, or by applying the construction above directly to the combined interval. The corresponding random proability measure is a Dirichlet process with parameters c and F^* and we write $F_0 \sim \mathscr{D}(F^*, c)$

Consider now the random probabilities

$$P_i = P\{T \in (a_{i-1}, a_i] | F_0\}, \qquad i = 1, \ldots, k.$$

It can be shown that $(P_1, \ldots, P_{k-1})' \sim D_{k-1}(\gamma_0 - \gamma_1, \ldots, \gamma_{k-1} - \gamma_k)$. Note that $X = (X_1, \ldots, X_r)' \sim D_r(\alpha_1, \ldots, \alpha_{r+1})$ denotes an r-dimensional Dirichlet distribution with density

$$f(x_1, \ldots, x_r) = \frac{\Gamma(\alpha)}{\prod_{i=1}^{r+1} \Gamma(\alpha_i)} \prod_{i=1}^{r+1} x_i^{\alpha_i - 1},$$

where r is a positive integer, $\alpha_i > 0$, $i = 1, \ldots, r+1$, $\alpha = \sum_{i=1}^{r+1} \alpha_i$, $x_{r+1} = 1 - x_1 - \cdots - x_r$, and $x_i \geq 0$, $i = 1, \ldots, r+1$. This distribution of $(P_1, \ldots, P_{k-1})'$ is a characterization of the Dirichlet process. The Dirichlet process actually satisfies a more general property. If $A_1, \ldots, A_{r+1}$ is any set of mutually exclusive and exhaustive events on $(0, \infty)$, the variates

$$P(T \in A_i | F_0), \qquad i = 1, \ldots, r$$

have an r-dimensional Dirichlet distribution with parameters

$$cF^*(A_1), \ldots, cF^*(A_{r+1}),$$

where $F^*(A_i)$ is the probability attached to A_i by the distribution with survivor function F^*.

Ferguson (1973) gives an interpretation of the parameters c and F^* of the Dirichlet process. For any fixed $t > 0$, it can be seen that

$$E[F_0(t)] = F^*(t) \quad \text{and} \quad \text{var}[F_0(t)] = c^{-1} F^*(t)[1 - F^*(t)]. \tag{11.33}$$

Thus, $F^*(t)$ is a guess at the true survivor function, and c is a measure of the precision of the guess.

A second process, similar to the Dirichlet process, is obtained if $-\log F_0(t)$ is generated by a gamma process. To describe the gamma process, consider the same partition $(a_{i-1}, a_i]$, $i = 1, \ldots, k$ as above and let $r_i = -\log(1 - q_i)$ have independent gamma distributions

$$r_i \sim G(\alpha_i - \alpha_{i-1}, c), \qquad i = 1, \ldots, k, \tag{11.34}$$

where $\alpha_i = -c \log F^*(a_i)$, $i = 1, \ldots, k$ and c and F^* have the same meanings as before. The notation $X \sim G(a, b)$ indicates that X has the gamma distribution with parameters a (shape) and b (scale) (see Section 2.2.4). The conventions adopted are that $G(0, c)$ is the distribution with unit mass at 0 and $G(\infty - \infty, c)$ or $G(\infty, c)$ is the distribution with unit mass at ∞. We shall write $-\log F_0 \sim \mathscr{G}(-c \log F^*, c)$ to denote this gamma process. From the additive property of the gamma distribution, it is easy to see that this process satisfies the consistency equations. Simple interpretations are also available for the prior parameters c and F^*. For a fixed $t > 0$, consider a partition $[0, t), [t, \infty)$. Then $r_1 = -\log F_0(t)$ and

$$E[\log F_0(t)] = \log F^*(t) \quad \text{and} \quad \text{var}[\log F_0(t)] = c^{-1} \log F^*(t).$$

Again, the specified F^* is an initial guess at F_0, and c specifies the weight attached to that guess.

In the following sections, the gamma process is used as a prior distribution in analyzing data arising from the proportional hazards model. Calculations could have been carried out equally well using the Dirichlet process (see, e.g., Wild and Kalbfleisch, 1981 and Hjort, 1990).

11.8.2 Estimation of β in the Cox Model with Gamma Process Prior

Let T be a random variable with corresponding covariate vector Z and conditional survivor function

$$F(t|Z) = P(T > t \mid Z, F_0) = F_0(t)^{\exp(Z'\beta)}. \tag{11.35}$$

This is the Cox or proportional hazards model with time-independent covariates. For the moment, consider the case of no censoring and suppose that $-\log F_0 \sim \mathcal{G}(-c \log F^*, c)$. We consider estimating β on the basis of data (t_i, Z_i), $i = 1, \ldots, n$.

For this purpose, we find the marginal probability density of $t_1, \ldots, t_n$ conditional on the Z_i's, F_0 having been eliminated, and interpret this as an (average or integrated) likelihood function for β. Now,

$$P(T_1 > t_1, \ldots, T_n > t_n \mid \beta, \mathcal{Z}, F_0) = \prod_{i=1}^{n} F_0(t_i)^{\exp(Z_i'\beta)}, \tag{11.36}$$

where $\mathcal{Z}$ is the design matrix with ith column Z_i. Without loss of generality, suppose that $t_1 \leq t_2 \leq \cdots \leq t_n$ and define $r_i = \log F_0(t_{i-1}) - \log F_0(t_i)$, $i = 1, \ldots, n + 1$, where $t_0 = 0$ and $t_{n+1} = \infty$. The t_i's play the role of the a_i's in the preceding section. Further,

$$r_i \sim G\{c[\log F^*(t_{i-1}) - \log F^*(t_i)], c\}, \qquad i = 1, \ldots, n + 1 \tag{11.37}$$

independently. Since $-\log F_0(t_i) = \sum_{j=1}^{i} r_j$, $i = 1, \ldots, n$, (11.36) implies that

$$P(T_1 > t_1, \ldots, T_n > t_n \mid \beta, \mathcal{Z}, r_1 \cdots r_n) = \exp\left(-\sum_{j=1}^{n} r_j A_j\right), \tag{11.38}$$

where

$$A_j = \sum_{l=1}^{n} \mathbf{1}(t_l \geq t_j) \exp(Z_l\beta), \qquad j = 1, \ldots, n. \tag{11.39}$$

Integrating (11.38) with respect to the distribution (11.37) of $r_1, \ldots, r_n$ gives

$$P(T_1 > t_1, \ldots, T_n > t_n \mid \beta, \mathcal{Z}) = \exp\left[c \sum_{j=1}^{n} B_i \log F^*(t_j)\right] \tag{11.40}$$

where $B_j = -\log[1 - \exp(Z_j\beta)/(c + A_j)]$.

The expression (11.40) is valid for any survivor function F^*, whether discrete, continuous, or mixed. To avoid problems with fixed discontinuities, however, we assume that $F^*(t) = \exp[-\Lambda^*(t)]$ is absolutely continuous. The multiple decrement function (11.40) is then absolutely continuous except along any hyperplane with $t_i = t_j$ for some $i \neq j$. Thus if there are no ties in the data $(t_1 < t_2 < \cdots < t_n)$, the PDF of $T_1, \ldots, T_n$ is computed by differentiation of (11.40) and yields

$$c^n \exp\left[-\sum cB_j\Lambda^*(t_j)\right]\prod_1^n[\lambda^*(t_i)B_i], \qquad (11.41)$$

where $\lambda^*(t) = d\Lambda^*(t)/dt$. The expression (11.41) can be interpreted as a likelihood function for β on the data $T_1 = t_1, \ldots, T_n = t_n$.

Right censoring is also easily accommodated since, again assuming no ties among the failure times, the appropriate likelihood is obtained by differentiating (11.40) with respect to those t_i's corresponding to observed failures. This gives

$$L(\beta) = \exp\left[-\sum_{i=1}^n cB_i\Lambda^*(t_i)\right]\prod_{i=1}^n[\lambda^*(t_i)B_i]^{\delta_i}, \qquad (11.42)$$

where $\delta_j = 0$ or 1 for censored or failure times t_j, respectively.

Two cases are of particular interest. If c is near 0, then to a first order approximation

$$L(\beta) \simeq K\prod_{i=1}^n\left[-\log\frac{1-\exp(Z_i'\beta)}{c+A_i}\right]^{\delta_i} \simeq K\prod_{i=1}^n\left[\frac{\exp(Z_i'\beta)}{\sum_{l=1}^n \mathbf{1}(t_l \geq t_i)\exp(Z_l'\beta)}\right]^{\delta_i} \quad (11.43)$$

The last term in (11.43) is proportional to the partial likelihood of β. Small values of c correspond to placing little weight on the prior guess $F^*(t)$. On the other hand

$$\lim_{c\to\infty} L(\beta) = \exp\left[-\sum_{i=1}^n\Lambda^*(t_i)\exp(Z_i'\beta)\right]\prod_{i=1}^n[\lambda^*(t_i)\exp(Z_i'\beta)],$$

which is the appropriate likelihood if it is assumed that $F_0(t) = F^*(t) = \exp[-\Lambda^*(t)]$. In effect, (11.43) gives a spectrum of likelihoods ranging from non-parametric situations (c near 0) to situations where $F_0(t)$ is assumed completely known. By allowing $F^*(t)$ to depend on one or more unknown parameters, the likelihood (11.42) corresponds to a Bayesian approach to the usual parametric analysis ($c \to \infty$). For example, if $F^*(t) = \exp(-\lambda t)$, an analysis based on (11.42) complements the exponential regression methods of Section 3.5. An examination of the likelihood for varying c can lead to an evaluation of how assumption dependent the analysis is.

The leukemia data of Gehan (1965a) reproduced in Table 11.3 can be used to illustrate this point. The covariate is treatment-group specified by the indicator

Table 11.3   Times of Remission in Weeks of Leukemia Patients [a]

Group 1 ($Z = 0.5$)	1, 1, 2, 2, 3, 4, 4, 5, 5, 8, 8, 8, 8, 11, 11, 12, 12, 15, 17, 22 $+ \epsilon^\dagger$, 23 $+ \epsilon^\dagger$
Group 0 ($Z = -0.5$)	6, 6, 6, 6*, 7, 9*, 10, 10*, 11*, 13, 16, 17*, 19*, 20*, 22, 23, 25*, 32*, 32*, 34*, 35*

[a],[†] These data were 22, 23 but were adjusted slightly in the positive direction to break the ties with items in group 0; *, censored.

variable $Z = -0.5, 0.5$. We select $\Lambda^*(t) = \lambda t$, where λ is unspecified. From (11.42), the joint integrated likelihood of λ and β for each c specified is

$$L(\beta, \lambda) = \exp\left(-c\lambda \sum t_j B_j\right) \prod (c\lambda B_i)^{\delta_i}.$$

Maximum likelihood equations are easily obtained and solved using a Newton–Raphson algorithm. For the data in Table 11.3, there are ties present, and these were broken at random in order to apply the results above.

Table 11.4 summarizes the estimation of β for various c values and gives the results for the exponential regression model $(c \to \infty)$ and for the proportional hazards model using partial likelihood. The entries var$(\hat{\beta})$ give the asymptotic variance of $\hat{\beta}$ based on the entry in the inverse of the information matrix for β and λ. The estimation of β is very stable over the range $0 < c < \infty$.

It should be noted that ties in the failure time data cause substantial difficulty in this analysis. The reason is that the gamma and Dirichlet process priors place all probability mass on the class of discrete distributions. As a consequence, even with a continuous parameter F^* in the prior, there is a positive probability of ties in the data, and a full analysis must deal with the fact that the joint survivor function (11.41) is not absolutely continuous in that it places positive probability mass on hyperplanes where two or more of the t_j's are equal. Perhaps the most satisfactory way to take account of ties in the data is to use a discrete model for the failure times, and this is discussed further in Section 11.8.4.

11.8.3   Posterior Distribution of Λ_0

In this section the posterior distribution of the underlying survivor function F_0 is obtained when a sample $(t_1, Z_1), \ldots, (t_n, Z_n)$ is obtained from the model (11.35).

Table 11.4   Estimation of β Based on the Integrated Likelihood (11.42) for the Data of Table 11.3

	Partial Likelihood	$c = 1$	$c = 5$	$c = 25$	$c = 125$	$c = \infty$
$\hat{\beta}$	1.652	1.606	1.512	1.461	1.490	1.580
var$(\hat{\beta})$	0.148	0.134	0.124	0.130	0.146	0.158
$\hat{\beta}[\text{var}(\hat{\beta})]^{-1/2}$	4.29	4.39	4.29	4.05	3.90	3.97

Again we assume the gamma process prior $-\log F_0(t) \sim \mathcal{G}(-c \log F^*, c)$ and will assume that F^* is absolutely continuous.

Consider again a partition $(a_{i-1}, a_i]$, $i = 1, \ldots, k$ with $a_0 = 0$ and $a_k = \infty$ and suppose that the data are (t_1, Z_1) with $n = 1$. The extension to general n follows easily from the result for $n = 1$. As before, $r_j = -\log(1 - q_j)$, where q_j is the hazard contribution of the jth interval $j = 1, \ldots, k$. Assume that $a_{i-1} < t_1 \le a_i$ and $r_{i1} = -\log(1 - q_{i1})$ and $r_{i2} = -\log(1 - q_{i2})$, where q_{i1} and q_{i2} are the hazard contributions of the intervals $(a_{i-1}, t_1]$ and $(t_1, a_i]$, respectively. Then

$$P(T_1 > t_1 \mid r, r_{i1}, Z_1) = \exp[-(r_1 + \cdots + r_{i-1} + r_{i1})e^{Z_1'\beta}] \qquad (11.44)$$

and

$$P(T_1 > t_1, r_j \le r_{0j}, j = 1, \ldots, k \mid Z_1) = H(t_1, r_{01}, \ldots, r_{0k}, Z_1)$$

is obtained by integrating (11.44) with respect to the independent gamma prior distributions of $r_1, \ldots, r_{i-1}, r_{i1}, r_{i2}, r_{i+1}, \ldots, r_k$ over the appropriate range. The posterior distribution of r given $T_1 = t_1$ is then specified by

$$P(r_j \le r_{0j}, j = 1, \ldots, k \mid T_1 = t_1, Z_1) = \frac{\partial H(t_1, r_{01}, \ldots, r_{0k}, Z_1)/\partial t_1}{\partial H(t_1, \infty, \ldots, \infty, Z_1)/\partial t_1}.$$

There does not appear to be a simple closed-form expression for this, but the probability laws can be characterized simply using moment generating functions. It can be shown that

$$M_r(\theta \mid T_1 = t_1) = E[\exp(\theta_1 r_1 + \cdots + \theta_k r_k) \mid T_1 = t_1] = \prod_{j=1}^{k} M_{r_j}(\theta_j \mid T_1 = t_1),$$

where

$$M_{r_j}(\theta_j \mid T_1 = t_1) = \begin{cases} \left(\dfrac{c_1}{c_1 - \theta_j}\right)^{\alpha_j - \alpha_{j-1}}, & j < i \\[2ex] \left(\dfrac{c}{c - \theta_j}\right)^{\alpha_j - \alpha_{j-1}}, & j > i \end{cases}$$

for $c_1 = c + \exp(Z_1'\beta)$. The moment generating function of r_i is

$$\left(\frac{c_1}{c_1 - \theta_i}\right)^{\alpha(t_1) - \alpha_{i-1}} \left(\frac{c}{c - \theta_i}\right)^{\alpha_i - \alpha_{t-1}} \frac{\log[(c_1 - \theta_i)/(c - \theta_i)]}{\log(c_1/c)}, \qquad (11.45)$$

where $\alpha_j = -c \log F^*(a_j), j = 1, \ldots, k$, and $\alpha(t_1) = -c \log F^*(t_1)$. Thus, r_i is distributed as the sum of three independent random variables X, Y, U, where X and

Y are gamma variables and $U \sim Q(c_1, c)$ with density

$$\frac{(1/u)(e^{-cu} - e^{-c_1 u})}{\log(c_1/c)},$$

the moment generating function of which is the last factor in (11.45). If t_1 corresponds to a right-censored time so that it is observed only that $T_1 > t_1$, the posterior distribution of the r_j's is the same as above, except that r_i is the sum of only the two gamma variables X and Y. In either case, whether T_1 is an observed or censored time, all finite-dimensional distributions of the posterior process have been obtained by the argument above, and characterization of the process is straightforward. This special case ($n = 1$) is covered in the next paragraph.

The generalization of these results to obtain the posterior distribution of F_0 given $(t_1, Z_1, \delta_1), \ldots, (t_n, Z_n, \delta_n)$ is straightforward if there are no ties among the failure times. Given $t_1 < t_2 < \cdots < t_n$, $-\log F_0(t)$ is an independent increments process. At t_i the increment is $\delta_i U_i$ where

$$U_i \sim Q(c + A_i, c + A_{i+1}), \qquad i = 1, \ldots, n,$$

and between $(t_{i-1}, t_i]$ increments occur as for the gamma process $\mathcal{G}(c \log F^*, c + A_i)$. This result is easily seen by inserting t_n first then $t_{n-1}, \ldots, t_1$. Insertion of t_i affects the process only at points $t \le t_i$.

If the loss function is squared error in $\log F_0(t)$, the posterior expectation of $\log F_0(t)$ provides the optimum Bayes estimate. This estimate is a log survivor function for a mixed distribution, and the estimator of the survivor function can be obtained by exponentiation. If the data satisfy $t_1 < t_2 < \cdots < t_n$, as above, and $t_{i-1} \le t < t_i$, the posterior distribution of $-\log F_0(t)$ is that of the sum of independent variables $X_1 + \delta_1 U_1 + \cdots + X_{i-1} + \delta_{i-1} U_{i-1} + D_i$, where $X_j \sim G\{c[\log F^*(t_{j-1}) - \log F^*(t_j)], c + A_j\}$, $U_j \sim Q(c + A_j, c + A_{j+1})$, $j = 1, \ldots, i - 1$, and $D_i \sim G\{c[\log F^*(t_{i-1}) - \log F^*(t)], c + A_i\}$. Now it can be seen that

$$E(U_j) = \frac{\exp(Z_j \beta)}{(c + A_j)(c + A_{j+1})} \log \frac{c + A_j}{c + A_{j+1}}$$

and the Bayes estimator is

$$E[-\log F_0(t) \mid \text{data}, \beta] = \sum_1^{i-1} [E(X_j) + E(U_j)] + E(D_i).$$

For c small, n large, and j moderate in size, $E(U_j) \simeq 1/A_j$, and for c small, $E(X_j) \simeq E(D_i) \simeq 0$. Thus, for t not too large,

$$E[-\log F_0(t) \mid \text{data}, \beta] \simeq \sum_{j \mid t_j \le t} \frac{\delta_j}{A_j}, \qquad (11.46)$$

where the right side is the Nelson–Aalen estimator.

These Bayes estimates for the survivor function are simple to compute. They amount to a smoothing of the usual step function estimates, although for $c < \infty$, jumps occur at each of the observed failure times.

As noted above, a serious deficiency of the gamma process, the Dirichlet process, and similar "neutral to the right" processes is that they place all probability mass on the class of dicrete distributions so that the resulting F_0 is a discrete survivor function with probability 1. Further, the prior parameter c indexes the degree of discreteness in that for $c \to 0$, the probability that two random observations chosen from F_0 are equal tends to 1. As a consequence, the increase in variance of $F_0(t)$ for c near zero is to a large extent accounted for by the fact that the realization F_0 will (with high probability) exhibit a very large jump at a random point in time. It is also this discreteness that accounts for the more complicated handling of ties, as mentioned earlier.

11.8.4 Ties and Discrete Models

If many ties are present in the data, it is best to use a discrete model as considered, for example, by Kalbfleisch and MacKay (1978a) and, more completely, by Burridge (1981). These discrete models also play prominenty in the book by Ibrahim et al. (2001). In this section we consider a discrete version of the gamma process and note the steps required to apply the Bayesian approach.

Discrete models have some appeal since failure times are always discrete due to measurement error. For example, the results in Table 11.3 arise because failure times are being recorded only to the nearest week. For the sake of simplicity, suppose that failures can occur only at positive integer values and, let F^* be a discrete survivor function with mass points on the integers. Consider a censored sample in which the individuals in D_i are observed to fail at time i and those in R_i are observed to survive past time i, $i = 1, 2, \ldots$. Under the model (11.29), the likelihood function can be written as

$$\prod_{i \geq 1} \left\{ \prod_{l \in D_i} [1 - \exp(-r_{0i} e^{z_i' \beta})] \exp\left[-\sum_{l \in R_i} r_{0i} e^{z_i' \beta} \right] \right\}, \qquad (11.47)$$

where $-\log F_0(t) = \sum_{i \leq t} r_{0i}$. Under the gamma process prior, $r_{0i} \sim G(cr_i^*, c)$, $i = 1, 2, \ldots$ independently, where $r_i^* = \log F^*(i - 1) - \log F^*(i)$. The likelihood for β in this case is obtained by integrating (11.47) with respect to this prior specification. If the ties are not too numerous, this integration is easily accomplished on multiplying out the ith factor in (11.47) and itegrating term by term. Burridge (1981) makes some remarks on computation more generally.

One can also obtain the posterior distribution of F_0 under the discrete model in a straightforward way. From the product form of (11.47) and the independent priors in the gamma process, it is easily seen that the r_{0i}'s have independent posterior distributions. The distribution of r_{0i} has density proportional to the product of the ith term in (11.47) and its prior gamma density.

11.9 SOME ANALYSES OF A PARTICULAR DATA SET

Nowinski et al. (1979) conducted a study to examine genetic and viral factors that may influence the development of spontaneous leukemia in AKR mice. A genetic cross was prepared between leukemia-prone AKR mice and leukemia-resistant C57BL/6 mice. Two hundred and four mice of the AKR × (C57BL/ 6 × AKR)F_1 backcross were examined for (1) the production of endogenous murine leukemia virus (MuLV), (2) the production of antiviral antibodies, (3) phenotype at the major histocompatibility complex (MHC), and (4) phenotype at the Gpd-1 region. The phenotype at the MHC was thought possibly to influence the immune response to viral proteins, whereas the Gpd-1 locus is closely linked (within 1 centimorgan) to an Fv-1 locus that may be related to the rate of spread of leukemia virus. Determinations of the Gpd-1 phenotype began midway through the study and are available on 100 mice. Sex and coat color were also recorded for each animal. The mice were followed over a 2-year period for mortality due to thymic leukemia, nonthymic leukemia, or other natural causes. The surviving mice were then sacrificed. The data from this experiment are given in data set V of Appendix A.

Product limit survival curves (1.13) for all natural causes of mortality and cumulative mortality plots for specific causes of death form an important part of the initial exploration of these data. Each prognostic factor in turn was used to divide the data into strata. The survival or cumulative mortality curves stratified on a specific factor were plotted on a single figure, and log-rank (1.21) and Gehan generalized Wilcoxon tests were carried out to test the hypothesis of equality of the survival curves.

Table 11.5 includes the results of log-rank and generalized Wilcoxon tests as applied to strata formed from each of the six factors mentioned above. Such tests were carried out for each of the mortality categories of thymic leukemia, nonthymic leukemia, and all natural causes. Corresponding to each factor (covariate) are the stratum definitions along with the number of mice known to belong to that stratum. The tabular entries are the number of deaths, the ratio of observed to "expected" deaths, the log-rank significance level, and the generalized Wilcoxon significance level.

A total of 67 mice died of thymic leukemia. The rank tests suggest strong associations between thymic leukemia mortality and each level of the virus, the level of viral antibody, and the Gpd-1 phenotype. Only 12 animals died of nonthymic leukemia, so that there is limited ability to detect associations between this mortality type and the genetic and viral factors. It does seem, however, that the strong association noted between virus levels and thymic leukemia mortality is not evident for nonthymic leukemia, suggesting different etiologies for the two diseases. A total of 115 animals died of natural causes. Not surprisingly, in view of the fact that more than half of these deaths were attributed to thymic leukemia, the factors virus level, antibody level, and Gpd-1 phenotype appear strongly related to all natural causes of mortality. In addition, the female mice may experience slightly lower natural mortality rates than do the males.

Table 11.5 Log-Rank and Generalized Wilcoxon Significance Tests to Identify Associations Between Various Factors and Mortality Caused by Thymic Leukemia, Nonthymic Leukemia, and All Natural Causes

Factor	Class	No. Mice	Thymic Leukemia O^a	Thymic Leukemia O/E^b	Nonthymic Leukemia O	Nonthymic Leukemia O/E	All Natural Causes O	All Natural Causes O/E
MHC	k	110	39	1.10	9	1.46	67	1.13
	b	93	28	0.88	3	0.51	47	0.86
	Log-rank significance level			0.36		0.10		0.15
	Generalized Wilcoxon			0.38		0.21		0.23
Antibody	<0.5	101	45	1.64	9	1.82	70	1.52
(% gp70	0.5–5.0	27	4	0.38	2	0.94	9	0.48
ppt.)	5.0–20.0	53	12	0.63	1	0.29	25	0.77
	>20.0	19	4	0.50	0^c	0.00	7	0.50
	Log-rank significance level			0.0002		0.09		<0.0001
	Generalized Wilcoxon			0.0001		0.08		<0.0001
Virus	$<10^{1.6}$	17	3	0.47	1	0.78	8	0.75
(PFU/mL)	$10^{1.6}$–$10^{3.0}$	35	1	0.07	2	0.71	9	0.39
	$10^{3.0}$–$10^{4.0}$	33	2	0.16	3	1.08	11	0.51
	$>10^{4.0}$	90	50	2.12	5	1.21	63	1.75
	Log-rank significance level			<0.0001		0.92		<0.0001
	Generalized Wilcoxon			<0.0001		0.93		<0.0001
Sex	Male	96	34	1.09	7	1.28	65	1.23
	Female	108	33	0.92	5	0.77	50	0.80
	Log-rank significance level			0.49		0.38		0.02
	Generalized Wilcoxon			0.79		0.43		0.07
Albino	c/c	92	29	0.91	5	0.84	46	0.83
	$c/+$	112	38	1.08	7	1.15	69	1.15
	Log-rank significance level			0.49		0.60		0.08
	Generalized Wilcoxon			0.46		0.68		0.16
Gpd-1	b/b	30	14	3.01	1	2.86	20	2.60
	b/a	70	5	0.35	1	0.61	15	0.55
	Log-rank significance level			<0.0001		0.23		<0.0001
	Generalized Wilcoxon			<0.0001		0.16		<0.0001

[a] O, number of deaths.
[b] O/E, ratio of the number of deaths to the expected number of deaths (calculated under a hypothesis of equality of mortality).
[c] This group excluded from the comparison.

The following practical statistical questions arise in the analyses of Table 11.5:

1. How should classes be formed?
2. Do the multiple significance tests affect the interpretation of the suggested associations?

3. What should be done if test statistics such as the log-rank and generalized Wilcoxon tests are in qualitative disagreement?

4. Are sample sizes adequate to permit the use of the asymptotic likelihood theory?

5. How are these or other inferences affected by the substantial fraction of missing data on Gpd-1 phenotype?

As usual, such practical problems are more difficult to answer precisely than are many of the associated theoretical questions. We provide here some discussion of these problems along with possible approaches to their resolution.

In regard to question 1, the four genetic factors are all binary, so that no question of class formation arises. Antibody and virus levels are at least partially quantitative. Antibody levels less than 5% (gp70 precipitate) could not be detected, as was the case with virus levels below $10^{1.6}$PFU/mL. Virus levels in excess of 10^4PFU/mL also could not be further specified. Beyond these restrictions, grouping could be done as desired. The formation of three or four classes on the basis of a quantitative factor would often provide adequate resolution without unduly compromising efficiency. Roughly equal sample sizes among classes is a sensible criterion, although natural division points may be present, and some limited imbalance in the initial groups may be preferable if censoring depends markedly on the factor under consideration. Of course, formation of groups on the basis of the mortality data observed itself would invalidate the corresponding tests. Table 11.5 defines four classes for each of the antibody and virus levels. Some collapsing of these classes, particularly for the nonthymic mortality, may be preferable.

Six significance tests are carried out for each of the three mortality categories in Table 11.5. If significance levels were based on k *independent* test statistics, the probability that one or more would show significance at the level α is, under an overall null hypothesis, $1 - (1 - \alpha)^k$, which for $k = 6$ and $\alpha = 0.05, 0.01, 0.001$, and 0.0001 has values 0.26, 0.06, 0.01, and 0.001, respectively. This point should be kept in mind if a large number of factors are simultaneously studied in an exploratory manner for association with failure time. Frequently, however, a study is designed to examine the prognostic value of one or a small number of factors (e.g., MHC and Gpd-1 in the current data set), and other variables are known or suspected risk factors that are included for model-building purposes or to avoid confounding. The multiple testing problem is then not severe in respect to the primary factors. The related multiple comparison problem involving tests to compare all pairs of mortality curves when the data are divided into $s > 2$ groups has been discussed by Koziol and Reid (1977). They consider both the log-rank and Gehan generalized Wilcoxon tests.

The log-rank and generalized Wilcoxon significance levels in Table 11.5 are generally in good agreement. Some relatively minor differences can be noted. For example, the comparison of male and female mortality from all natural causes has significance level 0.02 under the log-rank test and significance level 0.07 under the generalized Wilcoxon test. It is possible, however, that large differences in these significance levels occur (e.g., Prentice and Marek, 1979).

One approach to resolving a discrepant result among rank tests would permit the rank test to adapt to the data. For example, any of the censored data rank tests may be generalized by stratifying on time. The time axis may be divided into s (e.g., 2 or 3) strata and separate χ^2_{r-1} statistics may be calculated in each stratum, where r is the number of groups being compared. [See the discussion on time-dependent strata in Section 6.4.2 and in Peto et al. (1977) for the log-rank case.] The generalization of the log-rank test is precisely the score statistic based on a global test for $\beta = 0$ in a proportional hazards model when time-dependent indicator covariates are defined in each stratum for any $r - 1$ of the groups. Of course, for formal testing, the subdivision must not be based on previous examination of the failure times. An alternative statistic that would often be more efficient is the score statistic based on a test for $\beta = 0$ with indicator variables for $r - 1$ of the groups along with time-dependent covariates that are products of these indicator variables and simple functions of time, such as t or $\log t$. The latter tests would be sensitive for the detection of hazard ratios different from unity when such ratios are monotone and relatively smooth functions of time.

Problem 4 is concerned with sample sizes for the use of asymptotic results. This topic is of particular concern for the tests of nonthymic leukemia mortality because of the small number of deaths. In such situations it may be advisable to estimate the actual sampling distribution of the test statistic (under the null hypothesis) by simulation rather than by relying solely on asymptotic results. Very briefly, if censoring is independent of the covariates under consideration, a permutation approach to estimating the actual distribution of the test statistic is valid. From (7.18) the rank statistic can be written

$$\mathbf{v} = \sum_{i=1}^{k} [Z_{(i)} c_i + S_{(i)} C_i]$$

where $Z_{(1)}, \ldots, Z_{(k)}$ are covariate values corresponding to the uncensored failure times $t_{(1)}, \ldots, t_{(k)}$, and $S_{(i)}$ is the sum of covariate values for the m_i censored times in $[t_{(i)}, t_{(i+1)})$ with $t_{(0)} = 0, t_{(k+1)} = \infty$. The scores $(c_i, C_i), i = 1, \ldots, k$, are known for any particular rank test. Under the hypothesis $\beta = 0$ and the assumption of censorship independent of Z, the permutation distribution of $\mathbf{v}$ may be generated. Each step in the simulation simply involves a random assignment of the scores $\{c_i, C_i, \ldots, C_i; i = 1, \ldots, k\}$ to the $k + (m_1 + \cdots + m_k) Z$ values and the calculation of $\mathbf{v}$. A more complex simulation is required if the censoring depends on Z in an important manner. One possibility in this regard is to generate failure times at each specific Z from an exponential distribution with the actual censoring scheme at each particular Z approximated by a progressive type II censoring procedure.

Before problem 5 listed is addressed, some regression analyses for the mouse leukemia data are presented. Analyses relating to missing covariate data are then discussed for the rank tests and the regression methods.

The analyses presented in Table 11.5 provide no insight into the joint association between two or more covariates and mortality and only limited insight into the magnitude of relative risks associated with the covariates. Table 11.6 gives results of four applications of the Cox model to these data for the type-specific mortality

Table 11.6 Proportional Hazards Regression Analysis of Factors Related to Mortality in AKR $\times$ (B6 $\times$ AKR)F$_1$ Mice[a]

Factor (Code)	Thymic Leukemia Mortality				All Natural Mortality			
	Analysis 1 Coef. (S.E.)	Analysis 2 Coef. (S.E.)	Analysis 3 Coef. (S.E.)	Analysis 4 Coef. (S.E.)	Analysis 5 Coef. (S.E.)	Analysis 6 Coef. (S.E.)	Analysis 7 Coef. (S.E.)	Analysis 8 Coef. (S.E.)
MHC ($0 = k/k$; $1 = k/b$)	0.166 (0.294)	−1.179 (0.299)	n.i.	−0.122 (0.551)	−0.055 (0.232)	−0.333 (0.237)	n.i.	−0.206 (0.433)
Antibody (% gp70 ppt.) ($0 < 0.5$; $1 \geq 0.5$)	−0.996† (0.324)	−0.405 (0.388)	n.i.	−0.896 (0.582)	−0.607* (0.249)	−0.225 (0.263)	n.i.	−0.441 (0.425)
Virus (PFU/mL) ($1 < 10^{1.6}$; $0 \geq 10^{1.6}$)	n.i.	1.559 (0.825)	1.226 (1.238)	1.914 (1.313)	n.i.	0.776 (0.881)	0.503 (0.659)	1.236 (0.732)
Virus (PFU/mL) ($1 \leq 10^4$; $0 > 10^4$)	n.i.	−2.891‡ (0.603)	−1.672* (0.824)	−1.701* (0.824)	n.i.	−1.349‡ (0.272)	−0.132 (0.437)	−0.373 (0.447)
Sex ($0 = M$; $1 = F$)	−0.097 (0.290)	0.058 (0.301)	n.i.	−0.279 (0.504)	−0.392 (0.231)	−0.345 (0.238)	n.i.	−0.848* (0.380)
Albino ($0 = c/+$; $1 = c/c$)	0.121 (0.274)	0.275 (0.277)	n.i.	0.676 (0.481)	0.301 (0.217)	0.444* (0.221)	n.i.	0.413 (0.361)
Gpd-1 ($0 = b/b$; $1 = b/a$)	n.i.	n.i.	−1.510* (0.626)	−1.977† (0.636)	n.i.	n.i.	−1.598‡ (0.439)	−1.792‡ (0.414)

[a] n.i., not included in analysis; S.E., standard error.

* $p < .05$; † $p < .01$; ‡ $p < .001$

rate (7.1) for thymic leukemia as well as for all natural causes. There were too few nonthymic leukemia deaths to support such an analysis.

The left side of Table 11.6 gives the codes for seven indicator covariates defined from the six factors under study, and the tabular entries are the estimated regression coefficients and (asymptotic) standard errors corresponding to covariates included in a particular run. The first column under either mortality category indicates that the production of detectable viral antibody is associated with reduced mortality, particularly thymic leukemia mortality, after adjusting mortality rates for possible influences of MHC, sex, and coat color. The second column shows that this association is no longer close to significant when the mortality rates are permitted to vary among three virus-level categories. The association with antibody concentration is largely explained by the suppressed virus levels that occur in antibody-producing animals. Virus and antibody levels could be entered into these regression analyses in a quantitative manner with equal ease.

The initial analyses using log-rank and Wilcoxon tests provide valuable guidance for more refined model building. In addition, they are themselves useful summaries for many purposes. In particular, they are valuable aids in presenting the results of an analysis to nonstatisticians since these tests (log rank and Wilcoxon) can be described and explained at an intuitive level. Because of the close relationship between the log-rank and the Cox model (see Section 4.2.4), the main results of an analysis based on the latter can often be described with reasonable accuracy using log-rank or stratified log-rank tests. For example, the second log-rank test in Table 11.5 illustrates the strong association between detectable viral antibody and mortality. A stratified log-rank test of this same variable with strata defined by the four virus-level categories would illustrate the extent to which virus level accounts for this association. For the statistician, however, the proportional hazard presentation has the advantage of exhibiting these facts and others quickly and in an easily understood manner.

The final two analyses in Table 11.6, under either mortality category, are restricted to the 100 animals with known Gpd-1 phenotype. Analyses 3–4 and 7–8 indicate a much weaker virus–mortality association when taking account of Gpd-1 phenotype. This reflects a strong association between Gpd-1 and virus level. The significant Gpd-1 coefficients in the presence of virus-level covariates may also reflect a role for Gpd-1 in relation to mortality, beyond its association with virus level.

Useful supplementary information may be obtained by studying the relation between the covariates themselves. For example, a binary logistic model applied to whether or not virus levels exceed 10^4 PFU/mL shows strong ($p < 0.001$) associations between virus levels and both antibody production and Gpd-1. The same type of analysis with the presence or absence of detectable antibody as dependent variables showed simultaneous associations between antibody and each of MHC, virus level, and Gpd-1 phenotype.

A troublesome point in the analysis of Tables 11.5 and 11.6 is that one of the most important prognostic factors, Gpd-1 phenotype, is available on only about one-half of the experimental animals. Moreover, the probability of a missing Gpd-1 phenotype is related to the survival time of the animal.

Several authors have used these data to illustrate and compare missing data methods (see Section 11.5). Chen and Little present thymic leukemia analyses using data on all 204 mice. Their analysis included two covariates: Gpd-1 phenotype coded as 0 for b/b and 1 for b/a, and viral concentration coded as 0 for concentrations less than 10^4 and 1 otherwise. There were 100 mice having Gpd-1 phenotype available and 175 having virus concentration available. As noted above, none of the mice dying in the first 100 days have Gpd-1 determinations, so missingness clearly depends on the observed times (T) for the study animals. On the other hand, it seems unlikely that the availability of covariate measurements depends on the actual covariate values in this application, so a MAR assumption seems reasonable.

The left side of Table 11.7 show analyses of Gpd-1 and a binary $(0 \leq 10^4,$ $1 > 10^4)$ virus concentration covariate in relation to thymic leukemia mortality from Chen and Little (1999). Note the smaller standard error estimate for the virus coefficient estimate under the nonparametric maximum likelihood procedure compared to the complete case analysis, and the variability in the coefficient estimates themselves. The complete case and approximate partial likelihood (Lin and Ying, 1993b) methods do not apply here in view of the dependence of missingness rates for Gpd-1 on the observed time T.

To implement the inverse probability weighted and AIPW estimators Wang and Chen (2001) defined $\eta_i/\pi_i = 0$ for mice dying within the first 400 days of life. Wang and Chen also excluded mice having missing virus concentration data. The right side of Table 11.7 shows various analyses of these data, slightly revised from Wang and Chen (2001). These authors fit a binary logistic missingness model for Z_1(Gpd-1) using data on mice having $T \geq 400$. They observed a dependence of missingness on T but not on (δ, Z_2). They then used a local linear smoothing method, with a bandwidth of 200 days, to obtain estimates of the nonmissingness probability π as a function of T. These estimates also varied fairly strongly with T. They were used in the simple inverse probability weighted and the augmented inverse probability weighted estimates shown in Table 11.7. Regression calibration and estimated partial likelihood estimates are also given. Note the substantial variability of regression coefficient estimates and in corresponding standard error estimates from these various analyses. The nonparametric maximum likelihood, inverse probability weighted, and augmented inverse probability weighted estimates should all be appropriate here, assuming that a simple Cox model applies in this setting. As expected, comparison of the simple and augmented inverse probability weighted estimators suggests that inclusion of the augmentation yields a reduction in standard error for the estimated virus coefficient. These various analyses are fairly consistent in indicating joint negative and positive associations of Gpd-1 phenotype and virus concentration with the hazard rate for thymic leukemia mortality.

Table 11.7 Cox Model Analysis of Factors Related to Thymic Leukemia Mortality, with Provision for Missing Covariate Data[a]

| Factor (Code) | All Mice (Chen and Little, 1999) | | | | Mice Having Virus Conc. (Wang and Chen, 2001) | | | |
| | Gpd-1 $(0 - b/b; 1 - b/a)$ | | Virus $(0 \leq 10^4; 1 > 10^4)$ | | Gpd-1 $(0 - b/b; 1 - b/a)$ | | Virus $(0 \leq 10^4; 1 > 10^4)$ | |
Analysis Method	Coef.	S.E.	Coef.	S.E.	Coef.	S.E.	Coef.	S.E.
Complete case	-1.44	0.60	1.44	0.72	-1.46	0.57	1.22	0.65
Approx. partial likelihood	-1.21	0.62	2.14	0.59	-1.40	0.64	1.36	0.47
Nonparametric ML	-1.49	0.59	1.67	0.55	-1.44	0.52	1.11	0.44
Inverse prob. weighted					-1.50	0.61	1.17	0.68
Aug. inv. prob. weighted					-1.47	0.60	1.14	0.61
Regression calibration					-0.78	0.32	1.50	0.35
Estimated partial likelihood					-1.23	0.61	1.32	0.47

[a] S.E., standard error.

BIBLIOGRAPHIC NOTES

Paired data from the proportional hazards model were presented by Holt and Prentice (1974) in essentially the form given here. Related early work is described by Armitage (1959), Downton (1972), and Breslow (1975). Gross and Huber (1987) consider corresponding asymptotic distribution theory. See also Andersen et al. (1993, pp. 523–525). Wild (1983) considers the possibility of recovering interpair information by introducing a random effects (frailty) assumption for the baseline hazard functions (Exercise 11.2). See also related work by Woolson and Lachenbruch (1980) and by Clayton and Cuzick (1985).

The statistical literature is replete with methods for the analysis of binary data. Finney (1971) provides a detailed discussion of the probit model, as do Cox (1970) and Cox and Snell (1989) for the logit model. Additional detail on the methods described in Section 11.3 can be found in Prentice (1976a).

The cohort sampling methods of Section 11.4 are intimately linked to case–control studies. Basic results for case–control studies were given by Cornfield (1951) and Mantel and Haenszel (1959). Other early works focusing on disease occurrence as a binary outcome include Zelen (1971), Anderson (1972), Mantel (1973), Miettinen (1974), Fisher and Patil (1974), Breslow (1976), Prentice (1976b), and Prentice and Pyke (1979). See Breslow and Holubkov (1997), Chatterjee et al. (2002), Scott and Wild (1997), and Lawless et al. (1999) for more recent developments. Cohort sampling in the context of the Cox model and time-matched cases and controls was considered by Thomas (1977) and Prentice and Breslow (1978). Asymptotic distribution theory for nested case–control studies using martingale methods was given by Goldstein and Langholz (1992) and in a more general form by Borgan et al. (1995). Related work on properties and variants of nested case–control studies includes Oakes (1981), Breslow et al. (1983), Prentice (1986b), Lubin and Gail (1984), Robins et al. (1986, 1989) and Langholz and Borgan (1995). Samuelsen (1997), building on the work of Kalbfleisch and Lawless (1988), proposed inverse sampling probability weighted estimators as a means of improving the efficiency of nested case–control analyses. Breslow and Day (1980, 1987) present a detailed discussion of statistical aspects of case–control and cohort methods, respectively.

Cox model estimation under case–cohort sampling was considered by Prentice (1986a). Corresponding asymptotic distribution theory was given by Self and Prentice (1988). Score statistic variance estimators were proposed by Prentice (1986a), Wacholder et al. (1989), and Barlow (1994). Sampling variants were considered by Barlow (1994) and Borgan et al. (2000). Therneau and Li (1999) discuss computational aspects of case–cohort estimation. Sorensen and Andersen (2000) generalize case–cohort methods to allow competing risks. Langholz and Thomas (1990) and Wacholder (1991) discuss aspects of the choice of cohort sampling procedures. Chen and Lo (1999) consider inverse sampling probability weighted estimators under case–cohort sampling and demonstrate an efficiency advantage to the more comprehensive use of data on cases. Recently, Chen (2001) has proposed a local covariate averaging estimation approach for a broader class of sampling procedures that include case–cohort and nested case–control special cases. This estimation approach has potential for semiparametric efficiency with more complex computations.

Little and Rubin (1987) provide an extensive discussion of statistical approaches to the accommodation of missing covariate data in the analysis of various types of data. More specific references, in the context of failure time data under the Cox model, include Lin and Ying (1993b) and Zhou and Pepe (1995), both of which require fairly strong assumptions on the missingness mechanism. The nonparametric maximum likelihood approach of Chen and Little (1999) relaxes such assumptions but requires a correct specification of the covariate distribution and does not allow the censoring to depend on the missing covariates. The augmented inverse sampling probability method of Robins et al. (1994) and Wang and Chen (2001) avoids the latter restriction and evidently can yield consistent estimates if either the covariate distribution or missingness model is specified correctly. Both the nonparametric maximum likelihood and the augmented inverse sampling probability methods will typically be computationally intensive, especially if covariates are continuous or of high dimension.

These same approaches also apply to mismeasured covariate data with a validation subsample. With only a reliability covariate sample and a simple classical measurement model possible, approaches include regression calibration (Wang et al., 1997) and risk set regression calibration (Xie et al., 2001). These methods are simply applied and tend to be fairly efficient, but incorporate some asymptotic bias. A corrected score approach can yield consistent regression parameter estimates. Some applications of this approach (Nakamura, 1992; Buzas, 1998) impose distributional assumptions on the measurement errors, but work by Huang and Wang (2000) avoids any such assumption for regression parameter estimation. Likelihood-based approaches to this problem have also been considered (Zhong et al., 1996; Hu et al., 1998). The monograph by Carroll et al. (1995) provides a valuable discussion of measurement error estimation procedures in nonlinear models, with some discussion of the Cox model.

Early important work on sequential testing include Barnard (1946) and Wald (1947). In the comparative clinical trial context, the book by Armitage (1975) is particularly noteworthy. Key papers on group sequential testing in clinical trials include Pocock (1977), O'Brien and Fleming (1979), Fleming et al. (1984), and Wang and Tsiatis (1987). The papers by Slud and Wei (1982), Lan and DeMets (1983), and Kim and DeMets (1987a) relaxed the common information assumption of much of the earlier work. Two-sided testing boundaries were considered by Gould and Pecore (1982), Whitehead and Stratton (1983), and Emerson and Fleming (1989). Estimation at the termination of a clinical trial is discussed by Siegmund (1978, 1985), Tsiatis et al. (1984), Kim and DeMets (1987b), Whitehead (1986), and Emerson and Fleming (1990), and estimation on an ongoing basis is considered by Jennison and Turnbull (1984, 1989). Spiegelhalter et al. (1994), among others, consider Bayesian approaches to group sequential methodology. The book by Jennison and Turnbull (2000) provides a detailed account of group sequential methods with substantial emphasis on failure time endpoints.

There is a very large literature on nonparametric Bayesian methods and on the application of Bayesian techniques to the analysis of failure time data, especially in the Cox model. The book by Ibrahim et al. (2001) gives a very good summary with an extensive bibliography and bibliographic notes. In the treatment in this chapter, we have attempted only to indicate the type of results available from Dirichlet, gamma, and other "neutral the right" process priors. Many other related directions have also been explored.

Ferguson (1973) introduced the Dirichlet process and showed that many standard nonparametric techniques could be derived in this way. Doksum (1974) considered a wide class of prior distributions called *tail-free* or *neutral to the right* random probability measures, and both the Dirichlet process and the gamma process are special cases of these. Ferguson (1974) gives a survey of work done in this area. The application of these methods to survival data has been considered by Susarla and Van Ryzin (1976), who obtain the Kaplan–Meier estimator from a Dirichlet prior process; by Cornfield and Detre (1977), who consider a discrete gamma process (see also Kalbfleisch and MacKay, 1978a); and by Kalbfleisch (1978a). Ferguson and Phadia (1979) have considered the estimation of the survivor function with no covariates for a number of tail-free or neutral-to-the-right prior processes, including the gamma and Dirichlet cases. Wild and Kalbfleisch (1981) extended these analyses to the Cox model with time-independent covariates. This work is closely related to that of Doksum (1974) and Kalbfleisch (1978a). Hjort (1990) considers a beta process in which infitesimal increments in the hazard process are modeled as beta variables. He extends the analysis of the Cox model to include time-dependent covariates and extends the analyses to include estimation of intensity rates in Markov processes as discussed in Section 8.3. Burridge (1981) advocates the use of discrete models to avoid the singularities caused by the discreteness of the neutral-to-the-right processes. Many of the approaches advocated in Ibrahim et al. (2001) are also based on discretized or grouped models.

Mixed Dirichlet process priors have been considered and developed by many authors but were first introduced by Escobar (1994) and MacEachern (1994) (see also Escobar and West, 1995). As the name would suggest, the mixed Dirichlet process is a two-stage model for the observable T. It is assumed that T has some parametric distribution $F(t \mid \theta)$, conditional on a parameter or vector of parameters θ. It is then assumed that θ arises from some unknown distribution H which is the realization of a Dirichlet process. Generally, computation is achieved through Markov chain Monte Carlo simulation techniques. See, for example, Kleinman and Ibrahim (1998a,b) and MacEachern and Muller (1998). Mixed Dirichlet process priors in the context of failure time data have been considered by Doss (1994), Doss and Huffer (1998), and Doss and Narasimhan (1998).

EXERCISES AND COMPLEMENTS

11.1   Consider the data on skin graft survival given in Table 11.1 using the marginal regression methods of Section 10.2.2. Using the marginal regression model (10.5) with baseline hazard functions λ_{0j} restricted to be equal $(j = 1, 2, \dots)$ derive a test for the hypothesis of no association between quality of the tissue match and graft survival and estimate the corresponding regression parameter. Describe the interpretation of this regression parameter relative to the within-pair regression parameter of Section 11.2.1 (see Lee et al., 1992 and Cai and Prentice, 1997 for a discussion of marginal models having a common baseline hazard).

11.2 (*continuation*) Consider again data of Table 11.1 and the Weibull special case of the paired data Cox model (11.1), so that $\lambda_{0s}(t) = \lambda_s \eta t^{\eta-1}$. Suppose that the frailty parameters $\lambda_s, s = 1, \dots$ are independent gamma variates rescaled to have mean 1 and unknown variance θ. Derive an estimation procedure for the relative risk parameter β and test $\beta = 0$ using the skin graft data. What can be said about the efficiency of this regression parameter estimate compared to the maximum likelihood estimate from (11.3) if the modeling assumptions obtain? (Wild, 1983)

11.3 Suppose that the paired failure time Cox model (11.1) is extended to allow competing failure types $j = 1, \dots, m$ with the hazard function for failure type j in pair s given by $\lambda_{ojs}(t)\exp[Z(t)'\beta_j]$. Define r_{js} to be zero if the first failure in pair s is of type j, and $t_{s1} < t_{s2}$, and by 1 if the first failure in pair s is of type j and $t_{s2} < t_{s1}$. Derive a conditional likelihood function for $\{\beta_1, \dots, \beta_m\}$ under independent censorship and show that it is a product of factors like (11.3) over $j = 1, \dots, m$. Suppose further that $\lambda_{0js}(t) = \lambda_{0s}(t) \exp(\gamma_j)$ for each j. Derive a partial likelihood function for $(\beta_j, \gamma_j), j = 1, \dots, m$ under this proportional risk model. (Kalbfleisch and Prentice, 1980, Sec. 8.1.4)

11.4 Consider the mouse leukemia data (data set V, Appendix A) discussed in Section 11.9. Develop an analysis like analysis 2 of Table 11.4 by matching each mouse dying of thymic leukemia to m at-risk comparison mice, with control mice selected independently at each thymic leukemia mortality time. Examine how the analysis using estimating function (11.16) relates to that in Table 11.6 for various choices 1, 2, 5, and 10 for m (take all available controls if fewer than m are available). How large does m need to be for estimated standard errors for regression parameters to be close to those in Table 11.6? Repeat the analysis several times selecting new controls each time. Comment on the variation in the regression parameter estimate.

11.5 Suppose that n cases occur at time t under the model (11.14), and suppose that m controls are selected from individuals at risk at time t in a large, possibly conceptual cohort. Suppose further that the modeled covariate $Z(t)$ has a finite sample space $Z_1, \dots, Z_q$. Show that the induced distribution for $Z(t)$ given t and given case or control status j (0 for control, 1 for case) is

$$\exp(\alpha_i + Z_i'\beta_j) \bigg/ \sum_{\ell=1}^{q} \exp(\alpha_\ell + Z_i'\beta_j)$$

for $i = 1, \ldots, q$, where $\beta_1 = \beta$ and $\beta_0 = 0$ and where

$$\alpha_i = \log \{P[Z(t) = Z_i \mid (t, 0)]/P[Z(t) = Z_1 \mid (t, 0)]\}.$$

Compute the asymptotic distribution of the maximum likelihood estimator $\hat{\beta}$ from this induced logistic model. Compare $\hat{\beta}$ and its asymptotic distribution to that obtained from a direct application of the binary logistic model

$$P[j|t, Z(t)] = \exp\{[\gamma + Z(t)'\beta]j/[1 + \exp[\gamma + Z(t)'\beta]\},$$

$j = 0, 1$ to the case–control data, as if the data have been obtained prospectively. Show that this same result holds even if the sample space for $Z(t)$ is not finite. (Prentice and Pyke, 1979)

11.6 Reanalyze the mouse leukemia data (Appendix A, data set V) by replacing missing values for virus concentration and for Gpd-1 allele by the average of such values over the mice for which such data were available, and compare the results with the Chen and Little (1999) analyses on the left side of Table 11.7. Describe the advantages or disadvantages of this crude form of data imputation.

11.7 Derive a general expression for the conditional probability

$$P[N_i(t) = 1 \mid dN.(t) = 1; X_\ell(t), \ell \in \tilde{R}(t); \tilde{R}(u), 0 \le u \le t]$$

under model (11.14) with $\tilde{R}(u)$ comprised of the case at time t along with a random sample of subjects at risk at time t. Assume independent censorship and independent left truncation. Discuss circumstances under which this probability will equal (11.17). Contrast this probability with (11.17) under case–cohort sampling. (Borgan et al., 1995)

11.8 In the sequential testing setting of Section 11.7, show that the log-rank tests U_j are approximately independently normally distributed with mean $\beta I_j^{1/2}$ and variance I_j under model (11.28) with binary covariate and with β close to zero. (*Hint*: Apply a Taylor approximation in β to the score test at each inspection time.)

11.9 (*continuation*) Consider the O'Brien–Fleming stopping boundaries as applied to the standardized tests $\mid v_j \mid = \mid U_j/I_j^{1/2} \mid$ with $I_j = j\,I, j = 1, \ldots, K$ (equal information increments). Compute the type I error (α) for the sequential procedure under the normal approximation to the distribution of the log-rank tests with critical value given by $c(K/j)^{1/2}$ for the jth test, for $K = 3, 5,$ and 10. Describe how c can be chosen to yield a test of size α for a specified value of K. [*Hint*: The null hypothesis is rejected at test j if

$| v_\ell | < (K/\ell)^{1/2}$ for $\ell = 1, \ldots, j = 1$ and $| v_j | > (K/j)^{1/2}$ the probability of which can be computed for any $j = 1, \ldots, K$ under the null hypothesis. The type I error is the sum of these rejection probabilities.]

11.10   Verify that the mean and variance of $F_0(t)$ under the Dirichlet process are as stated in (11.33).

11.11   Verify the Kolmogorov consistency condition (11.32) for the Dirichlet process.

11.12   For the Dirichlet process, find the distribution of $r_i = -\log(1 - q_i)$, $i = 1, \ldots, k$ corresponding to the partition $(a_{i-1}, a_i]$, $i = 1, \ldots, k$. Show that the moment generating function $E[\exp(\theta r_i)]$ of r_i is $B(\gamma_{i-1} - \gamma_i, \gamma_i - \theta)/B(\gamma_{i-1} - \gamma_i, \gamma_i)$, where $B(a, b) = \Gamma(a + b)/[\Gamma(a)\Gamma(b)]$ is the beta function. Verify the consistency condition on the log scale using the generating function. That is show that the MGF of $r_i + r_{i+1}$ is of the same form and so verify that $-\log F_0(t)$ is an independent increments process.

11.13   (a) Suppose that the survivor function F_0 is a realization of a Dirichlet process with parameters c, F^* and that $T_1 = t_1$ is a single failure time observed from F_0. Show that given t_1, F_0 is a Dirichlet process with parameters c_1, F_1^* where $c_1 = c + 1$ and $(c + 1)F_1^*(t) = cF^*(t) + 1(t < t_1)$, $0 \leq t$. Thus, the effect of the observation t_1 is to place a fixed mass point at t_1 in the parameters of the process.

   (b) Describe the posterior distribution of $F_0(t)$ when independent failure times $t_1, \ldots, t_n$ are observed. Find the posterior expectation of $F_0(t)$ and show that as $c \to 0$, this yields the empirical survivor function.

   (c) Suppose that a censored sample is available from F_0. Show that the posterior expectation of $F_0(t)$ approaches the Kaplan–Meier estimate as $c \to 0$. (Susarla and Van Ryzin, 1976)

11.14   Apply the discrete model of Section 11.8.4 to the data of Table 11.3. For this purpose, suppose that $F^*(j) = \exp(-j\lambda)$, $j = 0, 1, \ldots$ and vary c as in Table 11.4. Compare the results to the continuous analysis.

Glossary of Notation

For ease of reference, the following gives a summary of some basic notation used throughout the book.

Vectors and matrices appear in regular italic type and are identified through context. All vectors are column vectors.

$T > 0$ and $\tilde{T} > 0$ denote a failure time random variable.

C denotes a censoring time.

$T \wedge C = \min(T, C)$.

$\mathbf{1}(A)$ is an indicator variable that takes value 1 if A is true and 0 otherwise.

P denotes probability.

$F(t) = P(T > t)$ is the survivor function of T.

$\bar{F}(t) = P(T \leq t)$ is the distribution function.

$\Lambda(t)$ denotes a cumulative hazard function (or process).

$\lambda(t)$ denotes a hazard function (or process) in the continuous case.

$\mathscr{P}$ denotes the product integral.

$F(t) = \mathscr{P}_0^t [1 - d\Lambda(u)]$.

$X(t^-) = \lim_{s \to t^-} X(s); \ X(t^+) = \lim_{s \to t^+} X(s)$.

p denotes the number of modeled covariates.

n denotes the sample size.

If $a = (a_1, \ldots, a_m)'$ is a vector, then $a^{\otimes 2} = aa'$.

$x = (x_1, x_2, \ldots)'$ denotes a basic fixed or time-independent basic covariate.

$x(t) = [x_1(t), x_2(t), \ldots,]'$ denotes a possibly time-dependent basic covariate. The covariate history is denoted $X(t) = \{x(u) : 0 \leq u < t\}$.

$Z(t) = [Z_1(t), \ldots, Z_p(t)]'$ is the (left-continuous) modeled covariate vector.

$\beta = (\beta_1, \ldots, \beta_p)'$ is a vector of regression parameters.

$N_i(t)$ counts number of *observed* failures or events in $[0, t]$ for individual i.

$\tilde{N}_i(t)$ counts number of *actual* failures or events in $[0, t]$ for i.

$Y(t)$ is the (left-continuous) at-risk process.

$$Y_i(t) = \begin{cases} 1 & \text{if the } i\text{th individual is being observed and at risk of failure at } t^-; \\ 0 & \text{otherwise.} \end{cases}$$

$R(t) = \{i : Y_i(t) = 1\}$ is the set of individuals at risk of failure at time t^-.

$N.(t) = \sum_{i=1}^{n} N_i(t)$, $Y.(t) = \sum_{i=1}^{n} Y_i(t)$.

$dN_i(t) = N_i(t^- + dt) - N_i(t^-)$.

$$d\Lambda(t) = \begin{cases} \Lambda(t) - \Lambda(t^-) & \text{if } t \text{ is a discontinuity point of } \Lambda \\ \lambda(t)\,dt & \text{if } d\Lambda(t)/dt = \lambda(t) \\ 0 & \text{otherwise.} \end{cases}$$

$\Delta N_i(t) = N_i(t) - N_i(t^-)$.

$\Delta\Lambda(t) = \Lambda(t) - \Lambda(t^-)$.

In Chapter 10, the notation $N_i(dt), \Lambda(dt), N_i(\Delta t)$, and so on, is used instead of $dN_i(t), d\Lambda(t), \ldots$. This allows a convenient notation for bivariate and multivariate distributions. For example, $N(dt_1, dt_2), \Lambda(\Delta t_1, \Delta t_2)$, etc.

$X \sim F$ means that the random variable X has distribution F.

$\overset{\mathscr{P}}{\rightarrow}$ denotes convergence in probability.

$\overset{\mathscr{D}}{\rightarrow}$ denotes convergence in distribution.

$\sigma\{\cdot\}$ denotes the σ-field of events generated by the random variables in $\{\cdot\}$.

$\{\mathscr{F}_t : t \geq 0\}$ denotes a filtration or history process.

$\mathscr{F}_t$ is the σ-algebra of events up to and including time t.

$\mathscr{F}_{t^-} = \lim_{s \to t^-} \mathscr{F}_t$ is the history up to but not including t.

$M(\cdot)$ denotes a mean zero process, often a martingale.

The remaining notation given here relates to the Cox or relative risk model.

$\lambda[t; X(t)] = \lambda_0(t) \exp[Z(t)'\beta]$.

$M_i(t) = \int_0^t \{dN_i(u) - Y_i(u) \exp[Z_i(u)'\beta]\, d\Lambda_0(u)\}$ is a mean 0 process; often a martingale.

$\langle M \rangle(t)$ denotes the predictable variation process of the martingale M.

$[M](t)$ denotes the optional variation process of the martingale M.

$\mathscr{F}_t = \sigma\{N_i(u), Y_i(u^+), X_i(u^+), i = 1, \ldots, n; 0 \leq u \leq t\}$ is the filtration.

$S^{(0)}(\beta, t) = \sum_{i=1}^{n} Y_i(t) \exp[Z_i(t)'\beta]$.

$S^{(1)}(t) = \sum_{i=1}^{n} Y_i(t) Z_i(t) \exp[Z_i(t)'\beta]$.

$S^{(2)}(t) = \sum_{i=1}^{n} Y_i(t) Z_i(t) Z_i(t)' \exp[Z_i(t)'\beta] = \sum_{i=1}^{n} Y_i(t) Z_i(t)^{\otimes 2} \exp[Z_i(t)'\beta]$.

$p_l(\beta, t) = Y_l(t) \exp[Z_l(t)\beta] / \sum_{i=1}^{n} Y_i(t) \exp[Z_i(t)'\beta]$.

$\mathscr{E}(\beta, t) = \sum_{i=1}^{n} Z_i(t) p_i(t) = S^{(1)}(\beta, t)/S^{(0)}(\beta, t)$.

$\mathscr{V}(\beta, t) = \sum_{i=1}^{n} [Z_i(t) - \mathscr{E}(\beta, t)][Z_i(t) - \mathscr{E}(\beta, t)]' p_i(t)$
$\qquad = S^{(2)}(\beta, t)/S^{(0)}(\beta, t) - \mathscr{E}(\beta, t)^{\otimes 2}$.

$$n^{-1}S^{(j)}(\beta, t) \xrightarrow{\mathscr{P}} s^{(j)}(\beta, t), \quad j = 0, 1, 2.$$

$$\mathscr{E}(\beta, t) \xrightarrow{\mathscr{P}} e(\beta, t) = s^{(1)}(\beta, t)/s^{(0)}(\beta, t).$$

$$\mathscr{V}(\beta, t) \xrightarrow{\mathscr{P}} v(\beta, t) = s^{(2)}(\beta, t)/s^{(0)}(\beta, t) - e(\beta, t)^{\otimes 2}.$$

The Cox model score function is

$$U(\beta, t) = \sum_{i=1}^{n} \int_0^t [Z_i(u) - \mathscr{E}(\beta, t)] \, dN_i(t)$$

$$= \sum_{i=1}^{n} \int_0^t [Z_i(u) - \mathscr{E}(\beta, u)] \, dM_i(u)$$

The observed information is

$$I(\beta) = \sum_{i=1}^{n} \int_0^{\infty} \mathscr{V}(\beta, u) \, dN_i(u)$$

$$= \int_0^{\infty} [S^{(2)}(\beta, u)/S^{(0)}(\beta, u) - S^{(1)}(\beta, u)S^{(1)}(\beta, u)'/S^{(0)}(\beta, u)^2] \, dN.(u)$$

$\mathscr{I}(\beta) = E[I(\beta)]$ is the Fisher information.

$\hat{\Lambda}_0(t) = \int_0^t \{S^{(0)}(\hat{\beta}, u)\}^{-1} \, dN.(u)$ is the Nelson–Aalen estimator.

$\hat{M}_i(t) = N_i(t) - \int_0^t Y_i u \exp[Z_i(u)'\hat{\beta}] \, d\hat{\Lambda}_0(t)$ is the martingale (or mean model) residual.

$\hat{U}_i = \int_0^{\infty} [Z_i(u) - \mathscr{E}(\hat{\beta}, u)] \, d\hat{M}_i(u)$ is the ith score residual.

$nI(\hat{\beta})^{-1} \left(\sum_{i=1}^{n} \hat{U}_i \hat{U}_i' \right) I(\hat{\beta})^{-1}$ is the robust variance estimate for $\hat{\beta}$.

APPENDIX A

Some Sets of Data

The following sets of data are used for examples and discussion at various places in the text.

Data Set I Veterans Administration Lung Cancer Trial [a]

t	x_1	x_2	x_3	x_4	t	x_1	x_2	x_3	x_4	t	x_1	x_2	x_3	x_4
Standard, squamous [b]					153	60	14	63	10	92	70	10	60	0
72	60	7	69	0	59	30	2	65	0	35	40	6	62	0
411	70	5	64	10	117	80	3	46	0	117	80	2	38	0
228	60	3	38	0	16	30	4	53	10	132	80	5	50	0
126	60	9	63	10	151	50	12	69	0	12	50	4	63	10
118	70	11	65	10	22	60	4	68	0	162	80	5	64	0
10	20	5	49	0	56	80	12	43	10	3	30	3	43	0
82	40	10	69	10	21	40	2	55	10	95	80	4	34	0
110	80	29	68	0	18	20	15	42	0					
314	50	18	43	0	139	80	2	64	0					
100 [c]	70	6	70	0	20	30	5	65	0	Standard, large				
42	60	4	81	0	31	75	3	65	0	177	50	16	66	10
8	40	58	63	10	52	70	2	55	0	162	80	5	62	0
144	30	4	63	0	287	60	25	66	10	216	50	15	52	0
25 [c]	80	9	52	10	18	30	4	60	0	553	70	2	47	0
11	70	11	48	10	51	60	1	67	0	278	60	12	63	0
					122	80	28	53	0	12	40	12	68	10
					27	60	8	62	0	260	80	5	45	0
Standard, small					54	70	1	67	0	200	80	12	41	10
30	60	3	61	0	7	50	7	72	0	156	70	2	66	0
384	60	9	42	0	63	50	11	48	0	182 [c]	90	2	62	0
4	40	2	35	0	392	40	4	68	0	143	90	8	60	0
54	80	4	63	10	10	40	23	67	10	105	80	11	66	0
13	60	4	56	0						103	80	5	38	0
123 [c]	40	3	55	0	Standard, adeno.					250	70	8	53	10
97 [c]	60	5	67	0	8	20	19	61	10	100	60	13	37	10

378

Data Set I   (*Continued*)

t	x_1	x_2	x_3	x_4
Test, squamous				
999	90	12	54	10
112	80	6	60	0
87[c]	80	3	48	0
231[c]	50	8	52	10
242	50	1	70	0
991	70	7	50	10
111	70	3	62	0
1	20	21	65	10
587	60	3	58	0
389	90	2	62	0
33	30	6	64	0
25	20	36	63	0
357	70	13	58	0
467	90	2	64	0
201	80	28	52	10
1	50	7	35	0
30	70	11	63	0
44	60	13	70	10
283	90	2	51	0
15	50.	13	40	10
Test, small				
25	30	2	69	0
103[c]	70	22	36	10

t	x_1	x_2	x_3	x_4
21	20	4	71	0
13	30	2	62	0
87	60	2	60	0
2	40	36	44	10
20	30	9	54	10
7	20	11	66	0
24	60	8	49	0
99	70	3	72	0
8	80	2	68	0
99	85	4	62	0
61	70	2	71	0
25	70	2	70	0
95	70	1	61	0
80	50	17	71	0
51	30	87	59	10
29	40	8	67	0
Test, adeno.				
24	40	2	60	0
18	40	5	69	10
83[c]	99	3	57	0
31	80	3	39	0
51	60	5	62	0
90	60	22	50	10
52	60	3	43	0

t	x_1	x_2	x_3	x_4
73	60	3	70	0
8	50	5	66	0
36	70	8	61	0
48	10	4	81	0
7	40	4	58	0
140	70	3	63	0
186	90	3	60	0
84	80	4	62	10
19	50	10	42	0
45	40	3	69	0
80	40	4	63	0
Test, large				
52	60	4	45	0
164	70	15	68	10
19	30	4	39	10
53	60	12	66	0
15	30	5	63	0
43	60	11	49	10
340	80	10	64	10
133	75	1	65	0
111	60	5	64	0
231	70	18	67	10
378	80	4	65	0
49	30	3	37	0

Source: Prentice (1973). For a discussion of these data, see Section 4.5.

[a] Data for lung cancer patients: days of survival (t), performance status (x_1), months from diagnosis (x_2), age in years (x_3), and prior therapy (x_4) (0, no prior therapy; 10, prior therapy).

[b] Standard therapy, squamous tumor cell type.

[c] Censored survival.

Data Set II   Clinical Trial in the Treatment of Carcinoma of the Oropharynx[a]

Case	Inst.	Sex	Trt. Gp.	Grade	Age	Cond.	Site	T	N	Entry Date	Status	Time
1	2	2	1	1	51	1	2	3	1	2468	1	631
2	2	1	2	1	65	1	4	2	3	2968	1	270
3	2	1	1	2	64	2	1	3	3	3368	1	327
4	2	1	1	1	73	1	1	4	0	5768	1	243
5	5	1	2	2	64	1	1	4	3	9568	1	916
6	4	1	2	1	61	1	2	3	0	10668	0	1823
7	4	1	1	2	65	1	2	4	3	10768	1	637
8	4	1	2	3	84	1	4	1	3	12068	1	235
9	6	1	1	2	54	2	1	3	3	13368	1	255

Data Set II (*Continued*)

Case	Inst.	Sex	Trt. Gp.	Grade	Age	Cond.	Site	T	N	Entry Date	Status	Time
10	3	1	1	2	72	2	4	2	2	15468	1	184
11	3	1	1	2	42	1	4	2	2	15468	1	1064
12	2	1	1	2	61	1	1	4	3	18268	1	414
13	3	1	2	1	71	1	2	3	1	18468	1	216
14	4	1	2	2	83	3	4	3	1	19068	1	324
15	2	1	1	3	43	1	2	4	3	20768	1	480
16	5	1	2	2	52	1	4	4	3	21768	1	245
17	4	2	1	3	68	1	4	2	3	22768	0	1565
18	6	1	2	2	69	1	1	3	0	23368	1	560
19	3	2	2	3	65	3	1	3	0	25968	1	376
20	5	1	1	2	58	1	2	4	3	28068	1	911
21	2	1	2	2	63	1	2	4	3	28068	1	279
22	4	1	1	2	59	3	2	4	3	28268	1	144
23	3	1	1	1	75	1	2	3	1	28268	1	1092
24	6	1	1	1	65	2	1	3	3	28968	1	94
25	4	1	2	3	41	1	2	4	3	29468	1	177
26	3	1	1	2	60	1	4	3	3	29868	0	1472
27	3	1	2	2	72	1	4	1	3	30468	1	526
28	5	1	2	2	51	1	1	4	3	30868	1	173
29	2	2	1	2	72	2	2	3	1	30868	1	575
30	6	1	1	2	49	1	4	3	2	31068	1	222
31	3	1	2	2	82	3	1	3	0	31868	1	167
32	2	2	1	2	64	1	1	2	3	32468	1	1565
33	4	2	2	2	57	2	2	4	3	33568	1	256
34	3	1	2	1	67	2	2	3	3	33368	1	134
35	6	1	2	2	65	2	1	3	0	33868	1	404
36	3	1	2	2	62	1	4	1	2	369	0	1495
37	2	1	1	2	49	1	4	1	3	769	1	162
38	5	1	1	2	60	1	4	3	3	969	1	262
39	3	1	1	2	75	2	2	3	3	1769	1	307
40	2	1	2	2	54	1	2	2	3	2469	1	782
41	3	1	2	2	59	1	4	2	2	2469	1	661
42	5	1	1	2	58	1	1	3	2	3569	1	546
43	3	1	2	2	50	1	1	4	0	4469	0	1766
44	2	1	1	1	60	1	1	3	0	4569	1	374
45	3	2	1	1	43	1	2	2	2	4969	0	1489
46	4	1	1	2	48	2	2	3	3	5169	0	1446
47	4	1	2	2	49	3	1	4	3	5669	1	74
48	3	1	1	1	44	1	1	3	1	2769	0	1609
49	2	1	1	1	77	1	1	4	1	8369	1	301
50	2	1	1	1	75	1	2	4	1	9369	1	328
51	3	1	1	1	54	1	1	3	3	11869	1	459
52	3	1	1	1	68	1	1	4	0	12569	1	446
53	6	1	2	3	58	1	4	3	2	12769	0	1644
54	2	1	2	3	66	1	2	4	3	12969	1	494

Data Set II (*Continued*)

Case	Inst.	Sex	Trt. Gp.	Grade	Age	Cond.	Site	T	N	Entry Date	Status	Time
55	3	1	2	1	47	1	1	3	2	13269	1	279
56	5	1	1	2	60	1	4	3	2	13569	1	915
57	2	1	1	2	66	1	4	4	2	14369	1	228
58	3	1	1	3	51	1	1	3	3	15569	1	127
59	2	2	1	1	49	1	1	3	0	15669	1	1574
60	6	1	1	1	50	1	2	4	0	16669	1	561
61	2	2	1	1	52	1	4	4	3	16769	1	370
62	2	1	2	2	40	1	4	4	3	17869	1	805
63	4	1	2	2	69	1	1	3	3	19969	1	192
64	5	1	2	2	56	1	2	1	3	20469	1	273
65	5	2	2	3	70	2	4	4	3	20469	0	1377
66	3	2	2	3	47	1	4	3	2	23069	1	407
67	3	1	1	3	46	1	2	3	1	24569	1	929
68	3	1	2	1	53	1	4	2	3	26669	1	548
69	3	1	2	1	67	1	4	3	1	27969	0	1317
70	3	2	2	1	68	1	4	3	1	26869	0	1317
71	2	1	1	2	90	1	4	3	3	28069	1	517
72	3	2	1	3	44	1	4	3	2	28969	0	1307
73	5	1	2	2	48	1	1	4	2	29069	1	230
74	4	1	1	2	67	1	2	3	1	30469	1	763
75	5	2	1	2	58	2	4	4	3	30469	1	172
76	4	1	2	2	69	1	1	3	2	32869	0	1455
77	4	1	2	2	75	1	4	3	0	32869	0	1234
78	6	1	2	3	58	1	2	3	3	33069	1	544
79	3	1	1	1	72	1	4	3	3	33269	1	800
80	6	1	1	2	72	1	1	3	0	33569	0	1460
81	6	1	1	3	70	1	4	2	3	33669	1	785
82	6	1	1	2	71	1	2	4	0	34469	1	714
83	1	1	2	2	55	1	1	3	1	35369	1	338
84	3	2	2	1	73	1	1	3	0	36369	1	432
85	1	1	2	2	50	1	4	3	3	870	0	1312
86	6	1	2	2	63	2	1	3	0	4270	1	351
87	2	2	1	1	58	1	2	1	3	4470	1	205
88	1	2	1	2	56	1	4	3	0	4870	0	1219
89	6	2	2	2	62	3	4	4	3	4970	1	11
90	3	2	1	1	55	1	4	2	2	5470	1	666
91	2	1	1	1	50	2	2	4	3	5770	1	147
92	3	1	2	1	77	1	4	2	2	7870	0	1060
93	1	2	1	2	67	1	2	3	2	8270	1	477
94	3	1	1	3	53	1	2	3	2	9670	0	1058
95	2	1	2	3	55	1	2	2	2	11070	0	1312
96	6	1	2	1	71	2	2	3	3	11870	1	696
97	2	1	1	1	65	1	1	4	3	12470	1	112
98	1	2	2	2	50	1	2	4	3	13170	1	308
99	5	1	2	2	61	2	4	4	3	14470	1	15

Data Set II   (*Continued*)

Case	Inst.	Sex	Trt. Gp.	Grade	Age	Cond.	Site	T	N	Entry Date	Status	Time
100	5	1	2	1	72	2	1	4	0	14670	1	130
101	4	1	1	1	51	1	2	3	0	15270	1	296
102	4	1	1	2	59	1	4	3	3	15870	1	293
103	2	1	2	2	56	1	1	4	0	16070	1	545
104	3	1	1	2	61	1	1	3	1	16670	0	1086
105	1	1	1	3	61	1	2	2	3	17470	0	1250
106	3	1	2	2	68	2	1	3	3	18770	1	147
107	5	2	1	2	71	2	1	3	3	18970	1	726
108	2	2	2	2	57	1	2	2	2	19070	1	310
109	2	1	1	2	72	1	4	3	1	20570	1	599
110	3	1	1	2	55	1	2	3	0	21170	0	998
111	4	2	2	3	61	1	2	2	2	21970	0	1089
112	5	1	1	1	47	1	4	4	1	23170	1	382
113	4	2	1	3	66	1	2	3	2	24370	0	932
114	1	1	2	2	52	2	4	4	3	25170	1	264
115	1	1	2	1	61	2	4	4	3	25470	1	11
116	5	1	1	1	66	2	2	4	3	25870	0	911
117	2	1	1	3	64	2	4	4	3	28570	1	89
118	5	1	1	2	73	1	1	4	0	28770	1	525
119	2	1	2	2	67	1	1	3	3	31670	0	532
120	3	1	1	2	68	1	1	2	3	32770	1	637
121	6	1	1	3	58	2	2	3	3	33370	1	112
122	6	2	1	1	68	1	4	3	3	33670	0	1095
123	1	1	1	3	85	2	4	2	2	34170	1	170
124	4	1	1	2	74	1	1	3	0	34270	0	943
125	5	1	1	2	53	1	1	4	0	34370	1	191
126	6	1	2	2	60	1	2	1	2	34470	0	928
127	3	1	1	3	58	1	2	3	2	35570	0	918
128	3	1	1	2	66	1	1	1	2	36270	0	825
129	1	1	1	2	58	2	2	2	3	1271	1	99
130	2	1	1	2	39	1	4	3	3	1571	1	99
131	2	1	1	1	54	1	1	4	1	1871	0	933
132	6	1	2	3	49	1	2	2	3	2271	1	461
133	6	1	2	2	52	1	1	3	1	2671	1	347
134	1	1	1	2	35	2	4	3	0	3371	1	372
135	5	1	2	3	44	1	2	4	3	4371	0	731
136	5	1	1	9	81	1	4	3	3	4971	1	363
137	1	1	2	2	74	2	1	3	0	6771	1	238
138	4	1	2	2	65	1	4	4	3	7571	0	593
139	2	2	1	2	66	2	4	4	3	7771	1	219
140	3	1	1	2	74	2	4	3	2	8871	1	465
141	1	2	2	3	90	0	2	3	0	10571	1	446
142	2	1	2	2	60	1	1	4	1	11371	1	553
143	5	1	2	2	63	1	1	4	1	15371	1	532
144	2	2	2	2	61	1	4	4	1	15471	1	154

Data Set II (*Continued*)

Case	Inst.	Sex	Trt. Gp.	Grade	Age	Cond.	Site	T	N	Entry Date	Status	Time
145	2	1	2	1	67	1	1	4	3	15971	1	369
146	4	1	1	2	88	1	2	3	0	16171	1	541
147	5	2	2	3	69	2	4	4	0	18371	1	107
148	2	1	2	2	46	1	1	4	1	18871	0	854
149	1	1	1	2	69	1	1	2	2	20171	0	822
150	6	2	1	2	48	1	2	3	0	20271	1	775
151	2	1	1	1	77	1	1	4	0	20271	1	336
152	6	1	2	3	69	1	2	1	3	20271	1	513
153	6	1	2	3	75	1	4	3	3	20971	0	914
154	5	1	1	2	71	1	4	4	3	21671	1	757
155	5	2	1	2	58	1	1	4	3	21871	0	794
156	3	1	2	2	66	2	2	3	2	22171	1	105
157	5	2	1	1	44	1	2	4	1	23771	0	733
158	1	2	2	2	59	1	1	3	1	25371	0	600
159	2	1	1	2	78	9	4	4	3	26371	1	266
160	2	2	2	1	58	2	1	4	3	27371	1	317
161	2	1	2	3	65	1	4	3	3	28071	1	407
162	6	2	2	2	53	2	4	3	2	28471	1	346
163	3	1	2	2	49	1	4	2	2	29471	1	518
164	1	1	2	2	65	1	2	3	2	29971	1	395
165	5	1	1	1	59	1	1	4	1	31471	1	81
166	6	2	2	2	79	1	4	2	2	31971	1	608
167	6	1	1	1	57	1	2	4	3	32171	0	760
168	6	1	1	1	54	1	2	4	0	32371	1	343
169	3	1	2	2	47	1	1	3	0	32671	1	324
170	1	1	1	2	68	1	1	4	1	33071	1	254
171	2	1	1	3	63	1	4	2	2	34071	0	751
172	2	1	1	3	72	1	2	3	3	34271	1	334
173	6	2	2	1	51	2	2	3	1	34771	1	275
174	5	2	2	1	43	1	2	3	3	1272	0	546
175	6	2	2	2	43	2	4	3	3	3572	1	112
176	1	1	2	2	65	4	4	2	3	4672	0	182
177	1	1	2	2	54	1	1	4	3	5472	1	209
178	6	1	2	3	50	1	2	4	3	5572	1	208
179	2	1	2	2	39	1	1	4	3	5672	1	174
180	2	1	1	1	46	1	1	4	0	5972	0	651
181	2	1	2	3	49	1	2	3	3	8072	0	672
182	1	2	2	2	52	2	1	3	2	8272	1	291
183	1	2	2	2	69	2	4	4	1	13671	0	723
184	4	2	1	2	55	1	2	3	0	14372	1	498
185	4	1	2	2	48	2	2	3	0	14372	0	276
186	1	1	1	2	20	1	4	2	3	15672	0	90
187	4	2	1	2	47	1	4	3	3	15772	1	213
188	5	2	2	2	67	2	4	3	0	20572	1	38
189	4	2	1	2	66	2	2	3	2	20772	1	128

Data Set II   *(Continued)*

Case	Inst.	Sex	Trt. Gp.	Grade	Age	Cond.	Site	T	N	Entry Date	Status	Time
190	2	1	1	1	60	1	2	3	0	20972	0	445
191	2	1	1	1	54	1	2	4	3	22772	1	159
192	5	1	2	2	54	1	1	3	3	24372	1	219
193	4	1	2	2	59	2	1	4	0	24872	1	173
194	5	1	2	3	47	1	1	3	3	27672	0	413
195	3	1	2	2	57	2	1	3	3	12371	1	274

Source: Radiation Therapy Oncology Group. For a discussion of these data, see Sections 1.1.2 and 4.5.
[a]*Definitions*

Sex	1 = male, 2 = female.
Treatment	1 = standard, 2 = test.
Grade	1 = well differentiated, 2 = moderately differentiated, 3 = poorly differentiated.
Age	In years at time of diagnosis.
Condition	1 = no disability, 2 = restricted work, 3 = requires assistance with self care, 4 = bed confined.
Site	1 = faucial arch, 2 = tonsillar fossa, 3 = posterior pillar, 4 = pharyngeal tongue, 5 = posterior wall.
T staging	1 = primary tumor measuring 2 cm or less in largest diameter; 2 = primary tumor measuring 2 to 4 cm in largest diameter, minimal infiltration in depth; 3 = primary tumor measuring more than 4 cm; 4 = massive invasive tumor.
N staging	0 = no clinical evidence of node metastases; 1 = single positive node 3 cm or less in diameter, not fixed; 2 = single positive node more than 3 cm in diameter, not fixed; 3 = multiple positive nodes or fixed positive nodes.
Date of entry	Day of year and year.
Status	0 = censored, 1 = dead.
Survival	In days from day of diagnosis.

Data Set III   **Transfusion-Related AIDS Data**[a]

Inf. Time	Inc. Time	Age	Inf. Time	Inc. Time	Age	Inf. Time	Inc. Time	Age	Inf. Time	Inc. Time	Age
23	27	4	62	22	38	53	12	2	49	38	59
27	28	57	69	16	34	34	32	26	61	26	42
25	34	46	82	4	46	21	46	30	56	32	66
33	29	54	56	30	26	48	21	70	46	42	67
50	13	2	73	13	54	64	6	2	50	38	69
3	61	29	41	46	68	67	4	29	78	11	61
28	38	46	67	20	2	60	13	69	22	67	59
19	48	61	65	23	66	53	20	56	62	27	4
31	37	4	57	31	69	45	29	62	58	32	56
17	53	33	64	24	62	46	28	69	70	21	53
49	22	57	46	43	66	56	19	70	37	55	80
19	53	52	19	70	17	42	33	62	38	54	68

Data Set III *(Continued)*

Inf. Time	Inc. Time	Age	Inf. Time	Inc. Time	Age	Inf. Time	Inc. Time	Age	Inf. Time	Inc. Time	Age
65	8	73	27	62	65	66	10	32	56	37	66
50	24	52	57	33	51	67	11	68	76	17	2
37	37	34	41	50	46	60	19	70	86	8	1
34	41	32	63	29	78	71	8	1	74	20	63
57	18	64	39	53	37	38	15	56	84	10	61
36	40	66	41	52	72	23	34	20	45	49	77
53	24	44	83	10	38	33	29	53	59	36	65
40	39	50	69	25	65	34	29	34	65	31	51
74	5	39	75	19	61	26	38	56	81	15	36
38	14	2	61	33	66	40	25	61	87	9	57
45	10	1	58	36	21	42	24	68	77	19	77
42	17	46	84	11	55	37	30	25	54	43	47
39	23	2	91	4	1	32	37	62	48	49	60
35	29	62	73	23	3	58	13	39	30	68	29
53	12	2	67	29	72	35	37	57	43	55	59
34	32	26	72	24	76	53	20	78	57	41	11
21	46	30	53	43	46	36	38	56	60	38	68
48	21	70	65	32	68	33	41	67	36	63	67
64	6	2	36	61	63	62	13	2	48	51	27
67	4	29	75	23	41	57	18	23	89	10	1
60	13	69	86	12	36	54	21	3	68	32	71
53	20	56	51	47	65	19	58	69	37	63	59
45	29	62	48	51	54	68	10	67	60	40	73
46	28	69	36	63	57	54	25	50	39	62	69
56	19	70	60	39	35	43	37	76	80	21	62
42	33	62	41	59	65	75	6	68	29	72	73
66	10	32	49	51	5	48	33	3	61	40	72
67	11	68	52	48	56	70	12	63	97	4	36
60	19	70	37	64	49	64	18	35	38	15	56
71	8	1	63	38	85	67	16	59	23	34	20
52	29	24	85	16	38	36	47	22	33	29	53
67	14	45	70	31	71	66	18	67	34	29	34
55	27	68	63	38	69	42	42	73	26	38	56
47	35	23	38	14	2	65	20	2	40	25	61
35	48	26	45	10	1	62	23	54	42	24	68
67	16	72	42	17	46	46	40	64	37	30	25
76	8	1	39	23	2	70	16	62	32	37	62
49	35	62	35	29	62	76	10	50	58	13	39
35	37	57	68	19	74	66	0	63	52	34	73
53	20	78	82	5	84	33	34	51	27	59	60
36	38	56	59	29	52	43	26	61	66	20	67
33	41	67	68	20	60	67	4	1	50	36	46
62	13	2	59	29	67	12	60	21	39	48	64
57	18	23	26	63	62	56	17	66	69	18	2

Data Set III (Continued)

Inf. Time	Inc. Time	Age	Inf. Time	Inc. Time	Age	Inf. Time	Inc. Time	Age	Inf. Time	Inc. Time	Age
54	21	3	26	63	48	12	62	58	57	31	41
19	58	69	75	15	68	55	19	68	76	12	29
68	10	67	59	32	32	57	18	61	80	8	1
54	25	50	57	34	78	64	11	1	62	27	80
43	37	76	75	17	52	28	48	42	76	13	66
36	18	65	73	19	81	45	32	60	76	14	70
48	10	1	50	43	4	54	25	58	8	83	66
45	17	39	29	64	67	42	37	82	58	33	33
48	15	63	29	65	53	23	27	4	58	34	65
43	21	67	42	52	64	27	28	57	72	20	77
53	12	46	41	53	58	25	34	46	79	14	71
66	0	63	83	12	55	33	29	54	17	76	78
33	34	51	85	10	1	50	13	2	65	29	72
43	26	61	47	49	51	3	61	29	65	29	58
67	4	1	79	17	67	28	38	46	83	11	4
12	60	21	85	11	59	19	48	61	80	15	2
56	17	66	72	24	44	31	37	4	59	36	53
12	62	58	17	80	54	17	53	33	64	32	3
55	19	68	57	40	66	49	22	57	36	60	66
57	18	61	45	53	71	19	53	52	29	67	69
64	11	1	76	22	68	65	8	73	67	29	65
28	48	42	75	23	65	50	24	52	61	36	28
45	32	60	41	57	61	37	37	34	29	68	60
54	25	58	72	27	70	34	41	32	40	58	28
42	37	82	55	44	51	57	18	64	66	32	65
61	20	68	37	63	67	36	40	66	71	27	68
46	35	65	71	29	44	53	24	44	53	46	37
68	13	2	60	40	49	40	39	50	60	39	55
33	49	71	85	15	73	74	5	39	58	41	70
15	68	6	22	79	33	68	13	63	82	18	60
35	48	51	45	56	64	62	19	58	63	37	63
69	14	61	60	41	74	58	23	64	78	22	41
22	62	70	69	32	3	56	26	71	68	33	71
15	69	30	36	18	65	41	42	62	12	89	38
75	10	81	48	10	1	71	12	71	47	54	53
56	30	73	45	17	39	45	38	78	84	17	49
75	11	59	48	15	63	74	10	42	77	24	70
49	37	4	43	21	67	29	55	54			
43	43	62	53	12	46	65	20	67			

[a]Definitions

Inf. time   Infection (transfusion) time in months with 1 = Jan. 1978.
Inc. time   Incubation time (time from infection to diagnosis of AIDS measured in months from time of infection.
Age   In years at time of infection.

For a discussion of these data, see Section 1.6.

Data Set IV   Stanford Heart Transplant Data

Patient Ident.	Time of Accept.	Age at Accept.	Surg.	Transpl. Time	Surv. Status	Surv. Time	Matching 1	2	3
1	0.12	30.84	0		1	50			
2	0.25	51.84	0		1	6			
3	0.27	54.30	0	1	1	16	2	0	1.11
4	0.49	40.26	0	36	1	39	3	0	1.66
5	0.61	20.79	0		1	18			
6	0.70	54.60	0		1	3			
7	0.78	50.87	0	51	1	675	4	0	1.32
8	0.84	45.35	0		1	40			
9	0.86	47.16	0		1	85			
10	0.86	42.50	0	12	1	58	2	0	0.61
11	0.87	47.98	0	26	1	153	1	0	0.36
12	0.96	53.19	0		1	8			
13	0.97	54.57	0	17	1	81	3	0	1.89
14	0.97	54.01	0	37	1	1387	1	0	0.87
15	0.99	53.82	1		1	1			
16	1.07	49.45	0	28	1	308	2	0	1.12
17	1.08	20.33	0		1	36			
18	1.09	56.85	0	20	1	43	3	0	2.05
19	1.13	59.12	0		1	37			
20	1.33	55.28	0	18	1	28	3	1	2.76
21	1.34	43.34	0	8	1	1032	2	0	1.13
22	1.46	42.78	0	12	1	51	3	0	1.38
23	1.53	58.36	0	3	1	733	3	0	0.96
24	1.57	51.80	0	83	1	219	3	1	1.62
25	1.57	33.22	0	25	0	1800	2	0	1.06
26	1.58	30.54	0		0	1401			
27	1.59	8.79	0		1	263			
28	1.68	54.02	0	71	1	72	2	0	0.47
29	1.79	50.43	0		1	35			
30	1.88	44.91	0	16	1	852	4	0	1.58
31	1.89	54.89	0		1	16			
32	1.91	64.41	0	17	1	77	4	0	0.69
33	2.16	48.90	0	51	0	1587	3	0	0.91
34	2.20	40.55	0	23	0	1572	2	0	0.38
35	2.31	43.47	0		1	12			
36	2.51	48.93	0	46	1	100	2	0	2.09
37	2.57	61.50	0	19	1	66	3	1	0.87
38	2.59	41.47	0	4.50	1	5	3	0	0.87
39	2.63	50.52	0	2	1	53			
40	2.65	48.48	1	41	0	1408	4	0	0.75
41	2.88	45.30	1	58	0	1322	2	0	0.98
42	2.89	36.44	0		1	3			
43	3.06	43.39	1		1	2			
44	3.16	42.58	1		1	40			

Data Set IV *(Continued)*

Patient Ident.	Time of Accept.	Age at Accept.	Surg.	Transpl. Time	Surv. Status	Surv. Time	Matching 1	2	3
45	3.26	36.18	0	1	1	45	1	0	0.0
46	3.28	48.61	1	2	1	996	2	0	0.81
47	3.34	47.10	0	21	1	72	3	0	1.38
48	3.35	56.04	0		1	9			
49	3.38	36.65	1	36	0	1142	4	0	1.35
50	3.38	45.89	1	83	1	980			
51	3.48	48.73	0	32	1	285	4	1	1.08
52	3.56	41.25	0		1	102			
53	3.75	47.34	0	41	1	188			
54	3.75	47.79	0		1	3			
55	3.85	52.45	0	10	1	61	2	0	1.51
56	3.92	38.74	0	67	0	942	4	0	0.98
57	3.95	41.26	0		1	149			
58	3.98	48.02	1	21	1	343	2	1	1.82
59	3.99	41.38	1	78	0	916	2	0	0.19
60	4.13	49.05	0	3	1	68	3	0	0.66
61	4.18	52.56	0		1	2			
62	4.19	39.35	0		1	69			
63	4.20	32.66	0	27	0	842	3	1	1.93
64	4.34	48.82	1	33	1	584	1	0	0.12
65	4.43	51.29	0	12	1	78	2	0	1.12
66	4.47	53.21	0		1	32			
67	4.48	19.55	0	57	1	285	3	0	1.02
68	4.52	45.24	0	3	1	68	3	1	1.68
69	4.67	47.99	0	10	0	670	2	0	1.20
70	4.71	53.00	0	5	1	30	3	1	1.68
71	4.80	47.41	0	31	0	620	3	0	0.97
72	4.87	26.73	0	4	0	596	3	1	1.46
73	4.95	56.33	0	27	1	90	3	1	2.16
74	4.97	29.17	0	5	1	17	1	0	0.61
75	5	52.18	0		1	2			
76	5.01	52.08	1	46	0	545	3	1	1.70
77	5.02	41.11	0		1	21			
78	5.09	48.70	0	210	0	515	3	0	0.81
79	5.17	53.78	0	67	1	96	2	0	1.08
80	5.18	46.44	1	26	0	482	3	0	1.41
81	5.28	52.89	0	6	0	445	4	1	1.94
82	4.08	29.20	0		0	428			
83	5.32	53.31	0	32	1	80	4	0	3.05
84	5.33	42.72	0	37	1	334	4	0	0.60
85	5.35	47.98	0		1	5			
86	5.42	48.92	0	8	0	397	3	1	1.44
87	5.47	46.25	0	60	1	110	2	0	2.25
88	5.49	54.36	0	31	0	370	3	0	0.68

Data Set IV   (*Continued*)

Patient Ident.	Time of Accept.	Age at Accept.	Surg.	Transpl. Time	Surv. Status	Surv. Time	Matching 1	2	3
89	5.51	51.05	0	139	1	207	4	1	1.33
90	5.51	52.03	1	160	1	186	3	1	0.82
91	5.53	47.59	0		1	340			
92	5.57	44.98	0	310	0	340	1	0	0.16
93	5.78	47.75	0	28	0	265	2	0	0.33
94	5.95	43.84	1	4	1	165	3	0	1.20
95	5.98	40.28	0	2	1	16			
96	6.01	26.65	0	13	0	180	2	0	0.46
97	6.14	23.62	0	21	0	131	3	1	1.78
98	6.20	28.63	0	96	0	109	4	1	0.77
99	6.23	49.83	0		1	21			
100	6.35	35.06	1	38	0	39	3	0	0.67
101	6.37	49.52	0		0	31			
102	6.47	40.39	0		0	11			
103	−.05	39.32	0		1	6			

Source: Crowley and Hu (1977). For a discussion of these data, see Sections 1.1.3 and 6.4.3.

[a]*Definitions*

Time of acceptance	Years since Jan. 1, 1967.
Matching	1 = number of mismatches, 2 = HLA-A2 mismatch, 3 = mismatch score.
Survival status	1 = dead, 0 = censored.
Previous surgery	1 = yes, 0 = no.
Survival time, transplant time	In days from acceptance.

Data Set V   Days Until Death and Cause of Death for Male Mice Exposed to 300 rads of Radiation

				Control Group					
Thymic lymphoma	159,	189,	191,	198,	200,	207,	220,	235,	245
(22%)	250,	256,	261,	265,	266,	280,	343,	356,	383
	403,	414,	428,	432					
Reticulum cell	317,	318,	399,	495,	525,	536,	549,	552,	554
sarcoma (38%)	337,	558,	571,	586,	594,	596,	605,	612,	621
	628,	631,	636,	643,	647,	648,	649,	661,	663
	666,	670,	695,	697,	700,	705,	712,	713,	738
	748,	753							
Other causes	40,	42,	51,	62,	163,	179,	206,	222,	228
(39%)	252,	249,	282,	324,	333,	341,	366,	385,	407
	420,	431,	441,	461,	462,	482,	517,	517,	524
	564,	567,	586,	619,	620,	621,	622,	647,	651
	686,	761,	763						

Data Set V   *(Continued)*

				Germ-Free Group					
Thymic lymphoma	158,	192,	193,	194,	195,	202,	212,	215,	229
(22%)	230,	237,	240,	244,	247,	259,	300,	301,	321
	337,	415,	434,	444,	485,	496,	529,	537,	624
	707,	800							
Reticulum cell	430,	590,	606,	638,	655,	679,	691,	693,	696
sarcoma (18%)	747,	752,	760,	778,	821,	986			
Other causes	136,	246,	255,	376,	421,	565,	616,	617,	652
(46%)	655,	658,	660,	662,	675,	681,	734,	736,	737
	757,	769,	777,	800,	807,	825,	855,	857,	864
	868,	870,	870,	873,	882,	895,	910,	934,	942
	1015,	1019							

Source: Hoel (1972). For a discussion of these data, see Sections 1.1.1 and 8.2.3.

Data Set VI   Mouse Leukemia Data [a]

T_1	T_2	J	δ	Z_1	Z_2	Z_3	Z_4	Z_5	Z_6
121175	122377	2	2	1	2	2	1	00.0	10000
121175	122377	3	2	1		2	2	07.4	02400
121175	060677	3	1	1		2	2	13.2	00000
121175	121476	2	1	1		2	2	00.0	10000
121175	122377	3	2	2	2	1	1	05.8	00000
121175	010577	2	1	2		1	1	00.0	08800
121175	111677	5	1	1	2	1	1	00.0	08000
121175	101477	1	1	2		1	2	00.0	10000
121175	122377	3	2	2	2	2	1	78.7	00000
121175	122377	3	2	2		2	2	05.3	00080
121175	122377	1	2	2	2	2	2	20.3	00000
121175	040777	1	1	1	1	2	2	04.7	10000
121175	081076	1	1	1		2	2		
121175	050577	1	1	2		1	1	00.0	10000
121175	081077	1	1	1	1	1	1	00.0	10000
121175	091277	2	1	1		1	1	00.0	00080
121175	021777	1	1	2		1	2	00.0	10000
121174	091476	1	1	2		1	2	00.0	
121175	031877	1	1	1	2	1	2	00.0	10000
121275	100676	1	1	2		2	1	00.0	10000
121275	122377	2	2	2	2	2	2	00.0	03600
121275	122377	5	2	2	2	2	2	07.5	02000
121275	062877	2	1	1		2	2	00.4	00000
121275	122377	3	2	2	2	2	2	09.7	00000
121275	081776	1	1	1		2	2	00.0	
121275	080676	3	1	1		1	1		

Data Set VI   (*Continued*)

T_1	T_2	J	δ	Z_1	Z_2	Z_3	Z_4	Z_5	Z_6
121275	123377	3	2	1	2	1	1	00.0	10000
121275	041877	3	1	1		1	2	00.0	
121275	122377	3	2	1	2	1	2	00.0	10000
121275	122377	3	2	2	1	2	1	06.1	10000
121275	081977	2	1	1	1	2	1	07.5	10000
121275	122377	3	2	1		2	1	01.7	01000
121275	122377	3	2	1	1	2	1	00.0	10000
121275	102076	1	1	1		1	1	00.0	10000
121275	122377	3	2	2		1	2	18.1	05200
121275	102077	3	1	1	2	1	2	00.0	0300
121275	041177	3	1	1		1	2	00.0	10000
121275	052477	5	1	2	1	1	2	01.6	10000
121275	122377	3	2	1	2	1	2	00.0	05600
121275	070677	1	1	2		1	2	00.0	10000
121575	083076	6	1	1		2	1	00.0	
121575	083076	6	1	2		2	2	00.0	
121575	083076	6	1	2		2	2	00.0	
121575	083076	6	1	1		2	2	00.0	
121575	100776	3	1	2		1	1	00.0	
121575	082677	2	1	2		1	2	00.0	10000
121575	091476	1	1	1		2	1	00.0	
121575	080476	1	1	2		2	1		
121575	082677	1	1	2	2	2	1	68.5	00000
121575	031877	1	1	1	1	2	2	00.0	10000
121575	110477	4	1	2		2	2	00.0	00000
121575	050577	1	1	2	1	2	2	49.7	10000
121575	122377	3	2	2		1	1	02.2	00000
121575	122176	1	1	1		1	1	00.0	10000
121575	122377	3	2	2	2	1	2	00.0	00800
121575	122377	3	1	1	1	1	2	00.0	00360
121575	091277	3	1	1		1	2	00.0	10000
121575	072577	6	1	2		2	1	05.0	00160
121575	072577	6	1	1		2	2	06.2	00480
121575	063077	3	1	1	1	1	1	08.6	00160
121575	101276	1	1	2		1	1	00.0	10000
121575	100477	1	1	2	1	1	1	00.0	10000
121575	100476	1	1	1		1	1	00.0	
121575	052077	1	1	2	1	1	1	06.2	10000
121575	100477	3	1	2	2	1	1	05.1	00000
121575	100477	3	1	1	2	1	1	00.0	02000
121575	072676	3	1	1		1	2	06.7	03100
121575	020977	1	1	1		1	2	00.0	10000
121675	072577	6	1	2		2	1	63.7	00280
121675	072577	6	1	2		2	1	17.7	10000
121675	072577	6	1	1		2	2	06.2	10000
121675	110876	3	1	1		1	1	00.0	

Data Set VI   (*Continued*)

T_1	T_2	J	δ	Z_1	Z_2	Z_3	Z_4	Z_5	Z_6
121675	122377	3	2	2		1	1	05.1	02300
121675	122377	3	2	1	2	1	2	00.0	06400
121675	122377	3	2	2	2	1	2	00.0	10000
121775	031877	1	1	2	1	2	1	00.0	10000
121775	081877	1	1	2		2	2	12.8	0040
121775	091477	6	1	1		2	2	47.5	00240
121775	120276	3	1	1		2	2	28.7	
121775	020878	2	2	2	2	2	2	74.7	00040
121775	122377	5	2	2	2	1	1	05.8	00080
121775	122377	2	2	1	2	1	1	00.0	01400
121775	031877	3	1	2	2	1	1	10.7	00000
121775	062077	1	1	2		1	1	06.6	10000
121775	060777	1	1	1	1	1	2	00.00	10000
121775	122377	3	2	2	2	1	2	04.6	00080
121775	102776	1	1	1		1	2	00.0	10000
122275	111577	3	2	1	2	2	1	00.0	10000
122275	111677	1	1	2	1	2	1	00.0	10000
122275	061477	1	1	1		2	2	00.0	10000
122275	111577	3	2	1	2	2	2	00.0	00960
122275	111577	3	2	1	2	2	2	00.0	00000
122275	092476	1	1	1		2	1	00.0	
122275	012677	1	1	1		1	2	00.0	10000
122275	111577	3	1	2	2	1	2	00.0	10000
122275	122377	1	2	1	2	1	2	00.0	04000
122275	042677	2	1	1		1	2	00.0	00100
122275	033077	1	1	1	1	1	2	00.0	10000
122275	010577	1	1	1		2	1	00.0	10000
122275	122377	3	2	2	2	2	1	14.4	00720
122275	070677	1	1	1		2	1	02.2	10000
122275	122377	5	2	1	2	2	1	00.0	02400
122275	122377	3	2	1	2	2	1	00.0	02000
122275	122377	3	2	2	1	2	2	09.2	10000
122275	102776	1	1	1		2	2	00.0	10000
122275	101477	3	1	1	2	1	1	00.0	00880
122275	122377	3	2	1	2	1	1	00.0	10000
122275	122377	3	2	2	1	1	1	12.5	10000
122275	052777	1	1	1		1	1	00.0	10000
122275	091377	2	1	1		1	2	01.3	01500
122275	122377	3	2	2		1	2	87.9	00000
010576	012778	1	1	1		2	1	55.5	00000
010576	072677	1	1	2		2	2	17.4	10000
010576	083077	3	1	2	2	2	2	14.3	00560
010576	020178	3	2	1		1	1	00.0	10000
010576	060777	1	1	1		1	2	06.7	10000
011276	020178	3	2	2	2	2	1	43.4	00000
011276	091376	1	1	2		2	1	00.0	

Data Set VI   (*Continued*)

T_1	T_2	J	δ	Z_1	Z_2	Z_3	Z_4	Z_5	Z_6
011276	020178	1	2	1	2	2	1	01.3	00360
011276	020178	2	2	1	2	2	1	02.9	03800
011276	100676	1	1	2		2	1	03.4	10000
011276	122977	2	1	1	2	2	2	00.0	02800
011276	020178	3	2	2	2	2	2	45.4	00000
011276	020178	3	2	2	2	1	1	06.1	01500
011276	020178	3	2	2	2	1	1	02.6	10000
011276	011277	3	1	2		1	2	00.0	
011276	120676	3	1	2		1	2	01.7	
011276	082377	3	1	1		1	2	00.0	10000
011576	020178	3	2	2	2	2	2	27.7	00040
012676	020178	3	2	1	1	2	1	07.1	00040
012676	010577	3	1	2		2	2	02.1	
012676	030877	1	1	2	1	2	2	05.8	10000
012676	102076	1	1	1		2	2	00.0	10000
012676	122977	2	2	1	2	2	2	04.4	00280
012676	020178	3	2	1	2	2	1	00.0	00280
012676	100477	4	1	2		2	2	05.8	10000
012676	012677	6	1	1		2	2	00.0	
012676	020178	1	2	1		2	2	03.3	00080
012676	120776	1	1	2		2	2	27.7	10000
012676	020178	3	2	2	2	1	1	04.0	00040
012676	020178	1	2	2		1	2	00.0	10000
012676	020178	3	2	2	2	1	2	86.6	00040
012676	021777	1	1	2		1	2	05.2	10000
012676	120877	1	2	1	2	2	1	00.0	09200
012676	020178	2	2	1	2	2	1	05.2	10000
012676	020178	3	2	2	2	2	1	06.0	00040
012676	031477	4	1	1		2	2	00.0	
012676	082076	1	1	2		2	2	00.0	
012676	122077	1	2	2		2	2	04.4	10000
012676	021777	1	1	1		1	1	00.0	00000
012676	022176	1	1	1		1	1	00.0	100000
012676	111176	1	1	2		1	2	00.0	10000
012676	020178	3	2	1	2	1	2	00.7	06000
012676	020178	1	2	1		2	1	05.1	00880
012676	020178	3	2	1	2	2	1	04.1	01800
012676	020178	3	2	2		2	1	23.1	00200
012676	011777	2	1	1		2	2	00.0	
012676	091477	3	1	1		2	2	49.5	00400
012676	020178	3	2	2	1	2	2	05.1	10000
012676	111677	2	1	1		1	1	00.0	10000
012676	010177	4	1	2		1	1	05.1	
012676	081177	3	1	2	1	1	1	08.3	10000
012676	020977	1	1	1		1	2	16.6	10000
012676	012778	5	2	1	2	1	2	00.0	01200

Data Set VI   (*Continued*)

T_1	T_2	J	δ	Z_1	Z_2	Z_3	Z_4	Z_5	Z_6
012676	112277	1	1	1		1	2	00.0	10000
012776	021777	1	1	2		2	1	06.4	10000
012776	111776	1	1	1		2	1	00.0	10000
012776	111176	1	1	2		2	2	05.5	10000
012776	092077	4	1	2		2	2	18.1	05200
012776	102676	1	1	1		2	2	00.0	
012776	020178	3	2	2	2	1	1	15.2	04000
012776	020178	2	2	2	2	1	1	02.8	10000
012776	062077	1	1	1		1	2	00.0	10000
012776	020878	3	2	2	2	1	2	00.0	10000
012776	033077	1	1	2	2	1	2	00.0	10000
020276	020878	3	2	2	2	2	1	11.5	00040
020276	100477	3	1	2		2	2	05.3	10000
020276	030877	1	1	1	1	2	2	17.3	10000
020276	010178	4	1	2	2	2	2	50.2	00240
020276	090676	1	1	1		2	2	00.0	
020276	020878	3	2	2	1	1	1	03.7	10000
020276	040777	1	1	2	2	1	1	00.0	10000
020276	050677	1	1	1	1	1	1	00.0	10000
020276	101876	2	1	2		1	1	00.0	10000
020276	101876	3	1			1	2		
020276	020878	3	2	1	2	2	1	00.0	10000
020276	050977	3	1	1		2	1	00.0	
020276	020878	3	2	2	1	2	1	05.7	10000
020276	020878	3	2	2	2	2	1	41.1	00040
020276	020878	3	2	1	2	2	2	01.2	03900
020276	050577	1	1	1		2	2	00.0	10000
020276	012177	1	1	1		1	2	00.0	10000
020276	090777	1	1	1	1	1	2	01.2	04200
020276	061677	1	1	1		1	2	00.0	10000
020276	072977	3	1	1	1	1	1	02.9	02900
020976	011078	1	1	1	2	2	1	09.5	09200
020976	020878	3	2	1	2	2	2	02.5	00600
020976	020878	3	2	2	2	2	2	08.6	00800
020976	020878	3	2	1	1	2	2	13.1	10000
020976	020878	3	2	1	2	2	2	00.0	10000
020976	020878	3	2	1	2	2	1	00.0	01000
020976	020878	3	2	2	1	2	1	13.8	10000
020976	020878	3	2	1	2	1	1	01.2	10000
020976	020878	3	2	1	2	1	2	01.6	02200

Source: Mortality data on AKR $\times$ (C57BL/6 $\times$ AKR)F$_1$ mice from the laboratories of Dr. Robert Nowinski, Fred Hutchinson Cancer Research Center, Seattle, Washington. For a discussion of these data, see Section 11.9.

[a]*Definitions*

Date of birth (T_1)	Month/day/year.
Date of death (T_2)	Month/day/year.

Disposition of disease at death (J)	1 = thymic leukemia, 2 = nonthymic leukemia, 3 = nonleukemic and no other tumors, 4 = unknown, 5 = other tumors, 6 = accidental death.
"Type" of death (δ)	1 = natural, 2 = terminated.
MHC phenotype (z_1)	1 = k, 2 = b, blank = unknown.
Gpd-1 phenotype (z_2)	1 = b/b, 2 = b/a, blank = unknown.
Sex (z_3)	1 = male, 2 = female.
Coat color (z_4)	1 = c/c, 2 = $c/+$.
Antibody level (z_5)	% gp 70 precipitate; 0 = if % gp 70 < 0.5, blank = unknown.
Virus level (z_6)	PFL/mL; 0 = if PFL/mL < $10^{1.6}$, 10000 = if PFU/mL $\geq 10^4$, blank = unknown.

APPENDIX B

Supporting Technical Material

This appendix contains three brief sections to complement the material in the book. Section B.1 describes the EM algorithm that formed the basis of the analysis of interval censored data in Section 3.8 and that has also been used widely in various contexts to fit models with missing or incomplete data as they arise, in particular, with censoring. In Section B.2 we give a brief intuitive introduction to Stieltjes (or Lebesgue–Stieltjes) integration. These materials are relevant to several of the chapters in this book. Finally, in Section B.3 we provide a short summary of statistical software for failure time data analysis.

B.1 EXPECTATION-MAXIMIZATION ALGORITHM

The expectation-maximization (EM) algorithm (Dempster et al. 1977) is a convenient and widely used tool to compute maximum likelihood estimates when the data have, or can be viewed as having, missing elements. For example, one can view a censored sample of failure times in this way in that the unobserved time to failure for the censored observations can be viewed as the missing data.

Suppose that the complete (partially unobserved) data are denoted by X and that the incomplete data are denoted $Y = g(X)$, where g is a many-to-one function that corresponds to some grouping or incomplete observation of the complete data. Suppose that the probability laws for X depend on an unknown parameter vector θ and that the log-likelihood function for θ given X, the complete data log likelihood, is $\ell_X(\theta)$. The log likelihood arising from the incomplete data is denoted $\ell_Y(\theta)$. In many instances, ℓ_Y is of complicated form, whereas ℓ_X is relatively much simpler and more easily maximized. In this case, the EM algorithm often provides a simple approach to finding the MLE of θ based on Y.

Let $\theta^{(0)}$ be an initial estimate of θ. The algorithm proceeds through an expectation step and then a maximization step:

1. Calculate $Q(\theta, \theta^{(0)}) = E[\ell_X(\theta) \mid Y; \theta^{(0)}]$.
2. Find $\theta^{(1)}$ to maximize $Q(\theta, \theta^{(0)})$.

396

In the next iteration, replace $\theta^{(0)}$ with $\theta^{(1)}$ and repeat steps 1 and 2 to convergence. In many instances, this algorithm converges to the value of θ that maximizes the log liklihood ℓ_Y based on the incomplete data. If ℓ_Y is multimodal, however, the algorithm may converge to a local maximum or it can converge to a saddle point. In some fairly simple problems, such as the interval censoring example of Section 3.9.1, the algorithm can be shown to be globally convergent. In most applications, however, one would wish to explore various starting values to be assured that the MLE has been obtained. It should be noted that the estimation of the variance of $\hat{\theta}$ generally involves a separate calculation of the observed information based on the incomplete likelihood ℓ_Y and cannot be computed directly by taking second derivatives of Q. Louis (1982) gives some alternative approaches to estimating the covariance matrix for $\hat{\theta}$.

It can readily be seen that the algorithm is monotone in that at each step, the incomplete log likelihood ℓ_Y cannot decrease. To show this, we note that

$$\ell_Y(\theta) = \ell_X(\theta) - \ell_{X|Y}(\theta),$$

where the second term on the right is the log likelihood arising from the conditional density of X given Y. Taking an expectation conditional on Y at $\theta = \theta^{(0)}$ gives

$$\ell_Y(\theta) = Q(\theta, \theta^{(0)}) - E[\ell_{X|Y}(\theta) \mid Y; \theta^{(0)}].$$

It now follows that

$$\ell_Y(\theta^{(1)}) - \ell_Y(\theta^{(0)}) = A + B,$$

where

$$A = Q(\theta^{(1)}, \theta^{(0)}) - Q(\theta^{(0)}, \theta^{(0)}) \geq 0$$

by the definition of $\theta^{(1)}$, and

$$B = E[\ell_{X|Y}(\theta^{(0)})|\theta^{(0)}] - E[\ell_{X|Y}(\theta^{(1)})|\theta^{(0)}].$$

Since $\ell_{X|Y}(\theta)$ is the log of the conditional density of X given Y, it follows from Jensen's inequality that $B \geq 0$, and the monotonicity follows.

From this, it is easily seen that the MLE $\hat{\theta}$ arising from ℓ_Y is a fixed point of the algorithm and, under differentiability conditions, is a solution to the *self-consistency equation*

$$\frac{\partial}{\partial \theta_*} Q(\theta_*, \theta)|_{\theta_* = \theta} = 0.$$

An example is provided by a simple multinomial experiment. The underlying complete data comprise the frequencies $X = (X_1, X_2, X_3)'$, which have a trinomial distribution with known index n and probabilities $\theta = (\theta_1, \theta_2, \theta_3)'$. We do not, however, observe the complete data. Rather, for the first n_1 individuals, we observe only whether or not outcome 1 occurs; for the next n_2, whether or not outcome 2 occurs; and for the final $n_3 = n - n_1 - n_2$, whether or not outcome 3 occurs. Denote the incomplete data by $Y = (Y_1, Y_2, Y_3)'$, where Y_j has a binomial distribution with parameters (n_j, θ_j), $j = 1, 2, 3$ independently. The complete data log likelihood is

$$\ell_X(\theta) = \sum_{j=1}^{3} X_j \log \theta_j,$$

where $\theta_i \geq 0$ and $\theta_1 + \theta_2 + \theta_3 = 1$. It follows that

$$Q(\theta, \theta^{(0)}) = \sum_{j=1}^{3} X_j^{(0)} \log \theta_j, \qquad (\text{B.1})$$

where $X_j^{(0)} = E[X_j \mid Y; \theta^{(0)}]$ and

$$X_1^{(0)} = Y_1 + (n_2 - Y_2) \frac{\theta_1^{(0)}}{\theta_1^{(0)} + \theta_3^{(0)}} + (n_3 - Y_3) \frac{\theta_1^{(0)}}{\theta_1^{(0)} + \theta_2^{(0)}},$$

with similar expressions for $X_2^{(0)}$ and $X_3^{(0)}$. The M step involves maximization of the imputed complete data likelihood (B.1) to give $\theta_j^{(1)} = X_j^{(0)}/n$, $j = 1, 2, 3$. This leads to a very simple iteration that converges to the MLE based on the incomplete data. With data $Y = (15, 8, 12)$ and $(n_1, n_2, n_3) = (24, 15, 20)$ and an initial value of $\theta^{(0)} = (0.2, 0.5, 0.3)$, for example, the algorithm converges in eight iterations to the estimates $\hat{\theta} = (0.4011, 0.2522, 0.3467)$, which are accurate to four significant digits. The EM algorithm in this example can be shown to converge to the MLE from any starting value in the interior of the parameter space. Newton's method can also be applied directly to the incomplete log likelihood ℓ_Y and is easily implemented. It converges more quickly and also gives, as a by-product, estimates of the covariance matrix of $\hat{\theta}$.

This example is for illustration only. The EM algorithm is most useful in much more complicated problems and often provides a simple and intuitively appealing algorithm. It can be very slow to converge, and as noted ealier, may not converge to the MLE. A more practical example of the usefulness of the algorithm is in the treatment of interval censored data in Section 3.9.1.

B.2 STIELTJES INTEGRATION

In combining discrete and continuous cases, or in writing score functions, estimating functions, test statistics, and estimators using counting process and martingale theory, we often encounter integrals of the form

$$\int_s^t f \, dG = \int_s^t f(u) \, dG(u).$$

This is called a *Stieltjes integral*, and in this section, we define the integral for the types of functions considered in this book and examine some of its properties. In this expression, the function f is referred to as the *integrand* and G as the *integrator*. For our purposes, the function G is a right-continuous function and differentiable at all but perhaps a finite or countable number of points, some of which may be jump discontinuities.

We take a rather informal approach to defining the Stieltjes integral. More complete definitions and derivations can be found, for example, in the books by Royden (1968) and Ash (1972). Consider first the case in which G is a nondecreasing function and suppose that $G : [0, \infty) \to [0, \infty)$ can be written as

$$G(t) = \int_0^t g(u)\, du + \sum_{0 < a_\ell \le t} g_\ell, \qquad 0 < t < \infty, \tag{B.2}$$

where g is a nonnegative (Riemann) integrable function and $g(u) = G'(u)$ at all $u \in [0, \infty)$ except possibly at a finite or countable set of points, and $g_\ell > 0$ for all ℓ. Note that G has jump discontinuities at positive values $a_1, a_2, \ldots$ and $g_\ell = \Delta G(a_\ell) = G(a_\ell) - G(a_\ell^-), \ell = 1, 2, \ldots$. It is easy to see that G is a right-continuous function with left-hand limits. For a given function $f : [0, \infty) \to \Re$, the Stieltjes integral from s to t $(0 < s \le t)$ of f with respect to G is

$$\int_s^t f(u)\, dG(u) = \int_{(s,t]} f(u)\, dG(u)$$

$$= \int_s^t f(u)g(u)\, du + \sum_{a_\ell \in (s,t]} f(a_\ell)g_\ell, \tag{B.3}$$

provided that the integral and the sum in the final expression exist. Note that by convention, the upper point t, but not the lower point s, is included in the range of integration.

In taking (B.3) to define the Stieltjes integral, we are using results from Lebesgue integration. More specifically, (B.3) arises from interpreting (B.2) as a Lebesgue integral, where G defines the measure on the nonnegative real line. The general arguments leading to (B.3) involve successive approximations of f over the interval $(s, t]$ with step functions f_n which converge pointwise to f at all $u \in (s, t]$. The approximation f_n is a step function with n component steps and we define

$$\int_s^t f_n\, dG = \sum_{j=1}^{k_n} f_{nj} G(B_{nj}),$$

where $B_{nj} = \{u \in (s, t] : f_n(u) = f_{nj}\}$, $f_{n1}, \ldots, f_{nk_n}$ are the distinct values taken by f_n, and $G(B)$ is the measure assigned to the set B by G. It is possible to approximate any Borel measurable function f in this way. The Stieltjes (or Lebesgue–Stieltjes) integral is then defined as

$$\int_s^t f \, dG = \lim_{n \to \infty} \int_s^t f_n \, dG,$$

which can be shown to be independent of the sequence of approximations f_n used and, in the case considered above, yields the result (B.3).

It is easily seen that $G(t) = \int_0^t dG(u)$, $t \geq 0$. Further, if $G(u)$ is continuous, then $\int_s^t f(u) \, dG(u) = \int_s^t f(u) g(u) \, du$. If G is a step function, so that $g'(u) = 0$, then $\int_s^t f(u) \, dG(u) = \sum_{a_\ell \in (s,t]} f(a_\ell) g_\ell$. Integrals over open intervals can be obtained as limits. For example,

$$\int_{(s,t)} f(u) \, dG(u) = \lim_{v \to t^-} \int_{(s,v]} f(u) \, dG(u).$$

Similarly, we obtain the integral over $[s, t)$ by taking a limit as $v \to s^-$ of the integral over (v, t).

In many instances, we consider integrals with respect to a function $G : [0, \infty) \to \Re$, where $G = G_1 - G_2$ and G_1 and G_2 are nondecreasing right-continuous functions of the type discussed above. If the integrals of f with respect to G_1 and G_2 exist, the Stieltjes integral of f with respect to G is defined to be

$$\int_s^t f \, dG = \int_s^t f \, dG_1 - \int_s^t f \, dG_2$$

$$= \int_s^t f(u)[g_1(u) - g_2(u)] \, du + \sum_{a \in (s,t]} f(a) \, \Delta G(a). \qquad (B.4)$$

Such integrals arise, in particular, as stochastic integrals with respect to martingales in Chapter 5 and elsewhere.

The usual formula for integration by parts can be extended to apply to Stieltjes integrals. Suppose that $F(t)$ and $G(t)$ are right-continuous functions of bounded variation on finite intervals and expressible as differences of nondecreasing right continuous functions as above. It can be shown that

$$\int_s^t F(u) \, dG(u) = F(u)G(u)|_s^t - \int_s^t G(u) \, dF(u) + \sum_{a \in (s,t]} \Delta F(a) \Delta G(a). \qquad (B.5)$$

Variations on this formula are also sometimes given. For example, it is sometimes useful to work with the left-continuous versions of F and G as the integrands,

$$\int_s^t F(u^-) \, dG(u) = F(u)G(u)|_s^t - \int_s^t G(u^-) \, dF(u) - \sum_{a \in (s,t]} \Delta F(a) \, \Delta G(a).$$

Since the sum on the right side of these expressions is zero when F and G have no jumps, these reduce to the usual formula for integration by parts in that special case. To establish these expressions, apply (B.4) to the left side of (B.5) to obtain

$$\int_s^t F(u)g(u)\,du + \sum_{a\in(s,t]} F(a)\,\Delta G(a)$$

$$= \int_{(s,t]} \left[F(s) + \int_{(s,u]} dF(v) \right] g(u)\,du + \sum_{a\in(s,t]} F(a)\,\Delta G(a)$$

$$= F(s)[G(t) - G(s)] + \int_{(s,t]} \left[\int_{[v,t]} g(u)\,du \right] dF(v) + \sum_{a\in(s,t]} F(a)\,\Delta G(a).$$

Some inspection and calculation show that this reduces to

$$F(s)[G(t) - G(s)] + \int_{(s,t]} \left[\int_{[v,t]} dG(u) \right] dF(v)$$

$$= F(s)[G(t) - G(s)] + \int_{(s,t]} [G(t) - G(v) + \Delta G(v)]\,dF(v)$$

which reduces to the right side of (B.5).

B.3   SOFTWARE FOR FAILURE TIME ANALYSES

In the first edition of this book we included Fortran programs for applying the Cox regression model as an appendix, since there was a dearth of commercially available software for these analyses in 1980. In contrast, a review by Goldstein and Harrell (1998) lists 14 commercially available software packages that have substantial capabilities for failure time data analysis, including some form of Cox regression, and this list is not exhaustive. The 14 packages listed all included Kaplan–Meier estimators, and most allow the convenient fitting of parametric regression models. Several packages allow left truncation as well as right censoring and include some model building and model-checking procedures for Cox regression. Some provide various options for handling tied data, and some allow interval censoring. At present there does not appear to be commercially available software for regression parameter estimation in the semiparametric accelerated failure time model, presumably because numerical aspects have only recently been well addressed. Readers are probably best advised to contact the authors of recent papers (e.g., Jin et al., 2001) for information on software availability. On the other hand, several packages include censored data rank tests pertinent to the accelerated failure time class. Also, much of the material on cohort sampling and on missing or mismeasured covariate data (Chapter 11) has yet to be included in software packages, although some packages provide simple options for accommodating missing data.

We will not a discuss the capabilities of commercial software packages for failure time data analysis comprehensively here, but rather, provide contact information and a few summary comments for some popular packages.

- [SAS (SAS Institute, Inc., Box 8000, Cary, NC 27511; (919)677-7000; http://www.sas.com].

  SAS failure time software includes PROC LIFETEST, which produces Kaplan–Meier curves and life-table survivor function estimators as well as selected tests for comparing survival curves; PROC LIFEREG, which has options to fit most of the parametric regression models described in Chapter 2, while allowing left truncation, right censoring, and interval censoring; and PROC PHREG, which fits the Cox model while allowing stratification and time-dependent covariates, various options for handling tied failure times, various approaches to model building, and providing martingale and other residuals.

- S-Plus [Insightful Corporation, 1700 Westlake Avenue North, Suite 500, Seattle, WA 98109-3044; (206)283-8802, http://www.insightful.com].

  The S-Plus functions SURVFIT and SURVDIFF compute Kaplan–Meier estimates and censored data rank tests to compare survival curves, respectively. The function SURVREG allows the application of various log-linear parametric models with left truncation, right censoring, or interval censoring. The function COXPH applies the Cox model while allowing step-function time-dependent covariates and time-dependent strata. It also computes various types of residuals; the score residuals, in particular, are convenient for including sandwich-type robust variance estimates as arise, for example, in marginal models for recurrent events. Certain frailty model generalizations are also allowed.

  Therneau and Grambsch (2000) provide illustrations and code for using SAS and S-Plus for failure time data applications, including, for example, recurrent event data analyses. They also provide helpful comments on the relative merits of the two packages, and they provide various SAS Macros and S functions for specialized calculations that are not included in the basic package. Also the recent volumes by Der and Everitt (2002) and Everitt (2002) provide current handbooks for the use of SAS and S-Plus, respectively, with specific chapters devoted to failure time methods.

- BMDP [SPSS Inc., 444 N. Michigan Ave., Chicago, IL 60611; (312)329-4000, http://www.statsol.ie/bmdp/bmdp.htm].

  BMDP program 1L provides Kaplan–Meier and life-table survivor function estimators and selected censored data rank tests. Program 2L applies selected parametric models in both the accelerated failure time class and the Cox regression model. The latter application accommodates time-dependent covariates and left truncation as well as right-censoring and selected residual and model-checking procedures.

- SPSS [SPSS Inc., 444 N. Michigan Ave., Chicago, IL 60611; (312)329-4000, http://www.spss.com].

   The SPSS procedure KM has features similar to BMDP 1L and SAS procedure LIFETEST; SPSS procedure COXREG fits the Cox model while allowing time-dependent covariates and selected residuals but evidently doesn't accommodate left truncation. SPSS doesn't include procedures for fitting parametric regression models.

Collett (1994) includes a chapter on computer software for failure time data analysis that includes a more detailed presentation and comparison of SAS, BMDP, and SPSS, as well as illustrations of the use of these packages. Collett also provides a brief discussion and references as to how the GLIM and Genstat packages can be used to fit the Cox model and certain other failure time models.

A number of other statistical packages will be mentioned briefly. Stata [Stata Corp., 702 University Drive East, College Station, TX 77849; (800)782-8272, http://www.stata.com] includes substantial failure time analysis capabilities, including the accommodation of left truncation and right censoring, recurrent events, and sandwich variance estimators for the Cox model and for a number of parametric models, in addition to Kaplan–Meier and related estimators. Epicure [HiroSoft International Corp., 1463 E. Republican Ave., Suite 103, Seattle, WA 98112; (206)328-5301, http://www.hirosoft.com] also has a broad range of capabilities for applying the Cox model and parametric regression models, as does Egret [Cytel Software Corp., 675 Massachusetts Ave., Cambridge, MA 02139; (617)667-2011, http://www.cytel.com]. Epilog Plus [Epicenter Software, P.O. Box 90073, Pasadena, CA 91109; (818)304-9487, http://icarus2.hsc.usc.edu/epicenter] includes a version of Cox regression but not parametric regression model fitting. It does, however, include some novel programs for failure time data analyses, including the application recursive partitioning and neural network methods and a mixture failure time model that allows a fraction of study subjects not to be at risk for failure. Also Limdep [Econometric Software Inc., 15 Gloria Place, Plainview, NY 11803; (516)938-5254, http://www.limdep.com] and Statistica [Statsoft, 2325 East 13th St., Tulsa, OK 74104; (918)583-4149, http://www.statsoftinc.com] provide additional functional packages for failure time data analysis, as does Spida (The Statistical Laboratory, Macquarie University, NSW 2109, Australia; 02-850-8792, http://www.efs.mq.edu.au/statlab/spida.html), although parametric regression models are evidently not included in Spida. As mentioned above, the list given here is not comprehensive concerning software for failure time analysis. Rather, this brief presentation is intended only to provide an orientation toward some of the most widely used and most flexible packages.

Bibliography

Aalen, O. O. (1975). Statistical inference for a family of counting processes. Ph.D. thesis, University of California, Berkeley.

— — — (1976). Nonparametric inference in connection with multiple decrements models. *Scand. J. Statist.*, **3**, 15–27.

— — — (1978a). Nonparametric estimation of partial transition probabilities in multiple decrement models. *Ann. Statist.*, **6**, 534–545.

— — — (1978b). Nonparametric inference for a family of counting processes. *Ann. Statist.*, **6**, 701–726.

Aalen, O. O. and Hoem, J. M. (1978). Random time changes for multivariate counting processes. *Scand. Actuar. J.*, 81–101.

Aalen, O. O. and Husebye, E. (1991). Statistical analysis of repeated events forming renewal processes. *Statist. Med.*, **10**, 1227–1240.

Aalen O. O. and Johansen, S. (1978). An empirical transition matrix for nonhomogeneous Markov chains based on censored observations. *Scand. J. Statist.*, **5**, 141–150.

Abramowitz, M. and Stegun, I. A. (eds.) (1965). *Handbook of Mathematical Functions.* New York: Dover.

Adichie, J. N. (1967). Estimates of regression parameters based on rank tests. *Ann. Math. Statist.*, **38**, 894–904.

Altshuler, B. (1970). Theory for the measurement of competing risks in animal experiments. *Math. Biosci.*, **6**, 1–11.

Andersen, P. K. (1982). Testing goodness-of-fit of Cox's regression and life model. *Biometrics*, **75**, 67–77; amendment: **40**, 1217 (1984).

Andersen, P. K. and Borgan, O. (1985). Counting process models for life history data: a review (with discussion). *Scand. J. Statist.*, **12**, 97–158.

Andersen, P. K., Borgan, O., Gill, R. D., and Keiding, N. (1982). Linear nonparametric tests for comparison of counting processes, with application to censored survival data (with discussion). *Internat. Statist. Rev.*, **50**, 219–258; amendment: **52**, 225 (1984).

— — — (1993). *Statistical Models Based on Counting Processes.* New York: Springer–Verlag.

Andersen, P. K. and Gill, R. D. (1982). Cox's regression model for counting processes: a large sample study. *Ann. Statist.*, **10**, 1100–1120.

Andersen, P. K. and Ronn, R. B. (1995). A nonparametric test for comparing two samples where all observations are either left- or right-censored. *Biometrics*, **51**, 323–329.

Anderson, J. A. (1972). Separate sample logistic discrimination. *Biometrika*, **59**, 19–35.

Anderson, T. W. and Ghurye, S. B. (1977). Identification of parameters by the distribution of a maximum random variable. *J. Roy. Statist. Soc. Sec. Ser. B*, **39**, 337–342.

Andrews, D. F. and Herzberg, A. M. (1985). *Data: A Collection of Problems from Many Fields for the Student and Research Worker*. New York: Springer-Verlag.

Andrews, D. F., Bickel, P. J., Hampel, F. R., Huber, P. J., Rogers, W. H., and Tukey, J. W. (1972). *Robust Estimates of Location: Survey and Advances*. Princeton, N.J.: Princeton University Press.

Anscombe, F. J. (1964). Normal likelihood functions. *Ann. Inst. Statist. Math. (Tokyo)*, **16**, 1–19.

Arjas, E. (1988). A graphical method for assessing goodness of fit in Cox's proportional hazards model. *J. Amer. Statist. Assoc.*, **83**, 204–212.

— — — (1989). Survival models and martingale dynamics (with discussion). *Scand. J. Statist.*, **16**, 177–225.

Arjas, E. and Haara, P. (1988). A note on the asymptotic normality in the Cox regression model. *Ann. Statist.*, **16**, 1133–1140.

Armitage, P. (1959). The comparison of survival curves. *J. Roy. Statist. Soc. Ser. A*, **122**, 279–292.

— — — (1975). *Sequential Medical Trials*. Oxford: Blackwell.

Armitage, P. and Doll, R. (1954). The age distribution of cancer and a multistage theory of carcinogenesis. *British J. Cancer*, **8**, 1–12.

— — — (1961). Stochastic models for carcinogenesis. In *Proceedings of the Fourth Berkeley Symposium in Mathematical Statistics*, **IV** (L. M. Le Cam et al., eds.). Berkeley, Calif.: University of California Press, pp. 19–38.

Arnold, B. and Brockett, P. (1983). Identifiability for dependent multiple decrement/competing risk models. *Scand. Actuar. J.*, **10**, 117–127.

Ash, R. B. (1972). *Real Analysis and Probability*. New York: Academic Press.

Atkinson, A. C. (1970). A method for discriminating between models (with discussion). *J. Roy. Statist. Soc. Ser. B*, **32**, 323–353.

Ayer, M., Brunk, H. D., Ewing, G. M., Reid, W. T., and Silverman, E. (1955). An empirical distribution function for sampling with incomplete information. *Ann. Math. Statist.*, **26**, 641–647.

Barlow, R. E., Bartholomew, D. J., Bremner, J. M., and Brunk, H. D. (1972). *Statistical Inference Under Order Restrictions*. New York: Wiley.

Barlow, W. E. (1994). Robust variance estimation for the case–cohort design. *Biometrics*, **50**, 1064–1072.

Barlow, W. E. and Prentice, R. L. (1988). Residuals for relative risk regression. *Biometrika*, **75**, 65–74.

Barnard, G. A. (1946). Sequential tests in industrial statistics. *J. Roy. Statist. Soc.*, **Suppl. 8**, 1–26.

Bartholomew, D. J. (1957). A problem in life testing. *J. Amer. Statist. Assoc.*, **52**, 350–355.

Basu, A. P. and Ghosh, J. K. (1978). Identifiability of the multinomial and other distributions under competing risks model. *J. Multivariate Anal.*, **8**, 413–429.

— — — (1980). Identifiability of distributions under competing risks and complementary risks models. *Comm. Statist. A Theory Methods*, **9**, 1515–1525.

Batchelor, J. R. and Hackett, M. (1970). HLA matching in treatment of burned patients with skin allografts. *Lancet*, **2**, 581–583.

Begun, J. M., Hall, W. J., Huang, W.-M., and Wellner, J. A. (1983). Information and asymptotic efficiency in parametric–nonparametric models. *Ann. Statist.*, **11**, 432–452.

Bennett, S. (1983). Analysis of survival data by the proportional odds model. *Statist. Med.*, **2**, 273–277.

Berkson, J. and Elveback, L. (1960). Competing exponential risks, with particular reference to the study of smoking and lung cancer. *J. Amer. Statist. Assoc.*, **55**, 415–428.

Berkson, J. and Gage, R. P. (1952). Survival curve for cancer patients following treatment. *J. Amer. Statist. Assoc.*, **47**, 501–515.

Bickel, P. J., Klassen, C. A., Ritov, Y. and Wellner, J. A. (1993). *Efficient and Adaptive Estimation for Semiparametric Models*. Baltimore, Md.: Johns Hopkins University Press.

Billingsley, P. (1961). *Statistical Inference for Markov Processes*. Chicago: University of Chicago Press.

Birnbaum, A. and Laska, E. (1967). Efficiency robust two–sample rank tests. *J. Amer. Statist. Assoc.*, **62**, 1241–1251.

Bliss, C. I. (1935). The calculation of the dosage–mortality curve. *Ann. Appl. Biol.*, **22**, 134–167.

Böhmer, P. E. (1912). Theorie der unabhängigen Warscheinlichkeiten. *Rapports, Mém. et Procés–verbaux 7ᵉ Congrès Internat. Act., Amsterdam*, **2**, 327–343.

Böhning, D., Schlattmann, P., and Dietz, E. (1996). Interval censored data: a note on the nonparametric maximum likelihood estimator of the distribution function. *Biometrika*, **83**, 462–466.

Borgan, O. (1984). Maximum likelihood estimation in parametric counting process models, with applications to censored failure time data. *Scand. J. Statist.*, **11**, 1–16.

Borgan, O., Goldstein, L., and Langholz, B. (1995). Methods for the analysis of sampled cohort data in the Cox proportional hazards model. *Ann. Statist.*, **23**, 1749–1778.

Borgan, O., Langholz, B., Samuelsen, S. O., Goldstein, L., and Pogoda, J. (2000). Exposure stratified case–cohort designs. *Lifetime Data Anal.*, **6**, 39–58.

Breslow, N. E. (1970). A generalized Kruskal–Wallis test for comparing K samples subject to unequal patterns of censorship. *Biometrika*, **57**, 579–594.

— — — (1974). Covariance analysis of censored survival data. *Biometrics*, **30**, 89–99.

— — — (1975). Analysis of survival data under the proportional hazards model. *Internat Statist. Rev.*, **43**, 45–58.

— — — (1976). Regression analysis of the log odds ratio: a method for retrospective studies. *Biometrics*, **32**, 409–416.

Breslow, N. E. and Cain, K. (1988). Logistic regression for two stage case–control data. *Biometrika*, **75**, 11–20.

Breslow, N. E. and Crowley, J. (1974). A large sample study of the life table and product limit estimates under random censorship. *Ann. Statist.*, **2**, 437–453.

Breslow, N. E. and Day, N. E. (1980). *Statistical Methods in Cancer Research: 1. The Design and Analysis of Case–Control Studies*. IARC Scientific Publications 32. Lyon, France: International Agency for Research on Cancer.

— — — (1987). *Statistical Methods for Cancer Research: 2. The Design and Analysis of Cohort Studies*. IARC Scientific Publications 82. Lyon, France: International Agency for Research on Cancer.

Breslow, N. E. and Holubkov, R. (1997). Maximum likelihood estimation of logistic regression parameters under two-phase, outcome-dependent sampling. *J. Roy. Statist. Soc. Ser. B*, **59**, 447–61.

Breslow, N. E. and Edler, L., and Berger, J. (1984). A two-sample censored-data rank test for acceleration. *Biometrics*, **40**, 1049–1062.

Breslow, N. E., Lubin, J. H., Marek, P., and Langholz, B. (1983). Multiplicative models and cohort analysis. *J. Amer. Statist. Assoc.*, **78**, 1–12.

Brown, B. M. (1971). Martingale central limit theorems. *Ann. Math. Statist.*, **42**, 59–66.

Buckley, J. and James, I. (1979). Linear regression with censored data. *Biometrika*, **66**, 429–436.

Burke, M. D. (1988). Estimation of a bivariate distribution function under random censorship. *Biometrika*, **75**, 379–382.

Burridge, J. (1981). Empirical Bayes analysis for survival time data. *J. Roy. Statist. Soc. Ser. B*, **43**, 65–75.

Buzas, J. S. (1998). Unbiased scores in proportional hazards regression with covariate measurement error. *J. Statist. Plann. Inference*, **67**, 247–257.

Byar, D. P. (1980). The Veterans Administration study of chemoprophylaxis of recurrent stage I bladder tumors: comparisons of placebo, pyridoxine, and topical thiotepa. In *Bladder Tumors and Other Topics in Urological Oncology* (M. Pavone–Macaluso, P. H. Smith, and F. Edsmyn, eds.). New York: Plenum, pp. 363–370.

Byar, D. P., Huse, R., and Bailar, J. C. III (1974). An exponential model relating censored survival data and concomitant information for prostrate cancer patients. *J. Nat. Cancer Inst.*, **52**, 321–326.

Cai, J. and Prentice, R. L. (1995). Estimating equations for hazard ratio parameters based on correlated failure time data. *Biometrika*, **82**, 151–164.

— — — (1997). Regression estimation using multivariate failure time data and a common baseline hazard function model. *Lifetime Data Anal.*, **3**, 197–213.

Cain, K. C. and Lange, N. T. (1984). Approximate case influence for the proportional hazards regression model with censored data. *Biometrics*, **40**, 493–499.

Campbell, G. (1981). Nonparametric bivariate estimation with randomly censored data. *Biometrika*, **68**, 417–422.

Campbell, G. and Földes, A. (1982). Large-sample properties of nonparametric bivariate estimators with censored data. In *Nonparametric Statistical Inference* (B. V. Gnredenko, M. L. Puri, and I. Vineze, eds.). Amsterdam: North–Holland. pp. 103–122.

Carroll, R. J., Ruppert, D., and Stefanski, L. A. (1995). *Nonlinear Measurement Error Models*. London: Chapman & Hall.

Chatterjee, N., Chen, H. Y., and Breslow, N. E. (2002). A pseudo-score estimator for regression problems with two phase sampling. *J. Amer. Statist. Assoc.*, **97**, to appear.

Chen, H. Y. and Little, R. J. A. (1999). Proportional hazards regression with missing covariates. *J. Amer. Statist. Assoc.*, **94**, 896–908.

Chen, K. (2001). Generalized case–cohort sampling. *J. Roy. Statist. Soc. ser. B*, **63**, 791–809.

Chen, K. and Lo, S.-H. (1999). Case–cohort and case–control analysis with Cox's model. *Biometrika* **86**, 755–764.

Chen, Y. Q. and Jewell, N. P. (2001). On a general class of semiparametric hazards regression model. *Biometrika*, **88**, 687–702.

Chen, Y. Q. and Wang, M.-C. (2000). Analysis of accelerated hazards model. *J. Amer. Statist. Assoc.*, **95**, 608–18.

Cheng, S. C., Wei, L. J., and Ying, Z. (1995). Analysis of transformation models with censored data. *Biometrika*, **82**, 835–845.

Chiang, C. L. (1960). A stochastic study of life table and its applications: I. Probability distribution of the biometric functions. *Biometrics*, **16**, 618–635.

— — — (1961). On the probability of death from specific causes in the presence of competing risks. In *Proceedings of the Fourth Berkeley Symposium in Mathematical Statistics*, **IV** (L. M. Le Cam et al., eds.). Berkeley, Calif.: University of California Press, pp. 169–180.

— — — (1968). *Introduction to Stochastic Processes in Biostatistics*. New York: Wiley.

— — — (1970). Competing risks and conditional probabilities. *Biometrics*, **26**, 767–776.

Clark, D. A., Stinson, E. B., Griepp, R. B., Schroeder, J. S., Shumway, N. E., and Harrison, D. C. (1971). Cardiac transplantation in men: VI. Prognosis of patients selected for cardiac transplantation. *Ann. Internal Med.*, **75**, 15–21.

Clayton, D. G. (1978). A model for association in bivariate life tables and its application in epidemiological studies of familial tendency in chronic disease incidence. *Biometrika*, **65**, 141–151.

Clayton, D. G. and Cuzick, J. (1985). Multivariate generalizations of the proportional hazards model (with discussion). *J. Roy. Statist. Soc. Ser. A*, **148**, 82–117.

Collett, D. (1994). *Modeling Survival Data in Medical Research*. New York: Chapman & Hall.

Conover, W. J. (1971). *Practical Nonparametric Statistics*. New York: Wiley.

Cornfield, J. (1951). A method of estimating comparative rates from clinical design: applications to cancer of the lungs, breast, and cervix. *J. Nat. Cancer Inst.*, **11**, 1269–1275.

— — — (1957). The estimation of the probability of developing a disease in the presence of competing risks. *Amer. J. Public Health*, **47**, 601–607.

— — — (1971). The university group diabetes program: a further statistical analysis of the mortality findings. *J. Amer. Med. Assoc.*, **217**, 1676–1687.

Cornfield, J. and Detre, K. (1977). Bayesian analysis of life tables. *J. Roy. Statist. Soc. Ser. B*, **39**, 86–94.

Cox, D. R. (1953). Some simple tests for Poisson variates. *Biometrika*, **40**, 354–360.

— — — (1959). The analysis of exponentially distributed life-times with two types of failure. *J. Roy. Statist. Soc. Ser. B*, **21**, 411–421.

— — — (1961). Tests of separate families of hypotheses. In *Proceedings of the Fourth Berkeley Symposium in Mathematical Statistics*, **I** (L. M. Le Cam et al., eds.). Berkeley, Calif.: University of California Press, pp. 105–123.

— — — (1962a). Further results on tests of separate families of hypotheses. *J. Roy. Statist. Soc. Ser. B*, **24**, 406–424.

— — — (1962b). *Renewal Theory*. London: Methuen.

— — — (1964). Some applications of exponential ordered scores. *J. Roy. Statist. Soc. Ser. B*, **26**, 103–110.

— — — (1970). *The Analysis of Binary Data*. London: Methuen.

— — — (1972). Regression models and life tables (with discussion). *J. Roy. Statist. Soc. Ser. B*, **34**, 187–220.

— — — (1973). The statistical analysis of dependencies in point processes. In *Symposium on Point Processes* (P. A. W. Lewis, ed.). New York: Wiley, pp. 55–66.

— — — (1975). Partial likelihood. *Biometrika*, **62**, 269–276.

Cox, D. R. and Hinkley, D. V. (1974). *Theoretical Statistics*. London: Chapman & Hall.

Cox, D. R. and Isham, V. (1980). *Point Processes*. London: Chapman & Hall.

Cox, D. R. and Lewis, P. A. (1966). *The Statistical Analysis of a Series of Events*. London: Methuen.

Cox, D. R. and Oakes, D. (1984). *Analysis of Survival Data*. London: Chapman & Hall.

Cox, D. R. and Snell, E. J. (1968). A general definition of residuals (with discussion). *J. Roy. Statist. Soc. Ser. B*, **30**, 248–275.

— — — (1989). *Analysis of Binary Data*, 2nd ed. London: Chapman & Hall.

Crowder, M. (2001). *Classical Competing Risks*. London: Chapman & Hall/CRC.

Crowley, J. (1974a). Asymptotic normality of a new nonparametric statistic for use in organ transplant studies. *J. Amer. Statist. Assoc.*, **69**, 1006–1011.

— — — (1974b). A note on some recent likelihoods leading to the log rank test. *Biometrika*, **61**, 533–538.

Crowley, J. and Hu, M. (1977). Covariance analysis of heart transplant data. *J. Amer. Statist. Assoc.*, **72**, 27–36.

Crowley, J. and Thomas, D. R. (1975). Large sample theory for the log-rank test. *Technical Report 415*. Department of Statistics, University of Wisconsin.

Crowley, J. J. and Storer, B. E. (1983). Comment on "A reanalysis of the stanford heart transplant data," by M. Aitkin, N. Laird, and B. Francis. *J. Amer. Statist. Assoc.*, **78**, 277–281.

Crump, K. F., Hoel, D. G., Langley, C. H., and Peto, R. (1976). Fundamental carcinogenic processes and their implications for low dose risk assessment. *Cancer Res.*, **36**, 2973–2979.

Cutler, S. J. and Ederer, F. (1958). Maximum utilization of the life table in analyzing survival. *J. Chronic Disease*, 699–712.

Cuzick, J. (1982). Rank tests for association with right censored data. *Biometrika*, **69**, 351–364.

— — — (1985). Asymptotic properties of censored linear rank tests. *Ann. Statist.*, **13**, 133–141.

— — — (1988). Rank regression. *Ann. Statist.*, **16**, 1369–1389.

Dabrowska, D. M. (1988). Kaplan–Meier estimate on the plane. *Ann. Statist.*, **16**, 1475–1489.

Dabrowska, D. M. and Doksum, K. A. (1988). Partial likelihood in transformation models with censored data. *Scand. J. Statist.*, **15**, 1–24.

D'Agostino, R. B., Lee, M. L. T., Belander, A. J., Cupples, L. A., Anderson, K., and Kannel, W. B. (1990). Relation of pooled logistic regression to time dependent Cox regression analysis: the Framingham heart study. *Statist. Med.*, **9**, 1501–1515.

David, H. A. (1970). *Order Statistics*. New York: Wiley.

David, H. A. and Moeschberger, M. (1978). *Theory of Competing Risks*. London: Griffin.

Dempster, A. P., Laird, N. M., and Rubin, D. B. (1977). Maximum likelihood estimation from incomplete data via the EM algorithm (with discussion). *J. Roy. Statist. Soc. Ser. B*, **39**, 1–38.

Der, G. and Everitt, B. S. (2002). *A Handbook of Statistical Analyses Using SAS*, 2nd ed. Boca Raton, Fla.: Chapman & Hall/CRC.

Dewanji, A. and Kalbfleisch, J. D. (1986). Nonparametric analysis of survival/sacrifice experiments. *Biometrics*, **42**, 325–341.

Diabetic Retinopathy Study Research Group (1981). Diabetic retinopathy study. *Invest. Ophthalmol. Vis. Sci.*, **21**, 149–226.

Dinse, G. E. and Lagakos, S. W. (1983). Regression analysis of tumour prevalence data. *Appl. Statist.*, **32**, 236–248.

Doksum, K. A. (1974). Tailfree and neutral random probability measures and their posterior distributions. *Ann. Probab.*, **2**, 183–201.

Dollard, J. D. and Friedman, C. N. (1979). *Product Integration with Applications to Differential Equations*. Reading, Mass.: Addison-Wesley.

Doss, H. (1994). Bayesian nonparametric estimation for incomplete data via successive substitution sampling. *Ann. Statist.*, **22**, 1763–1786.

Doss, H. and Huffer, F. (1998). Monte Carlo methods for Bayesian analysis of survival data using mixtures of Dirichlet priors. *Technical Report*. Department of Statistics, Ohio State University.

Doss, H. and Narasimhan, B. (1998). Dynamic display of changing posterior in Bayesian survival analysis. In *Practical Nonparametric and Semiparametric Bayesian Statistics* (D. Dey, P. Muller, and D. Sinha, eds.). New York: Springer-Verlag, pp. 63–84.

Downton, F. (1972). Contribution to the discussion of paper by D. R. Cox. *J. Roy. Statist. Soc. Ser. B*, **34**, 202–205.

Duffy, D. L., Martin, N. G., and Matthews, J. D. (1990). Appendectomy in Australian twins. *Amer. J. Human Genet.*, **47**, 590–592.

Dumonceaux, R. and Antle, C. E. (1973). Discrimination between lognormal and Weibull distributions. *Technometrics*, **15**, 923–926.

Efron, B. (1967). The two sample problem with censored data. In *Proceedings of the Fifth Berkeley Symposium in Mathematical Statistics*, **IV**. New York: Prentice–Hall, pp. 831–853.

— — — (1977). Efficiency of Cox's likelihood function for censored data. *J. Amer. Statist. Assoc.*, **72**, 557–565.

Elveback, L. (1958). Estimation of survivorship in chronic disease; the actuarial method. *J. Amer. Statist. Assoc.*, **53**, 420–440.

Emerson, S. S. and Fleming, T. R. (1989). Symmetric group sequential designs. *Biometrics*, **45**, 905–923.

— — — (1990). Parameter estimation following group sequential hypothesis testing. *Biometrika*, **77**, 875–892.

Epstein, B. and Sobel, M. (1953). Life testing. *J. Amer. Statist. Assoc.*, **48**, 486–502.

Escobar, L. A. and Meeker, W. Q. (1992). Assessing influence in regression analysis with censored data. *Biometrics*, **48**, 507–528.

Escobar, M. D. (1994). Estimating normal means with a Dirichlet process prior. *J. Amer. Statist. Assoc.*, **89**, 268–277.

Escobar, M. D. and West, M. (1995). Bayesian density estimation and inference using mixtures. *J. Amer. Statist. Assoc.*, **90**, 578–588.

Everitt, B. S. (2002). *A Handbook of Statistical Analyses Using S-Plus*, 2nd ed. Boca Raton, Fla.: Chapman & Hall/CRC.

Fan, J., Hsu, L., and Prentice, R. L. (2000). Dependence estimation over a finite bivariate failure time region. *Lifetime Data Anal.*, **6**, 343–355.

Farewell, V. T. and Prentice, R. L. (1977). A study of distributional shape in life testing. *Technometrics*, **19**, 69–76.

Feigl, P. and Zelen, M. (1965). Estimation of exponential survival probabilities with concomitant information. *Biometrics*, **21**, 826–838.

Feller, W. (1971). *An Introduction to Probability Theory*, 2nd ed., Vol. II. New York: Wiley.

Ferguson, T. S. (1973). A Bayesian analysis of some nonparametric problems. *Ann. Statist.*, **1**, 209–230.

— — — Prior distributions on spaces of probability measures. *Ann. Statist.*, **2**, 615–629.

Ferguson, T. S. and Phadia, E. G. (1979). Bayesian nonparametric estimation based on censored data. *Ann. Statist.*, **7**, 163–186.

Fine, J. P., Ying, Z., and Wei, L. J. (1998). On the linear transformation model for censored data. *Biometrika*, **85**, 980–986.

Finkelstein, D. M. (1986). A proportional hazards model for interval–censored failure time data. *Biometrics*, **42**, 845–854.

Finkelstein, D. M. and Schoenfeld, D. A. (1994). Analyzing survival in the presence of an auxiliary variable. *Statist. Med.*, **13**, 1747–1754.

Fisher, L. and Kanarek, P. (1974). Presenting censored survival data when censoring and survival times may not be independent. In *Reliability and Biometry: Statistical Analysis of Life Length* (F. Prochan and R. J. Serfling, eds.). Philadelphia: SIAM, pp. 303–326.

Fisher, L. and Patil, K. (1974). Matching and unrelatedness. *Amer. J. Epidemiol.*, **100**, 347–349.

Fisher, R. A. (1922). On the mathematical foundations of theoretical statistics. *Philos. Trans. Roy. Soc. (London) Ser. A*, **222**, 309–368.

— — — (1925). Theory of statistical estimation. *Proc. Cambridge Philos. Soc.*, **22**, 700–725.

Fisher, R. A. and Yates, F. (1938). *Statistical Tables for Biological, Agricultural and Medical Research*. London: Oliver & Boyd (1st ed. 1938, 6th ed. 1963).

Fleming, T. R. and Harrington, D. P. (1991). *Counting Processes and Survival Analysis*. New York: Wiley.

Fleming, T. R., Harrington, D. P., O'Brien, P. C. (1984). Designs for group sequential tests. *Control. Clin. Trials*, **5**, 348–361.

Fraser, D. A. S. (1968). *The Structure of Inference*. New York: Wiley.

Freedman, L. S., Anderson, G., Kipnis, V., Prentice, R. L., Wang, C. Y., Rossouw, J. R., Wittes, J., and DeMets, D. L. (1996). Approaches to monitoring the results of long–term disease prevention trials: example of the Women's Health Initiative. *Control. Clin. Trials*, **17**, 509–525.

Freireich, E. O. et al. (1963). The effect of 6-mercaptopmine on the duration of steroid induced remission in acute leukemia. *Blood*, **21**, 699–716.

Fygenson, M. and Ritov, Y. (1994). Monotone estimating functions for censored data. *Ann. Statist.*, **22**, 732–46.

Gail, M. H. (1972). Does cardiac transplantation prolong life? A reassessment. *Ann. Internal Med.*, **76**, 815–817.

— — — (1975). A review and critique of some models used in competing risk analysis. *Biometrics*, **31**, 209–222.

Gail, M. H., Santner, T. J., and Brown, C. C. (1980). An analysis of comparative carcinogenesis experiments based on multiple times to tumor. *Biometrics*, **36**, 255–266.

Gail, M. H., Lubin, J. H., and Rubenstein, L. V. (1981). Likelihood calculations for matched case–control studies and survival studies with matched death times. *Biometrika*, **68**, 703–707.

Gastwirth, J. L. (1970). On robust rank tests. In *Nonparametric Techniques in Statistical Inference* (M. L. Puri, ed.). London: Cambridge University Press.

Gehan, E. A. (1965a). A generalized Wilcoxon test for comparing arbitrarily singly–censored samples. *Biometrika*, **52**, 203–223.

— — — (1965b). A generalized two-sample Wilcoxon test for doubly censored data. *Biometrika*, **52**, 650–652.

— — — (1969). Estimating survivor functions from the life table. *J. Chronic Diseases*, **21**, 629–644.

Gentleman, R. and Geyer, C. J. (1994). Maximum likelihood for interval censored data: consistency and computation. *Biometrics*, **81**, 618–623.

Gentleman, R. and Vandal, A. C. (2001). Computational algorithms for censored data problems using intersection graphs. *J. Comput. Graph. Statist.*, **10**, 403–421.

— — — (2002). Graph theoretic aspects of bivariate censored data. *Canad. J. Statist.*, **32**, to appear.

Ghosh, D. and Lin, D. Y. (2000). Nonparametric analysis of recurrent events and death. *Biometrics*, **56**, 554–562.

Gilbert, J. P. (1962). Random censorship. Ph.D. thesis, University of Chicago.

Gilbert, J. P. et al. (1975). Report of the committee for the assessment of biometric aspects of controlled trials of hypoglycemic agents. *J. Amer. Med. Assoc.*, **231**, 583–608.

Gill, R. D. (1980). *Censoring and Stochastics Integrals*. Math Centre Tract 124. Amsterdam: Math. Centrum.

Gill R. D. and Johansen, S. (1990). A survey of product integration with a view toward application in survival analysis. *Ann. Statist.*, **18**, 1501–1555.

Gill, R. D., van der Laan, M. J., and Wellner, J. A. (1995). Inefficient estimators of the bivariate survival function for three models. *Ann. Inst. H. Poincaré Prob. Statist.*, **31**, 547–597.

Glasser, M. (1967). Exponential survival with covariance. *J. Amer. Statist. Assoc.*, **62**, 561–568.

Glidden, D. V. (2000). A two-stage estimator of the dependence parameter for the Clayton–Oakes model. *Lifetime Data Anal.*, **6**, 141–156.

Goggins, W. B. Finkelstein, D. M., Schoenfeld, D. A., and Zaslavsky, A. M. (1998). A Markov chain Monte Carlo EM algorithm for analyzing interval-censored data under the Cox proportional hazards model. *Biometrics*, **54**, 1498–1507.

Goldstein, L. and Langholz, B. (1992). Asymptotic theory for nested case–control sampling in the Cox regression model. *Ann. Statist.*, **20**, 1903–1928.

Goldstein, R. and Harrell, F. (1998). Survival analysis, software. In *Encyclopedia of Biostatistics* (P. Armitage and T. Colton, eds.), Vol. 6. New York: Wiley, pp. 4461–4466.

Gompertz, B. (1825). On the nature of the function expressive of the law of human mortality. *Philos. Trans. Roy. Soc. (London)*, **115**, 513–583.

Gould, A. L. and Pecore, V. J. (1982). Group sequential methods for clinical trials allowing for early acceptance of H_0 and incorporating costs. *Biometrika*, **69**, 75–80.

Gray, R. J. (1994). A kernel method for incorporating information on disease progression in the analysis of survival. *Biometrika*, **81**, 527–539.

Greenwood, M. (1926). The natural duration of cancer. *Reports on Public Health and Medical Subjects*, **33**, London: Her Majesty's Stationery Office, pp. 1–26.

Groenboom, P. and Wellner, J. A. (1992). *Nonparametric Maximum Likelihood Estimators for Interval Censoring and Deconvolution.* Boston: Birkhauser.

Gross, A. J. and V. A. Clark (1975). *Survival Distributions: Reliability Applications in the Biomedical Sciences.* New York: Wiley.

Gross, S. T. and Huber, C. (1987). Matched pair experiments: Cox and maximum likelihood estimation. *Scand. J. Statist.*, **14**, 27–41.

Hagar, H. W. and Bain, L. J. (1970). Inferential procedures for the generalized gamma distribution. *J. Am. Statist. Assoc.*, **65**, 1601–1609.

Hájek, J. (1969). *A Course in Nonparametric Statistics.* San Francisco: Holden-Day.

Hájek, J. and Sǐdák, Z. (1967). *Theory of Rank Tests.* New York: Academic Press.

Hampel, F. (1974). The influence curve and its role in robust estimation. *J. Amer. Statist. Assoc.*, **69**, 383–393.

Hanley, J. A. and Parnes, M. N. (1983). Nonparametric estimation of a multivariate distribution in the presence of censoring. *Biometrics*, **39**, 129–139.

Harrington, D. P. and Fleming, T. R. (1982). A class of rank test procedures for censored survival data. *Biometrika*, **69**, 133–143.

Harter, H. L. (1967). Maximum likelihood estimation of the parameters of four parameter generalized gamma population from complete and censored samples. *Technometrics*, **9**, 159–165.

Hartley, H. O. and Sielken, R. L., Jr. (1977). Estimation of "safe doses" in carcinogenesis experiments. *Biometrics*, **33**, 1–30.

Haybittle, J. L. (1971). Repeated assessment of results in clinical trials of cancer treatment. *British J. Radiol.*, **44**, 793–797.

Heitjan, D. F. (1993). Ignorability and coarse data: some biomedical examples. *Biometrics.* **49**, 1099–1109.

Heitjan, D. F. and Rubin, D. B. (1991). Ignorability and coarse data. *Ann. Statist.*, **19**, 2244–2253.

Hettmansperger, T. P. (1984). *Statistical Inference Based on Ranks.* New York: Wiley.

Hettmansperger, T. P. and McKean, J. W. (1977). A robust alternative based on ranks to least squares in analyzing linear models. *Technometrics*, **19**, 275–284.

Hjort, N. L. (1985). Discussion of the paper by P. K. Andersen and O. Borgan. *Scand. J. Statist.*, **12**, 141–150.

— — — (1990) Nonparametric Bayes estimators based on beta processes in models of life history data. *Ann. Statist.*, **18**, 1259–1294.

Hodges, J. L. and Lehmann, E. L. (1963). Estimates of location based upon rank tests. *Ann. Math. Statist.*, **34**, 598–611.

Hoel, D. G. (1972). A representation of mortality data by competing risks. *Biometrics*, **28**, 475–488.

Hoel, D. G. and Walburg, H. E. (1972). Statistical analysis of survival experiments. *J. Nat. Cancer Inst.*, **49**, 361–372.

Hogg, R. V. (1974). Adaptive robust procedures: a partial review and some suggestions for future applications and theory (with discussion). *J. Amer. Statist. Assoc.*, **69**, 909–927.

Hollander, M. and Wolfe, D. A. (1973). *Nonparametric Statistical Methods*. New York: Wiley.

Holford, T. R. (1976). Life tables with concomitant information. *Biometrics*, **32**, 587–598.

Holt, J. D. and Prentice, R. L. (1974). Survival analysis in twin studies and matched pair experiments. *Biometrika*, **61**, 17–30.

Hougaard, P. (1986). Survival models for heterogeneous populations derived from stable distributions. *Biometrika*, **73**, 387–396.

— — — (2000). *Analysis of Multivariate Survival Data*. New York: Springer–Verlag.

Howard, S. V. (1972). Contribution to discussion of paper by D. R. Cox. *J. Roy. Statist. Soc. Ser. B*, **34**, 210–211.

Hsu, L. and Prentice, R. L. (1996). A generalization of the Mantel–Haenszel test to bivariate failure time data. *Biometrika*, **83**, 905–911.

Hu, P., Tsiatis, A. A., and Davidian, M. (1988). Estimating the parameters in the Cox model when covariate variables are measured with error. *Biometrics*, **54**, 1407–1419.

Huang, J. (1996). Efficient estimation for the proportional hazards model with interval censoring. *Ann. Statist.*, **24**, 540–568.

Huang, J. and Wellner, J. A. (1995). Asymptotic normality of the NPMLE of linear functionals for interval censored data: case I. *Statist. Neerlandica*, **49**, 153–163.

Huang, Y. and Wang, C. Y. (2000). Cox regression with accurate covariates unascertainable: a nonparametric correction approach. *J. Amer. Statist. Assoc.*, **45**, 1209–1219.

Huber, P. J. (1972). Robust statistics: a review. *Ann. Math. Statist.*, **43**, 1041–1067.

— — — (1973). Robust regression: asymptotics, conjectures and Monte Carlo. *Ann. Statist.*, **1**, 799–821.

Huster, W. S., Brookmeyer, R., and Self, S. G. (1989). Modeling paired survival data with covariates. *Biometrics*, **45**, 145–156.

Ibrahim, J. G., Chen, M.-H., and Sinha, D. (2001). *Bayesian Survival Analysis*. New York: Springer–Verlag.

— — — (1984). Maximum likelihood estimation in the multiplicative intensity model: a survey. *Internat. Statist. Rev.*, **52**, 193–207.

Jacobsen, M. (1982). *Statistical Analysis of Counting Processes*. Lecture Notes in Statistics 12. New York: Springer–Verlag.

Jacod, J. (1975). Multivariate point processes: predictable projection, Radon–Nikodym derivatives, representation of martingales. *Z. Wahrsch. Verw. Gebiete*, **31**, 235–253.

Jennison, C. and Turnbull, B. W. (1984). Repeated confidence intervals for group sequential clinical trials. *Control. Clin. Trials*, **5**, 33–45.

— — — (1989). Interim analyses: the repeated confidence interval approach (with discussion). *J. Roy. Statist. Soc. Ser. B*, **51**, 305–361.

— — — (2000). *Group Sequential Methods with Applications to Clinical Trials*. Boca Raton, Fla.: Chapman & Hall CRC.

Jewell, N. P. and Kalbfleisch, J. D. (2002). Maximum likelihoood estimation of ordered multinomial parameters. Unpublished manuscript.

Jewell, N. P. and van der Laan, M. (1995). Generalizations of current status data with applications. *Lifetime Data Anal.*, **1**, 101–110.

Jin, Z., Lin, D. Y., Wei, L. J., and Ying, Z. (2002). Rank–based inference for the accelerated failure time model. Revised for *Biometrika*.

Johansen, S. (1983). An extension of Cox's regression model. *Internat. Statist. Rev.*, **51**, 165–174.

Johnson, N. L. and Kotz, S. (1970a). *Distributions in Statistics: Continuous Univariate Distributions*, Vol. 1. Boston: Houghton Mifflin.

— — — (1970b). *Distributions in Statistics: Continuous Univariate Distributions*, Vol. 2, Boston: Houghton Mifflin.

Johnson, R. A. and Mehrotra, K. G. (1972). Locally most powerful rank tests for the two-sample problem with censored data. *Ann. Math. Statist.*, **43**, 823–831.

Jurečková, J. (1969). Asymptotic linearity of a rank statistic in regression parameter. *Ann. Math. Statist.*, **40**, 1889–1900.

— — — (1971). Nonparametric estimate of regression coefficients. *Ann. Math. Statist.*, **42**, 1328–1338.

Kalbfleisch, J. D. (1974). Some efficiency calculations for survival distributions. *Biometrika*, **61**, 31–38.

— — — (1978a). Nonparametric Bayesian analysis of survival time data. *J. Roy. Statist. Soc. Ser. B*, **40**, 214–221.

— — — (1978b). Likelihood methods and nonparametric tests. *J. Amer. Statist. Assoc.*, **73**, 167–170.

Kalbfleisch, J. D. and Lawless, J. F. (1988). Likelihood analysis of multi–state models for disease incidence and mortality. *Statist. Med.*, **7**, 147–160.

— — — (1989). Inference based on retrospective ascertainment: an analysis of the data on transfusion–related AIDS. *J. Amer. Statist. Assoc.*, **84**, 360–372.

Kalbfleisch, J. D. and MacKay, R. J. (1978a). Some remarks on a paper by Cornfield and Detre. *J. Roy. Statist. Soc. Ser. B*, **40**, 175–177.

— — — (1978b). Censoring and the immutable likelihood. *Technical Report 78-09.* Department of Statistics, University of Waterloo.

— — — (1979). On constant-sum models for censored survival data. *Biometrika*, **66**, 87–90.

Kalbfleisch, J. D. and McIntosh, A. A. (1977). Efficiency in survival distributions with time dependent covariables. *Biometrika*, **64**, 47–50.

Kalbfleisch, J. D. and Prentice, R. L. (1973). Marginal likelihoods based on Cox's regression and life model. *Biometrika*, **60**, 267–278.

— — — (1980). *The Statistical Analysis of Failure Time Data*. New York: Wiley.

Kalbfleisch, J. D. and Sprott, D. A. (1970). Application of likelihood methods to models involving a large number of parameters (with discussion). *J. Roy. Statist. Soc. Ser. B*, **32**, 175–208.

Kalbfleisch, J. D., Lawless, J. F., and Robinson, J. A. (1991). Methods for the analysis of prediction of warranty claims. *Technometrics*, **33**, 273–285.

Kaplan, E. L. and Meier, P. (1958). Nonparametric estimation from incomplete observations. *J. Amer. Statist. Assoc.*, **53**, 457–481.

Kay, R. (1977). Proportional hazard regression models and the analysis of censored survival data. *J. Roy. Statist. Soc. Ser. C*, **26**, 227–237.

— — — (1979). Some further asymptotic efficiency calculations for survival data regression models. *Biometrika*, **66**, 91–96.

Keiding, N. (1991). Age specific incidence and prevalence: a statistical perspective (with discussion). *J. Roy. Statist. Soc., Ser. A*, **154**, 371–412.

Kellerer, A. M. and Chmelevsky, D. (1983). Small–sample properties of censored data rank tests. *Biometrics*, **39**, 675–682.

Kim, K. and DeMets, D. L. (1987a). Design and analysis of group sequential tests based on the type I error spending rate function. *Biometrika*, **74**, 149–154.

— — — (1987b). Confidence intervals following group sequential tests in clinical trials. *Biometrics*, **43**, 857–864.

Kimball, A. W. (1958). Disease incidence estimation in populations subject to multiple causes of death. *Bull. Internat. Inst. Statist.*, **36**, 193–204.

— — — (1969). Models for the estimation of competing risks from grouped data. *Biometrics*, **25**, 329–337.

Klein, J. P. (1991). Small sample moments of some estimators of the variance of the Kaplan–Meier and Nelson–Aalen estimators. *Scand. J. Statist.*, **18**, 333–340.

Klein, J. P. and Moeschberger, M. (1997). *Survival Analysis*. New York: Springer–Verlag.

Kleinman, K. P. and Ibrahim, J. G. (1998a). A semiparametric Bayesian approach to the random effects model. *Biometrics*, **54**, 921–938.

— — — (1998b). A semi–parametric Bayesian approach to generalized linear mixed models. *Statist. Med.*, **17**, 2579–2596.

Koul, H., Susarla, V., and Van Ryzin, J. (1981). Regression analysis with randomly right-censored data. *Ann. Statist.*, **9**, 1276–1288.

Koziol, J. and Reid, N. (1977). On multiple comparison among K samples subject to unequal patterns of censorship. *Comm. Statist., A Theory Methods*, **6**(12), 1149–1164.

Lagakos, S. W. (1977). A covariate model for partially censored data subject to competing causes of failure. *J. Roy. Statist. Soc.*, **27**, 235–241.

— — — (1981). The graphical evaluation of explanatory variables in proportional hazards regression models. *Biometrika*, **68**, 93–98.

Lagakos S. W. (1988). The loss in efficiency from misspecifying covariates in proportional hazards regression models. *Biometrika*, **75**, 156–160.

Lagakos, S. W., Sommer, C. J., and Zelen, M. (1978). Semi-Markov models for partially censored data. *Biometrika*, **65**, 311–317.

Lagakos, S. W., Barraj, L. M., and De Gruttola, V. (1988). Nonparametric analysis of truncated survival data with application to AIDS. *Biometrika*, **75**, 515–523.

Lai, T. L. and Ying, Z. (1991a). Rank regression methods for left truncated and right censored data. *Ann. Statist.*, **19**, 531–554.

— — — (1991b). Large sample theory of a modified Buckley–James estimator for regression analysis with censored data. *Ann. Statist.*, **19**, 1370–1402.

Lan, K. K. G. and DeMets, D. L. (1983). Discrete sequential boundaries for clinical trials. *Biometrika*, **70**, 659–663.

— — — (1989). Group sequential procedures: calendar versus information time. *Statist. Med.*, **8** 1191–1198.

Langholz, B. and Borgan O. (1995). Counter–matching: a stratified nested case–control sampling method. *Biometrika*, **82**, 69–79.

Langholz, B. and Thomas, D. C. (1990). Nested case–control and case–cohort methods of sampling from a cohort: a critical comparison. *Amer. J. Epidemiol.*, **131**, 169–176.

Latta, R. B. (1981). A Monte Carlo study of some two-sample rank tests with censored data. *J. Amer. Statist. Assoc.*, **76**, 713–719.

Lawless, J. F. (1973). Conditional versus unconditional confidence intervals for the parameters of the Weibull distribution. *J. Amer. Statist. Assoc.*, **68**, 665–669.

— — — (1978). Confidence interval estimation for the Weibull and extreme value distributions. *Technometrics*, **20**, 355–364.

— — — (1982). *Statistical Models and Methods for Lifetime Data*. New York: Wiley.

— — — (1987). Regression methods for Poisson process data. *J. Amer. Statist. Assoc.*, **82**, 808–815.

Lawless, J. F. and Nadeau, C. (1995). Some simple and robust methods for the analysis of recurrent events. *Technometrics*, **37**, 158–168.

Lawless, J. F., Nadeau, C., and Cook, R. J. (1997). Analysis of the mean and rate functions for recurrent events. In *Proceedings of the First Seattle Symposium in Biostatistics: Survival Analysis* (D. Y. Lin and T. R. Fleming, eds.). New York: Springer–Verlag, pp. 37–49.

Lawless, J. F., Kalbfleisch, J. D., and Wild, C. J. (1999). Semi-parametric methods for response-selective and missing data problems in regression. *J. Roy. Statist. Soc. Ser. B*, **61**, 413–438.

Le Cam, L. (1970). On the assumptions used to prove asymptotic normality of maximum likelihood estimates. *Ann. Math. Statist.*, **41**, 802–828.

Lee, E. T. (1980). *Statistical Methods for Survival Data Analysis*. Belmont, Calif.: Lifetime Learning Publications.

Lee, E. T., Desu, M. M., and Gehan, E. A. (1975). A Monte Carlo study of the power of some two-sample tests. *Biometrika*, **62**, 425–432.

Lee, E. W., Wei, L. J., and Amato, D. A. (1992). Cox-type regression analysis for large numbers of small groups of correlated failure time observations. In *Survival Analysis: State of the Art* (J. P. Klein and P. K. Goel, eds.). Norwell, Mass.: Kluwer Academic, pp. 237–247.

Lehmann, E. L. (1953). The power of rank tests. *Ann. Math. Statist.*, **24**, 23–43.

— — — (1959). *Testing Statistical Hypotheses*. New York: Wiley.

— — — (1975). *Nonparametrics: Statistical Methods Based on Ranks*. San Francisco: Holden-Day.

Lenglart, E. (1977). Relation de domination entre deux processus. *Ann. Inst. H. Poincaré*, **13**, 171–179.

Leurgans, S. (1983). Three classes of censored data rank tests: strengths and weaknesses under censoring. *Biometrika*, **70**, 651–658.

— — — (1984). Asymptotic behavior of two-sample rank tests in the presence of random censoring. *Ann. Statist.*, **12**, 572–589.

Li, G. (1995a). Nonparametric likelihood ratio estimation of probabilities for truncated data. *J. Amer. Statist. Assoc.*, **87**, 120–127.

— — — (1995b). On nonparametric likelihood ratio estimation of survival probabilities for censored–data. *Statist. Probab. Lett.*, **25**, 95–104.

Li, G., Hollander, M., McKeague, I. W., and Yang, J. (1996). Nonparametric likelihood ratio confidence bands for quantile functions from incomplete survival data. *Ann. Statist.*, **24**, 628–640.

Lin, D. Y. (1991a). Goodness of fit for the Cox regression model based on a class of parameter estimators. *J. Amer. Statist. Assoc.*, **86**, 725–728.

— — — (1991b). Nonparametric sequential testing in clinical trials with incomplete multivariate observations. *Biometrika*, **78**, 123–131.

— — — (1994). Cox regression analysis of multivariate failure time data: the marginal approach. *Statist. Med.*, **13**, 2233–2247.

Lin, D. Y. and Geyer, C. J. (1992). Computational methods for semiparametric linear regression with censored data. *J. Comput. Graph. Statist.*, **1**, 77–90.

Lin, D. Y. and Ying, Z. (1993a). A simple nonparametric estimator of the bivariate survival function under univariate censoring. *Biometrika*, **80**, 573–581.

— — — (1993b). Cox regression with incomplete covariate measurements. *J. Amer. Statist. Assoc.*, **88**, 1341–1349.

— — — (1995). Semiparametric inference for the accelerated life model with time-dependent covariates. *J. Statist. Plann. Inference*, **44**, 47–63.

Lin, D. Y., Wei, L. J., and Ying, Z. (1998). Accelerated failure time models for counting processes. *Biometrika*, **85**, 605–618.

Lin, D. Y., Sun, W., and Ying, Z. (1999). Nonparametric estimation of the gap time distribution for serial events with censored data. *Biometrika*, **86**, 59–70.

Lin, D. Y., Wei, L. J., Yang, I., and Ying, Z. (2000). Semiparametric regression for the mean and rate functions of recurrent events. *J. Roy. Statist. Soc. Ser. B*, **62**, 711–730.

Lin, D. Y., Wei, L. J., and Ying, Z. (2001). Semiparametric transformation models for point processes. *J. Amer. Statist. Assoc.*, **96**, 620–628.

Lin, J. S. and Wei, L. J. (1992). Linear regression for multivariate failure time observations. *J. Amer. Statist. Assoc.*, **87**, 1091–1097.

Lininger L, Gail, M. H., Green, S. B., and Byar, D. P. (1979). Comparison of four tests for equality of survival curves in the presence of stratification and censoring. *Biometrika*, **66**, 419–428.

Little, R. J. A. and Rubin, D. B. (1987). *Statistical Analysis of Missing Data*. New York: Wiley.

Louis, T. A. (1981). Nonparametric analysis of an accelerated failure time model. *Biometrika*, **68**, 381–390.

— — — (1982). Finding the observed information matrix when using the EM algorithm. *J. Roy. Statist. Soc. Ser. B*, **44**, 226–233.

Lubin, J. H. and Gail, M. H. (1984). Biased selection of controls for case–control analysis of cohort studies. *Biometrics*, **40**, 63–75.

MacEachern, S. N. (1994). Estimating normal means with a conjugate style Dirichlet process prior. *Comm. Statist. A Theory Methods*, **23**, 727–741.

MacEachern, S. N. and Muller, P. (1998). Estimating mixture of Dirichlet process models. *J. Comput. Graph. Statist.*, **7**, 223–238.

Makeham, W. M. (1860). On the law of mortality and the construction of annuity tables. *J. Inst. Actuaries (London)*, **8**.

— — — (1874). On an application of the theory of the composition of decremental forces. *J. Inst. Actuaries (London)*, **18**, 317–322.

Mann, N. R., Schafer, R. E., and Singpurwalla, N. D. (1974). *Methods for Statistical Analysis of Reliability and Life Data*. New York: Wiley.

Mantel, N. (1963). Chi-square tests with one degree of freedom: extensions of the Mantel–Haenszel procedure. *J. Amer. Statist. Assoc.*, **58**, 690–700.

— — — (1966). Evaluation of survival data and two new rank order statistics arising in its consideration. *Cancer Chemother. Rep.*, **50**, 163–170.

— — — (1973). Synthetic retrospective studies and related topics. *Biometrics*, **29**, 479–486.

Mantel, N. and Bryan, W. R. (1961). Safety testing for carcinogens. *J. Nat. Cancer Inst.*, **27**, 455–470.

Mantel, N. and Byar, D. P. (1974). Evaluation of response time data involving transient states: an illustration using heart transplant data. *J. Amer. Statist. Assoc.*, **69**, 81–86.

Mantel, N. and Haenszel, W. (1959). Statistical aspects of the analysis of data from retrospective studies of disease. *J. Nat. Cancer Inst.*, **22**, 719–748.

Mantel, N. and Myers, M. (1971). Problems of convergence of maximum likelihood iterative procedures in multiparameter situations. *J. Amer. Statist. Assoc.*, **66**, 484–491.

Mantel, N., Bohidar, N. R., Brown, C. C., Ciminera, J. L., and Tukey, J. W. (1975). An improved Mantel–Bryan procedure for "safety" testing of carcinogens. *Cancer Res.*, **35**, 865–872.

Marshall, A. W. and Olkin, I. (1967). A multivariate exponential distribution. *J. Amer. Statist. Assoc.*, **62**, 30–44.

McKean, J. W. and Hettmansperger, T. P. (1978). A robust analysis of the general linear model based on one step *r*-estimates. *Biometrika*, **65**, 571–579.

McKnight, B. and Crowley, J. J. (1984). Tests for differences in tumor incidence based on animal carcinogenesis experiments. *J. Amer. Statist. Assoc.*, **79**, 639–648.

McLeish, D. L. (1974). Dependent central limit theorems and invariance principles. *Ann. Probab.*, **2**, 620–628.

Mehrotra, K. C., Michalek, J. E., and Mihalko, D. (1982). A relationship between two forms of linear rank procedures for censored data. *Biometrika*, **69**, 674–676.

Miettinen, O. S. (1974). Confounding and effect modification. *Amer. J. Epidemiol.*, **100**, 350–353.

Miller, R. G. (1976). Least squares regression with censored data. *Biometrika*, **63**, 449–464.

Moeschberger, M. L. (1974). Life tests under competing causes of failure. *Technometrics*, **16**, 39–47.

Moeschberger, M. L. and David, H. A. (1971). Life tests under competing causes of failure and the theory of competing risks. *Biometrics*, **27**, 909–933.

Moran, P. A. P. (1959). *The Theory of Storage*. London: Methuen.

— — — (1971). Maximum likelihood estimation in non-standard conditions. *Proc. Cambridge Philos. Soc.*, **70**, 441–450.

Moreau, T., O'Quigley, J., and Mesbah, M. (1985). A global goodness-of-fit statistic for the proportional hazards model. *Appl. Statist.*, **34**, 212–218.

Murphy, S. A.(1995). Likelihood ratio-based confidence intervals in survival analysis. *J. Amer. Statist. Assoc.*, **90**(432), 1399–1405.

Murphy, S. A., Rossini, A. J., and van der Vaart, A. W. (1997). Maximum likelihood estimation in the proportional odds model. *J. Amer. Statist. Assoc.*, **92**, 968–976.

Murray, S. and Tsiatis A. A. (1996). Nonparametric approach to incorporating prognostic longitudinal covariate information in survival estimation. *Biometrics*, **52**, 137–151.

— — — (2001). Using auxiliary time-dependent covariates to recover information in nonparametric testing with censored data. *Lifetime Data Anal.*, **7**, 125–141.

Myers, M., Hankey, B. F., and Mantel, N. (1973). A logistic–exponential model for use with response-time data involving regressor variables. *Biometrics*, **29**, 257–269.

Nakamura, T. (1992). Proportional hazards model with covariates subject to measurement error. *Biometrics*, **48**, 829–838.

Nelson, W. B. (1969). Hazard plotting for incomplete failure data. *J. Qual. Technol.*, **1**, 27–52.

— — — (1970). Statistical methods for accelerated life test data—the inverse power law model. *General Electric Corporate Research and Development T15 Report 71-C-011.*

— — — (1972). Theory and applications of hazard ploting for censored failure data. *Technometrics*, **14**, 945–965.

— — — (1988). Graphical analysis of system repair data. *J. Qual. Technol.*, **20**, 24–35.

— — — (1995). Confidence limits for recurrence data—applied to cost or number of product repairs. *Technometrics*, **37**, 147–157.

Nelson, W. B. and Hahn, G. J. (1972). Linear estimation of a regression relationship from censored data: 1. Simple methods and their applications (with discussion). *Technometrics*, **14**, 247–276.

Nielsen, G. G., Gill, R. D., Andersen, P. K., and Sorenson, T. I. A. (1992). A counting process approach to maximum likelihood estimation in frailty models. *Scand. J. Statist.*, **19**, 25–43.

Nowinski, R. C., Brown, M., Doyle, T., and Prentice, R. L. (1979). Genetic and viral factors influencing the development of spontaneous tumors in AKR mice. *Virology*, **96**, 186–204.

Oakes, D. (1977). The asymptotic information in censored survival data. *Biometrika*, **64**, 441–448.

— — — (1981). Survival times: aspects of partial likelihood (with discussion). *Internat. Statist. Rev.*, **49**, 235–264.

— — — (1989). Bivariate survival models induced by frailties. *J. Amer. Statist. Assoc.*, **84**, 487–493.

— — — (2001). *Biometrika* centenary: survival analysis. *Biometrika*, **88**, 99–142.

O'Brein, P. C. and Fleming, T. R. (1979). A multiple testing procedure for clinical trials. *Biometrics*, **35**, 549–556.

Odell, P. M., Anderson, K. M., and D'Agostino, R. B. (1992). Maximum likelihood estimation for interval-censored data using a Weibull–based accelerated failure time model. *Biometrics*, **48**, 951–959.

O'Quigley, J. and Pessione, F. (1989). Score tests for homogeneity of regression effect in the proportional hazards model. *Biometrics*, **45**, 135–144.

Owen, A. B. (2001). *Empirical Likelihood.* Boca Raton, Fla.: Chapman & Hall/CRC.

Parr, V. B. and Webster, J. T. (1965). A method for discriminating between failure density functions used in reliability predictions. *Technometrics*, **7**, 1–10.

Parzen, M. I., Wei, L. J., and Ying, Z. (1994). A resampling method based on pivotal estimating functions. *Biometrika*, **81**, 341–350.

Peace, K. E. and Flora, R. E. (1978). Size and power assessments of tests of hypotheses on survival parameters. *J. Amer. Statist. Assoc.*, **73**, 129–132.

Peña, E. and Hollander, M. (2001). Models for recurrent events in reliability and survival analysis. In *Mathematical Reliability* (T. Mozzuchi, N. Singpurwalla, and R. Soyer, eds.). Norwell, Mass.: Kluwer Academic. To appear.

Pepe, M. S. and Cai, J. (1993). Some graphical displays and marginal regression analyses for recurrent failure times and time dependent covariates. *J. Amer. Statist. Assoc.*, **88**, 811–820.

Peterson, A. V. (1975). Nonparametric estimation in the competing risks problem. Ph.D. thesis, Department of Statistics, Stanford University.

— — — (1976). Bounds for a joint distribution function with fixed sub-distribution functions: application to competing risks. *Proc. Nat. Acad. Sci. U.S.A.*, **73**, 11–13.

Peterson, A. V., Kealey, K. A., Mann, S. L., Marek, P. M., and Sarason, I. G. (2000). Hutchinson smoking prevention project: long–term randomized trial in school-based tobacco use prevention—results on smoking. *J. Nat. Cancer Inst.*, **92**, 1979–1991.

Peto, R. (1972a). Rank tests of maximal power against Lehmann-type alternatives. *Biometrika*, **59**, 472–475.

— — — (1972b). Contribution to the discussion of paper by D. R. Cox. *J. Roy. Statist. Soc. Ser. B*, **34**, 205–207.

Peto, R. and Lee, P. (1973). Weibull distributions for continuous carcinogenesis experiments. *Biometrics*, **29**, 457–470.

Peto, R. and Peto, J. (1972). Asymptotically efficient rank invariant test procedures (with discussion). *J. Roy. Statist. Soc. Ser. A*, **135**, 185–206.

Peto, R., Pike, M. C., Armitage, P., Breslow, N. E., Cox, D. R., Howard, S. V., Mantel, N., McPherson, K., Peto, J., and Smith, P. G. (1976). Design and analysis of randomized clinical trials requiring prolonged observation of each patient: I. Introduction and design. *British J. Cancer*, **34**, 585–612.

— — — (1977). Design and analysis of randomized clinical trials requiring prolonged observation of each patient: 2. Analysis and examples. *British J. Cancer*, **35**, 1–39.

Pettitt, A. N. (1982). Proportional odds model for survival data and estimates using ranks. *Appl. Statist.*, **33**, 169–175.

Pike, M. C. (1966). A method of analysis of certain class of experiments in carcinogenesis. *Biometrics*, **22**, 142–161.

— — — (1970). A note on Kimball's paper "Models for the estimation of competing risks from grouped data." *Biometrics*, **26**, 579–581.

Pocock, S. J. (1997). Group sequential methods in the design and analysis of clinical trials. *Biometrika*, **64**, 191–199.

Pons. O. and Turckheim, E. (1991). Tests of independence of bivariate censored data based on the empirical joint hazard function. *Scand. J. Statist.*, **18**, 21–37.

Prentice, R. L. (1973). Exponential survivals with censoring and explanatory variables. *Biometrika*, **60**, 279–288.

— — — (1974). A log-gamma model and its maximum likelihood estimation. *Biometrika*, **61**, 539–544.

— — — (1975). Discrimination among some parametric models. *Biometrika*, **62**, 607–614.

— — — (1976a). Use of the logistic model in retrospective studies. *Biometrics*, **32**, 599–606.

— — — (1976b). A generalization of the probit and logit models for dose response curves. *Biometrics*, **32**, 761–768.

— — — (1978). Linear rank tests with right censored data. *Biometrika*, **65**, 167–179.

— — — (1982). Covariate measurement errors and parameter estimation in a failure time regression model. *Biometrika*, **69**, 331–342.

— — — (1986a). A case-cohort design for epidemiologic cohort studies and disease prevention trials. *Biometrika*, **73**, 1–11.

— — — (1986b). On the design of synthetic case-control studies. *Biometrics*, **42**, 301–310.

— — — (1999). Nonparametric maximum likelihood estimation of the bivariate survivor function. *Statist. Med.*, **18**, 2517–2527.

Prentice, R. L. and Breslow, N. E. (1978). Retrospective studies and failure time models. *Biometrika*, **65**, 153–158.

Prentice, R. L. and Cai, J. (1992). Covariance and survivor function estimation using censored multivariate failure time data. *Biometrika*, **79**, 495–512.

Prentice, R. L. and Gloeckler, L. A. (1978). Regression analysis of grouped survival data with application to breast cancer data. *Biometrics*, **34**, 57–67.

Prentice, R. L. and Hsu, L. (1997). Regression on hazard ratios and cross ratios in multivariate failure time analysis. *Biometrika*, **84**, 349–363.

Prentice, R. L. and Kalbfleisch, J. D. (1979). Hazard rate models with covariates. *Biometrics*, **35**, 25–39.

— — — (2002). Mixed continuous and discrete Cox models. *Lifetime Data Analysis*, **8**, in press.

Prentice, R. L. and Marek, P. (1979). A qualitative discrepancy between censored data rank tests. *Biometrics*, **35**, 861–868.

Prentice, R. L. and Pyke, R. (1979). Logistic disease incidence models and case-control studies. *Biometrika*, **66**, 403–411.

Prentice, R. L. and Self, S. G. (1983). Asymptotic distribution theory for Cox-type regressions models with general relative risk form. *Ann. Statist.*, **11**, 804–813.

Prentice, R. L. and Shillington, E. R. (1975). Regression analysis of Weibull data and the analysis of clinical trials. *Utilitas Math.*, **8**, 257–276.

Prentice, R. L., Williams, B. J., and Peterson, A. V. (1981). On the regression analyses of multivariate time data. *Biometrika*, **68**, 373–379.

Prentice, R. L., Kalbfleisch, J. D., Peterson, A. V. Jr., Flournoy, N., Farewell, V. T., and Breslow, N. E. (1978). The analysis of failure times in the presence of competing risks. *Biometrics*, **34**, 541–554.

Pruitt, R. C. (1991). Strong consistency of self-consistent estimators: general theory and an application to bivariate survival analysis. *Technical Report 543*. University of Minnesota.

Puri, M. L. (ed.) (1970). *Nonparametric Techniques in Statistical Inference*. London: Cambridge University Press.

Puri, M. L. and Sen, P. K. (1971). *Nonparametric Methods in Multivariate Analysis*. New York: Wiley.

Rebolledo, R. (1979). La méthode des martingales appliqué a l'étude de la convergence en loi de processus. *Mem. Soc. Math. France*, **62**.

— — — (1980). Central limit theorems for local martingales. *Z. Wahrsch. Verw. Gebiete*, **51**, 269–286.

Reid, N. (1994). A conversation with Sir David Cox. *Statist. Sci.*, **9**, 439–55.

Reid, N. and Crepeau, H. (1985). Influence functions for proportional hazards regression. *Biometrika*, **72**, 1–9.

Ritov, Y. (1990). Estimation in a linear regression model with censored data. *Ann. Statist.*, **18**, 303–328.

Robins, J. M. and Finkelstein, D. (2000). Correcting for non-compliance and dependent censoring in an AIDS clinical trial with inverse probability of censoring weighted (IPCW) log-rank tests. *Biometrics*, **56**, 779–788.

Robins, J. M. and Tsiatis, A. A. (1992). Semiparametric estimation of an accelerated failure time model with time-dependent covariates. *Biometrika*, **79**, 311–319.

Robins, J. M. and Rotnitsky, A. (1992). Recovery of information and adjustment for dependent censoring using surrogate markers. In *AIDS Epidemiology: Methodological Issues* (N. Jewell, K. Dietz, and V. Farewell, eds.). Boston: Birkhauser, pp. 297–331.

Robins, J. M., Gail, M. H., and Lubin, J. H. (1986). More on "Biased selection of controls for case-control analysis of cohort studies." *Biometrics*, **42**, 273–299.

Robins, J. M., Prentice, R. L., and Blevins, D. (1989). Designs for synthetic case-control studies in open cohorts. *Biometrics*, **45**, 1103–1116.

Robins, J. M., Rotnitsky, A., and Zhao, L. P. (1994). Estimation of regression coefficients when some regressors are not always observed. *J. Amer. Statist. Assoc.*, **89**, 846–866.

Robins, J. M., Rotnitzky, A., and van der Laan, M. (2000). Commentary on "Profile Likelihood" by S. A. Murphy and A. W. van der Vaart. *J. Amer. Statist. Assoc.*, **95**, 477–482.

Rossini, A. and Tsiatis, A. A. (1996). A semiparametric proportional odds regression model for the analysis of current status data. *J. Amer. Statist. Assoc.*, **91**, 713–721.

Royden, H. L. (1968). *Real Analysis*, 2nd ed. London: Macmillan.

Samuelsen, S. O. (1997). A pseudolikelihood approach to the analysis of nested case–control studies. *Biometrika*, **84**, 379–394.

Sarhan, A. E. and Greenberg, B. G. (eds.) (1962). *Contributions to Order Statistics*. New York: Wiley.

Satten, G. A. (1996). Rank-based inference in the proportional hazards model for interval censored data. *Biometrika* **83**, 355–370.

Satten, G. A. and Datta, S. (2000). The Kaplan–Meier estimator as an inverse-probability-of-censoring weighted average. *Amer. Statist.*, **55**, 207–210.

Savage, I. R. (1956). Contributions to the theory of rank order statistics—the two sample case. *Ann. Math. Statist.*, **27**, 590–615.

— — — (1957). Contributions to the theory of rank order statistics: the "trend" case. *Ann. Math. Statist.*, **28**, 968–977.

Schoenfeld, D. (1982). Partial residuals for the proportional hazards regression model. *Biometrika*, **69**, 239–241.

Scott, A. J. and Wild, C. J. (1997). Fitting regression models to case–control data by maximum likelihood. *Biometrika*, **47**, 497–510.

Seal, H. L. (1954). The estimation of mortality and other decremental probabilities. *Skand. Akt.*, **37**, 137–162.

Self, S. G. and Grossman, E. A. (1986). Linear rank tests for interval-censored data with application to PCB levels in adipose tissue of transformer repair workers. *Biometrics*, **42**, 521–530.

Self, S. G. and Prentice, R. L. (1988). Asymptotic distribution theory and efficiency results for case–cohort studies. *Ann. Statist.*, **16**, 64–81.

Shih, J. H. and Louis, T. A. (1995). Inferences on the association parameter in copula models for bivariate survival data. *Biometrics*, **51**, 1384–1399.

Shorack, G. R. (2000). *Probability for Statisticians*. New York: Springer-Verlag.

Shorack, G. R. and Wellner, J. A. (1986). *Empirical Processes*. New York: Wiley. Corrections and changes: *Technical Report 167*, Department of Statistics, University of Washington (1989).

Siegmund, D. (1978). Estimation following sequential tests. *Biometrika*, **65**, 341–349.

— — — (1985). *Sequential Analysis*. New York: Springer-Verlag.

Sinha, D., Tanner, M. A., and Hall, W. J. (1994). Maximizing the marginal likelihood from grouped survival data. *Biometrika*, **81**, 53–60.

Slud, E. V. and Wei, L. J. (1982). Two-sample repeated significance tests based on the modified Wilcoxon statistic. *J. Amer. Statist. Assoc.*, **77**, 862–868.

Snyder, D. L. (1975). *Random Point Processes*. New York: Wiley.

Sorensen, P. and Andersen, P. K. (2000). Computing the Cox model for case–cohort designs. *Biometrika*, **87**, 49–59.

Spiegelhalter, D. J., Freedman, L. S. and Parman, M. K. B. (1994). Bayesian approaches to clinical trials (with discussion). *J. Roy. Statist. Soc. Ser. A*, **157**, 357–416.

Sprott, D. A. (1975). Application of maximum likelihood methods for finite samples. *Sankhya B*, **37**, 259–270.

Sprott, D. A. and Kalbfleisch, J. D. (1969). Examples of likelihoods and comparison with point estimates and large sample approximations. *J. Amer. Statist. Assoc.*, **64**, 468–484.

Stacy, E. W. (1962). A generalization of the gamma distribution. *Ann. Math. Statist.*, **33**, 1187–1192.

Stacy, E. W. and Mihram, G. A. (1965). Parameter estimation for a generalized gamma distribution. *Technometrics*, **7**, 349–358.

Storb, R. H., Deeg, J., Farewell, V., Doney, K., Appelbaum, F., Beatty, P., Bensinger, W., Buckner, C. D., Clift, R., Hansen, J., Hill, R., Longton, G., Lum, L., Maring, P., McGuffin, R., Sanders, J., Singer, J., Stewart, P., Sullivan, K., Witherspoon, R., and Thomas, E. D. (1986). Marrow transplantation for severe aplastic anemia: methotrexate alone compared with a combination of methotrexate and cyclosporine for prevention of acute graft-versus-host disease. *Blood*, **68**, 119–125.

Struthers, C. A. (1984). Asymptotic properties of linear rank tests with censored data. Ph.D. thesis, Statistics and Actuarial Science Department, University of Waterloo.

Sun, J. (1996). A nonparametric test for interval-censored failure time data with application to AIDS studies. *Statist. Med.*, **15**, 1387–1395.

Sun, J. and Kalbfleisch, J. D. (1993). The analysis of current status data on point processes. *J. Amer. Statist. Assoc.*, **88**, 1449–1454.

Susarla, V. and Van Ryzin, J. (1976). Nonparametric Bayesian estimation of survival curves from incomplete observations. *J. Amer. Statist. Assoc.*, **71**, 897–902.

Tarone, R. and Ware, J. (1977). On distribution-free tests for equality of survival distributions. *Biometrika*, **64**, 156–160.

Temkin, N. R. (1978). An analysis for transient states with application to tumor shrinkage. *Biometrics*, **34**, 571–580.

Therneau, T. M. and Grambsch, P. M. (2000). *Modeling Survival Data: Extending the Cox Model*. New York: Springer-Verlag.

Therneau, T. M. and Hamilton, S. A. (1997). rhDNase as an example of recurrent event analysis. *Statist. Med.*, **16**, 2029–2047.

Therneau, T. M. and Li, H. (1999). Computing the Cox model for case–cohort designs. *Lifetime Data Anal.*, **5**, 99–112.

Therneau, T. M., Grambsch, P., and Fleming. T. (1990). Martingale based residuals for survival models. *Biometrika*, **77**, 147–160.

Thomas, D. C. (1977). Addendum to "Methods for cohort analysis: appraisal by application to asbestos mining" by F. D. D. Liddell, J. C. McDonald, and P. C. Thomas. *J. Roy. Statist. Soc. Ser. A*, **140**, 469–491.

Thomas, D. R. (1969). Conditional locally most powerful rank tests for the two-sample problem with arbitrarily censored data. *Technical Report 7*, Department of Statistics, Oregon State University.

Thomas, D. R. and Grunkemeier, G. L. (1975). Confidence interval estimation of survival probabilities for censored data. *J. Amer. Statist. Assoc.*, **70**, 865–871.

Thomas, E. D., Storb, R., Clift, R. A., Fefer, A., Johnson, F. L., Neiman, P. E., Lerner, K. G., Glucksberg, H., and Buckner, C. D. (1975a). Bone-marrow transplantation (first of two parts). *N. Engl. J. Med.*, **292**, 832–843.

— — — (1975b). Bone-marrow transplantation (second of two parts). *N. Engl. J. Med.*, **292**, 895–902.

Thompson, M. E. and Godambe, V. P. (1974). Likelihood ratio vs. most powerful rank test: a two sample problem in hazard analysis. *Sankhya*, **36**, 13–40.

Thompson, W. A. (1977). On the treatment of grouped observations in life studies. *Biometrics*, **33**, 463–470.

Tsai, W. Y. and Crowley, J. (1998). A note on nonparametric estimators of the bivariate survival function under univariate censoring. *Biometrika*, **85**, 573–580.

Tsai, W. Y., Leurgans, S., and Crowley, J. (1986). Nonparametric estimator of a bivariate survival function in the presence of censoring. *Ann. Statist.*, **14**, 1351–1365.

Tsiatis, A. A. (1975). A nonidentifiability aspect of the problem of competing risks. *Proc. Nat. Acad. Sci.*, **72**, 20–22.

— — — (1978). A large sample study of the estimate for the integrated hazard function in Cox's regression model for survival data. *Technical Report 562*. Department of Statistics, University of Wisconsin, Madison.

— — — (1981). A large sample study of Cox's regression model. *Ann. Statist.*, **9**, 93–108.

— — — (1990). Estimating regression parameters using linear rank tests for censored data. *Ann. Statist.*, **18**, 354–372.

Tsiatis, A. A., Rosner, G. L., and Mehta, C. R. (1984). Exact confidence intervals following a group sequential test. *Biometrics*, **40**, 797–803.

Turnbull, B. W. (1974). Nonparametric estimation of a survivorship function with doubly censored data. *J. Amer. Statist. Assoc.*, **69**, 169–173.

— — — (1976). The empirical distribution function with arbitrarily grouped censored and truncated data. *J. Roy. Statist. Soc. Ser. B*, **38**, 290–295.

Turnbull, B. W., Brown, B. W., and Hu, M. (1974). Survivorship analysis of heart transplant data (A). *J. Amer. Statist. Assoc.*, **69**, 74–80.

Van der Laan, M. J. (1996). Efficient estimation in the bivariate censoring model and repairing NPMLE. *Ann. Statist.*, **24**, 596–627.

— — — (1997). Nonparametric estimators of the bivariate survival function under random censoring. *Statist. Neerlandica*, **51**, 178–200.

Van der Vaart, A. W. and Wellner, J. A. (1996). *Weak Convergence and Empirical Processes with Applications to Statistics.* New York: Springer-Verlag.

Van der Waerden, B. L. (1953). Ein neuer Test fur das Problem der zwei Stichproben. *Math. Ann.*, **126**, 93–107.

Wacholder, S. (1991). Practical considerations in choosing between the case–cohort and nested case–control designs. *Epidemiology*, **2**, 155–158.

Wacholder, S., Gail, M. H., Pee, D., and Brookmeyer, R. (1989). Alternative variance and efficiency calculations for the case–cohort design. *Biometrika*, **76**, 117–123.

Wald, A. (1947). *Sequential Analysis.* New York: Wiley.

Wang, C. Y. and Chen, H. Y. (2001). Augmented inverse probability weighted estimator for Cox missing covariate regression. *Biometrics*, **57**, 414–419.

Wang, C. Y., Hsu, L., Feng, Z. D., and Prentice, R. L. (1997). Regression calibration in failure time regression. *Biometrics*, **53**, 131–145.

Wang, M.-C. and Chang, S.-H. (1999). Nonparametric estimation of a recurrent survival function. *J. Amer. Statist. Assoc.*, **94**, 146–153.

Wang, M.-C., Qin, J., and Chiang, C.-T. (2001). Analyzing recurrent event data with informative censoring. *J. Amer. Statist. Assoc.*, **96**, 1057–1065.

Wang, S. K. and Tsiatis, A. A. (1987). Approximately optimal one-parameter boundaries for group sequential trials. *Biometrics*, **43**, 193–200.

Wang, Y. and Leung, D. H. Y. (1997). Bias reduction via resampling for estimation following sequential tests. *Sequential Anal.*, **16**, 249–267.

Wei, C. C. G. and Tanner, M. A. (1990). A Monte Carlo implementation of the EM algorithm and the poor man's data augmentation algorithms. *J. Amer. Statist. Assoc.*, **85**, 699–704.

Wei, L. J. and Gail, M. H. (1983). Nonparametric estimation of a scale-change with censored observations. *J. Amer. Statist. Assoc.*, **78**, 382–388.

Wei, L. J., Lin, D. Y., and Weissfeld, L. (1989). Regression analysis of multivariate incomplete failure time data by modeling marginal distributions. *J. Amer. Statist. Assoc.*, **84**, 1065–1073.

Wei, L. J., Ying, Z. and Lin. D. Y. (1990). Linear regression analysis of censored survival data based on rank tests. *Biometrika*, **77**, 845–852.

Whitehead, J. (1986). On the bias of maximum likelihood estimation following a sequential test. *Biometrika*, **73**, 573–581.

Whitehead, J. and Stratton, I. (1983). Group sequential clinical trials with triangular continuation regions. *Biometrics*, **39**, 227–236.

Whittemore, A. S. (1981). The efficiency of synthetic retrospective studies. *Biometrical J.*, **23**, 73–78.

Wild, C. J. (1983). Failure time models with matched data. *Biometrika*, **70**, 633–641.

Wild, C. J. and Kalbfleisch, J. D. (1981) A note on a paper by Phadia and Ferguson. *Ann. Statist.*, **9**, 1061–1065.

Williams, J. S. and Lagakos, S. W. (1977). Models for censored survival analysis: constant sum and variable sum models. *Biometrika*, **64**, 215–224.

Women's Health Initiative Study Group (1998). Design of the Women's Health Initiative clinical trial and observational study. *Control. Clin. Trials*, **19**, 61–109.

Woolson, R. F. and Lachenbruch, P. A. (1980). Rank tests for censored matched pairs. *Biometrika*, **67**, 597–606.

Wu, C. O. (1995). Estimating the real parameter in a two-sample proportional odds model. *Ann. Statist.* **23**, 376–395.

Xie, S. X., Wang, C. Y., and Prentice, R. L. (2001). A risk set calibration method for failure time regression by using a covariate reliability sample. *J. Roy. Statist. Soc. Ser. B*, **63**, 855–870.

Yang, S. and Prentice, R. L. (1999). Semiparametric inference in the proportional odds regression model. *J. Amer. Statist. Assoc.*, **94**, 125–136.

Ying, Z. (1993). A large sample study of rank estimation for censored regression data. *Ann. Statist.*, **21**, 76–99.

Zelen, M. (1971). The analysis of several 2×2 contingency tables. *Biometrika*, **38**, 129–137.

Zhong, M., Sen, P. K., and Cai, J. (1996). Cox regression model with mismeasured covariate or missing covariate. *ASA Proceedings of the Biometrics Section*, pp. 323–328.

Zhou, H. and Pepe, M. (1995). Auxiliary covariate data in failure time regression analysis. *Biometrika*, **82**, 139–149.

Zhou, H. and Wang, C. Y. (2000). Failure time regression with continuous covariates measured with error. *J. Roy. Statist. Soc. Ser. B*, **62**, 657–665.

Zippin, C. and Armitage, P. (1966). Use of concomitant variables and incomplete survival information in the estimation of an exponential survival parameter. *Biometrics*, **22**, 665–672.

Author Index

429

Subject Index

BERRY, CHALONER, and GEWEKE · Bayesian Analysis in Statistics and Econometrics: Essays in Honor of Arnold Zellner

BERNARDO and SMITH · Bayesian Theory

BHAT and MILLER · Elements of Applied Stochastic Processes, *Third Edition*

BHATTACHARYA and JOHNSON · Statistical Concepts and Methods

BHATTACHARYA and WAYMIRE · Stochastic Processes with Applications

BILLINGSLEY · Convergence of Probability Measures, *Second Edition*

BILLINGSLEY · Probability and Measure, *Third Edition*

BIRKES and DODGE · Alternative Methods of Regression

BLISCHKE AND MURTHY · Reliability: Modeling, Prediction, and Optimization

BLOOMFIELD · Fourier Analysis of Time Series: An Introduction, *Second Edition*

BOLLEN · Structural Equations with Latent Variables

BOROVKOV · Ergodicity and Stability of Stochastic Processes

BOULEAU · Numerical Methods for Stochastic Processes

BOX · Bayesian Inference in Statistical Analysis

BOX · R. A. Fisher, the Life of a Scientist

BOX and DRAPER · Empirical Model-Building and Response Surfaces

*BOX and DRAPER · Evolutionary Operation: A Statistical Method for Process Improvement

BOX, HUNTER, and HUNTER · Statistics for Experimenters: An Introduction to Design, Data Analysis, and Model Building

BOX and LUCEÑO · Statistical Control by Monitoring and Feedback Adjustment

BRANDIMARTE · Numerical Methods in Finance: A MATLAB-Based Introduction

BROWN and HOLLANDER · Statistics: A Biomedical Introduction

BRUNNER, DOMHOF, and LANGER · Nonparametric Analysis of Longitudinal Data in Factorial Experiments

BUCKLEW · Large Deviation Techniques in Decision, Simulation, and Estimation

CAIROLI and DALANG · Sequential Stochastic Optimization

CHAN · Time Series: Applications to Finance

CHATTERJEE and HADI · Sensitivity Analysis in Linear Regression

CHATTERJEE and PRICE · Regression Analysis by Example, *Third Edition*

CHERNICK · Bootstrap Methods: A Practitioner's Guide

CHILÈS and DELFINER · Geostatistics: Modeling Spatial Uncertainty

CHOW and LIU · Design and Analysis of Clinical Trials: Concepts and Methodologies

CLARKE and DISNEY · Probability and Random Processes: A First Course with Applications, *Second Edition*

*COCHRAN and COX · Experimental Designs, *Second Edition*

CONGDON · Bayesian Statistical Modelling

CONOVER · Practical Nonparametric Statistics, *Second Edition*

COOK · Regression Graphics

COOK and WEISBERG · Applied Regression Including Computing and Graphics

COOK and WEISBERG · An Introduction to Regression Graphics

CORNELL · Experiments with Mixtures, Designs, Models, and the Analysis of Mixture Data, *Third Edition*

COVER and THOMAS · Elements of Information Theory

COX · A Handbook of Introductory Statistical Methods

*COX · Planning of Experiments

CRESSIE · Statistics for Spatial Data, *Revised Edition*

CSÖRGÖ and HORVÁTH · Limit Theorems in Change Point Analysis

DANIEL · Applications of Statistics to Industrial Experimentation

DANIEL · Biostatistics: A Foundation for Analysis in the Health Sciences, *Sixth Edition*

*DANIEL · Fitting Equations to Data: Computer Analysis of Multifactor Data, *Second Edition*

*Now available in a lower priced paperback edition in the Wiley Classics Library.

DAVID · Order Statistics, *Second Edition*
*DEGROOT, FIENBERG, and KADANE · Statistics and the Law
DEL CASTILLO · Statistical Process Adjustment for Quality Control
DETTE and STUDDEN · The Theory of Canonical Moments with Applications in
    Statistics, Probability, and Analysis
DEY and MUKERJEE · Fractional Factorial Plans
DILLON and GOLDSTEIN · Multivariate Analysis: Methods and Applications
DODGE · Alternative Methods of Regression
*DODGE and ROMIG · Sampling Inspection Tables, *Second Edition*
*DOOB · Stochastic Processes
DOWDY and WEARDEN · Statistics for Research, *Second Edition*
DRAPER and SMITH · Applied Regression Analysis, *Third Edition*
DRYDEN and MARDIA · Statistical Shape Analysis
DUDEWICZ and MISHRA · Modern Mathematical Statistics
DUNN and CLARK · Applied Statistics: Analysis of Variance and Regression, *Second
    Edition*
DUNN and CLARK · Basic Statistics: A Primer for the Biomedical Sciences,
    Third Edition
DUPUIS and ELLIS · A Weak Convergence Approach to the Theory of Large Deviations
*ELANDT-JOHNSON and JOHNSON · Survival Models and Data Analysis
ETHIER and KURTZ · Markov Processes: Characterization and Convergence
EVANS, HASTINGS, and PEACOCK · Statistical Distributions, *Third Edition*
FELLER · An Introduction to Probability Theory and Its Applications, Volume I,
    Third Edition, Revised; Volume II, *Second Edition*
FISHER and VAN BELLE · Biostatistics: A Methodology for the Health Sciences
*FLEISS · The Design and Analysis of Clinical Experiments
FLEISS · Statistical Methods for Rates and Proportions, *Second Edition*
FLEMING and HARRINGTON · Counting Processes and Survival Analysis
FULLER · Introduction to Statistical Time Series, *Second Edition*
FULLER · Measurement Error Models
GALLANT · Nonlinear Statistical Models
GHOSH, MUKHOPADHYAY, and SEN · Sequential Estimation
GIFI · Nonlinear Multivariate Analysis
GLASSERMAN and YAO · Monotone Structure in Discrete-Event Systems
GNANADESIKAN · Methods for Statistical Data Analysis of Multivariate Observations,
    Second Edition
GOLDSTEIN and LEWIS · Assessment: Problems, Development, and Statistical Issues
GREENWOOD and NIKULIN · A Guide to Chi-Squared Testing
GROSS and HARRIS · Fundamentals of Queueing Theory, *Third Edition*
*HAHN · Statistical Models in Engineering
HAHN and MEEKER · Statistical Intervals: A Guide for Practitioners
HALD · A History of Probability and Statistics and their Applications Before 1750
HALD · A History of Mathematical Statistics from 1750 to 1930
HAMPEL · Robust Statistics: The Approach Based on Influence Functions
HANNAN and DEISTLER · The Statistical Theory of Linear Systems
HEIBERGER · Computation for the Analysis of Designed Experiments
HEDAYAT and SINHA · Design and Inference in Finite Population Sampling
HELLER · MACSYMA for Statisticians
HINKELMAN and KEMPTHORNE: · Design and Analysis of Experiments, Volume 1:
    Introduction to Experimental Design
HOAGLIN, MOSTELLER, and TUKEY · Exploratory Approach to Analysis
    of Variance
HOAGLIN, MOSTELLER, and TUKEY · Exploring Data Tables, Trends and Shapes

*Now available in a lower priced paperback edition in the Wiley Classics Library.

*Now available in a lower priced paperback edition in the Wiley Classics Library.

KOTZ and JOHNSON (editors) · Encyclopedia of Statistical Sciences: Supplement Volume

KOTZ, READ, and BANKS (editors) · Encyclopedia of Statistical Sciences: Update Volume 1

KOTZ, READ, and BANKS (editors) · Encyclopedia of Statistical Sciences: Update Volume 2

KOVALENKO, KUZNETZOV, and PEGG · Mathematical Theory of Reliability of Time-Dependent Systems with Practical Applications

LACHIN · Biostatistical Methods: The Assessment of Relative Risks

LAD · Operational Subjective Statistical Methods: A Mathematical, Philosophical, and Historical Introduction

LAMPERTI · Probability: A Survey of the Mathematical Theory, *Second Edition*

LANGE, RYAN, BILLARD, BRILLINGER, CONQUEST, and GREENHOUSE · Case Studies in Biometry

LARSON · Introduction to Probability Theory and Statistical Inference, *Third Edition*

LAWLESS · Statistical Models and Methods for Lifetime Data

LAWSON · Statistical Methods in Spatial Epidemiology

LE · Applied Categorical Data Analysis

LE · Applied Survival Analysis

LEE and WANG · Statistical Methods for Survival Data Analysis, *Third Edition*

LePAGE and BILLARD · Exploring the Limits of Bootstrap

LEYLAND and GOLDSTEIN (editors) · Multilevel Modelling of Health Statistics

LIAO · Statistical Group Comparison

LINDVALL · Lectures on the Coupling Method

LINHART and ZUCCHINI · Model Selection

LITTLE and RUBIN · Statistical Analysis with Missing Data

LLOYD · The Statistical Analysis of Categorical Data

MAGNUS and NEUDECKER · Matrix Differential Calculus with Applications in Statistics and Econometrics, *Revised Edition*

MALLER and ZHOU · Survival Analysis with Long Term Survivors

MALLOWS · Design, Data, and Analysis by Some Friends of Cuthbert Daniel

MANN, SCHAFER, and SINGPURWALLA · Methods for Statistical Analysis of Reliability and Life Data

MANTON, WOODBURY, and TOLLEY · Statistical Applications Using Fuzzy Sets

MARDIA and JUPP · Directional Statistics

MASON, GUNST, and HESS · Statistical Design and Analysis of Experiments with Applications to Engineering and Science

McCULLOCH and SEARLE · Generalized, Linear, and Mixed Models

McFADDEN · Management of Data in Clinical Trials

McLACHLAN · Discriminant Analysis and Statistical Pattern Recognition

McLACHLAN and KRISHNAN · The EM Algorithm and Extensions

McLACHLAN and PEEL · Finite Mixture Models

McNEIL · Epidemiological Research Methods

MEEKER and ESCOBAR · Statistical Methods for Reliability Data

MEERSCHAERT and SCHEFFLER · Limit Distributions for Sums of Independent Random Vectors: Heavy Tails in Theory and Practice

*MILLER · Survival Analysis, *Second Edition*

MONTGOMERY, PECK, and VINING · Introduction to Linear Regression Analysis, *Third Edition*

MORGENTHALER and TUKEY · Configural Polysampling: A Route to Practical Robustness

MUIRHEAD · Aspects of Multivariate Statistical Theory

MURRAY · X-STAT 2.0 Statistical Experimentation, Design Data Analysis, and Nonlinear Optimization

MYERS and MONTGOMERY · Response Surface Methodology: Process and Product Optimization Using Designed Experiments, *Second Edition*

MYERS, MONTGOMERY, and VINING · Generalized Linear Models. With Applications in Engineering and the Sciences

NELSON · Accelerated Testing, Statistical Models, Test Plans, and Data Analyses

NELSON · Applied Life Data Analysis

NEWMAN · Biostatistical Methods in Epidemiology

OCHI · Applied Probability and Stochastic Processes in Engineering and Physical Sciences

OKABE, BOOTS, SUGIHARA, and CHIU · Spatial Tesselations: Concepts and Applications of Voronoi Diagrams, *Second Edition*

OLIVER and SMITH · Influence Diagrams, Belief Nets and Decision Analysis

PANKRATZ · Forecasting with Dynamic Regression Models

PANKRATZ · Forecasting with Univariate Box-Jenkins Models: Concepts and Cases

*PARZEN · Modern Probability Theory and Its Applications

PEÑA, TIAO, and TSAY · A Course in Time Series Analysis

PIANTADOSI · Clinical Trials: A Methodologic Perspective

PORT · Theoretical Probability for Applications

POURAHMADI · Foundations of Time Series Analysis and Prediction Theory

PRESS · Bayesian Statistics: Principles, Models, and Applications

PRESS and TANUR · The Subjectivity of Scientists and the Bayesian Approach

PUKELSHEIM · Optimal Experimental Design

PURI, VILAPLANA, and WERTZ · New Perspectives in Theoretical and Applied Statistics

PUTERMAN · Markov Decision Processes: Discrete Stochastic Dynamic Programming

*RAO · Linear Statistical Inference and Its Applications, *Second Edition*

RENCHER · Linear Models in Statistics

RENCHER · Methods of Multivariate Analysis, *Second Edition*

RENCHER · Multivariate Statistical Inference with Applications

RIPLEY · Spatial Statistics

RIPLEY · Stochastic Simulation

ROBINSON · Practical Strategies for Experimenting

ROHATGI and SALEH · An Introduction to Probability and Statistics, *Second Edition*

ROLSKI, SCHMIDLI, SCHMIDT, and TEUGELS · Stochastic Processes for Insurance and Finance

ROSENBERGER and LACHIN · Randomization in Clinical Trials: Theory and Practice

ROSS · Introduction to Probability and Statistics for Engineers and Scientists

ROUSSEEUW and LEROY · Robust Regression and Outlier Detection

RUBIN · Multiple Imputation for Nonresponse in Surveys

RUBINSTEIN · Simulation and the Monte Carlo Method

RUBINSTEIN and MELAMED · Modern Simulation and Modeling

RYAN · Modern Regression Methods

RYAN · Statistical Methods for Quality Improvement, *Second Edition*

SALTELLI, CHAN, and SCOTT (editors) · Sensitivity Analysis

*SCHEFFE · The Analysis of Variance

SCHIMEK · Smoothing and Regression: Approaches, Computation, and Application

SCHOTT · Matrix Analysis for Statistics

SCHUSS · Theory and Applications of Stochastic Differential Equations

SCOTT · Multivariate Density Estimation: Theory, Practice, and Visualization

*SEARLE · Linear Models

SEARLE · Linear Models for Unbalanced Data

SEARLE · Matrix Algebra Useful for Statistics

SEARLE, CASELLA, and McCULLOCH · Variance Components

SEARLE and WILLETT · Matrix Algebra for Applied Economics

SEBER · Linear Regression Analysis

*Now available in a lower priced paperback edition in the Wiley Classics Library.

SEBER · Multivariate Observations

SEBER and WILD · Nonlinear Regression

SENNOTT · Stochastic Dynamic Programming and the Control of Queueing Systems

*SERFLING · Approximation Theorems of Mathematical Statistics

SHAFER and VOVK · Probability and Finance: It's Only a Game!

SMALL and McLEISH · Hilbert Space Methods in Probability and Statistical Inference

SRIVASTAVA · Methods of Multivariate Statistics

STAPLETON · Linear Statistical Models

STAUDTE and SHEATHER · Robust Estimation and Testing

STOYAN, KENDALL, and MECKE · Stochastic Geometry and Its Applications, *Second Edition*

STOYAN and STOYAN · Fractals, Random Shapes and Point Fields: Methods of Geometrical Statistics

STYAN · The Collected Papers of T. W. Anderson: 1943–1985

SUTTON, ABRAMS, JONES, SHELDON, and SONG · Methods for Meta-Analysis in Medical Research

TANAKA · Time Series Analysis: Nonstationary and Noninvertible Distribution Theory

THOMPSON · Empirical Model Building

THOMPSON · Sampling, *Second Edition*

THOMPSON · Simulation: A Modeler's Approach

THOMPSON and SEBER · Adaptive Sampling

TIAO, BISGAARD, HILL, PEÑA, and STIGLER (editors) · Box on Quality and Discovery: with Design, Control, and Robustness

TIERNEY · LISP-STAT: An Object-Oriented Environment for Statistical Computing and Dynamic Graphics

TSAY · Analysis of Financial Time Series

UPTON and FINGLETON · Spatial Data Analysis by Example, Volume II: Categorical and Directional Data

VAN BELLE · Statistical Rules of Thumb

VIDAKOVIC · Statistical Modeling by Wavelets

WEISBERG · Applied Linear Regression, *Second Edition*

WELSH · Aspects of Statistical Inference

WESTFALL and YOUNG · Resampling-Based Multiple Testing: Examples and Methods for p-Value Adjustment

WHITTAKER · Graphical Models in Applied Multivariate Statistics

WINKER · Optimization Heuristics in Economics: Applications of Threshold Accepting

WONNACOTT and WONNACOTT · Econometrics, *Second Edition*

WOODING · Planning Pharmaceutical Clinical Trials: Basic Statistical Principles

WOOLSON and CLARKE · Statistical Methods for the Analysis of Biomedical Data, *Second Edition*

WU and HAMADA · Experiments: Planning, Analysis, and Parameter Design Optimization

YANG · The Construction Theory of Denumerable Markov Processes

*ZELLNER · An Introduction to Bayesian Inference in Econometrics

ZHOU, OBUCHOWSKI, and McCLISH · Statistical Methods in Diagnostic Medicine

*Now available in a lower priced paperback edition in the Wiley Classics Library.